GEOCHEMISTRY

GEOCHEMISTRY

Editor

Engelbert Martz

scitus
academics

Geochemistry

Edited by **Engelbert Martz**

This Book is Printed on Acid-Free Paper. ♾

ISBN: 9781681178967

Library of Congress Control Number: 2017948925

SCITUS Academics LLC,
616, Corporate Way, Suite 2, 4766,
Valley Cottage, NY 10989

www.scitusacademics.com

Printed in United States of America.

Preface

Geochemistry is the branch of Earth Science that applies chemical principles to deepen an understanding of the Earth system and systems of other planets. Geochemists consider Earth composed of discrete spheres — rocks, fluids, gases and biology — that exchange matter and energy over a range of time scales. An effective management of geological and environmental issues requires the understanding of recent research in minerals, soil, ores, rocks, water, sediments. In this work, it is proposed to provide an overview of major developments that have occurred in a number of areas of geochemistry in recent years. This will include developments in reservoir geochemistry such as the use of high-resolution gas chromatography for reservoir continuity studies and high-temperature gas chromatography for characterization of wax deposits. A brief overview of recent developments in biomarker geochemistry will be provided on exploration geochemistry along with a discussion on the use of various pyrolysis techniques for the purposes of artificial maturation or characterization of the insoluble organic matter in source rocks or asphaltenes in oils. It presents the main aspects of coalbed gas geochemistry and current research advances. Coals are both sources and reservoirs of large amounts of gas that has received increasing attention in recent years as a largely untapped potential energy resource.

This book brings together the knowledge from a variety of topics within the field of geochemistry. It aims to characterize geochemical conditions prior to mine construction, operation, and reclamation/closure. The audience for this book consists of a multitude of scientists such as physicists, geologists, technologists, petroleum engineers, volcanologists, geochemists and government agencies.

Table of Contents

CHAPTER 1

Geochemistry and Stable Isotopes of Travertine from Jordan Valley and Dead Sea Areas

Khalil M. Ibrahim[1], Issa M. Makhlouf[1], Ali R. El Naqah[2] and Sana' M. Al-Thawabteh[1]

[1]*Department of Earth and Environmental Sciences, Hashemite University, P.O. Box 150459, Zarqa 13115, Jordan*

[2]*Faculty of Natural Resources and Environment, Hashemite University, P.O. Box 150459, Zarqa 13115, Jordan*

ABSTRACT

Travertine deposits in Deir Alla, Suwayma, and Az Zara areas were investigated. Mineralogy, geochemistry, stable isotopes and age dating indicate the presence of low-Mg calcite, with minor quartz components. The variable isotope ($\delta^{13}C$ and $\delta^{18}O$) signatures indicate dependence on water temperature and water/rock isotopic exchange. In contrast, the high $\delta^{13}C$ values in some travertine samples reflect $^{12}CO_2$ degassing processes, increased input of ^{13}C-enriched groundwater, and the presence of surface and groundwater hydrological systems. The high $\delta^{18}O$ values may be attributed to evaporation effects and low water temperature during the formation of localized travertine. The age of travertine is the Late Pleistocene.

Keywords: travertine; geochemistry; stable isotopes; carbon dating; Dead Sea; Jordan Valley

1. INTRODUCTION

Travertine refers to all non-marine carbonate precipitates formed at or near terrestrial springs, rivers, lakes, and caves [1], when CO_2-rich, Ca-bearing fluids are exposed to low pressure conditions at surface conditions [2]. Travertine formation is controversial since it has been attributed to both biotic and abiotic processes [3,4]. The terms travertine and tufa are sometimes used synonymously [5]. Calcite precipitation occurs due to CO_2-loss, and associated pH-reduction during fluid equilibration with the atmosphere [6,7]. Travertine precipitation seems to be controlled by a combination of several factors, such as: (i) a fast decrease of the hydrostatic pressure and the pCO_2, resulting in a rapid degassing process; and (ii) microbes and algae activity [6]. The term travertine must be retained for continental carbonates mainly composed of calcium carbonate deposits produced from non-marine, supersaturated calcium bicarbonate-rich waters, typically hydrothermal in origin [5]. Based on the CO_2 interaction with the groundwater, Pentecost [8] subdivided travertine into meteogene and thermogene deposits. CO_2 in the meteogene deposits is developed at a shallower origin, as a result of organic activity in the soil [9]. Thermogene deposits of CO_2 have a deeper origin either from magmatic degassing or from decarbonation processes [10]. Such deposits are typical of tectonically active areas where geothermal heat flux (endogenic or volcanic) is high [11]. The meteogene travertine is usually referred to as "tufa", especially that which contains remains of micro and macrophytes, invertebrates and bacteria [11,12]. It is usually derived from lower water temperature, and lower contents of dissolved inorganic carbon (Dissolved inorganic carbon (DIC) < 10 mmol·L^{-1}), CO_2 partial pressure ($pCO_2 < 0.1$ atm) and higher pH values (7–8) [10]. Meteogene usually displays low deposition rates (<10 mmol·cm^{-2} year^{-1}) and shows a negative carbon isotopic composition (δ ^{13}C vs. PDB between −12 and 0‰). However, thermogene travertine displays higher deposition rates and higher (δ ^{13}C vs. PDB) values [5,10]. They are characterized chiefly by high depositional rates, regular bedding and fine lamination, low porosity, low permeability and an inorganic crystalline fabric [5]. The main distinctive characteristics of travertine and tufa are detailed in Capezzuoli et al. [5].

Travertine and tufa rocks are exposed in the eastern margins of the Dead Sea and Jordan Valley. Some investigations focused on their geotechnical properties as a decoration stone [13]. However, geochemical and mineralogical studies of such rocks are limited [14,15]. The present study aims to elucidate the mineralogy, geochemistry, stable isotope implications and age determination of the travertine and tufa formation along the eastern margins of the Dead Sea and Jordan Valley.

1.1. Location

The study area is located along the margins of the Dead Sea (Figure 1). Most of the travertine outcrops are located in the eastern Jordan Valley and the Dead Sea Transform Fault system. Three main travertine outcrops were studied including: Deir Alla, Suwayma and Az Zara (Figure 1). Representative travertine, tufa and water samples were collected from these areas.

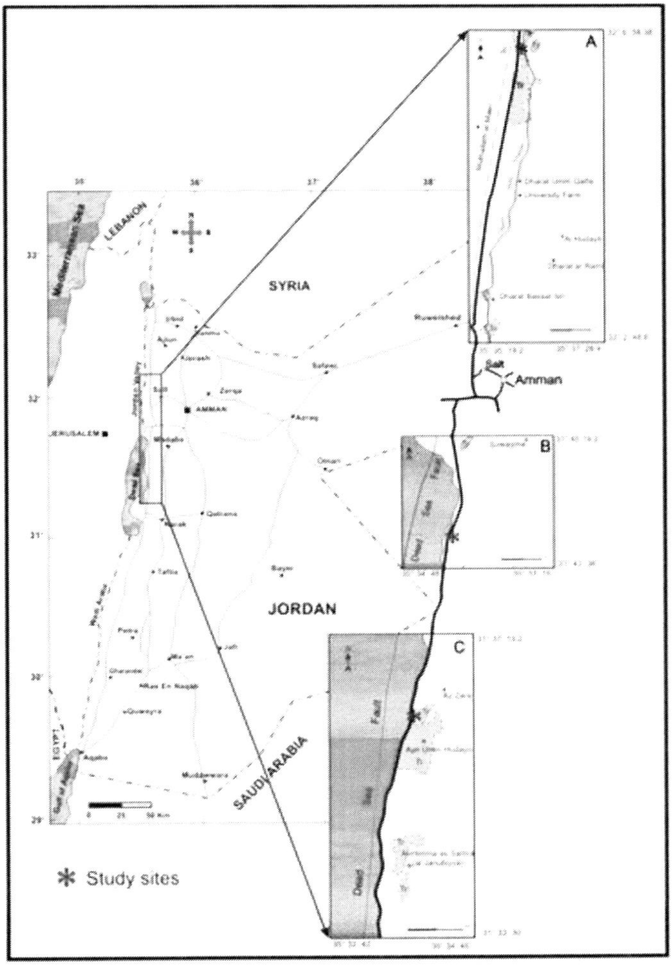

Figure 1. Location map to show the studied travertine outcrops.

1.2. Geological Setting

The travertine is exposed in a highly tectonized area associated with the main Dead Sea Transform Fault (Figure 2). It is an 1100 km long sinistral fault system that connects the Gulf of Aqaba-Red Sea spreading system to the convergence zone of the Taurus–Zagrous Mountains [16]. This fault is a left lateral transform plate boundary, separating the Arabian plate in the east and the Palestine-Sinai sub plate (part of the African plate) in the west [17]. It has been active since ca. 13–18 Ma ago with movement continuing today [17,18]. Most of the faults in the study area are related to the stress fields associated with the transform boundary.

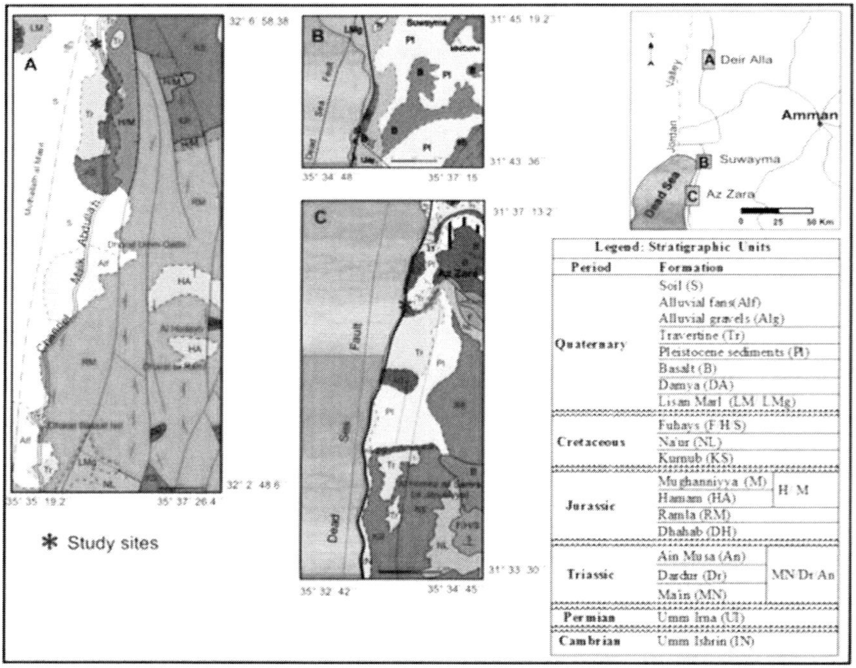

Figure 2. Geological maps of the studied travertine outcrops.

The stratigraphic sequence in the studied areas, as illustrated in the simplified geological maps in Figure 2, comprises a wide range of rock units including Cambrian, Permian, Triassic, Jurassic and Cretaceous sandstones, shale and carbonates (Figure 2). Quaternary deposits occur widely in the form of marls, siltstone, basalts, travertine, fluvial and alluvial sediments.

The studied travertine facies are composed of crystalline crust, shrub, paper-thin raft, coated gas bubble, reed, lithoclasts, pebbly travertine, and palaeosols with a tufa carbonate facies of macrophyte encrustation deposits, bryophyte build-ups, biomicrites and green and grey marls (Figure 3). The Deir Alla travertine is formed in a terraced and smooth slope depositional setting, where it has been subjected to meteoric diagenesis.

The inactive thermogene Suwayma travertine is deposited in fissure-ridge and depression depositional setting, with meteogene and burial diagenetic processes [19]. Both localities are characterized by hard crystalline travertine with low organic content. Az Zara travertines were deposited under self-built channels' depositional settings. Az Zara active tufa was formed in a predominantly fluvio-palaustrine environment, which is friable and rich in organics and clays.

1.3. Hydrology and Hydrogeology

The physiographic situation of the Dead Sea and Jordan Valley displays morphotectonic depressions (Figure 1). Therefore, most of the rainfall-infiltrated water emerges once again in the form of springs, and the general flow direction is due west, towards the Dead Sea and Jordan Valley. The water temperature of some springs is moderate (18–21 °C). The discharge amount is seasonal such that it increases during the winter and drops in the summer [20].

The groundwater around the Dead Sea is found in two aquifer complexes; the first is the upper limestone aquifer complex (Upper Cretaceous), and the second is the sandstone aquifer complex (Lower Cretaceous). The total amount of groundwater associated with the upper aquifer is around 87 million m^3/year [20], and half of it discharges on the surface through springs. The total discharge of the lower aquifer is around 90 million m^3/year, and the water is mostly thermal and mineralized [21]. The Jordan Valley floor consists of recent alluvial fan sediments inter-tonguing with Pleistocene salty, clayey sediments of the Lisan Formation. The available groundwater in this area ranges from 18–20 million m^3/year [22].

Figure 3. Selected field photos of the studied travertine: (**A**) root casts, downward tapering, rare lateral roots, smooth tube-like, ranging from 0.5 to 7.0 cm in length and from 0.1 to 3.5 cm in diameter preserved in porous travertine from Deir Alla; (**B**) tiny travertine from Deir Alla with irregular elliptical cavities are partly to completely infilled by secondary materials; (**C**) interconnected root traces on the upper bedding plain indicating the plants types of Az Zara area; (**D**) stalagmites and stalactites features (red circle), which are formed near Az Zara hot spring; (**E**) wavy laminae resembles stromatolite in Suwayma travertine, bedding plains indicate three stages of travertine precipitation; (**F**) the mound of travertine standing in an erosional relief in Suwayma. The hammer is 28 cm long.

Springs issued east of Deir Alla are alkaline with prevailing bicarbonate in summer, and alkaline with prevailing chlorides in winter. Fe oxides precipitate as soon as the water mixes with the Zarqa River water [23]. The number of thermal springs in Ma'in area is approximately 109, with temperature exceeding 34 °C. All such thermal springs are classified as hyperthermal [24]. Sixty-four hyperthermal springs are located in the northern part of Wadi Zarqa Ma'in forming the "Hammamt Ma'in", with 19 million m^3/year total discharge into the Dead Sea. The rest of the hyperthermal springs are aligned in a north–south trend at the Az Zara area, on the eastern shore of the Dead Sea. The total discharge into the Dead Sea from the Az Zara area is 17 million m^3/year [23].

2. MATERIALS AND METHODS

More than 80 travertine samples were collected from three investigated areas representing both vertical and horizontal sections. The mineralogy of 18 samples was determined by X-ray diffraction (XRD), using a Phillips X' Pert Pro PW 3040/80 diffractometer (Lelweg 1, 7602, EA Almelo, The Netherlands). Twenty-two samples were selected for major and trace elements determination by X-ray fluorescence (XRF)—using a Phillips MagiX PW 2424 diffractometer (Lelweg 1, 7602, EA Almelo, The Netherlands). Radiogenic and stable isotope measurements were also carried out. The method used for $\delta^{18}O$ measurements is CO_2 equilibrium/Delta Plus XP Isotope Ratio Mass Spectrometer HDO (ThermoFisher Scientific, Walton, MA, USA) using a platinum catalyst modified from (International Atomic Energy Agency). Measurements of ^{14}C and $\delta^{13}C$ are done by a benzene synthesis line and Liquid Scintillation Counter. The method is modified by (IAEA) technical procedure #25. The measurements of the ($^{13}C/^{12}C$) isotopes are made relative to the Vienna Pee Dee Belemnite standard (VPDB), and the water oxygen is based on a Vienna Standard Mean Ocean Water (VSMOW).

The ratios of ($^{13}C/^{12}C$) and ($^{18}O/^{18}O$) are expressed in the δ notation. The values of $\delta^{18}O$ are interchangeable according to such equations [8]: $\delta^{18}O$ (VSMOW) = [1.03088 $\delta^{18}O$ (VPDB) + 30.88] and $\delta^{18}O$ (VPDB) = [0.97008 $\delta^{18}O$ (VSMOW) – 29.94].

The calculation of travertine age was performed by the equation set by Vogel [25] and used by Thorpe et al. [26]. The decay rate constant (λ) is assumed to equal 1.209×10^{-4} year^{-1}. The value dilution factor is 0.85 ± 0.05 [8,26].

3. RESULTS AND DISCUSSION

Petrographic investigation reported the presence of eight travertine and four tufa lithofacies in the studied sites. The description of these lithofacies is summarized in Table 1.

Table 1. Summary of travertine and tufa lithofacies in the studied travertine outcrops.

Locality	Lithofacies	Description
Deir Alla & Suwayma Travertine	Crystalline crust	Dense, crudely fibrous, elongated calcite crystals (ray crystals)
	Shrub	Small bush-like growths, on horizontal-semi-horizontal surfaces
	Rafts	Thin, delicate and brittle crystalline layers
	Lithoclast	Penecontemporaneous, angular to subangular travertine fragments in different size, formed by erosion of the upper slope and collapse of waterfalls and terrace cliffs
	Coated gas bubble	Gas bubbles are coated by rapid precipitation of calcium carbonate
Deir Alla & Suwayma Travertine	Reed	Rich in molds of reed and coarse grass
	Pebbly	Rounded pebbles of limestone, chert and basalt
	Palaeosol	Exposure of travertine surface to rainwater, subaerial desiccation, and biological activities associated with soil formation
Az Zara Tufa	Macrophyte encrustation	Calcified stems, twigs and leaves that represent rushes, reeds and unidentifiable bushes.
	Bryophyte build-ups	Laminated deposits that commonly form asymmetrical mounds or small domes, up to 1.75 m thick or stacked sequences up to 2 m high.
	Biomicrites	Tabular layers, tens of centimeters thick, mostly at the top of the tufa succession. They consist of massive biomicrites or, in some places, biosparites.
	Marls	Massive lenticular beds that are associated with macrophyte palisades. They contain plant debris.

3.1. Mineralogy

Semiquantitative determination of mineral phases was reported by the X-ray diffraction. In the present work, components other than calcium carbonate may be defined as accessories. X-ray diffraction of travertine from Deir Alla, Suwayma, and Az Zara indicates close similarity. They are predominantly composed of calcite (Figure 4). Deir Alla and Suwayma samples occasionally contain a very low percent of quartz and Fe-oxides. Sample (D26a) shows a frequent amount of aragonite (Figure 4). Traces of gypsum, dolomite and halite may also occur.

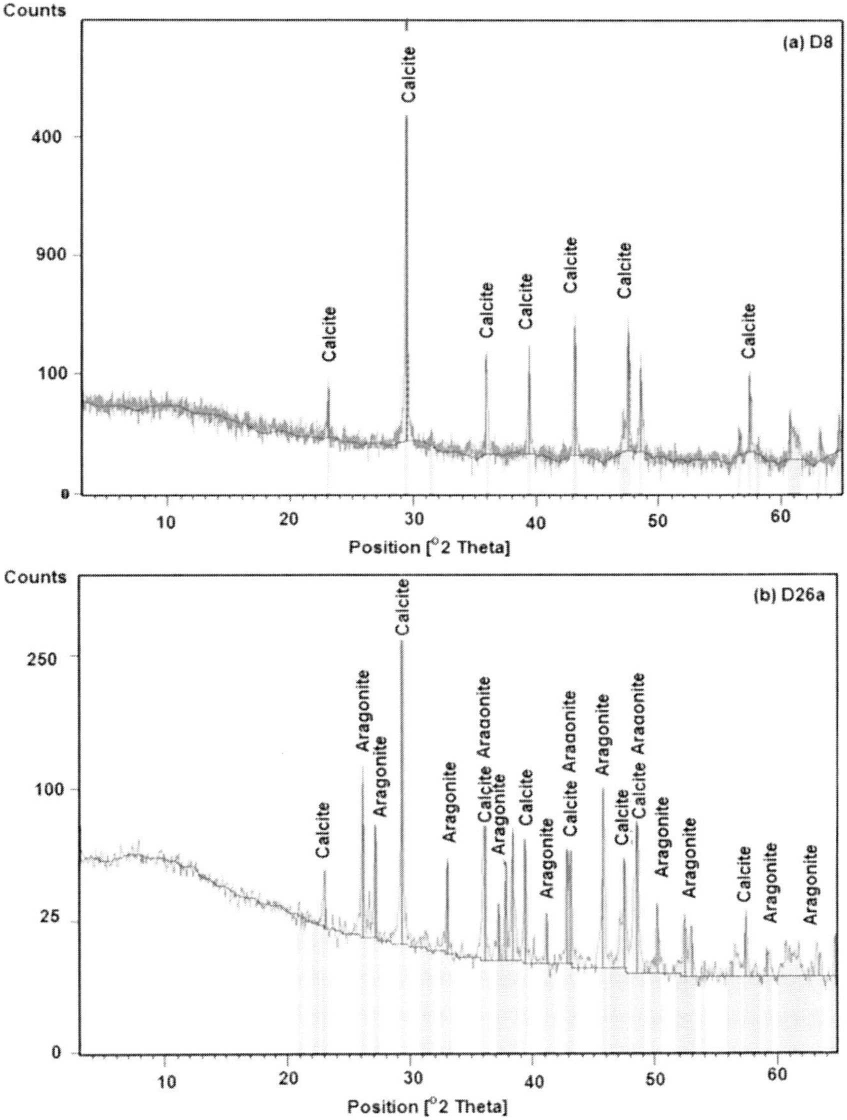

Figure 4. Selected X-ray diffractograms of the studied travertines (**a**) D8, (**b**) D26a.

Petrographic study by Al-Thawabteh [19] indicated that most of the Fe present in travertine and tufa is autochthonous, i.e., formed or grown in situ and has not been transported. Precipitation of hydrated Fe-oxides occurs when groundwater containing ferrous ions encounters the atmosphere. In contrast, quartz enters the travertine realm as a detrital mineral, derived from the soil and bedrocks, and are therefore considered as allochthonous minerals.

Calculations of the theoretical mineralogical composition including $MgCO_3$ percent are presented in Table 2. In Deir Alla, the detrital quartz grains are rare compared to Suwayma and Az Zara (Table 2). It is believed that they are derived from sandstone when hot groundwater emerges onto the surface. In addition, Deir Alla travertine includes different types of clay minerals that vary from 0.19–1.47 wt % (Table 2). In thermogene travertine of Suwayma and Az Zara, the presence of clay (2.82 wt % and 0.85 wt %, respectively) may be due to the hydrothermal alteration of the host rock.

Table 2. Calculated theoretical mineralogical composition of travertine and tufa samples (wt %).

No.	D3	D8	D11	D18	D23b	D26a	D26b	D29
Clay	0.8	0.7	0.4	1.5	0.4	0.7	1.3	0.2
Dol	0.5	0.7	0.5	0.7	0.4	0.9	0.5	0.4
Cc	98.4	94.3	97.2	93.0	94.3	93.7	98.5	98.8
Qz	nd	nd	nd	0.1	0.1	0.1	nd	nd
Mgs	0.4	0.4	0.3	0.5	0.3	0.8	0.4	0.3
Others	2.4	4.3	1.8	4.7	4.8	4.4	1.8	0.7

No.	S1d	S10	S19	S27	S32	S34	S38	S39	S45	S47b
Clay	9.2	2.2	4.5	1.4	1.2	1.8	1.8	1.3	2.2	2.8
Dol	2.1	2.3	2.7	1.3	3.0	3.4	2.0	2.2	2.5	2.9
Cc	72.0	90.7	88.9	94.4	92.3	90.7	93.8	93.1	90.7	87.8
Qz	3.7	0.5	1.5	0.2	0.4	1.1	0.4	0.3	0.9	1.2
Mgs	1.9	1.59	1.99	0.9	1.9	2.2	1.3	1.4	1.7	1.9
Others	12.9	4.0	4.4	2.7	3	3.0	2.3	2.7	3.5	5.3

No.	Z1	Z1*	Z3	Z4
Clay	0.8	0.5	1.0	1.1
Dol	2.5	1.8	1.5	2.2
Cc	91.8	88.7	93.5	94.2
Qz	0.8	1.8	0.1	0.2
Mgs	1.8	1.1	1.0	1.4
Others	3.9	7.1	3.7	2.1

Z1 and Z1* are tufa samples. (nd: is not detected; D: Deir Alla; S: Suwayma; Z: Az Zara; Dol: dolomite; Cc: calcite and/or aragonite; Qz: quartz; Mgs: magnesite).

3.1.1. $MgCO_3$

The calculated $MgCO_3$ represents all of Mg in the studied travertine. The Mg is usually incorporated in the calcite, aragonite, and dolomite crystals, or it may form magnesite. In Deir Alla travertine, the calculated $MgCO_3$ content ranges from 0.3 wt % to 0.8 wt %, with a mean value of 0.4 wt %, whereas in Suwayma travertine, the range is from 0.9 wt % to 2.2 wt %, with an average value of 1.7%; meanwhile, the calculated $MgCO_3$ content of Az Zara travertine and tufa ranges from 1.0% to 1.8%, with a mean value of 1.3% (Table 2). This indicates that the calcite in such travertines and tufas is invariably of low Mg calcite [27] (Table 2). Calcite that contains less than 8% of $MgCO_3$ has been defined as low-Mg calcite, whereas calcite, with more than 8–11% of $MgCO_3$, has been defined as high-Mg calcite [27].

3.1.2. Calcium Carbonate ($CaCO_3$)

Deir Alla travertine consists mainly of calcium carbonate with an arithmetic average value of 95.5% (Table 2, shown as cc). According to Pentecost [8], meteogene deposits have a high average of $CaCO_3$ content with a median of about 92.9%, and an upper range of 99.3%. In addition, Pentecost [8] indicated that active meteogene deposits have a lower $CaCO_3$content (mean 91.2%, No. of samples = 28) than their inactive meteogene counterparts (mean 94.8%, No. of samples = 22). This might be due to the loss of organic matter in the older deposits. Consequently, the Deir Alla travertine can be classified as inactive super-ambient meteogene deposits based on the $CaCO_3$ content [8].

The mean calcium carbonate abundance in Suwayma travertine is about 89.18% (Table 2). The $CaCO_3$ content of thermogene travertine is broadly similar to that of meteogene [8]. Therefore, the Suwayma travertine can be classified as thermogene deposits based on its $CaCO_3$ content. Az Zara travertine mainly consists of calcium carbonate with a mean of 92% (Table 2). Az Zara travertines and tufa can be classified as inactive and active thermogenic deposits, respectively. Az Zara deposits are localized and are commonly associated with regions of Quaternary basaltic outcrops and recent tectonic activities related to the Dead Sea Transform Fault System.

In general, calcium carbonates in travertine are formed of various combinations of aragonite and calcite polymorphs [28]. No attempts were carried out in this study to investigate the controls of calcium carbonate polymorph precipitation. The factors that control calcite and aragonite precipitation in travertine are still open to debate [28,29,30,31]. The precipitation has been variously attributed to water composition, water temperature, growth inhibitors (e.g., Mg/Ca ratio), and/or CO_2degassing and saturation levels [29]. Aragonite may co-precipitate with calcite at a high CO_2 content and rapid CO_2degassing, irrespective of the Mg/Ca ratio [30], or aragonite may precipitate from low levels of CO_2 degassing, provided the Mg/Ca ratio is high enough to inhibit calcite precipitation [30].

3.2. Geochemistry

3.2.1. CaO and MgO

The XRF results indicate that CaO wt % in the travertine ranges from 48.38% to 55.8% (Table 3). The relationship between CaO% and SiO_2% is presented in Figure 5a supported by the negative correlation coefficient of −0.82 in the correlation coefficient matrix (Figure 6). The inverse relationship between these two oxides (Figure 5a) indicates that they are involved in two different phases. A similar conclusion can be noted in Figure 5b, which exhibits the relationship between CaO and Al_2O_3.

It is evident that the studied travertine samples have low Mg content (Table 3), which is less than 1.0 wt % MgO with a mean value of 0.19 wt %. Consequently, calcites of such travertine are invariably of low Mg. The high value of Mg in some Suwayma travertine samples (Table 3) may have resulted from the input of Mg-enriched solutions that could lead to slight dolometization; i.e., formation of diagenetic dolomite. Mg forms a solid solution series with calcite and a wide range of compositions are possible. Figure 6 shows the negative correlation between CaO and MgO, which indicates that Mg substitutes Ca in calcite crystals. Previous studies by Pentecost [8] show that a few meteogene travertines are composed of calcite, which exceed 1% of Mg by weight, and yield a mean value of 0.29 wt %. Thermogene travertine consisting of calcite has similar Mg concentration similar meteogene travertine [8].

3.2.2. Fe_2O_3 and MnO

The Fe_2O_3 content of Deir Alla samples ranges from 0.24% to 3.18% (Table 3). In Suwayma travertine, Fe_2O_3 wt % ranges from 0.47% to 4.72%, and Az Zara travertine and tufa is between 0.18% and 0.88%. Such percents are for total iron, which include the Fe precipitated with calcite and that associated with detrital minerals. According to Pentecost [8], the mean Fe wt % for the meteogene and thermogene travertines is 0.18% to 0.28%, respectively, which is far less than the mean value of the studied travertines (0.83%). This is most probably related to the presence of allochthonous Fe-minerals within the travertine.

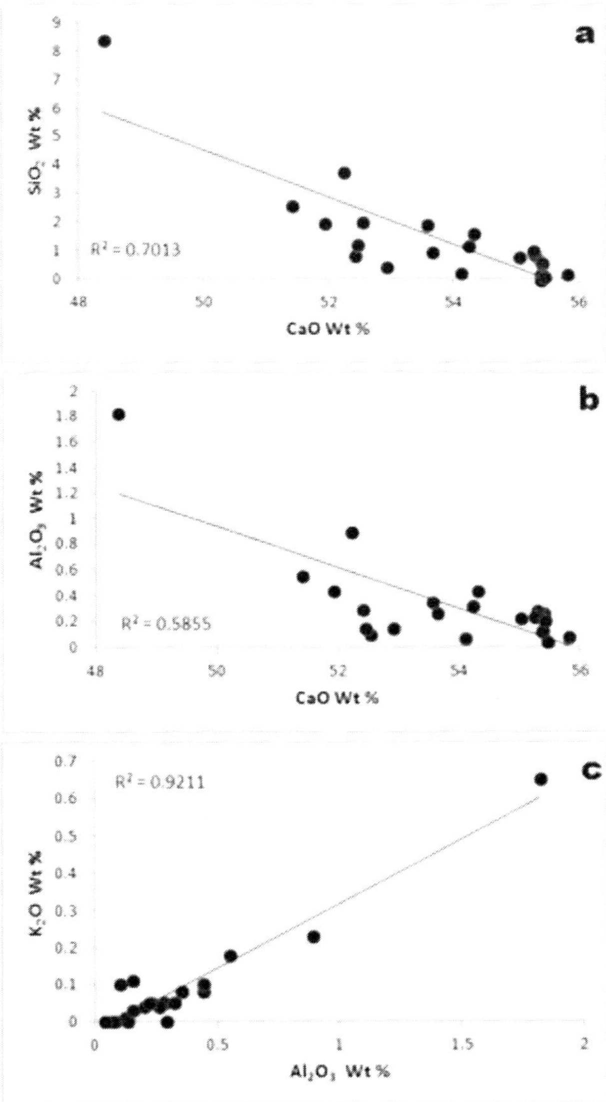

Figure 5. Variation diagrams of: (**a**) CaO versus SiO_2; (**b**) CaO versus Al_2O_3; and (**c**) Al_2O_3 versus K_2O.

	SiO_2	Al_2O_3	TiO_2	Fe_2O_3	MgO	CaO	MnO	Na_2O	K_2O	P_2O_5	SO_3	Sr	Zr	Mo	Rb	Ba	Cl	Br
SiO_2	1																	
Al_2O_3	0.97	1																
TiO_2	0.90	0.92	1															
Fe_2O_3	0.63	0.64	0.71	1														
MgO	0.62	0.55	0.53	0.18	1													
CaO	-0.82	-0.74	-0.83	-0.64	-0.56	1												
MnO	0.00	-0.15	-0.05	-0.13	-0.10	-0.23	1											
Na_2O	0.13	-0.10	-0.08	-0.10	0.13	-0.35	0.63	1										
K_2O	0.98	0.96	0.88	0.63	0.55	-0.78	-0.02	0.12	1									
P_2O_5	0.17	0.21	0.06	-0.03	0.36	-0.12	-0.08	-0.11	0.10	1								
SO_3	0.14	0.05	0.13	-0.13	0.66	-0.35	0.09	0.34	0.12	0.31	1							
Sr	-0.02	-0.02	0.12	-0.11	0.07	-0.36	0.17	0.14	-0.07	-0.01	0.36	1						
Zr	-0.03	-0.19	-0.19	-0.15	-0.16	0.05	0.54	0.62	-0.06	-0.14	-0.09	-0.14	1					
Mo	0.23	0.22	0.18	0.24	-0.03	-0.16	-0.17	0.08	0.25	0.04	-0.12	-0.41	-0.02	1				
Rb	0.82	0.82	0.81	0.77	0.38	-0.69	-0.06	-0.03	0.87	-0.04	0.07	-0.09	-0.10	0.31	1			
Ba	0.09	0.06	0.04	-0.15	0.38	-0.08	0.09	-0.01	0.00	0.49	0.24	-0.04	-0.03	0.26	-0.09	1		
Cl	0.13	-0.09	-0.11	-0.13	0.15	-0.30	0.60	0.98	0.12	-0.11	0.35	0.08	0.63	0.09	-0.03	-0.02	1	
Br	0.06	-0.17	-0.18	-0.13	0.06	-0.21	0.65	0.96	0.04	-0.10	0.21	-0.05	0.72	0.14	-0.09	0.01	0.96	1

Figure 6. Correlation coefficient matrix for the chemical components in the studied travertine.

The mean value of MnO content in Deir Alla and Suwayma travertines is about 0.14% and 0.03%, respectively, while the mean value for meteogene and thermogene travertines is 0.0188% and 0.091%, respectively [8]. MnO in Deir Alla and Suwayma travertines is most probably incorporated into calcium carbonate. The high percent of Mn in D26a sample is zoned in aragonite (Table 3). The low percent of MnO content reflects a high rate of travertine deposition in such areas [29]. Consequently, the precipitation rate in Suwayma was higher than that in Deir Alla. High percent of MnO was also recorded in Z1* tufa sample (Table 3), due to the presence of plants, especially mosses, which contain manganese in their internal structure, similar to those described by Obeidat [15]. Such MnO content indicates a slow rate of deposition of tufa [32]. Caboi et al. [33] found that Mn is mainly associated with calcite, while Fe is associated with the detrital minerals.

Pentecost [8] observed that when Fe levels are high, Mn tends to precipitate with it, but when Fe is low, Mn is incorporated into calcium carbonate. The selected travertine samples define a negative correlation between CaO and MnO (Figure 6). This might indicate that Mn is incorporated into calcium carbonate and replaces Ca.

Table 3. X-ray Fluorescence results of major and minor elements of travertine and tufa samples.

Chemical Constituents	D3	D8	D11	D18	D23b	D26a	D26b	D 29	S1d	S10	S19
	wt %										
CaO	55.38	55.38	55.81	52.42	54.1	52.92	55.39	55.45	48.38	54.31	52.22
SiO_2	0.22	nd	0.22	0.85	0.24	0.44	0.82	0.09	8.39	1.85	3.77
Al_2O_3	0.12	0.13	0.08	0.29	0.07	0.15	0.28	0.04	1.82	0.44	0.89
TiO_2	nd	nd	nd	0.06	nd	0.05	nd	nd	0.28	nd	0.13
Fe_2O_3	0.28	1.3	0.8	2.13	3.81	0.24	0.4	0.55	4.72	0.47	1.83
MgO	0.17	0.21	0.18	0.28	0.14	0.27	0.19	0.12	0.9	0.74	0.91
MnO	nd	nd	nd	0.03	0.03	1.93	nd	0.01	0.08	0.08	0.08
Na_2O	nd	nd	nd	0.04	0.04	0.13	nd	0.04	0.13	0.02	0.03
K_2O	0.01	nd	nd	nd	nd	0.03	0.04	nd	0.85	0.08	0.23
P_2O_5	dl	0.01	dl	0.02	0.01	0.01	0.01	dl	0.01	0.19	0.08
SO_3	0.02	0.02	0.02	nd	0.01	0.07	0.02	0.01	0.03	0.11	0.08
	mg/kg										
Sr	842	423	539	4189	833	3919	1200	788	511	1061	898
Zr	31	nd	nd	nd	nd	nd	nd	nd	nd	nd	nd
Mo	307	357	349	377	285	287	378	379	389	350	321
Rb	nd	nd	nd	nd	nd	nd	nd	nd	87	nd	nd
Ba	nd	447	nd	378	nd	2825	nd	nd	1035	8339	758
Cl	nd	nd	nd	nd	nd	88	nd	40	188	31	12
Br	nd	nd	nd	nd	nd	nd	nd	nd	nd	nd	nd
Chemical Constituents	S27	S32	S34	S38	S39	S45	S47b	Z1	Z1*	Z3	Z4
	wt %										
CaO	55.29	55.28	53.57	54.23	53.84	51.93	51.41	52.54	52.45	55.41	55.03
SiO_2	0.89	1.04	1.95	1.21	1	2.01	2.81	2.03	1.24	0.81	0.82
Al_2O_3	0.28	0.23	0.35	0.32	0.28	0.44	0.55	0.1	0.15	0.2	0.22
TiO_2	nd	nd	0.08	0.06	0.09	0.08	0.11	nd	nd	nd	nd
Fe_2O_3	0.83	0.47	0.87	0.91	1.81	0.53	3.43	0.88	0.18	0.17	0.47
MgO	0.42	0.9	1.03	0.82	0.89	0.79	0.92	0.54	0.74	0.47	0.85
MnO	0.03	0.04	nd	nd	0.04	0.12	0.15	2.1	0.03	nd	nd
Na_2O	nd	nd	0.02	0.02	0.05	0.28	0.04	1.17	0.73	nd	0.02
K_2O	0.05	0.05	0.08	0.05	0.04	0.1	0.18	0.1	0.11	0.04	0.05
P_2O_5	dl	0.01	0.03	0.02	0.02	0.02	0.03	0.01	0.01	0.01	0.01
SO_3	0.02	0.04	0.09	0.06	0.12	0.14	0.14	0.07	0.18	0.07	0.1
	mg/kg										
Sr	588	555	1008	802	980	1811	7987	2213	897	585	837
Zr	nd	nd	nd	nd	nd	nd	nd	33	nd	nd	nd
Mo	343	234	385	410	328	373	228	378	358	308	331
Rb	nd	nd	nd	nd	nd	48	nd	nd	nd	nd	nd
Ba	547	535	8341	10,968	1019	1159	1812	nd	2955	822	1305
Cl	nd	nd	24	nd	39	23	297	877	1545	45	338
Br	nd	nd	nd	nd	nd	nd	nd	54	134	nd	nd

Z1 and Z1* are tufa samples. nd: is not detected; dl: detection limit; D: Deir Alla; S: Suwayma; Z: Az Zara.

3.2.3. Alkali Metals

The alkali metals are believed to occupy interstitial positions in the calcite lattice with a frequency of occurrence: Li > Na > K > Rb [34]. However, in aragonite, they substitute for Ca^{2+} in the crystal structure. Therefore, they are more easily co-precipitated with aragonite than with calcite [34]. It is noted that some Na concentrations in the samples are associated with Cl to form halite. In Deir Alla travertine, Na and K contents average 0.085% and 0.022%, respectively, and, similar to those found in the super-ambient travertines of Matlock Bath, UK [35]. Therefore, Deir Alla travertine is considered as super-ambient meteogene travertine, especially when the water temperature rises above ambient issuing hot springs (>37 °C) [8]. Figure 5c shows that K is correlated with Al, with a positive 0.98 correlation coefficient (Figure 6). This may indicate that they were found during the same mineral phases, such as clay minerals (potassium aluminum silicates). This is supported by the positive 0.98 correlation coefficient with SiO_2 (Figure 6).

3.2.4. Sr

The Sr content of Deir Alla samples ranges from 423 to 4189 mg/kg. In Suwayma travertine, Sr ranges from 511 to 7987 mg/kg, whereas Az Zara travertine and tufa Sr ranges from 585 to 2213 mg/kg (Table 3). The range of Sr concentration in the meteogene travertine is from 9 to >2930 mg/kg, and in the thermogene travertine is 20–14,000 mg/kg. In general, the Sr proportion in calcite is less than that in aragonite [8]. The transformation of aragonite to calcite during meteoric water circulation in depth leads to precipitation of calcite with less Sr content [8].

There is probably no single universal factor that controls calcite and aragonite precipitation in spring systems. For example, calcite was precipitated directly from waters with temperatures of >90 °C in Kenya [29] and New Zealand [36]. According to Fouke et al. [1], only aragonite precipitates at water temperatures >44 °C, whereas calcite and aragonite precipitate at 30–43 °C, while calcite alone forms when temperature falls below 30 °C. The Sr proportion increases in high-temperature travertine facies [1]. Therefore, the low temperature Deir Alla travertines have low Sr concentrations and are commonly composed of pure calcite. Some samples that contain a high percent of Sr may reflect their high aragonite content, especially those in D26a samples that showed aragonite recrystallization under the microscope [19]. High Sr content in Suwayma travertine reflects deposition under high water temperature like aragonite, before its complete transformation into calcite. The association of aragonite and calcite in Az Zara samples indicates their formation under high water temperatures (25–45 °C), before the complete transformation of aragonite into calcite.

Kinsman [37], Veizer and Demovic [38] considered that the diagenetic solution is one of the main factors that control the distribution of Sr in the carbonate rocks. In addition, Cipriani et al. [39,40] found an appropriate correlation between Sr content and porosity, which means that calcite must contain less Sr as porosity decreases. Deir Alla travertine was affected by diagenetic processes, such as their tight lithification, high compaction, low pore spaces, and good cementation. In contrast, Suwayma travertine was affected much less by diagenetic processes, as indicated by their lose lithification, low compaction, high pore spaces, and weak cementation. A high Sr value was recorded in the Z1 sample of the spongy tufa that was slightly affected by diagenetic processes, as compared to hard travertine samples Z3 and Z4.

The negative correlation between Ca and Sr abundances in travertine samples indicates that some Ca was replaced by Sr in the calcite lattice. Ichikuni [41] observed that the partition coefficients of Sr are higher in both aragonite and calcite, if a small amount of Mn has substituted for Ca in the lattice.

3.3. Stable Isotopes

3.3.1. Background Information

The ($^{13}C/^{12}C$) and ($^{18}O/^{18}O$) ratios undergo discrete change during biogeochemical cycling, allowing inferences about their origin and past history. Such ratios provide an expanded scale for the comparative small differences observed [42]. For travertine, fractionation of carbon and oxygen isotopes is of particular interest. They provide information about the carbon dioxide source, the physio-chemical conditions of the precipitation event (rate and temperature), and the influence of metabolic processes. In favorable circumstances, they provide information about the temperature at which deposition occurred, and assist in radiometric dating [42].

Stable oxygen isotope ratios are of great interest to geochemists and carbonate sedimentologists. As the lithosphere water exchanges oxygen atoms with the dissolved carbon dioxide, and where equilibrium is established, the difference between the ratio of oxygen isotopes in carbonates and depositing water can be used to estimate past water temperatures, assuming that the fluid source has not changed through time. Information can also be obtained on rates of water evaporation with time [8,42].

The carbon stable isotopes also undergo fractionation during chemical reactions. The fractionations of $\delta^{13}C$ occur between gas phase CO_2 and the dissolved carbonate species. The CO_2 gas is slightly depleted in ^{13}C, whilst bicarbonates and carbonates are enriched by about 9% at 20 °C [8].

Carbonates have $\delta^{18}O$ (VSMOW) values ranging from about 0 to +35%. Their interpretation is generally less straightforward than carbon isotope values because carbonate oxygen readily exchanges with oxygen in water molecules [8]. Biotical and geochemical processes such as photosynthesis and temperature decrease during flow process, evaporation, mixing from soil profile or surface waters will exert an influence on the final oxygen isotope value for any carbonate sample [43,44].

3.3.2. Deir Alla Travertine

The results indicate that the $\delta^{13}C$ isotope values in the eight samples from Deir Alla are between −1.45 and 2.99 (‰VPDB) (Table 4), while the $\delta^{18}O$ values vary from −8.13 to 8.17 (‰VPDB). Deir Alla samples are described as super-ambient meteogene travertines [8]. The range of stable carbon isotope value in such a travertine type is from −12 to +2 (‰VPDB), and the median of stable oxygen isotope is −7.92 (‰VPDB) [8]. The bivariate distribution of $\delta^{18}O$ and $\delta^{13}C$ (‰VPDB) values in Deir Alla samples (Figure 7) shows four different travertine lithofacies that are: shrub, coated bubble, paper-raft + reed, and crystalline crust lithofacies. The stable isotope values are variable in the different sub-environments, due to possible variations in conditions (i.e., water temperature, flow velocity and distance from spring, surface area and biological effects). Figure 7 indicates that Deir Alla travertine is enriched in ^{13}C. The high $\delta^{13}C$ values in some travertine facies (as shown in D26b and D29 samples) can be explained by a combination of several processes: (i) increased CO_2degassing processes related to travertine formation [45]; and (ii) increased input from heavier thermal-spring water ^{13}C-enriched groundwater. The cooling process of spring water during the flow tract is responsible for the loss of $\delta^{18}O$ from the water system by evaporation; as a result, the $\delta^{18}O$ value will increase. This case was observed only in the D3 sample.

3.3.3. Suwayma Travertine

The $\delta^{13}C$ isotope values in the six samples of Suwayma travertine vary between −0.18 and 1.28 (‰VPDB), while $\delta^{18}O$ values vary from −5.38 to 8.12 (‰VPDB) as shown in Table 4. Suwayma travertine is considered as thermogene travertine. The stable carbon isotope in such travertine type ranges from −1 to +10 (‰VPDB), and the median of stable oxygen isotope is −0.95 (‰VPDB) [8].

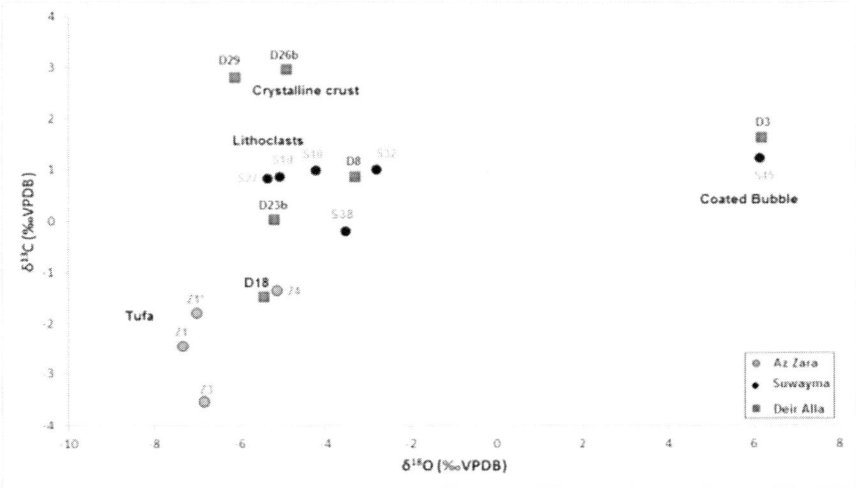

Figure 7. Bivariate plot of δ^{18}O and δ^{13}C (‰VPDB) values of the studied travertine samples.

Table 4. Values of δ^{18}O and δ^{13}C in the studied samples.

Sample	Facies	δ^{18}O (‰VPDB)	δ^{13}C (‰VPDB)
D3	coated bubble	8.17	1.88
D8	paper-raft + reed	−3.32	0.89
D18	Shrub	−5.48	−1.45
D23b	paper-raft + reed	−5.22	0.05
D26b	crystalline crust + palaeosol	−4.93	2.99
D29	crystalline crust	−8.13	2.83
S1d	lithoclast + pebbly	−5.1	0.89
S10	Lithoclast	−4.25	1.01
S27	Lithoclast	−5.38	0.85
S32	Lithoclast	−2.84	1.03
S38	Reed	−3.58	−0.18
S45	coated bubble	8.12	1.28
Z1	macrophyte tufa	−7.34	−2.45
Z1*	macrophyte tufa	−7.02	−1.78
Z3	coated bubble travertine	−8.83	−3.52
Z4	lithoclast + paper-raft travertine	−5.14	−1.34

VPDB: Vienna Pee Dee belemnite, standard value for C isotopes.

The $\delta^{18}O$ and $\delta^{13}C$ (%VPDB) bivariate distribution of Suwayma travertines shows three different lithofacies: lithoclast, reed, and coated bubble (Figure 7). Such results explain the enrichment of $\delta^{13}C$, which depends on the depositional environment fluctuations. The $\delta^{13}C$ is likely enriched due to the increase of degassing processes that are associated with the minor microbiologic and photosynthetic activities. The occurrence of ^{13}C-enriched carbonate in the Suwayma slope indicates the presence of ^{13}C-enriched surface water and groundwater that might be washed-out during hydrological open systems. It is quite evident that the low $\delta^{18}O$ values are related to the poor evaporation effects and high water temperature during travertine formation. Additionally, the $\delta^{18}O$ depletion in the thermogene water was displayed by the isotopic exchange with the host rock [46].

3.3.4. Az Zara Tufa

The $\delta^{18}O$ and $\delta^{13}C$ values in the Az Zara tufa deposits are distinctively low compared with the other areas (Figure 7). This may indicate different isotopic signatures. The average values of carbon and oxygen stable isotopes are −2.3‰ ± 0.2‰ VPDB and −8.58‰ ± 0.7‰ VPDB, respectively.

One per mill (1%) drop in the carbonate $\delta^{18}O$ value is equivalent to about a 4.5 °C temperature rise under equilibrium conditions [8]. The $\delta^{18}O$ compositions reflect the short stagnancy time of water. The lack of evaporation or continuous recharge could be the reason in such a fluvio-palustrine depositional environment. The low $\delta^{18}O$ values may also have resulted from isotopic exchange between the circulating thermal waters and the country rock and/or water adsorbed in the clay minerals.

The occurrence of ^{13}C-poor carbonate in the Az Zara area indicates the presence of ^{13}C-poor surface and groundwater that might be washed-out when the hydrological systems are closed. The variations in $\delta^{13}C$ are minimal (−3.52‰ to −1.34‰), reflecting similar carbon sources for all facies. The carbon dioxide is believed to originate from Upper Cretaceous limestone decarbonation reaction as a result of the Karst process; however, a magmatic component may also be significant and originates from the Pleistocene basaltic activity, which is dominant within two of the studied sites (Figure 2), and probably records the influence of meteoric and soil-derived CO_2.

3.4. Travertine Dating

3.4.1. Background Information

According to Srdoč et al. [47,48], if travertine is to be dated using ^{14}C, then the sources of carbon must be known. The ^{14}C dating method is not reliable for thermogene travertine because their carbon content is fossil (dead carbon/inorganic carbon). More reliable dates may be obtained by using a U-Series date [49,50] or by using the Electron Spin Resonance method [50]. For instance, where cosmogenic carbon is the only source, then dating is straightforward and the level of ^{14}C in the sample gives a direct measure of age [8]. The decay of isotopes ^{14}C has been used most extensively, and can date deposits up to 30 Ka. The isotope ^{14}C has a half-life of 5730 years and forms continuously in the atmosphere by the interaction of cosmic rays with molecules of nitrogen.

3.4.2. Age of Travertine

The previous studies did not provide any absolute ages of travertine in Jordan. The present work applied the radiocarbon method for dating Deir Alla and Suwayma travertines. Two rock samples were selected for radiocarbon dating. The travertine age calculations unfortunately yielded unreliable ages, and these indicate an age of 32.84 ± 0.005 Ka (Late Pleistocene).

4. CONCLUSIONS

The mineralogical compositions of the total travertine samples consist mainly of calcite with varying amounts of quartz (i.e., detrital quartz). Moreover, some aragonites are found in Deir Alla and Suwayma travertines. Based on geochemical studies, low Mg calcite is the main travertine component in all locations examined here. Accordingly, the Deir Alla travertines were classified as inactive super-ambient meteogene deposits. The Suwayma and Az Zara travertines were classified as inactive thermogene. Moreover, Az Zara tufa was classified as an active thermogene deposits.

The Sr abundances indicate pure calcite in the Deir Alla travertines and the diagenetic effect to be more than that recorded in Suwayma and Az Zara travertines. The Suwayma travertine was deposited as aragonite that completely transformed into calcite. Moreover, Az Zara travertine and tufa were deposited under high temperature water as the component of aragonite and calcite. Such aragonites are also completely transformed into calcite. The high $\delta^{13}C$ values in the travertine facies are related to evaporation effects, CO_2 degassing processes, increased input of ^{13}C-enriched groundwater, and the presence of surface and groundwater transportation medium in hydrological open systems. The high $\delta^{18}O$ values may be attributed to evaporation effects and low water temperature that prevailed during travertine formation. The travertine formed in the Late Pleistocene.

ACKNOWLEDGMENTS

The authors thank the Deanship of Research and Graduate Studies of Hashemite University very much for funding this work through the project entitled "Genesis and palaeo-environmental significance of hot spring travertines in Jordan". A huge thanks is due to Engneer Qais Al Qaisi, the Director of Travco Company (Amman, Jordan), who granted fieldwork access to study the travertine at the company site at Deir Alla. In addition, earnest thanks are due to Al al-Bayt University for the XRF analysis and the Water Authority for the radiogenic and stable isotope measurements. A special thanks is also extended to the anonymous reviewers of this article for their critical comments and suggestions.

AUTHOR CONTRIBUTIONS

Khalil Ibrahim, Sana' M. Al-Thawabteh, Issa M. Makhlouf and Ali R. El Naqah conducted the field work; Sana' M. Al-Thawabteh performed the mineral and chemical analysis; Khalil Ibrahim and Sana' M. Al-Thawabteh analyzed the data; and Khalil Ibrahim wrote the paper.

CONFLICTS OF INTEREST

The authors declare no conflict of interest.

REFERENCES

1. Fouke, B.W.; Farmer, J.D.; Des Marais, D.J.; Pratt, L.; Sturchio, N.C.; Burns, P.C.; Discipulo, M.K. Depositional facies and aqueous-solid geochemistry of travertine-depositing hot springs (Angel Terrace, Mammoth Hot Springs, Yellowstone National Park, USA). J. Sediment. Res. 2000, 70, 565–585.

2. Jamtveit, B.; Hammer, Ø.; Andersson, C.; Dysthe, D.K.; Heldmann, J.; Vogel, M.L. Travertines from the Troll thermal springs, Svalbard. Norwegian. J. Geol. 2006, 86, 387–395.
3. Fouke, B.W. Hot-spring systems geobiology: Abiotic and biotic influences on travertine formation at Mammoth hot springs, Yellowstone National Park, USA. Sedimentology 2011, 58, 170–219.
4. Zhu, T.; Dittrich, M. Carbonate precipitation through microbial activities in natural environment, and their potential in biotechnology: A review. Front. Bioeng. Biotechnol. 2016, 4, 4. [PubMed]
5. Capezzuoli, E.; Candin, A.; Pedley, M. Decoding tufa and travertine (fresh water carbonates) in the sedimentary record: The state of the art. Sedimentology 2014, 61, 1–21.
6. Rodrigo-Naharro, J.; Delgado, A.; Herrero, M.J.; Granados, A.; Pérez del Villar, L. Current travertines precipitation from CO2-rich groundwaters as an alert of CO2 leakages from a natural CO2 storage at Gañuelas-Mazarrón Tertiary Basin (Murcia, Spain). Inf. Técnicos Ciemat 2013, 1279, 1–53.
7. Dreybrodt, W.; Buhmann, D.; Michaelis, J.; Usdowski, E. Geochemically controlled calcite precipitation by CO2outgassing: Field measurements of precipitation rates in comparison to theoretical predictions. Chem. Geol. 1992, 97, 285–294.
8. Pentecost, A. Travertine, 1st ed.; Springer: London, UK, 2005; ISBN 978-1-4020-3606-4.
9. Pentecost, A.; Vile, H.A. A review and reassessment of travertine classification. Géogr. Phys. Quat. 1994, 48, 305–314.
10. Alessandro, D.W.; Giammanco, S.; Bellomo, S.; Parello, F. Geochemistry and mineralogy of travertine deposits of the SW flank of Mt. Etna (Italy): Relationships with past volcanic and degassing activity. J. Volcanol. Geotherm. Res. 2007, 165, 64–70.
11. Ford, T.D.; Pedley, H.M. A review of tufa and travertine deposits of the world. Earth Sci. Rev. 1996, 41, 117–175.
12. Arenas, C.; Gutiérrez, F.; Osàcar, C.; Sancho, C. Sedimentology and geochemistry of fluvio-lacustrine tufa deposits controlled by evaporate solution subsidence in the central Ebro Depression, NE Spain. Sedimentology 2000, 47, 883–909.
13. Jasser, D. Investigation of Deir Alla Travertine; Natural Resources Authority: Amman, Jordan, 1984.
14. Khoury, H.; Salameh, E.; Udluft, P. On the Zerqa Ma'in (thermal kallirrhoes) travertine/Dead Sea (hydrochemistry, geochemistry and isotopic composition). Neues Jahrb. Geol. Palaontol. Mh 1984, H8, 472–484.
15. Obeidat, O. 1992. Geochemistry, Mineralogy, and Petrography of Travertine of Deir Alla and Zerqa Ma'in Hot Springs. Master's Thesis, Yarmouk University, Irbid, Jordan, 1992.
16. Ibrahim, K.M.; Moh'd, B.K.; Masri, A.I.; Al-Taj, M.M.; Musleh, S.M.; Alzughoul, K.A. Volcanotectonic evolution of central Jordan: Evidence from the Shihan volcano. J. Afr. Earth Sci. 2014, 100, 541–553.
17. Smit, J.; Brun, J.P.; Cloetingh, S.; Ben-Avraham, Z. The rift-like structure and asymmetry of the Dead Sea Fault. Earth Planet. Sci. Lett. 2010, 290, 74–82.
18. Ferry, M.; Meghraoui, M.; Abou Karaki, N.; Al-Taj, M.; Khalil, L. Episodic behavior of the Jordan Valley section of the Dead Sea fault inferred from a 14-kyr long integrated catalogue of large earthquakes. Bull. Seism. Soc. Am. 2011, 101, 39–87.
19. Al-Thawabteh, S.M. 2008. Sedimentology, Geochemistry, and Petrographic Study of Travertine Deposits along the Eastern Side of the Jordan Valley and Dead Sea Areas. Master's Thesis, Hashemite University, Zarqa, Jordan, 2008.
20. Salameh, E.; Bannayan, H. Water Resources of Jordan: Present Status and Future Potentials; Friedrich Ebert Shiftung: Amman, Jordan, 1993.
21. Salameh, E.; Udluft, P. The Hydrodynamic Pattern of Central Jordan; Geol Jb, C38: Hannover, Germany, 1985.

22. Water Authority of Jordan. Open Files of the Water Authority of Jordan; Water Authority of Jordan: Amman, Jordan, 1991.
23. Salameh, E. Curative Water in Jordan; WRSC Bull, 7th Issue; University of Jordan: Amman, Jordan, 1988.
24. Abu Ajamieh, M. The Geothermal Resources of Zarqa Ma'in and Zara; Natural Recourses Authority: Amman, Jordan, 1980.
25. Vogel, J.C. Carbon-14 dating of groundwater. In Isotopes Hydrology; IAEA: Vienna, Austria, 1970.
26. Thorpe, P.M.; Holyoak, D.T.; Preece, R.C.; Willing, M.J. Validity of corrected 14C dates from calcareous tufa. In Actes Colloq. l'AGF Form. Carbonatées Externs Travertines Paris; Springer: London, UK, 1981; pp. 151–156.
27. Greensmith, J.T. Petrology of the Sedimentary Rocks, 6th ed.; George Allen and UNWIN: London, UK, 1978; ISBN 0 04 552011 9.
28. Kawano, J.; Shimobayashi, N.; Miyake, A.; Kitamura, M. Precipitation diagram of calcium carbonate polymorphs: Its construction and significance. J. Phys. Condens. Matter 2009, 21, 425102–425107. [PubMed]
29. Jones, B.; Renaut, R.W. Calcareous spring deposits in continental settings. In Carbonates in Continental Settings: Facies, Environments, and Processes; Alonso Zarza, A.M., Taner, L.H., Eds.; Elsevier Science: Amsterdam, The Netherlands, 2010; Volume 61, pp. 177–224.
30. Brian, J. Review of Calcium Carbonate Polymorph Precipitation in Spring Systems. Sediment. Geol. 2017, 353, 64–75.
31. Özkul, M.; Kele, S.; Gökgöz, A.; Shen, C.C.; Jones, B.; Baykara, M.O.; Fórizs, I.; Nemeth, T.; Chang, Y.W.; Alcicek, M.C. Comparison of the Quaternary travertine sites in the Denizli Extensional Basin based on their depositional and geochemical data. Sediment. Geol. 2013, 294, 179–204.
32. Rao, C.P. Petrography, trace elements and oxygen and carbon isotopes of Grodon group carbonates (Ordovician), Florentine Valley, Tasmania, Australia. J. Sediment. Geol. 1990, 80, 221–231.
33. Caboi, R.; Cidu, R.; Fanfani, L.; Zuddas, P.; Zuddas, P.P. Geochemistry of Funtana Maore travertines (Central Sardinia, Italy). Miner. Petrogr. Acta 1991, 34, 77–93.
34. Okumura, M.; Kitano, Y. Coprecipitation of alkali metal ions with calcium carbonate. Geochim. Cosmochim. Acta 1986, 50, 49–58.
35. Pentecost, A. The origin and development of the travertines and associated thermal waters at Matlock Bath, Derbyshire. Proc. Geol. Assoc. 1999, 110, 217–232.
36. Jones, B.; Renaut, R.W.; Rosen, M.R. High-temperature (>90 °C) calcite precipitation at Waikite Hot Springs, North Island, New Zealand. J. Geol. Soc. 1998, 153, 481–498.
37. Kinsman, D.J. Interpretation of Sr+2 concentration in carbonate minerals and rocks. J. Sediment. Pterol. 1989, 39, 488–508.
38. Veizer, J.; Demovic, R. Strontium as a tool in facies analysis. J. Sediment. Pterol. 1974, 44, 93–115.
39. Cipriani, N.; Ercoli, A.; Malesani, P.; Vannucci, S. I travertini di Rapolano Term (Siena). Mem. Soc. Geol. Ital. 1972, 11, 31–48.
40. Cipriani, N.; Malesani, P.; Vannucci, S. I travertine dell'Italia centrale. Boll. Serv. Geol. Ital. 1977, 98, 85–115.
41. Ichikuni, M. Partition of strontium between calcite and solution: Effect of substitution by manganese. Chem. Geol.1973, 11, 315–319.
42. Sun, H.; Liu, Z. Wet-dry seasonal and spatial variations in the δ13C and δ18O values of the modern endogenic travertine at Baishuitai, Yunnan, SW China and their paleoclimatic and paleoenvironmental implications. Geochim. Cosmochim. Acta 2010, 74, 1018–1029.
43. Chafetz, H.S.; Guidry, S.A. Bacterial shrubs, crystal shrubs, and ray-crystal crusts: Bacterially induced vs. a biotic mineral precipitation. Sediment. Geol. 1999, 128, 57–74.

44. Guo, L.; Andrews, J.; Riding, R.; Dennis, P.; Dresser, Q. Possible microbial effects on stable carbon isotopes in hot-spring travertines. J. Sediment. Res. 1996, 66, 468–473.
45. Turi, B. Stable isotope geochemistry of travertines. In Handbook of Environmental Isotope Geochemistry; Fritz, B.P., Fontes, L.J.C., Eds.; Elsevier: Amsterdam, The Netherlands, 1986; ISBN 9781483289830.
46. Fritz, P. Der Isotopengehalt der Mineralwasser von Stuttgart und Umgebung und ihrer ittel pleistozaenen Travertin-Ablagerungen. Jber. Mitt. Oberrhein. Geol. Ver. 1968, 50, 53–58.
47. Srdoč, D.; Chafetz, H.; Utech, N. Radiocarbon dating of travertine deposits, Arbuckle Mountains, Oklahoma. Radiocarbon 1989, 31, 819–828.
48. Nishikawaa, O.; Furuhashia, K.; Masuyamaa, M.; Ogatab, T.; Shiraishia, T.; Shenc, C. Radiocarbon dating of residual organic matter in travertine formed along the Yumoto Fault in Oga Peninsula, northeast Japan: Implications for long-term hot spring activity under the influence of earthquakes. Sediment. Geol. 2012, 243, 181–190.
49. Lovea, A.J.; Shanda, P.; Karlstromd, K.; Crosseyd, L.; Rousseau-Gueutina, P.; Priestleya, S.; Wholinge, D.; Fultonf, S.; Keppela, M. Geochemistry and travertine dating provide new insights into the hydrogeology of the Great Artesian Basin, South Australia. Procedia Earth Planet. Sci. 2013, 7, 521–524.
50. Lebatard, A.; Alcicek, M.C.; Rochette, P.; Khatib, S.; Vialet, A.; Boulbes, N.; Bourles, D.L.; Demory, F.; Guipert, G.; Mayda, S.; et al. Dating the Homo erectus bearing travertine from Kocabas, (Denizli, Turkey) at least 1.1 Ma. Earth Planet. Sci. Lett. 2014, 390, 8–18.

CHAPTER 2

Geochemistry of Rare Earth Elements in Sediment of Central India

Bharat Lal Sahu[1], Keshaw Prakash Rajhans[1], Shobhana Ramteke[1], Khageshwar Singh Patel[1*], Irena Wysocka[2], Irena Jaron[2]

[1] *School of Studies in Chemistry/Environmental Science, Pt. Ravishankar Shukla University, Raipur, India*

[2]*Polish Geological Institute, Warsaw, Poland*

ABSTRACT

The rare earth elements (REE) are of great interests due to wide industrial applications. In the present work, geochemistry of REE in the contaminated pond sediment of Ambagarh Chowki, central India is described. The concentration ($n = 24$) of La, Ce, Pr, Nd, Eu, Sm, Gd, Tb, Dy, Ho, Er, Tm, Yb, Lu, Th and U was ranged from 16 - 119, 41 - 163, 4.1 - 28, 16 - 99, 0.7 - 2.7, 3.2 - 18, 2.7 - 16, 0.4 - 2.3, 2.2 - 13, 0.4 - 2.4, 1.1 - 6.1, 0.2 - 0.8, 0.9 - 4.8, 0.13 - 0.70, 6 - 25 and 0.46 - 1.86 mg/kg with mean value of 42 ± 9, 89 ± 13, 9.4 ± 1.9, 34 ± 7, 1.2 ± 0.2, 6.3 ± 1.2, 5.4 ± 1.0, 0.8 ± 0.1, 4.2 ± 0.8, 0.8 ± 0.2, 2.0 ± 0.4, 0.3 ± 0.1, 1.6 ± 0.3, 0.24 ± 0.04, 11 ± 2 and 1.14 ± 0.13 mg/kg, respectively. The concentration ratio of ∑LREE/∑HREE, La/Y and Th/U are discussed. The spatial distribution and sources of the REE are described.

Keywords: Rare Earth Element, Sediment, Geochemistry, Sources

1. INTRODUCTION

The rare earth elements (REE) including actinide have wide industrial applications in several fields i.e. nuclear energy as fuel and modifier, organic synthesis as catalyst, ceramics as coloring agents, X-ray tubes as phosphors, etc. [1] [2] . The distribution of REE in various environmental materials i.e. soil and sediments was reported [3] - [13] . Recently, the health hazards of the REE were highlighted [14] [15] . The environment of the Ambagarh Chowki block, central India is seriously contaminated with As and heavy metals [16] . In this work, the geochemistry of the REE in pond sediments of Ambagarh Chowki block, Rajnandgaon district, central India is described.

2. MATERIALS AND METHODS

2.1. Study Area

Arsenic and other heavy metals are deposited in the environment of Ambagarh Chowki block, Rajnandgaon district, Chhattisgarh, India over area »3000 km^{-2}. The Ambagarh Chowki block falls in a tribal belt with population of ≈0.1 million. This block covers 155 villages and towns which suffering with serious health hazards due to contamination of the environment with toxic elements. In this work, 24 ponds lie in different 11 villages (locations) of Ambagarh Chowki block were selected for the investigation of the REE contamination (Figure 1).

2.2. Sampling

The composite surface sediment sample (0 - 10 cm) was collected from five points in the equal ratio from each pond as prescribed in literature [17] . For the depth profile studies, the samples at the depth profile of 0 - 10, 10 - 20 and 20 - 30 cm of the Bhadsena pond was collected.

2.3. Analysis

The samples were dried, crushed and sieved out the particles of size ≤ 1 mm. A 0.25 g of each sediment sample was digested with 8 ml aqua regia in the closed microwave vessel for 15 min. An aliquot of the extract was diluted with deionized water for the further analysis. The *Thermo Fisher* ICP-OES and ICP-MS (Polish Geological Institute, NRI, Central

Chemical Laboratory, and Warsaw) were employed for analysis of the elements. The NCS DC 73382 CRM sediment sample was used for the quality control. The background levels of the REE reported in the upper earth crust were used for computation of the enrichment factor (E_f) by using following equation [18] .

$$E_f = \{[X_s]/[Al_s]\} / \{[X_e]/[Al_e]\}$$

where, X_S, Al_S, X_e and Al_e are concentrations of metal and Al in the soil and earth crust, respectively.

3. RESULTS AND DISCUSSION

The sediment was found to be colored from light brown (LBr) to dark brown (DBr), Table 1. The pH value of the extract was ranged from 6.5 - 8.1 with mean value of 7.2 ± 0.2.

3.1. Concentration of Elements

The concentrations of 49 elements in 24 pond sediments are shown in Tables 1-4. The concentration of elements i.e. Fe, Al, Ca, Mg, K, Mn, Ti, Na, P, S, Ba, V, Zn, Cu, As, Ni, Cr, Co, Sr, Pb, Rb, Ga, Li, Be, Mo, Ag, Cd, Sn, Sb, Cs, Bi, Th, U, Sc, Y, La, Ce, Pr, Nd, Eu, Sm, Gd, Tb, Dy, Ho, Er, Tm, Tb and Lu was ranged from 25,400 - 65,900, 11,800 - 34,700, 1500 - 16,100, 1500 - 30,200, 1200 - 4500, 256 - 1660, 62 - 735, 90 - 830, 90 - 570, 50 - 790, 88 - 304, 29 - 166, 32 - 100, 28 - 89, 10 - 53, 19 - 340, 24 - 1298, 9 - 76, 10 - 69, 10 - 40, 12 - 46, 7 - 15, 5 - 18, 0.9 - 3.4,0.14 - 1.32, 0.07 - 2.36, 0.05 - 0.39, 0.8 - 2.6, 0.27 - 0.97, 0.5 - 2.2, 0.10 - 0.30, 6.1 - 25, 0.5 - 1.9, 3.0 - 25, 11 - 61, 16 - 119, 41 - 163, 4.0 - 28, 16 - 99, 0.7 - 2.7, 3.2 - 18, 2.7 - 16, 0.4 - 2.3, 2.2 - 13, 0.4 - 2.4, 1.1 - 6.1, 0.15 - 0.83, 0.9 - 4.8 and 0.13 - 0.70 mg/kg with mean value of 44842 ± 5133, 23813 ± 2819, 5133 ± 1418, 6850 ± 2581, 2708 ± 243, 759 ± 158, 254 ± 79, 355 ± 93, 268 ± 45, 285 ± 81, 196 ± 21, 84 ± 16, 63 ± 7, 52 ± 6, 22 ± 5, 75 ± 30, 177 ± 118, 30 ± 7, 26 ± 6, 23 ± 3, 26 ± 3, 11 ± 1, 12 ± 2, 1.6 ± 0.2, 0.63 ± 0.14, 0.67 ± 0.23, 0.13 ± 0.03, 1.5 ± 0.2, 0.44 ± 0.05, 1.2 ± 0.2, 0.15 ± 0.03, 11 ± 2, 1.1 ± 0.1, 11 ± 2, 20 ± 4, 42 ± 8, 42 ± 8, 89 ± 13, 9 ± 2, 34 ± 7, 1.2 ± 0.2, 6.3 ± 1.2, 5.4 ± 1.0, 0.77 ± 0.15, 4.2 ± 0.8, 0.8 ± 0.2, 2.0 ± 0.4, 0.27 ± 0.05, 1.6 ± 0.3 and 0.24 ± 0.04 mg/kg, respectively. Among them, the highest concentration of Fe was marked. They occurred in following decreasing order in the sediment: Fe < Al < Mg < Ca < K < Mn < Na < S < P < Ti < Ba < Ce < Cr < V < Ni < Zn < Cu < La < < Nd < Co < Sr » Rb < Pb < As < Y < Li < Ga » Th » Sc < Pr < Sm < Gd < Dy < Er < Yb » Be < Sn < Eu » Cs < U < Tb » Ho < Ag < Mo < Sb < Tm < Lu < Cd » Bi.

3.2. Concentration of REE

The concentration of the LREE (i.e. La, Ce, Pr, Nd, Eu and Sm) and HREE (i.e. Gd, Tb, Dy, Ho, Er, Tm, Yb and Lu) was ranged from 82 - 431 and 8 - 46 mg/kg with mean value of 182 ± 30 and 15 ± 3 mg/kg, respectively. The ratio of LREE/HREE and Lu/Y was found to be in the range of 9 - 18 and 14 - 56 with mean value of 12 ± 1 and 27 ± 4, respectively. The concentration of REE and LREE/HREE and Lu/Y ratio were found to be higher than values reported in river sediments of the other regions of the World [3] - [11] .

Table 1. Concentration of major element in sediments, mg/kg.

S. No.	Pond location	Color	Fe	Al	Ca	Mg	K	Mn	Ti	Na
1	Murethitola-1	LBr	62,700	31,400	5800	6300	2900	778	735	160
2	Murethitola-2	LBr	64,900	34,700	7100	6800	3000	956	586	540
3	Murethitola-3	RBr	32,700	17,200	1800	1700	2100	447	166	90
4	Netamtola-1	DB	54,700	32,200	9400	14,300	2500	892	394	630
5	Netamtola-2	B	48,000	32,200	12,500	9200	4000	1474	140	830
6	Netamtola-3	LB	53,800	30,000	16,100	9300	1300	959	455	740
7	Kaudikasa-1	DBr	49,800	30,600	7600	10,800	4000	1109	264	490
8	Kaudikasa-2	RBr	49,500	19,200	2000	2800	4500	467	254	180
9	Kaudikasa-3	LB	25,400	19,200	4700	4800	2400	256	64	220
10	Ghorda-1	Br	45,600	29,400	3700	7700	2900	1660	173	130
11	Ghorda-2	LB	41,000	28,600	5800	30,200	1600	1148	556	100
12	Ghorda-3	LBr	57,700	34,300	7600	17,900	1200	671	571	150
13	Metepar-1	Br	43,600	26,700	4800	6600	2000	856	105	110
14	Metepar-2	LBr	59,000	23,400	3400	4900	2200	878	313	730
15	Metepar-3	LBr	65,900	25,100	3600	5000	2900	540	320	270
16	Bhadsena	LBr	38,000	21,500	2100	2500	3000	350	68	340
17	Sonsaitola-1	LB	28,000	18,600	4800	3700	3200	694	62	280
18	Sonsaitola-2	LBr	28,000	14,100	2500	1800	3400	388	92	330
19	Sonsaitola-3	DBr	32,500	14,800	2500	2200	3800	435	98	600
20	Meregaon	Br	50,000	19,600	3100	2500	2200	1393	185	250
21	Joratarai	DBr	53,100	23,600	3300	5100	2700	853	133	360
22	Thailitola-1	LBr	26,900	11,800	2500	2600	2500	285	125	230
23	Thailitola-2	RBr	33,500	14,700	1500	1500	1700	369	163	140
24	Jadutola	LB	31,900	18,600	5000	4200	3000	350	73	630

The concentration of Th and U in the sediments (n = 24) was ranged from 6.1 - 25 and 0.46 - 1.86 mg/kg with mean value of 11 ± 2 and 1.14 ± 0.13 mg/kg, respectively. The Th/U ratio was ranged from 6 - 28 with mean value of 11 ± 2. The concentration of actinide elements in the studied area was found to be comparable to values reported in soil of other regions of the country [12] [13] (Figure 2).

Table 2. Chemical characteristics of metal in pond sediment, mg/kg.

S. No.	P	S	Ba	V	Zn	Cu	As	Ni	Cr	Co	Sr	Pb
1	300	80	217	144	76	64	40	51	45	30	29	17
2	380	380	259	160	88	71	17	58	64	44	39	24
3	160	70	129	59	42	53	33	29	39	13	12	20
4	430	790	179	86	90	89	16	133	301	41	42	24
5	420	540	304	87	80	71	12	85	136	39	54	22
6	170	70	150	166	53	49	12	43	56	33	69	12
7	340	600	264	96	88	73	20	108	232	42	30	27
8	320	320	212	48	100	48	53	30	61	18	16	40
9	220	400	179	50	50	41	15	55	120	18	21	23
10	160	360	300	79	52	47	12	196	635	76	17	39
11	570	90	218	67	54	36	10	340	1298	62	26	22
12	330	110	161	93	65	63	15	194	658	30	28	10
13	200	70	222	95	53	46	19	51	51	28	35	19
14	170	190	204	142	60	56	17	48	45	38	27	15
15	180	180	211	152	68	56	18	49	45	26	27	13
16	330	310	188	55	65	37	12	31	33	13	13	32
17	220	300	213	50	40	35	28	38	50	20	25	19
18	290	470	140	38	54	37	28	23	29	12	12	22
19	220	370	154	47	44	53	30	28	30	14	14	17
20	170	50	221	77	45	42	11	37	47	34	18	27
21	340	300	209	98	87	59	25	77	133	36	22	31
22	150	180	118	29	48	28	34	22	24	11	10	23
23	90	60	88	48	32	32	45	19	30	9	10	24
24	280	550	165	52	67	50	11	49	82	25	21	37

3.4. Spatial Variations and Sources

The sediment samples were collected from 24 ponds which situated in 11 locations (i.e. villages) of Ambagarh Chowki block. The content of the REE in the pond sediments of 11 locations are shown in Figure 3. The highest content of the REE in the pond sediment of Bhadsena was observed. The REE content in sediment was found to increase as the depth profile was increased from 0 - 30 cm, may be due to geogenic origin and less binding with organic components of the sediment, Figure 4.

Table 3. Chemical characteristics of minor element in pond sediment, mg/kg.

S. No.	Rb	Ga	Li	Be	Mo	Ag	Cd	Sn	Sb	Cs	Bi	Th	U
1	32	14	14	1.57	0.67	0.15	0.12	1.4	0.46	2.18	0.3	13.2	1.64
2	23	15	13	1.35	0.40	0.69	0.26	1.4	0.35	1.41	0.1	7.2	0.77
3	27	8	8	1.47	0.85	0.13	0.05	1.4	0.43	0.98	0.3	15.7	1.36
4	12	11	17	1.16	0.27	0.73	0.39	1.2	0.51	0.52	0.1	6.1	0.65
5	22	12	14	1.98	0.40	0.92	0.19	2.1	0.4	0.95	0.1	10.2	1.36
6	15	10	11	1.06	0.24	0.07	0.05	1.1	0.42	1.00	0.1	6.5	1.01
7	29	12	17	1.55	0.52	1.60	0.21	1.7	0.52	1.23	0.1	8.0	1.14
8	46	11	9	1.68	1.32	0.16	0.11	1.3	0.59	1.63	0.2	9.3	1.12
9	20	7	9	1.29	0.26	1.70	0.16	1.5	0.27	0.97	0.1	10.1	1.04
10	39	13	18	2.55	0.92	0.56	0.14	2.6	0.50	2.09	0.1	14.1	1.42
11	15	9	18	0.86	0.18	0.26	0.06	1.0	0.35	1.19	0.1	7.6	0.74
12	14	12	15	1.39	0.14	0.19	0.05	2.0	0.3	1.08	0.1	8.0	0.46
13	28	10	14	1.47	0.20	0.13	0.06	1.3	0.34	1.69	0.1	10.1	1.34
14	21	12	11	1.1	0.68	0.78	0.13	1.1	0.38	1.08	0.1	7.3	1.15
15	31	13	12	1.14	0.81	0.69	0.10	1.3	0.39	1.32	0.2	8.7	1.01
16	44	12	12	3.37	0.84	0.6	0.10	2.6	0.43	1.46	0.2	24.9	1.86
17	25	8	9	1.58	0.39	0.19	0.06	1.4	0.34	1.1	0.1	11.5	1.08
18	28	9	7	2.07	0.84	1.07	0.14	1.8	0.45	0.8	0.2	14.4	1.17
19	31	9	7	2.08	0.95	0.79	0.10	1.9	0.45	1.05	0.3	13.3	1.17
20	22	10	9	1.32	0.96	0.52	0.06	1.3	0.36	1.02	0.1	10.4	0.95
21	27	11	14	1.33	1.19	1.06	0.19	1.5	0.97	1.26	0.2	7.9	1.2
22	23	7	5	1.28	0.63	0.75	0.08	0.8	0.42	0.83	<0.1	16.9	1.33
23	24	8	7	1.61	1.04	0.07	0.05	1.0	0.49	1.08	0.1	15.1	0.84
24	29	9	9	2.24	0.53	2.36	0.19	1.9	0.38	0.91	0.1	13.2	1.64

Table 4. Concentration of lanthanide in pond sediment, mg/kg.

S. No.	Sc	Y	La	Ce	Pr	Nd	Eu	Sm	Gd	Tb	Dy	Ho	Er	Tm	Yb	Lu
1	18.0	21.0	40.0	66.0	8.7	32.0	1.3	5.9	5.2	0.8	4.1	0.8	2.1	0.3	1.6	0.2
2	25.0	21.0	32.0	73.0	7.3	27.0	1.3	5.3	4.7	0.7	4.1	0.8	2.2	0.3	1.8	0.3
3	8.0	23.0	56.0	99.0	12.9	47.0	1.6	8.6	7.0	0.9	4.9	0.9	2.2	0.3	1.5	0.2
4	18.0	13.0	21.0	49.0	4.8	18.0	0.8	3.4	3.1	0.4	2.6	0.5	1.4	0.2	1.2	0.2
5	12.0	22.0	37.0	93.0	8.3	30.0	1.1	5.8	5.0	0.8	4.4	0.8	2.2	0.3	1.8	0.3
6	16.0	13.0	16.0	43.0	4.1	16.0	0.8	3.3	2.9	0.5	2.6	0.5	1.4	0.2	1.2	0.2
7	14.0	18.0	33.0	77.0	7.5	27.0	1.1	5.2	4.5	0.7	3.6	0.7	1.8	0.2	1.4	0.2
8	10.0	22.0	47.0	94.0	10.8	40.0	1.5	7.2	6.1	0.9	4.5	0.8	2.0	0.3	1.4	0.2
9	7.0	15.0	29.0	74.0	6.3	22.0	0.8	4.2	3.7	0.6	3.1	0.6	1.6	0.2	1.3	0.2
10	11.0	23.0	50.0	157.0	10.5	37.0	1.2	7.0	6.1	0.9	4.7	0.9	2.4	0.3	1.8	0.3
11	8.0	11.0	22.0	64.0	4.6	17.0	0.7	3.2	2.7	0.4	2.2	0.4	1.1	0.2	0.9	0.1
12	23.0	13.0	31.0	41.0	6.4	22.0	0.8	3.9	3.2	0.5	2.6	0.5	1.3	0.2	1.1	0.2
13	11.0	18.0	28.0	84.0	7.1	26.0	1.0	5.2	4.4	0.7	3.7	0.7	1.8	0.3	1.5	0.2
14	15.0	16.0	23.0	55.0	5.7	22.0	1.0	4.2	3.9	0.6	3.4	0.6	1.7	0.2	1.3	0.2
15	18.0	20.0	28.0	53.0	6.9	26.0	1.2	5.1	4.7	0.7	3.9	0.8	2.0	0.3	1.5	0.2
16	9.0	61.0	119.0	163.0	28.3	99.0	2.7	18.0	16.0	2.3	13.0	2.4	6.1	0.8	4.8	0.7
17	7.0	21.0	40.0	91.0	9.0	33.0	1.1	6.3	5.5	0.8	4.3	0.8	2.1	0.3	1.7	0.2
18	5.0	28.0	65.0	126.0	14.2	51.0	1.4	9.2	7.9	1.1	6.0	1.1	2.8	0.4	2.2	0.3
19	6.0	24.0	54.0	111.0	11.9	43.0	1.3	7.9	6.6	0.9	5.0	0.9	2.4	0.3	1.9	0.3
20	9.0	18.0	45.0	108.0	9.8	36.0	1.3	6.4	5.4	0.7	3.8	0.7	1.7	0.2	1.2	0.2
21	12.0	18.0	32.0	73.0	7.5	28.0	1.2	5.5	4.9	0.7	3.7	0.7	1.8	0.2	1.4	0.2
22	3.0	15.0	59.0	110.0	11.7	41.0	0.9	6.9	5.4	0.7	3.3	0.6	1.5	0.2	1.1	0.2
23	5.0	14.0	53.0	99.0	10.8	38.0	1.0	6.7	5.2	0.6	3.1	0.5	1.4	0.2	1.1	0.2
24	5.0	20.0	56.0	131.0	11.1	39.0	1.1	6.9	5.8	0.8	4.2	0.8	1.9	0.3	1.5	0.2

The correlation matrix of the REE is summarized in Table 5. All REE inclusive of Eu, Th and U are correlated well in the sediment, indicating origin from the similar sources likely to geogenic origin. They were partially correlated with elements i.e. Be, K, Rb, Cs, Mo, As, Sb, Bi, Sn, and Ag. Their negative correlations with elements i.e. Na, Mg, Ca, Ba, Sr, Al, Ga, V, Cr, Fe, Mn, Co, Ni, Cu, Zn, Cd and Pb were observed.

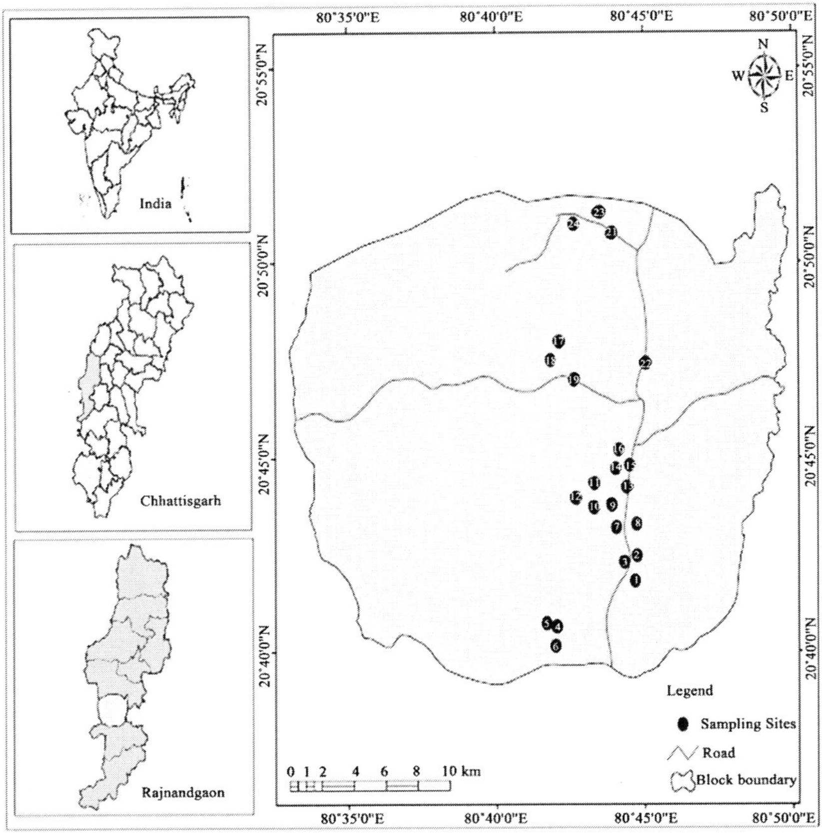

Figure 1. Representation of sampling locations (i.e. Murethitola-1, Murethitola-2, Murethitola- 3, Netamtola- 1, Netamtola-2, Netamtola-3, Kaudikasa-1, Kaudikasa-2, Kaudikasa-3, Ghorda-1, Ghorda-2, Ghorda-3, Metepar-1, Metepar-2, Metepar-3, Bhadsena, Sonsaitola-1, Sonsaitola-2, Sonsaitola-3, Meregaon, Joratarai,Thailitola-1,Thailitola-2 and Jadutola) in Ambagarh Chowki.

3.5. Enrichment

The E_f value for Sc, Y, La, Ce, Pr, Eu, Nd, Sm, Gd, Tb, Dy, Ho, Er, Tm, Yb, Lu, Th and U in the sediment was ranged from 0.8 - 4.2, 1.5 - 9.4,1.7 - 13, 1.4 - 12, 1.7 - 13, 1.6 - 9,1.7 - 12, 1.6 - 12, 1.9 - 13, 1.6 - 11,1.6 - 11, 1.6 - 9, 1.4 - 9, 1.4 - 9, 1.4 - 8, 1.1 - 7, 1.1 - 11 and 0.4 - 3 with mean value of 2.5 ± 0.4, 4 ± 1, 6 ± 1, 5 ± 1, 5 ± 1, 4 (± 1, 5 ± 1, 5 ± 1, 5 ± 1, 4 ± 1, 4 ± 1, 4 ± 1, 3 ± 1, 3 ± 1, 3 ± 1, 3 ± 1, 4 ± 1 and 1.7 ± 0.3, respectively. They were moderately enriched in the sediment. Their highest enrichments were marked in sediment of the Bhadsena village.

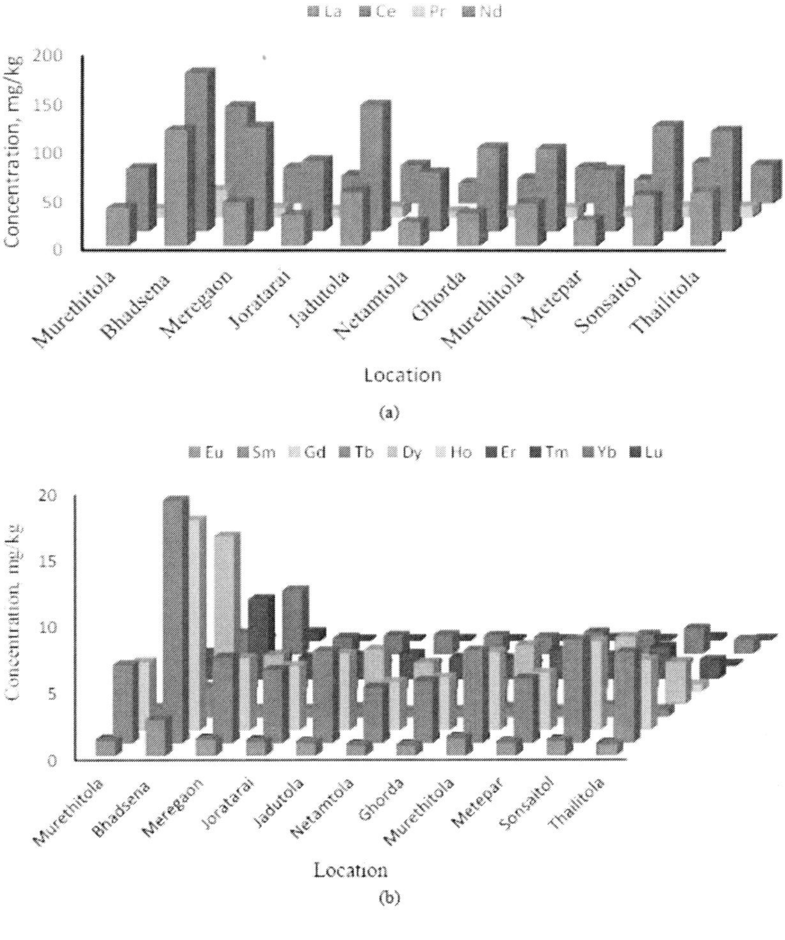

Figure 2. Spatial variation of REE concentration in sediment of Ambagarh Chowki.

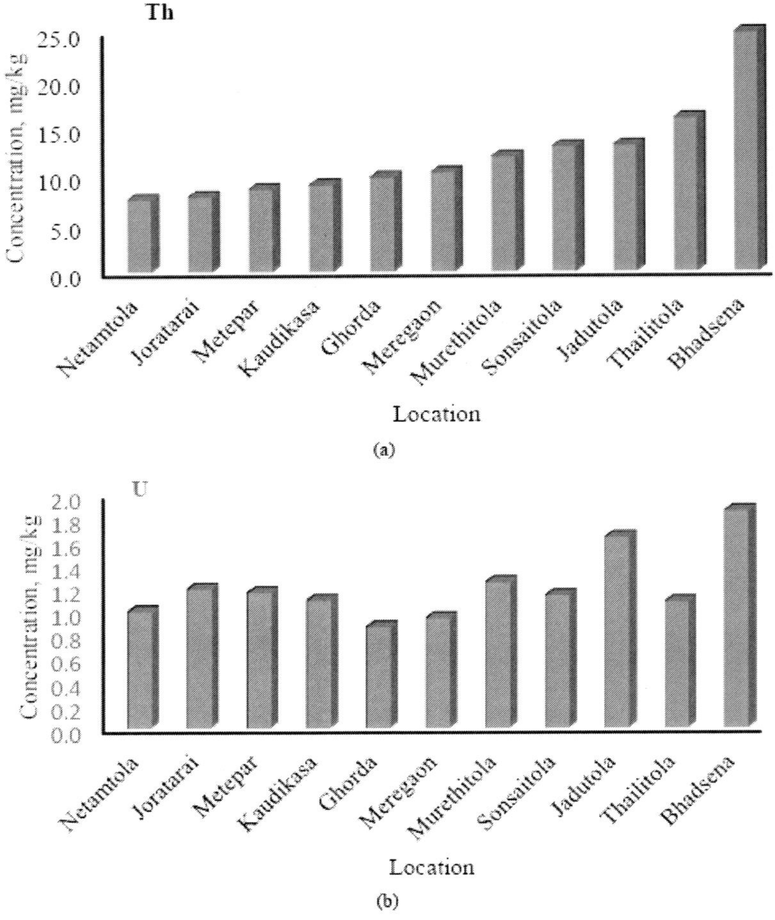

Figure 3. Spatial variation in concentration of actinide in sediment.

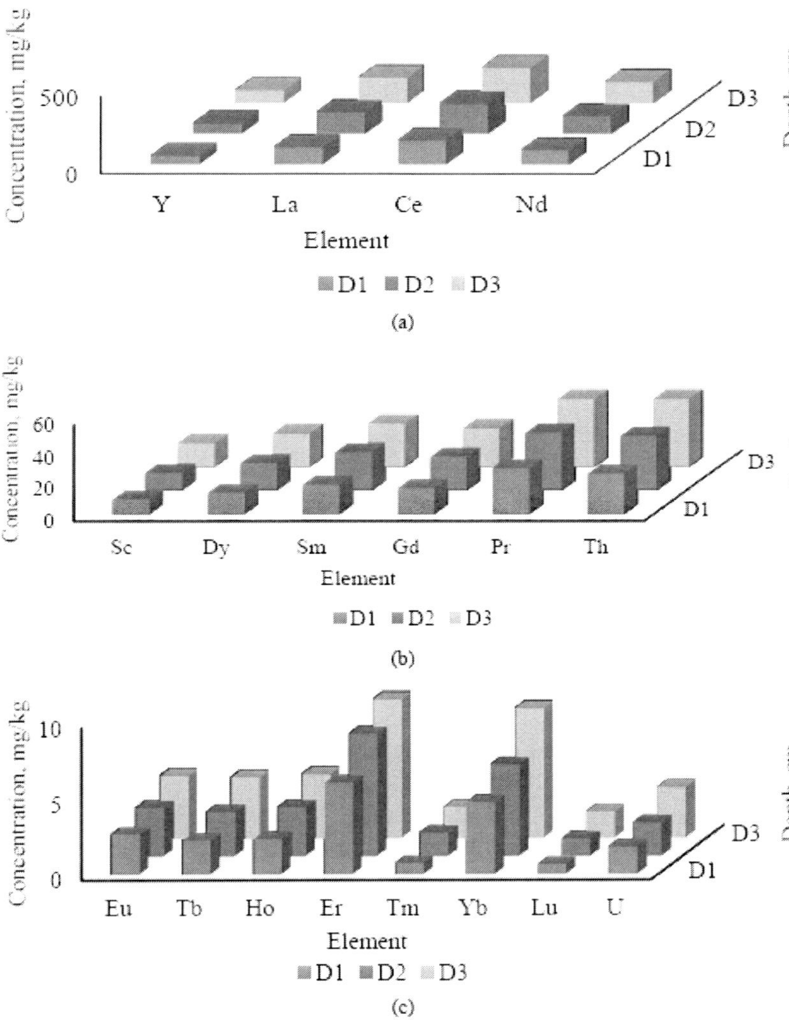

Figure 4. Depth profile studies of elements in Bhadsena pond sediment. D1 = 0 - 10 cm, D2 = 10 - 20 cm, D3 = 20 - 30 cm.

Table 5. Correlation matrix of REE.

	La	Ce	Pr	Nd	Eu	Sm	Gd	Tb	Dy	Ho	Er	Tm	Yb	Lu
La	1.00													
Ce	0.84	1.00												
Pr	0.99	0.80	1.00											
Nd	0.99	0.80	1.00	1.00										
Eu	0.84	0.63	0.90	0.91	1.00									
Sm	0.97	0.78	0.99	1.00	0.93	1.00								
Gd	0.95	0.76	0.98	0.99	0.94	1.00	1.00							
Tb	0.92	0.72	0.96	0.96	0.95	0.98	0.99	1.00						
Dy	0.89	0.69	0.94	0.94	0.94	0.96	0.98	1.00	1.00					
Ho	0.87	0.67	0.92	0.93	0.94	0.95	0.97	0.99	1.00	1.00				
Er	0.85	0.65	0.91	0.91	0.93	0.94	0.96	0.99	1.00	1.00	1.00			
Tm	0.83	0.64	0.89	0.89	0.90	0.92	0.94	0.98	0.99	0.99	1.00	1.00		
Yb	0.81	0.61	0.87	0.87	0.89	0.91	0.93	0.97	0.98	0.99	0.99	1.00	1.00	
Lu	0.83	0.63	0.89	0.89	0.90	0.92	0.94	0.97	0.98	0.99	1.00	1.00	1.00	1.00

4. CONCLUSION

The REE were moderately enriched in all ponds and their enrichments were found to decrease as the mass number of the REE was increased. The concentration of all REE in the sediment was increased vertically, may be due to origin from the rock weathering. Their highest contents were marked in the sediment of Bhadsena village, Rajnandgaon, India.

ACKNOWLEDGEMENTS

We are thankful to the Alexander von Humboldt Foundation, Bonn for financial support to one of author: KSP. Our sincere thanks are also extended to the UGC, New Delhi for awarding Rajiv Gandhi Scholarship to students: BLS, KPR and SR.

CITE THIS PAPER

Bharat Lal Sahu, Keshaw Prakash Rajhans, Shobhana Ramteke, Khageshwar Singh Patel, Irena Wysocka, Irena Jaron (2016) Geochemistry of Rare Earth Elements in Sediment of Central India. *Journal of Environmental Protection*,**07**,705-714.

REFERENCES

1. Cotton, S. (2006) Lanthanide and Actinide Chemistry. Wiley, Hoboken.
2. Ramesh, R., Ramanathan, Al., Ramesh, S., Purvaja, R. and Subramanian, V. (2000) Distribution of Rare Earth Elements and Heavy Metals in the Surficial Sediments of the Himalayan River System. Geochemical Journal, 34, 295-319.
3. Bounouira, H., Choukri, A., Elmoursli, R.C., Hakam, O.K. and Chakiri, S. (2007) Distribution of the Rare Earth Elements in the Sediments of the Bouregreg River (Morocco) Using the Instrumental Neutron Activation Analysis (INAA). Journal of Applied Sciences and Environmental Management, 11, 57-60.
4. Piper, D.Z. and Bau, M. (2013) Normalized Rare Earth Elements in Water, Sediments, and Wine: Identifying Sources and Environmental Redox Conditions. American Journal of Analytical Chemistry, 4, 69-83.
5. Antonina, A.N., Shazili, N.A.M., Kamaruzzaman, B.Y., Ong, M.C., Rosnan, Y. and Sharifah, F.N. (2013) Geochemistry of the Rare Earth Elements (REE) Distribution in Terengganu Coastal Waters: A Study Case from Redang Island Marine Sediment. Open Journal of Marine Science, 3, 154-159.
6. Obaje, S.O., Akpoborie, I.A., Ugbe, F.C. and Onugba, A. (2015) Rare Earth and Trace Elements Distribution in Sediments of River Gora, Minna Area, North-Central Nigeria: Implication for Provenance. Earth Science Research, 4, 103- 112.
7. Araújo, M.F., Corredeira, C. and Gouveia, A. (2007) Distribution of the Rare Earth Elements in Sediments of the Northwestern Iberian Continental Shelf. Journal of Radioanalytical and Nuclear Chemistry, 271, 255-260.
8. Xu, Y.Y., Song, J.M., Duan, L.Q., Li, X.G., Zhang, Y. and Sun, P.Y. (2010) Environmental Geochemistry Reflected by Rare Earth Elements in Bohai Bay (North China) Core Sediments. Journal of Environmental Monitoring, 12, 1547- 1555.
9. He, J., Lü, C.W., Xue, H.X., Liang, Y., Bai, S., Sun, Y., Shen, L.L., Mi, N. and Fan, Q.Y. (2010) Species and Distribution of Rare Earth Elements in the Baotou Section of the Yellow River in China. Environmental Geochemistry and Health, 32, 45-58.
10. Lee, M., Bae, W., Chung, J., Jung, H.S. and Shim, H. (2008) Seasonal and Spatial Characteristics of Seawater and Sediment at Youngil Bay, Southeast Coast of Korea. Marine Pollution Bulletin, 57, 325-334.
11. Tranchida, G., Oliveri, E., Angelone, M., Bellanca, A., Censi, P., D'Elia, M., Neri, R., Placenti, F., Sprovieri, M. and Mazzola, S. (2011) Distribution of Rare Earth Elements in Marine Sediments from the Strait of Sicily (western Mediterranean Sea): Evidence of Phosphogypsum Waste Contamination. Marine Pollution Bulletin, 62, 182-191.
12. Maharana, M., Krishnan, N. and Sengupta, D. (2010) Spatial Distribution of Gamma Radiation Levels in Surface Soils from Jaduguda Uranium Mineralization Zone, Jharkhand. Journal of Medical Physics, 35, 235-241.
13. Manigandan, P.K. and Natrajan, K.K. (2014) Activity Concentrations of Natural Radionuclides in Soils of Rainforest Sites in Western Ghats. International Journal of Students Research in Technology and Management, 2, 103-108.
14. Pagano, G., Guida, M., Tommasi, F. and Oral, R. (2015) Health Effects and Toxicity Mechanisms of Rare Earth Elements-Knowledge Gaps and Research Prospects. Ecotoxicological and Environmental Safety, 115, 40-48.
15. Pagano, G., Aliberti, F., Guida, M., Oral, R., Siciliano A., Trifuoggi, M. and Tommasi, F. (2015) Rare Earth Elements in Human and Animal Health: State of Art and Research Priorities. Environmental Research, 142, 215-220.
16. Patel, K.S., Shrivas, K., Brandt, R., Jakubowski, N., Corns, W. and Hoffmann, P. (2005) Arsenic Contamination in Water, Soil, Sediment and Rice of Central India. Environmental Geochemistry Health, 27, 131-145.
17. IAEA-TECDOC 1360 (2003) Collection and Preparation of Bottom Sediment Samples for Analysis of Radionuclides and Trace Elements. 34, 1-130.

18. Rudnick, R.L. and Gao, S. (2003) Vol. 3: The Crust, 3.01—The Composition of the Continental Crust. In: Holland, H.D. and Turekian, K.K., Eds., Treatise on Geochemistry, Elsevier-Pergamon, Oxford, 1-64.

CHAPTER 3

Continental Flood Basalts and Rifting: Geochemistry of Cenozoic Yemen Volcanic Province

M. A. Mattash[1,2], L. Pinarelli[3], O. Vaselli[3,4], A. Minissale[3], M. Al-Kadasi[5], M. N. Shawki[6], F. Tassi[3,4]

[1]*HCCGSR, Mukalla, Yemen*

[2]*Ministry of Oil and Minerals, Geological Survey and Mineral Resources Board, Aden, Yemen*

[3]*CNR, Institute of Geosciences and Earth Resources, Florence, Italy*

[4]*Department of Earth Sciences, University of Florence, Florence, Italy*

[5]*Department of Geology, Sana'a University, Sana'a, Yemen*

[6]*Department of Geology, Taiz University, Taiz, Yemen*

ABSTRACT

Rift formation is a crucial topic in global tectonics. The Yemen rift-related area is one of the most important provinces, being connected to the rifting processes of the Gulf of Aden, the Red Sea and Afar Triangle. In this paper, a review of the Yemen volcanic province and its relations with the Red Sea rifting are presented. Tertiary continental extension in Yemen resulted in the extrusion of large volumes of effusive rocks. This magmatism is divided in the Oligo-Miocene Yemen Trap Series (YTS) separated by an unconformity from the Miocene-Recent Yemen Volcanic Series (YVS). Magmas of the YTS were erupted during the synrift phase and correlate with the first stage of sea-floor spreading of the Red Sea and the Gulf of

Aden (30 - 15 Ma), whereas the magmas of the YVS were emplaced during the post rift phase (10 - 0 Ma). A continental within plate character is recognized for both the YTS and YVS basalts. The YTS volcanic rocks are contemporaneous with, and geochemically similar to, the Ethiopian rift volcanism, just as the volcanic fields of the YVS are geochemically alike to most of the Saudi Arabian volcanics. YTS and YVS have analogous SiO_2 ranges, but YVS tend to have, on average, higher alkalis and MgO contents than YTS. Fractional crystallization processes dominate geochemical variations of both series. Primitive magmas (MgO > 7.0%) are enriched in incompatible elements and LREEs with respect to primitive mantle, but YVS are more enriched than YTS. To first order, the different geochemical patterns agree with different degrees of partial melting of an astenospheric mantle source: 25% - 30% of partial melting for YTS and 10% - 3% for YVS. Secondly, the higher degree of enrichment in incompatible elements of YVS reflects also greater contribution of a lithospheric mantle component in their source region.

Keywords: Continental Flood Basalts; Continental Rifting; Geochemistry; Yemen; Cenozoic

1. INTRODUCTION

The Tertiary continental magmatism of Yemen was associated to the early opening phases of the Red Sea and the Gulf of Aden, and was part of the Afro-Arabian rift system (AARS), which included the Ethiopian Rift and the Afar Triangle (**Figure 1**(a)).

During Late Eocene to recent, the geological evolution of North Africa was dominated by the development of the Red Sea-Gulf of Aden-East Africa Rift System that resulted in the separation of the Arabian plate. The opening of narrow, elongated oceanic domains along the Gulf of Aden and the Red Sea from Miocene times resulted in the separation and northwards drift of the Arabian plate, allowed by sinistral movement along the Levant-Dead Sea fault zone. Rifting initiated in the Early Oligocene in several small, en echelon E-W to ESE-WNW trending basins in the Gulf of Aden province. By the OligoceneMiocene transition, rifting had spread to Afar and throughout the Red Sea system. The onset of continental rifting began ~22 Ma ago, and encompassed the whole length of the present-day Red Sea basin and Gulf of Aden [1]. Strong magmatic activity predated and accompanied rift tectonics, favouring extension by weakening the lithosphere. Oceanic spreading followed advanced continental rifting about 5 Ma ago with volcanic (mainly basaltic) activity [1].

In the late Oligocene-early Miocene through to the present time, large volumes of flood basalts emplaced at discrete eruptive centres along the western margin of the Arabian plate from the Gulf of Aden to the Mediterranean. These plateau basalts are concentrated on the Arabian side of the Red Sea without matching counterparts on the Nubian plate (**Figure 1**(a)), and represent one of the largest areas of predominantly alkali-olivine basalts in the world.

In Yemen, Oligo-Miocene to Recent (pre-/syn-rift) volcanic complexes were emplaced in the western and southwestern parts (**Figure 1**(b)).The area of this volcanism, which included Continental Flood Basalts, occupied approximately one-tenth of the total area of the country, and approximately 28% of the total area of the Arabian Plate volcanic rocks. The Yemen flood volcanics are characterized by: 1) large volumes of magmatic activity; 2) large-scale crustal extension; 3) mildly alkaline basalts and 4) bimodal distribution of basic and acid magma products. Volcanic products from Yemen are akin to coeval volcanic rocks from Djibouti, Ethiopia, and some parts of the Kenyan rift.

This work presents a review of petrological data for the rift-related volcanic rocks of Yemen.

2. THE YEMEN CONTINENTAL FLOOD BASALTS

The Cenozoic Yemen volcanic province can be divided into: 1) the late Oligocene-early Miocene Yemen Trap Series (YTS), separated by an unconformity from 2) the late Miocene-Recent Yemen Volcanic Series (YVS).

2.1. Yemen Trap Series (YTS)

The YTS represent the lowest part of the Cenozoic Yemen volcanic province, and mainly overly the Cretaceous Tawilah Group sandstones and the Paleogene lateritic paleosols, and in some cases the metamorphic basement rocks. They had been developed during the Oligoceneearly Miocenic preand sin-rift phases.

The YTS consists of thick bimodal volcanics, including alkaline to transitional basalts and peralkaline rhyolites and their associated ignimbrites, tuffs and rhyolitic obsidian flows, which cover an area of about 3000 km^2. The volume of acidic to basic volcanic products is greater than

0.5. Thickness of the YTS varies from >2000 m in the west down to hundreds to tens of meters in the east. The older units of this series are intruded by gabbro, alkali granite and syenite bodies. The majority of the felsic rocks are of peralkaline type (A/CNK > 1) with a comenditic character (normative-quartz > 20%) and are very poor in normative-mafic minerals.

Quoted ages for the YTS volcanism range from 31 to 16 Ma (see review in [1]). Within this large period, the peak of activity occurred between 30.9 and 26.5 Ma. In addition, some pre-Oligocenic ages [e.g. 2] could suggest early beginning of the volcanism in an Eocenic pre-Trap phase of magmatic activity, in agreement with Cenozoic igneous activity in the neighbouring Arabia and Gulf of Aden.

2.2. Yemen Volcanic Series (YVS)

The YVS was firstly defined by [3-5] on the basis of both geochronological and geochemical evidence. They were generated and developed through the post-rift stage (Miocene to Recent), and are separated by an unconformity from the older YTS. Volcanic cones, domes, sheets and lava flows are the typical occurrences of this series. They are composed mainly of basaltic lavas, stratified basic tuffs and agglomeratic pyroclastics, and less common differentiated rock-types, except the Al-Lisi volcano (5 km E of Dhamar city), which is entirely composed of rhyolitic lava flows.

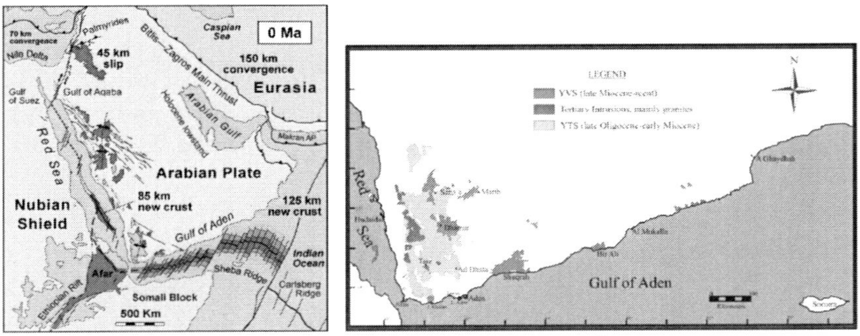

(a) (b)

Figure 1. (a) Generalized map of the Afro-Arabian rift system, comprising the Read Sea, the Gulf of Aden, the Ethiopian Rift and the Afar triangle. (b) Sketch map of the southwestern part of Yemen showing the main Tertiary magmatic rocks.

Published K-Ar ages for the YVS volcanism range from 11.3 to 0.04 Ma ([2] and references therein).

There are eight well-known volcanic fields situated along the Gulf of Aden coast. They are mainly composed of individual volcanoes characterized by central vent eruptions. There are geological, chronological and geochemical differences between those outcropping Westand Eastof Aden [4]. The former (West-of-Aden) are older (10 - 5 Ma), include the Island of Perim, Jabal Kharaz, Jabal Al-Birkah, Ras Imran, Little Aden, and Aden, and are stratoid volcanoes characterized by transitional-mildly alkaline basalt to peralkaline rhyolite. The latter (East-ofAden) are younger (5 - 0 Ma); include Shuqrah (Al Urkoob-Ahwar), Bir Ali, Ataq, as well as smaller basaltic fields in Hadramawt and Al-Mahrah in the farther east (Qusaier, Er-Raidah, Musayna'h-Hadhathem, HesayThamnoon, and the southern part of Wadi Al-Masilah), and the inland volcanic fields (Marib, Amran, Dhamar). They are low cones (basalts and basic pyroclastics), mostly characterized by an alkaline affinity.

3. GEOCHEMISTRY OF THE YEMEN FLOOD VOLCANISM

The following sections present a review of geochemical and isotope data of the Oligocene-Recent volcanic rocks from Yemen [2,3,6-19], along with their age equivalent from Saudi Arabia and Ethiopia [20-23].

The chemical characterization of the Yemen flood volcanism is illustrated in the total alkali versus silica (TAS) diagram of **Figure 2**. The volcanic rocks from northern YTS (**Figure 2**(a)) range in silica from 40% to 75%, witha marked bimodal distribution (SiO_2 < 50% or >60%) and alkaline character in the mafic part of the sequence. The southern YTS (**Figure 2**(b)) have only mafic lithotypes with the same SiO_2 range of mafic northern YTS, but tend to be lower in alkalies and their rock types are mainly transitional olivine basalt.

The East-of-Aden YVS (**Figure 2**(a)) have only mafic lithotypes with alkaline connotation comparable to the mafic northern YTS, whereas the West-of-Aden YVS (**Figure 2**(b)), show instead a continuous variation of silica from basalt to rhyolite, and have transitional character comparable to the southern YTS.

On the basis of several tectonic discrimination diagrams, not shown, the YTS and YVS basalts have within continental plate (WPB) character.

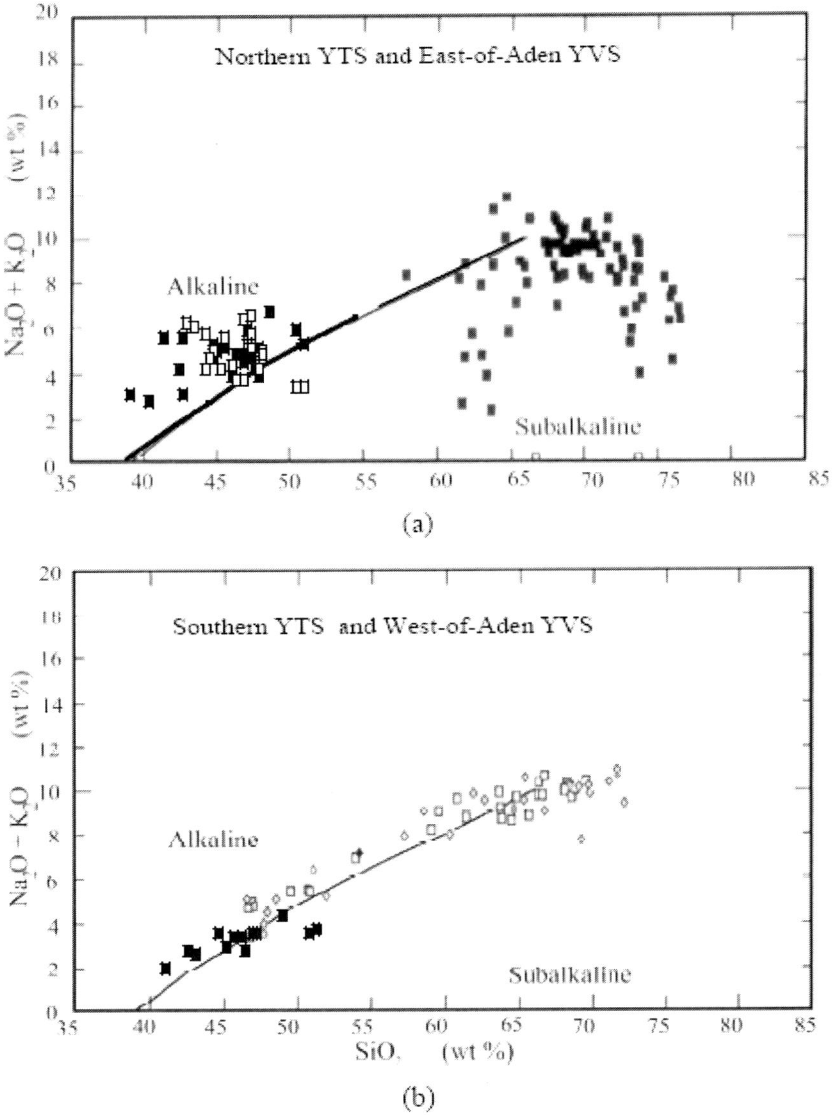

Figure 2. Total alkalis vs. silica diagram showing (a) the Northern YTS (closed squares) along with the East-of Aden YVS (open squares). (b) the Southern YTS (closed squares) along with the West-of-Aden YVS (open squares = Aden; open diamonds = Little Aden).

Binary diagrams of major elements versus MgO are reported in **Figure 3**. Despite the considerable scatter for most elements, the YTS and YVS have similar ranges of MgO, CaO and P_2O_5, even if YVS tend to have, on average, higher MgO contents (7.0% vs. 6.6, respectively). Al_2O_3, Na_2O and K_2O are higher, and TiO_2 lower, in the YVS than in the YTS. There is similarity in major element contents and distributions between the YTS and their age equivalent series in Ethiopia, and between the YVS and their age equivalent alkaline basalts in Saudi Arabia. Indeed, the Ethiopian basalts mainly overlap to the YTS, with rough positive correlation of CaO and rough negative correlations of Na_2O, K_2O, and P_2O_5 versus MgO. And the Saudi Arabian basalts tend to trace the YVS distributions, with rough positive correlation of CaO, rough negative correlation of Na_2O, and bell shaped patterns of K_2O, and Al_2O_3, versus MgO. The marked variation in MgO suggests olivine fractionation throughout the sequences.

Fractionating ol + plag could generate the decline in CaO observed with decreasing MgO, but this is inconsistent with the lack of decrease in Al_2O_3 observed with decreasing MgO in most but the evolved samples. Thus, this decrease in CaO also requires significant clinopyroxene fractionation. In summary, differentiation trends indicate crystallization of ol + cpx throughout the fractionation sequence, with plagioclase also a significant fractionating phase only in the most evolved part of the series (MgO < 6 - 7 wt%).These conclusions agree with quantitative fractionation modelling [e.g. 12].

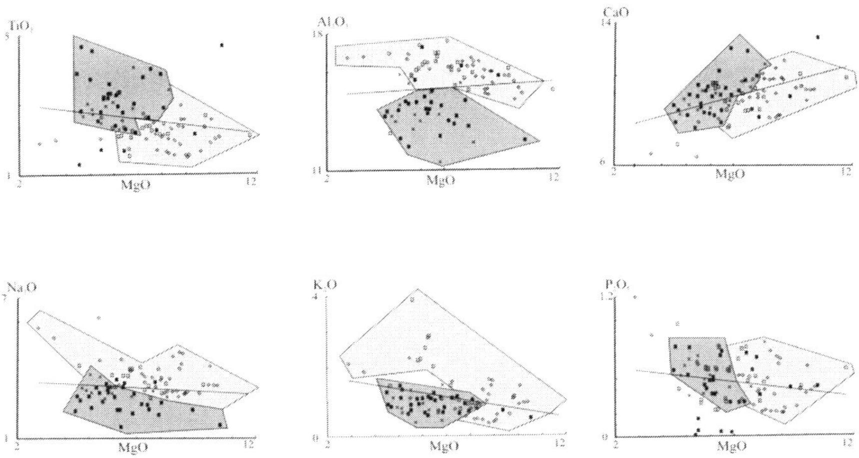

Figure 3. Major oxides versus MgO for YTS (closed squares) and YVS (open). The fields of Ethiopian (dark grey) and Saudi Arabian (light grey) basalts are shown for comparison.

The West-of-Aden YVS, which show a continuous variation of lithotypes from basalt to rhyolite, were considered separatelyin the major elements vs. Differentiation Index plots of **Figure 4**. The samples show welldefined negative correlations of TiO_2, MgO, CaO, and FeO, and a tight positive correlation of alkalies, with DI. These variations, taken together, point to olivine + clinopyroxene fractionation.

Trace element normalized concentrations of two representative primitive basalts from YTS and YVS were reported in **Figure 5**. The YTS are enriched with respect to normalizing primitive mantle, with increase of Large Ion Lithophile Elements (LILE) over High Field-Strength Elements (HFSE), and pronounced negative spikes of K and Pb. The pattern of the YVS is similar to that of YTS, but with even higher LILE/HFSE.

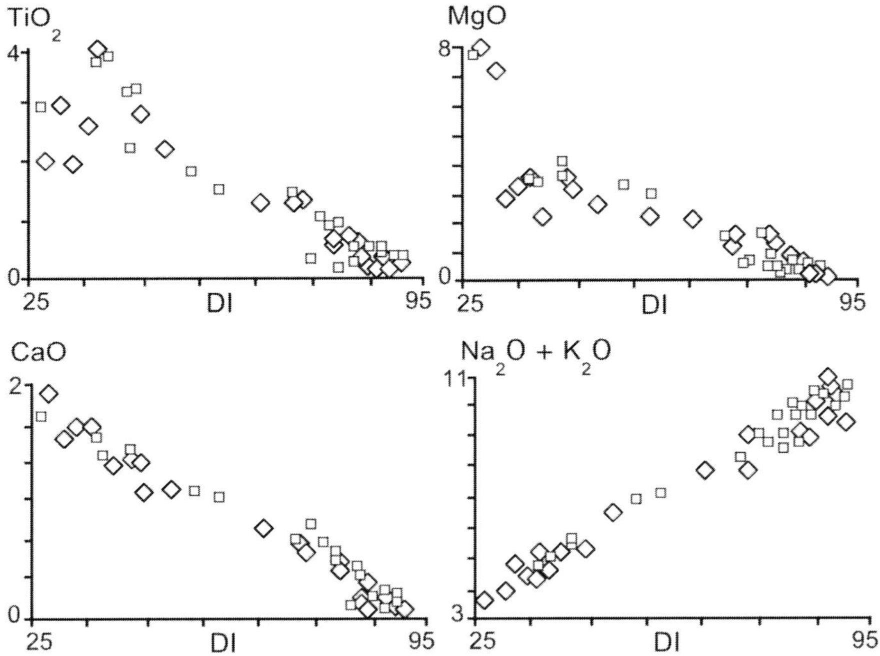

Figure 4. Major oxides versus DI (Differentiation Index) for the West-of-Aden YVS, which show a continuous variation of lithotypes from basalt to rhyolite. Open squares = Aden; open diamonds = Little Aden. DI is the normative weight ratio: (Qz + Or + Ab + Ne)/Total).

4. DISCUSSION

The Cenozoic Yemen volcanic rocks are rift-related, and their formation and evolution is the result of synand post-rift phases of the formation and development of the Gulf of Aden and the Red Sea. Magmas of the YTS were erupted during the syn-rift phase and correlate with the first stage of sea-floor spreading of the Red Sea and the Gulf of Aden (30 - 15 Ma), whereas magmas of the YVS were ascended during the post-rift phase (10 - 0 Ma). These differences in tectonic settings are reflected in analogue differences in geochemical characterization of the two rock series. YTS are bimodal, less alkaline, less enriched in incompatible elements and LREEs with respect to the YVS. The most primitive magma types in the YTS are transitional to alkali olivine basalts, in the YVS only alkali olivine basalt.

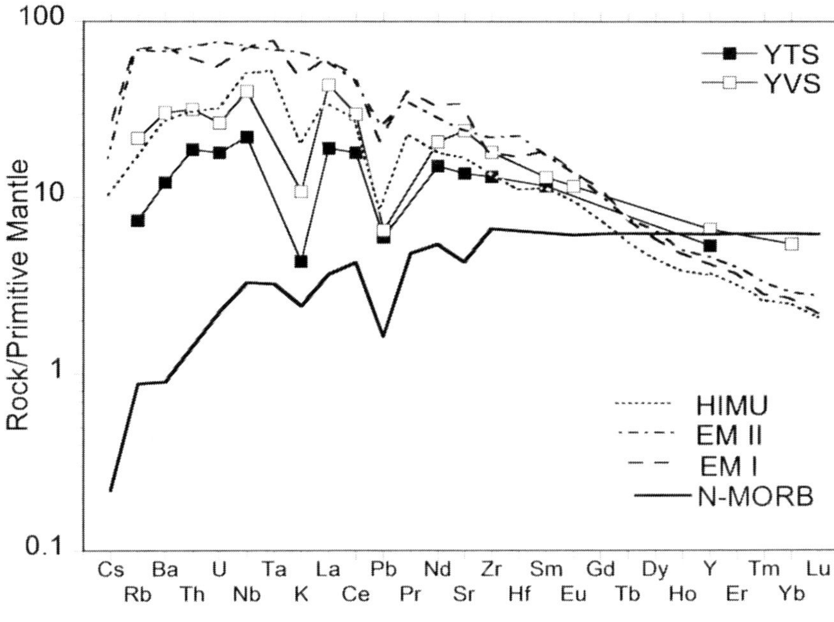

Figure 5. Spider diagrams displaying the normalized concentrations of trace elements ordered by increasing compatibility. Normalizing values after [24]. A representative pattern for MORB [24], and examples of HIMU (St. Helena, [25]), EM I (Tristan da Cunha, [25]), and EM II (Society Islands, [26]), are also reported for comparison.

The geodynamic setting influences the magma-generation process in the rift zones. In high-volcanicity rifts, the rates of crustal extension are higher, corresponding to higher degrees of lithospheric attenuation and hence more asthenospheric penetration into the lithosphere [27]. In this context, mafic magmas are derived by partial melting of asthenospheric mantle ascending beneath the rift axis, and possibly mixed with partial melts from the lithospheric mantle during their rise to the surface. The greater is the crustal extension rates and mantle upwelling, the higher are the partial melting degrees of the asthenosphere. Regarding the Yemen volcanism, major element geochemistry denoted the effects of fractional crystallization processes. In order to clarify the relative role of partial melting and fractional crystallization, trace element discrimination diagrams, availing the different behaviour of moderately and strongly incompatible elements, were used (**Figure 6**). The ratios of highly to mildly incompatible elements (Th/Yb and Th/Hf in Figures 6(a) and (b)) remain approximately constant during closed system fractional crystallization, whereas they decrease for increasing melting degrees. The distribution of the Yemen volcanic rocks suggests that YTS can be the result of higher degrees of partial melting of the mantle source with respect to YVS. This is confirmed by calculations made with equilibrium partial melting model [28], starting from an astenospheric mantle composition [29- 32], which indicate 25% - 30% of partial melting for YTS and 10% - 3% for YVS (**Figure 6**). However, departure of the YTS trend from the partial melting trend (Figures 6(c) and (d)) corroborates a consistent role of fractional crystallization in the more differentiated part of YTS.

As regards the nature of the mantle source, previous studies identified several distinct mantle domains con tributing to magma genesis in the AARS. To first order, these components include a mantle plume, broadly thought to consist of 1) an HIMU (High m) [e.g. 33] or FOZO (Focus Zone) [e.g. 15] component; 2) a depleted mantle reservoir, interpreted as the source region for N-MORB, and 3) one or two enriched (EMI = Enriched Mantle I - EMII = Enriched Mantle II) mantle components [15,33 among others]. In the spider diagrams of **Figure 5**, progressively higher contents of alkalies, incompatible elements and LREEsare evident passing from YTS to YVS. This could agree with alarger contribution of enriched mantle components in the source of the YVS with respect to the YTS.

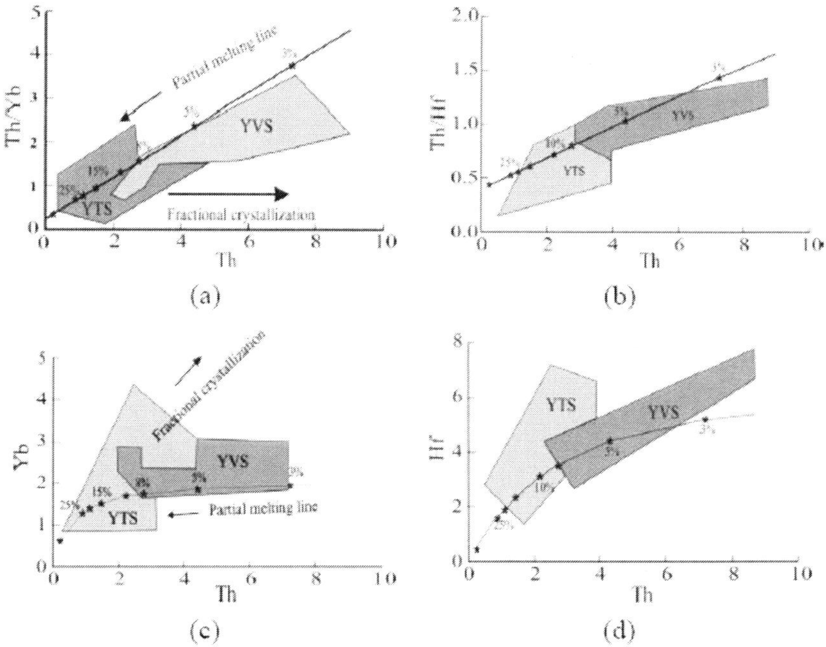

Figure 6. (a) Th/Yb versus Th; (b) Th/Hf versus Th; (c) Yb versus Th; (d) Hf versusTh reporting the fields of YTS and YVS mafic rocks, along with the result of calculations made with equilibrium partial melting model, starting from an astenospheric mantle composition. Numbers marked by asterisks along the solid lines are the degrees of partial melting.

Published Sr, Nd, Pb, and O isotope studies pointed out that part of the YTS and YVS underwent assimilation of heterogeneous Pan-African crust. This is consistent with the high heat flow that characterizes active rift zones, where elevated crustal temperatures give rise to the generation of acidic magmas by crustal melting, or mixing of crustal and mantle partial melts. On this basis, higher degrees of crustal contamination could be hypothesized as the agent of the enrichment in incompatible elements and LREEs of YVS with respect to YTS. To ascertain this hypothesis, the $(La/Sm)_N$ versus Ba/Nb patterns of the YTS and YVS, along with the Red Sea MORB, and the main mantle and crustal components, were reported in the plot of **Figure 7**. To minimize the effects of fractional crystallization, only samples with MgO > 7% were considered. It can be seen that crustal rocks have much higher Ba/Nb than all mantle components, whilst have $(La/Sm)_N$ higher than those of N-MORB, but similar to

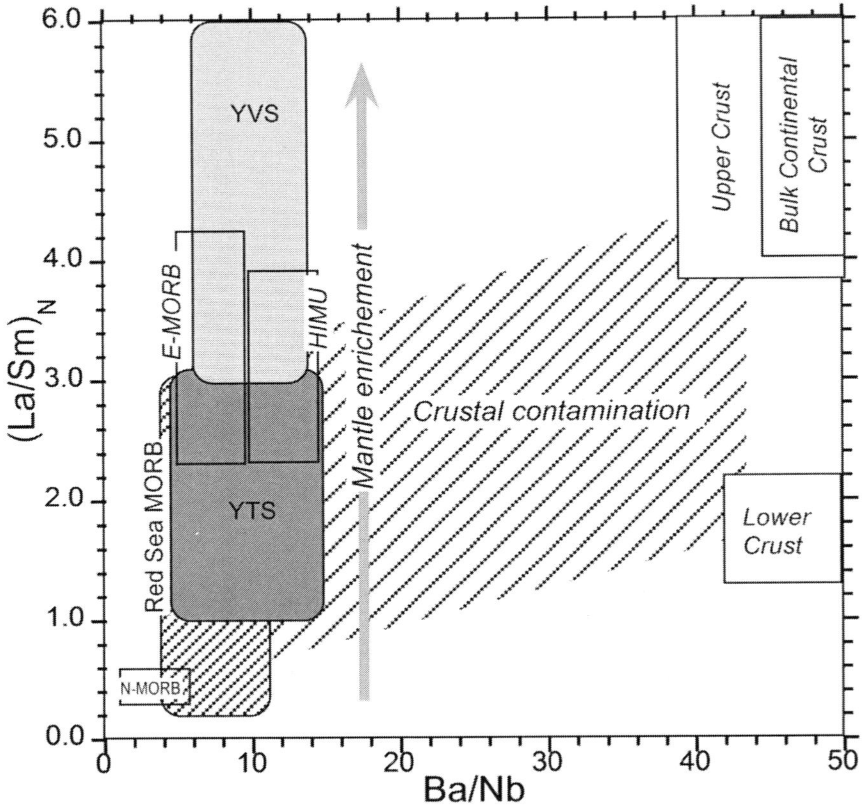

Figure 7. $(La/Sm)_N$ versus Ba/Nb plot reporting: YTS and YVS, Red Sea MORB [2 and references therein], N-MORB [24], E-MORB and HIMU [25,34], Upper Crust, Lower Crust, Bulk Continental Crust [35].

those of mantle sources like E-MORB or HIMU. Thus crustal contamination paths should afford a simultaneous increase of both Ba/Nb and $(La/Sm)_N$ (**Figure 7**). The ranges of Ba/Nb of basalts from both YTS and YVS show instead restricted variations, comprised in the range of mantle values, and do not show the increases expected for crustal contamination. As regards to the $(La/Sm)_N$, it is interesting to note that YTS are similar to the Red Sea MORB, and are displaced towards enriched components with respect to N-MORB. The YVS are still more enriched than YTS, and their $(La/Sm)_N$ encompass those of E-MORB, but without a concomitant increase of Ba/Nb, denoting a process of source enrichment rather than magma contamination.

Schilling [33] suggested, on the basis of combined NdSr-Pb isotopes, that the Afar mantle plume material may consist mainly of HIMU component and explained the isotope variations of the Cenozoic basalts from the Gulf of Aden by three components mixing including Afar plume material (HIMU), depleted mantle (DM) and an hybrid enriched mantle (EM I-EMII). However, the typical HIMU signatures are not found in the Cenozoic Yemen volcanics [15]. Indeed, in the $^{143}Nd/^{144}Nd$ versus $^{87}Sr/^{86}Sr$ plot (**Figure 8**), the Yemen flood volcanic rocks appear to plot within the mixing area of three components: Depleted mantle, Enriched Lithospheric Mantle (ELM, as defined by [15]), and C, an intermediate component between DM and HIMU. Although assimilation of heterogeneous crust has the major control on large-scale isotope variability, the least contaminated samples of the Yemen flood volcanics have radiogenic isotope ratios that approach those inferred for the Afar plume (field circled with dashed line in **Figure 8**).

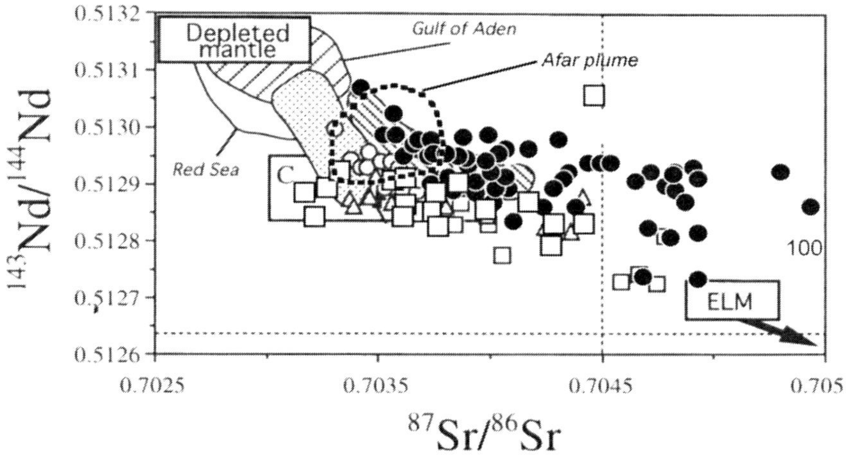

Figure 8. $^{143}Nd/^{144}Nd$ versus $^{87}Sr/^{86}Sr$ plot reporting YTS (closed symbols) and YVS (open symbols). Shaded areas outline data of the Red Sea [36-39], Gulf of Aden, and Afar plume [33,40]. C: Common component defined by [41]. ELM: Enriched Lithospheric Mantle defined by [33].

5. SUMMARY AND CONCLUSIONS

In summary, the following conclusions can be drawn:

1) The Cenozoic Yemen flood volcanic province can be divided in: Oligo-Miocenic bimodal transitional to alkaline Yemen Trap Series, YTS, and

Miocene-Recent alkaline Yemen Volcanic Series, YVS. Magmas of the YTS were erupted during the synrift phase and correlate with the first stage of sea-floor spreading of the Red Sea and the Gulf of Aden (30 - 15 Ma), whereas magmas of the YVS were ascended during the post rift phase, consistent with the sea-floor spreading model (10 - 0 Ma).

2) The YTS are contemporaneous with, and geochemically similar to, the Ethiopian Trap Series, whereas the more alkaline YVS are geochemically similar to the Saudi Arabian volcanics.

3) A continental within plate character is recognized for both the YTS and YVS basalts. The two rock series have similar SiO_2 ranges, but YVS tend to have, on average, higher alkalis and MgO contents than YTS. Fractional crystallization processes dominate geochemical variations of both series.

4) Primitive magmas (MgO > 7.0%) from both series are enriched in incompatible elements and LREEs with respect to primitive mantle, but YVS are more enriched than YTS. To first order, the different geochemical patterns agree with different degrees of partial melting of the mantle source: 25% - 30% of partial melting for YTS and 10% - 3% for YVS. Secondly, the higher degrees of enrichment in incompatible elements of YVS reflect also a greater contribution of an enriched component (EMORB) to the mantle source.

5) The large-scale Sr-Nd-Pb-O isotope variability of the Cenozoic Yemen flood volcanism is mostly accounted by crustal contamination. However, the least contaminated samples have radiogenic isotope ratios that approach those inferred for the Afar mantle plume.

6. ACKNOWLEDGEMENTS

M. A. Mattash, L. Pinarelli, O. Vaselli,, A. Minissale, M. Al-Kadasi, M. N. Shawki, F. Tassi, We are deeply indebted to Gianluca Bianchini and an anonymous referee for their useful and invaluable suggestions. Particular thanks are due to Gyorgy Buda, University of Budapest, Hungary for providing all the possibilities to make this work successful. We also acknowledge Professors of the University of Idaho, USA, for their continuous help and cooperation.

REFERENCES

1. B. Bosworth, P. Huchon and K. McClay, "The Red Sea and Gulf of Aden basins," Journal of African Earth Science, Vol. 43, No. 1-3, 2005, pp. 334-378.

2. M. A. Mattash, L. Pinarelli, O. Vaselli, A. Minissale, M. C. Jaimes-Viera, M. Al-Kadasi, M. N. Shawki and F. Tassi, "Geochemical Evolution of Southern Red Sea and Yemen Flood Volcanism: Evidence for Mantle Heterogeneity," Arabian Journal of Geosciences, in press.
3. M. A. Mattash, "Study of the Cenozoic Volcanic Rocks and Their Associated Intrusive Rock in Yemen in Relation to Rift Development," PhD Dissertation, Hungarian Academy of Sciences and Eötvös L. Univ., Budapest, 1994.
4. M. A. Mattash and K. Balogh, "K-Ar Radiometric Age Data on Cenozoic Volcanics and Their Associated Tertiary Intrusions from Yemen," Acta Mineralogica-Petrographica, Vol. 35, 1994, pp. 83-92.
5. Z. R. Beydoun, M. L. As-Saruri, H. El-Nakhal, I. N. AlGanad, R. S. Baraba, A. O. Nani and M. H. Al-Aawah, "International Lexicon of Stratigraphy," Republic of Yemen IUGS Publication, London, Vol. III, 1998, p. 245.
6. M. L. Kabesh, A. K. A. Salem and M. H. A. Higazy, "Contributions to the Petrochemistry and Geochemistry of Some Quaternary Basaltic Rocks, Northern and Southern Yemen," Acta Mineralogica-Petrographica, Vol. 26, 1984, pp. 171-186.
7. USGS (United States Geological Survey), "Geology of Dhamar Area," Internal Report for the Geological Survey of Yemen, 1985, p. 32.
8. S. Chiesa, L. Civetta, M. De Fino, L. La Volpe and G. Orsi, "The Yemen Trap Series: Genesis and Evolution of a Continental Flood Basalt Province," Journal Volcanological and Geothermal Research, Vol. 36, No. 4, 1989, pp. 337-350.
9. M. A. Mattash and Gy. Buda, "The Evolution of Yemen Basalts in Relation to Rift Development and the Chemistry of Their Mineral Constituents," Bulletin of the Geological Society of Greece, Vol. 28, No. 2, 1992, pp. 499- 514.
10. K. G. Cox, I. G. Gass and D. I. J. Mallick, "The Peralkaline Volcanic Suite of Aden and Little Aden, South Arabia," Journal of Petrology, Vol. 11, No. 3, 1970, pp. 433- 461.
11. J. A. Baker, M. F. Thirlwall and M. A. Menzies, "Sr-NdPb Isotope and Trace Element Evidence for the Crustal Contamination of Plume-Derived Flood Basalts: Oligocene Flood Volcanism in the Western Yemen," Geochimica et Cosmochimica Acta, Vol. 60, No. 14, 1996, pp. 2559-2581.
12. J. A. Baker, M. A. Menzies, M. F. Thirlwall and C. G. Macpherson, "Petrogenesis of Quaternary Intraplate Volcanism, Sana's, Yemen: Implications for Plume-Lithosphere Interaction and Polybaric Melt Hybridization," Journal of Petrology, Vol. 38, No. 10, 1997, pp. 1359- 1390.
13. M. T. S. Heikal, E. M. M. Lebda, Y. Orihashi and A. Habtoor, "Petrogenetic Evolution of Basaltic Lavas from Balhaf-Bir Ali Plio-Quaternary Volcanic Field, Arabia Sea, Republic of Yemen," Arabian Journal of Geosciences.
14. J. A. Baker, C. G. Macpherson, M. F. Menzies, M. F. Thirlwall, M. Al-Kadasi and D. P. Mattey, "Resolving Crustal and Mantle Contributions to Continental Flood Volcanism, Yemen; Constraints from Mineral Oxygen Isotope Data," Journal of Petrology, Vol. 41, No. 12, 2000, pp. 1805-1820.
15. Y. Orihashi, K. Nagao, A. Al-Jailani and B. B. Hanan, "Primordial Helium Isotope Signature from Plio-Quarternary Alkaline Basalts in Yemen," The Island Arc, Vol. 10, 2001, pp.145-157.
16. R. G. Coleman and A. V. McGuire, "Magma Systems Related to the Red Sea opening," Tectonophysics, Vol. 150, No. 1-2, 1988, pp. 77-100.
17. M. Menzies, K. Gallagher, A. Yelland and A. J. Hurford, "Volcanic and Non Volcanic Rifted Margins of the Red Sea and Gulf of Aden: Crustal Cooling and Margin Evolution in Yemen," Geochimica et Cosmochimica Acta, Vol. 61, No. 12, 1997, pp. 2511-2527.
18. Y. Orihashi "Evolution of the Afar Mantle Plume," Ph.D. Dissertation, Hokkaido University, Hokkaido, 1997.
19. J. A. Baker, M. A. Menzies, M. F. Thirlwall and C. G. Macpherson, "Petrogenesis of Quaternary Intraplate Volcanism, Sana's, Yemen: Implications for Plume-Lithosphere Interaction and Polybaric Melt Hybridization," Journal of Petrology, Vol. 38, 1997, pp. 1359-1390.

20. V. E. Camp and M. J. Roobol, "The Arabian Continental Alkali Basalt Province: Part I. Evolution of Harrat Rahat, Kingdom of Saudi Arabia," Bulletin of the Geological Society of America, Vol. 101, No. 1, 1989, p. 71.
21. V. E. Camp, M. J. Roobol and P. R. Hooper, "The Arabian Continental Alkali Basalt Province: Part II. Evolution of Harrats Khayber, Ithnayn and Kura, Kingdom of Saudi Arabia," Geological Society of America Bulletin, Vol. 103, No. 3, 1991, pp. 363-391.
22. V. E. Camp, M. J. Roobol and P. R. Hooper, "The Arabian Continental Alkali Basalt Province: Part III. Evolution of Harrat Kishb, Kingdom of Saudi Arabia," Geological Society of America Bulletin, Vol. 104, No. 4, 1992, pp. 379-396.
23. B. Zanettin, "Evolution of the Ethiopian Volcanic Province," In: Atti Della Nazionale Dei Lincei Rome, Memorie, Classe Scienze Matematiche Fisiche e Naturali dell'Accademia dei Lincei, series 9, Vol. 1, 1992, pp. 153-181.
24. S. S. Sun and W. F. McDonough, "Chemical and Isotopic Systematics of Oceanic Basalts: Implications for Mantle Composition and Processes," In: A. D. Saunders and M. J. Norry, Eds., Magmatism in the Ocean Basins, Geological Society of London, London, Special Publication, Vol. 42, 1989, pp. 313-345.
25. M. Willbold and A. Stracke, "Trace Element Composition of Mantle End-Members: Implications for Recycling of Oceanic and Upper and Lower Continental Crust," Geochemistry, Geophysics, Geosystems, Vol. 7, No. 4, 2006.
26. E. H. Hauri and S. R. Hart, "Rhenium Abundances and Systematics in Oceanic Basalts," Chemical Geology, Vol. 139, No. 1-4, 1997, pp. 185-205.
27. F. Barberi, R. Santacroce and J. Varet, "Chemical Aspects of Rift Magmatism," In: G. Pálmason, Ed., Continental and Oceanic Rifts, American Geophysical Union, Washington DC, 1982, pp. 223-258.
28. D. M. Shaw, "Trace Element Fractionation during Anatexis," Geochimica et Cosmochimica Acta, Vol. 34, No. 2, 1970, pp. 237-243.
29. G. G. Goles, "Trace Elements in Ultramafic Rocks," In: P. J. Wyllite, Ed., Ultramafic and Related Rocks, John Wiley, New York, 1967, pp. 352-362.
30. W. L. Griffin and V. R. Murthy, "Distribution of K, Rb, Sr, and Ba in Some Minerals Relevant to Basalt Genesis," Geochemica et Cosmochimica Acta, Vol. 33, No. 11, 1969, pp. 1389-1414.
31. D. M. Shaw, "Development of the Early Continental Crust, Part I. Use of Trace Element Distribution Coefficient Models for the Proto-Archean Crust," Canadian Journal of Earth Sciences, Vol. 9, 1972, pp. 1577-1595.
32. F. A. Frey, D. H. Green and S. D. Roy, "Integrated Models of Basalt Petrogenesis: A Study of Quartz Tholeiites to Olivine Melilites in the Southeastern Australia Using Geochemical and Experimental Petrologic Data," Journal of Petrology, Vol. 19, No. 3, 1978, pp. 463-513.
33. J.-G. Schilling, R. H. Kingsley, B. B. Hanan and B. L. McCully, "Nd-Sr-Pb Isotopic Variations along the Gulf of Aden: Evidence for Mantle Plume Continental Lithosphere," Journal of Geophysical Research, Vol. 97, No. B7, 1992, pp. 10927-10966.
34. A. H. Halliday, D. C. Lee, S. Tommasini, G. R. Davies, C. Paslick, J. G. Fitton and D. E. James, "Incompatible Trace Elements in OIB and MORB and Source Enrichment in the Sub-Oceanic Mantle," Earth Planetary Science Letters, Vol. 133, No. 3-4, 1995, pp. 379-395.
35. R. L. Rudnick and S. Gao, "Composition of the Continental Crust," Treatise on Geochemistry, Vol. 3, 2003, pp. 1- 64.
36. R. Altherr, F. Henjes-Kunst and V. Baumann, "Asthenosphere versus Lithosphere as Possible Sources for Basaltic Magmas Erupted during Formation of the Red Sea: Constraints from Sr, Pb and Nd Isotopes," Earth Planetary Science Letters, Vol. 96, No. 3-4, 1990, pp. 269-286.
37. R. Altherr, F. Henjes-Kunst, H. Puchelt and A. Bauman, "Volcanic Activity in the Red Sea Axial Trough-evidence for A Large Mantle Diapir?" Tectonophysics, Vol. 150, No. 1-2, 1988, pp. 121-133.

38. J. A. Barrat, B. M. Jahn, J. L. Joron, B. Auvray and H. Hamdi, "Mantle Heterogeneity in Northeastern Africa: Evidence from Nd Isotopic Composition and Hygromagmaphyle Element Geochemistry of Basaltic Rocks from the Gulf of Tadjoura and Southern Red Sea Regions," Earth Planetary Science Letters, Vol. 101, No. 2-4, 1990, pp. 233-247.
39. N. W. Rogers, "The Isotope and Trace Element Geochemistry of Basalts from the Volcanic Islands of the Southern Red Sea," In: H. M. Prichard, T. Alabaster, N. B. Haris and C. R. Neary, Eds., Magmatic Processes and Plate Tectonics, Geological Society of London, London, Special Publication, Vol. 76, 1993, pp. 455-467.
40. Vidal, C. Deniel, P. J. Vellutini, P. Piguet, C. Coulon, J. Vincent and J. Audin, "Changes of Mantle Sources in the Course of a Rift Evolution: The Afar Case," Geophysical Research Letters, Vol. 18, No. 10, 1991, pp. 1913-1916.
41. B. B. Hanan and D. W. Graham, "Lead and Helium Isotope Evidence from Oceanic Basalts for a Common Deep Source of Mantle Plumes," Science, Vol. 272, No. 5264, 1996, pp. 991-995.

CHAPTER 4

Geochemistry of Major Oxides in Host Rocks in Vizianagarm Manganese Ores Belt (A.P.), India

Farhat Nasim Siddiquie, Mohd Shaif*

Department of Geology, Aligarh Muslim University, Aligarh

ABSTRACT

The abundance, distribution trends and significance of the major oxides in the host rocks in Vizianagarm Manganese Ores Belt (A.P.) (between N latitude 18°12' and 18°30' and E longitudes 83°20' and 83°45'), 15 samples of host rocks from different localities of the area under study were collected and analyzed for major oxides. We describe here in major oxides geochemistry of host rocks and manganese ore deposits associated with Precambrian Khondalite and Charnockite in Vizianagarm Manganese Ores Belt (A.P.): 1) Preponderance of SiO_2 over Al_2O_3; 2) Dominance of K_2O and CaO over Na_2O; 3) Abnormally high concentration of phosphorus and a positive relationship of P_2O_5 with CaO and Ti contents; 4) Manganese increases with increases of iron, lime and soda and vice versa, 5) CaO increases with the increases of Al_2O_3, Ti, K_2O and vice versa. High P_2O_5content in these manganese ores appears to be the result of precipitation from secondary manganese rich solutions containing dissolved phosphorus from the P_2O_5 enriched host rocks. Another source of P_2O_5 may be the associated granitic and pegmatitic intrusions. Elements like K, Na, Ca, Mg, Co, Ni, Pb and Zn etc. appear to be mostly concentrated in the Mn-minerals viz. psilomelane, cryptomelane, hollandite and pyrolusite and related secondary phases [1] and [2] . Stratigraphically, the study area includes within a thick succession of Precambrian Group belonging to the Khondalite and Charnockite Groups of Dharwar Supergroup, that form a part of Eastern Ghat Complex of India. The manganiferous rocks that have been encountered in the Vizianagarm Manganese Ores Belt (A.P.) India are known as Kodurites.

Keywords: Vizianagarm, Manganese Ores, Major Oxides, Khondalite, Charnockite, A.P., Geochemistry, Eastern Ghat, Dharwar Supergroup

1. INTRODUCTION

A large number of manganese deposits occur in the Precambrian Eastern Ghats complex of Andhra Pradesh, India. These are dominantly low to medium grade metamorphic deposits with high iron, phosphorus and silica. These manganese bands (deposits) are generally syngenetic and syntectonic with metasedimentry country rock of Khondalite, quartzite and calc-granulites [3] .

Manganese ores are widely distributed throughout the Peninsular India. Stratigraphically, these are mainly confined to the Precambrian terrain and are encountered in the Sausar Group and Aravalli Supergroup of central and western India, Iron ore Group and Gangpur Group of eastern India. In Andhra Pradesh, there are major occurrences of manganese ores in Srikakulam-Visakhapatnam Manganese ores belt. The deposit is in this belt over an area of 50 kms in length and 20 km in width. The study area lies 150 km NE of Visakhapatnam and includes Garividi, Garbham and Chipurupalle areas (Figure 1). Systematic geochemical study of these manganese ores in the present study area was not carried out by earlier workers. Fifteen samples of host rocks were selected from different localities of the study area. They were analysed for number of major oxides and trace elements.

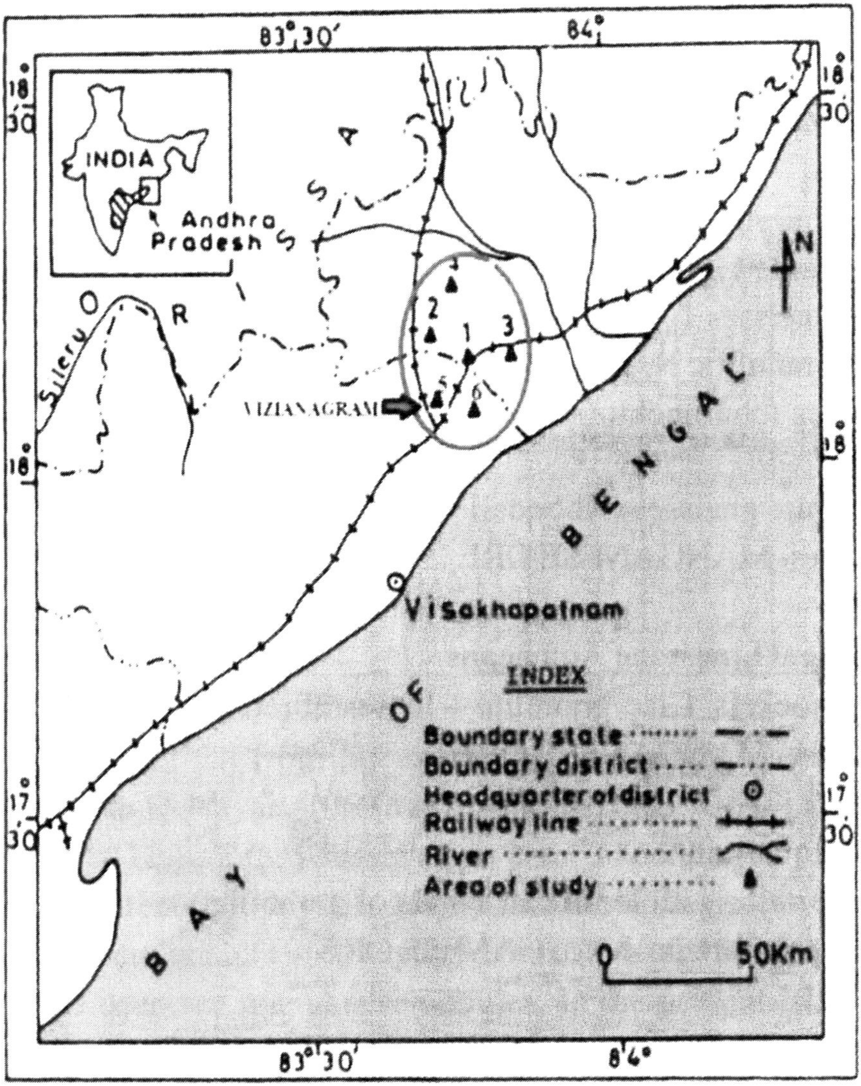

Figure 1. Location map of (1) Garividi, (2) Garbham (3) Chipurupalle manganese deposits, Vizianagram Manganese Ores Belt, (A.P.), India.

2. GENERAL GEOLOGY AND STRATIGRAPHY

The manganiferous rocks that have been encountered in the Srikakulam-Visakhapatnam manganese belt of Andhra Pradesh belongs to Kodurite Series which is typically exposed at Koduru (18°16': 83°36') a locality about 80 km north of Visakhapatnam, the rock series which constitutes an integral part of the Eastern Ghat complex of India. The rock series is composed of an intimate assembalges of metamorphosed argillaceous (garnet-sillimanite gneisses) as at Chipurupalle, Rayagada and Garbham, Arenaceous (Quartzite and garnetiferous quratzite) as at Garbham, Koduru, Perapi and Avagudem and calcareous sediments (calc-granulites and gneisses) as at Koduru and Gadabavalasa areas, this sequences belonging to granulite facies.

The manganese ores in the area occur primarily in association with calc-granulites, less often with quartzite and still less with the quartz garnet ± sillimanite graphite gneiss. According to [4] who made radiometric determination of the Precambrain rocks of India, the Eastern Ghat Complex was included in the Precambrain-II (2500 - 3000 m.y.) of the classification of Indian Precambrain rocks. The calc-granulite, quartzite and Khondalites being syn-sedimentary rock bodies, are conformable and co-folded with the associated lithic units such as garnetiferous quartzite, gneisses and garnet-sillimanite gneisses, etc. [5] . Subsequently, granitic and migmatitic activities of different intensities played their respective roles on these metasediments transmuting their earlier textural and mineralogical characters. However, their impact on quartzite and manganese ore is minimal, probably because these units are considered to be the resisters during migmatisation [6] . At places hypersthene gneisses (Charnockite) occur as intrusive bodies in the synsedimentary rock units. All these rocks have in turn been intruded by granite, pegmatite and quartz veins [7] .

[8] recongized several rock types ranging from ultrabasic to acidic in the mineralogical composition of Kodurite and ascribed them to differentiation prior to intrusion. [9] indicated that the Kodurite series might be due to hybridism on a large scale in which a granitic intrusion bodily assimilated the manganese of the Khondalites. [10] observed that by and large the manganese ore occurs associated either with quartzite (East Garbham) or with calc-granulite (Sadanandapuram).

However, the geological association in Saibaba main, Saibaba Central, Jai Bhavani-Kottakarra quaries is not clear owing to the highly weathered nature of the rocks. [11] suggested the following stratigraphic succession of the area under study and consider the granites to be intrusive bodies into the Khondalite and Kodurite series.

4	Granites	Peninsular and Crystalline Complex
3	Charnockites	
2	Kodurites ··· Hybrid	
1	Calc-granulites, crystalline limestone and associated manganese ores Garnet-Sillimanite gneisses Quartzite-Garnetiferous, Feldspathic and graphite-bearing	Khondalite Series

These rocks at places exhibit relict banding. The Garbham main quarry and the Srinavasa quarry at Garbham show weathered calc-granulite-quatzo-feldslpathic rock underlying quartzite. The rock formation and associated manganese ore deposits belonging to the Khondalite and Charnockite suit of metasediments are of the Archaean age [7] . The generalised stratigraphic sequence of the rock formations of Srikakulam area which has been suggested by [7] and supported by [12] is presented as follows:

Granite, pegmatitie and quartz vein

Hypersthese-gneisses (Charnockites?)

Khondalite Series:
- Garnet-sillimanite gneisses (khondalites)
- Garnet-granulites – Ore horizon I
- Calc-granulites
- Garnet-granulites – Ore horizon II
- Garnet-sillimanite gneissess (khondalites)
- Coarse grained quartzites)
- Feldspathic quartzite
- Garnetiferous quartzite – Ore horizon III
- Feldspathic quartzites
- Coarse grained quartzites
- Garnet-sillimanite gneisses (Khondalites)
- Biotite-gneisses

Crystalline algal limestone belonging to Knondalite Group (Archaean) [13] - [17] and gritty sandstone (Miocene?) in the form of rubble and laterite capping are also observed.

The present authors finds no evidence to differ from the suggestion made by [7] and [18] , regarding the stratigraphic succession in Srikakulam except that the occurrence of crystalline algal limestone and shales (red and green) which did not find place in their classification. The modified sequence is presented [13] - [17] as follows:

Some of the earlier workers have also made an attempts to classify and establish the sequence of rock formations encountered at Garividi, Garbham and Chipurupalle areas (Table 1 and Figure 2). They are as follows:

Garividi area [10]

Soil

Laterite-pisolites and concretions of

Mn-ore and calc-tufa

Pegmatite and quartz veins

Granites/garnetiferous biottie gneiss

Garnet-sillimanite gneiss (with or without graphite)

Calc-granulite and lenticles of crystalline limestone

Manganese ore and garnet granulites

Quartz-feldspar rock with occasional incidence of apatite

Garnet-sillimanite gneiss interlayered with quartzite and occasional lenticles of crystalline limestone

Garham Area [10]

Soil

Laterite-calc tufa calcreta

Pegmatite {
- pegmatite and quartz veins
- Garnetranite gneiss
- Garnetiferous and non-garnetiferous quartzo-feldspathic rock
- Biotite-gneiss

Charnockite

Garnet-sillimanite gneiss with or without graphite

Calc-granulite with manganese ore associated with bands and pockets of lithomarge and highly kaolinised quartzo-feldspathic apatite rock, the latter in the majority of cases seen underlying the Mn-ore zone.

Calc-granulite highly weathered with intermixture of quartz-feldspar-apatite rock

Manganese ore with chert and lemonite quartzite

Garnet-sillimanite gneiss (with or without graphite) interlayered with quartzite

Chirpurupalle area [20]

Archaean	Recent laterite and loamy alluvium
	Granite gneisses
	Manganese rich rocks, " Kodurites"
	Khondalite series :
	Calc-granulites
	Quartz-garnet-sillimanite gneisses
	Quartzites

Table 1. Geological succession of the Garividi area, Srikakulam-Visakhapatnam belt (A.P.).

Age	Group	Rock types
		Alluvium and soil
Recent		Laterite-limonite and ochre
Pleistocene (?)		Sandstone
Miocene (?)		Pegmatite and quartz vein
		Granite and granite gneisses
	Charnockite Group ←	Hypersthene gneisses (porphyro-blastic) (Charnockite) granulitic hypersthene diorite and granite gneisses (Charnockite).
		Garnet sillimanite gneisses (Khondalite)
Archaeans ←		Garnet granulite—MAGANESE ORE HORIZON – I
	Khondalite Group ←	* Crystalline algal limestone Archaeans
		* Green shale pockets, calc-granulite + lenses of red shale
		Garnet granulites—MANGANESE ORE HORIZON II
		Coarse grained quartzite
		Feldspethic quartzite with lenticular bands of garnetiferous quartzite—MANGANESE ORE HORIZON III
		Coarse grained quartzite
		Garnet-sillimanite-gneisses (Khondalite)
		Biotite gneisses

Two horizons were not reported by earlier workers. reported for the first time [15] and [19]

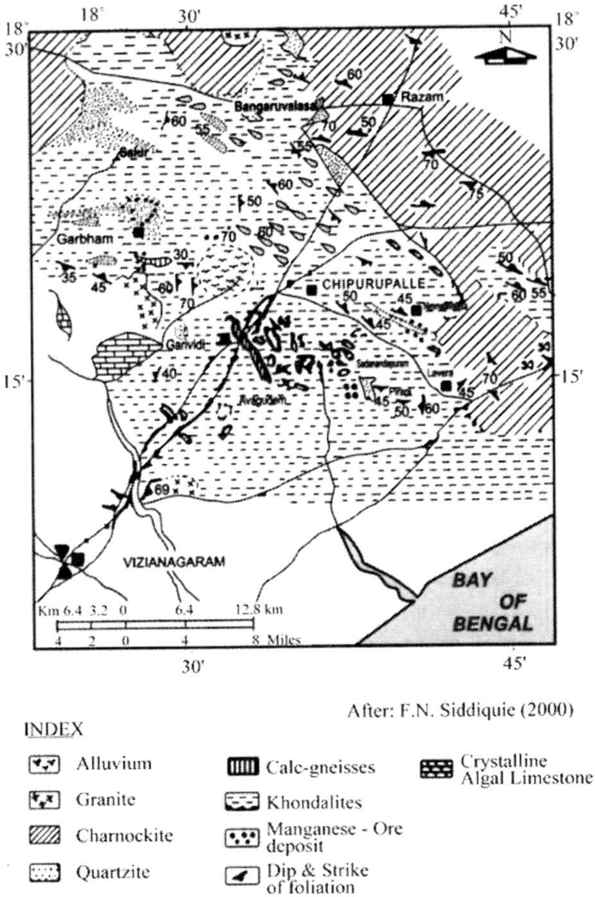

Figure 2. A generalised geological map of Garividi, Garbham and Chipurupalle manganese deposits, Vizianagram district (A.P.), India.

3. MAJOR OXIDES IN HOST ROCKS

General Statement

In the area of study, the host rocks for manganese mineralization, are mainly Khondalite and Charnockite Group of rocks.

Fifteen samples of host rocks were selected out of a large number of samples collected for their chemical analysis out of which ten samples were of Khondalites and remaining five of Charnockites. The samples were analysed chemically for the determination of their major oxides—SiO_2, Al_2O_3, TiO_2, MnO, Fe_2O_3, FeO, P_2O_5, MgO, CaO, Na_2O, K_2O and H_2O^+ and some trace elements which are discussed [21] . The relative variation trends of various major oxides are shown in Figure 3. The analytical data are presented in Table 2.

4. DISTRIBUTION OF THE MAJOR OXIDES

The abundance and variation range with average (wt%) of different major oxides of Khondalites and Charnockites are given in (Table 2 and Table 3) and the distribution of various major oxides are discussed as below:

4.1. Silica (SiO_2)

The quantitative variation trend of silica in the host rocks from 47.84 percent to 70.45 percent and gives an average value of 61.81 percent (Table 3, Figure 3). Khondalite, Charnockite rocks have wide variation of silica (Table 3). [22] reported 58.04% to 74.38% SiO_2 from Srikakulam area, [11] reported 58.4% from Garividi, [23] reported 67.10% and 69.90% respectively from Charnockites of Tamil Nadu and Karnataka. The histogram (Figure 4(a)) represents frequency percent distribution, showing polymodal nature of SiO_2 in the host rocks. The maxima is at 65 - 70 percent class. The typical mode of SiO_2 falls between 65 - 70 percent.

4.2. Alumina (Al_2O_3)

The quantitative variation trend of Alumina in the host rocks of the study area ranges 6.32 percent to 17.42 percent and gives an average value of 13.07 percent (Table 3).

Khondalite and charnockite rocks show wide variation of Al_2O_3 (Table 3). [22] reported 9.65% to 20.64% Al_2O_3 from Srikakulam area. [24] reported 14.08% from feldspar garnetiferous granite-gneiss of Peddapantu- palli, and [23] reported 14.10% and 15.10% from Charnockites of Tamil Nadu and Karnataka. The histogram (Figure 4(b)) represents frequency percent distribution, showing polymodal nature of Al_2O_3. The maxima is at 12 - 14 and 16 - 18 percent classes.

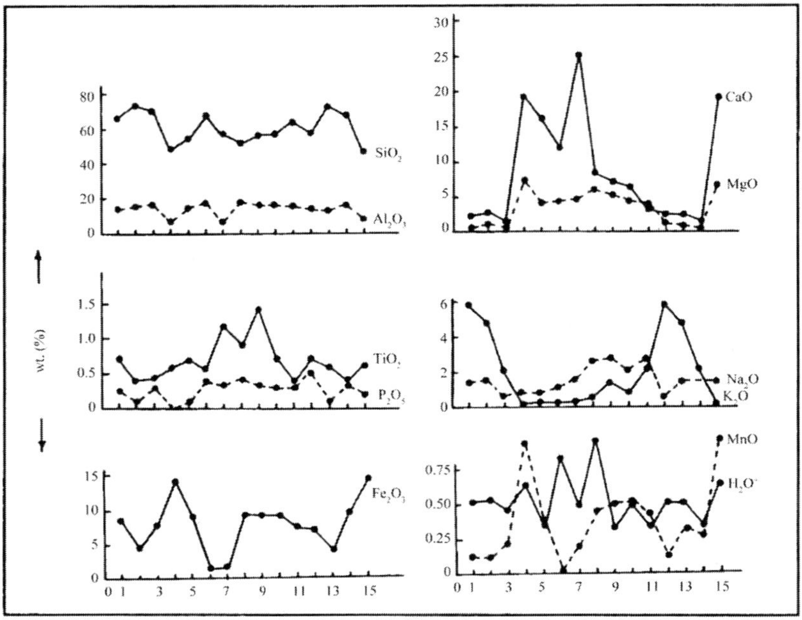

Figure 3. Variation diagrams of major oxides in Khondalite and Charnockite Groups of Garividi, Garbham and Chipurupalle areas, Srikakulam-Visakhapatnam Belt (A.P.).

The quantitative variation trend of manganese monoxide in the Khondalite and Charnockite rocks ranges from 0.12 percent to 0.94 percent and gives an average value of 0.36 percent (Table 3).

[22] reported 0.09% to 3.70% from the Srikakulam area whereas Khondalites of Sri Lanka consist 0.18% MnO [25] , [23] reported 0.05% and 0.02% from Charnockites of Tamil Nadu and Karnatka.

The histogram (Figure 4(c)) represents frequency percent distribution, showing unimodal positively skewed nature of MnO. The maxima is at 0.1 to 0.3 percent class and the typical mode of MnO also lies in the same class.

4.4. Ferric Oxide (Fe_2O_3)

The quantitative variation trend of Fe_2O_3 in the host rocks ranges from 0.44 percent to 3.36 percent and gives an average value of 1.45 percent (Table 3).

[22] reported 0.86% to 3.76% from Srikakulam area. [24] reported 1.22% from feldspar garnetiferous granite-gneiss of Peddapantupalli area, [23] reported 4.36% and 2.36% from the Charnockite of Tamil Nadu and Karnataka (Table 3 and Table 4).

Table 2. Distribution of major oxides of Khondalite and Charnockite Groups of Garividi, Garbham and Chipurupalle areas, Srikakulam-Visakhapatnam Belt (A.P.).

	KD1	KD2	KD3	KD4	KD5	KD6	KD7	KD8	KD9	KD10	CK11	CK12	CK13	CK14	CK15
SiO_2	69.35	70.3	70	48.84	56.1	68.46	58.29	52.24	56.1	59	64.16	69.09	70.45	68	47.84
Al_2O_3	13.09	13.45	16.75	6.42	13.4	10.94	6.32	17.42	16.34	16.02	15.66	13.35	13.12	16.4	7.42
MnO	0.12	0.12	0.21	0.94	0.36	0.08	0.21	0.4	0.5	0.52	0.38	0.12	0.3	0.26	0.9
Fe_2O_3	1.21	1.16	3.36	1.3	1.06	0.44	0.54	2.2	1.48	2.22	2.07	1.21	1.64	3.36	2.3
FeO	4.52	3.64	4.6	13.11	7.19	1.2	1.45	7.48	6.7	5.74	4.25	3.16	4.59	12.11	7.2
P_2O_5	0.24	0.1	0.3	n.d.	0.1	0.42	0.36	0.45	0.35	0.33	0.32	0.51	0.1	0.35	0.14
TiO_2	0.73	0.4	0.47	0.6	0.79	0.58	1.18	0.9	1.45	0.72	0.44	0.72	0.6	0.48	0.6
CaO	2.37	2.42	1.25	19.83	16.2	12.44	25.61	8.86	7.32	6.77	3.2	2.33	2.4	1.25	19.83
MgO	0.86	1.09	0.86	7.9	4.14	4.17	4.58	5.8	5.03	4.6	3.12	1.9	1.11	0.86	6.9
Na_2O	1.52	1.6	0.34	0.67	0.7	1.2	1.6	2.48	2.56	2.08	2.84	0.56	1.4	2.16	1.67
K_2O	5.94	4.96	2.16	0.02	0.24	0.16	0.2	0.64	1.42	0.94	2.2	5.9	4.97	2.09	0.02
H_2O^-	0.51	0.53	0.49	0.62	0.35	0.8	0.5	0.96	0.3	0.5	0.3	0.52	0.52	0.34	0.62
Total	100.46	99.77	100.79	100.25	100.63	100.89	100.76	99.55	100.33	100.4	100.43	100.46	99.77	100.14	100.35

KD = Khondalite; CK = Charnokite; n.d. = Not Detectable.

Table 3. Range of variation and average oxide % of Khondalite and Charnockite Groups of Garividi, Garbham and Chipurupalle areas, Srikakulam-Visakhapatnam Belt (A.P.).

Oxides	Minimum	Maximum	Average
SiO_2	47.84	70.45	61.81
Al_2O_3	6.32	17.42	13.07
TiO_2	0.12	0.94	0.36
MnO	0.44	3.36	1.45
Fe_2O_3	1.20	13.11	5.79
FeO	0.10	0.51	0.27
P_2O_5	0.40	1.45	0.70
CaO	1.25	25.61	8.80
MgO	0.86	7.90	3.52
Na_2O	0.34	2.84	1.55
K_2O	0.02	5.94	2.12
H_2O^-	0.30	0.96	0.52

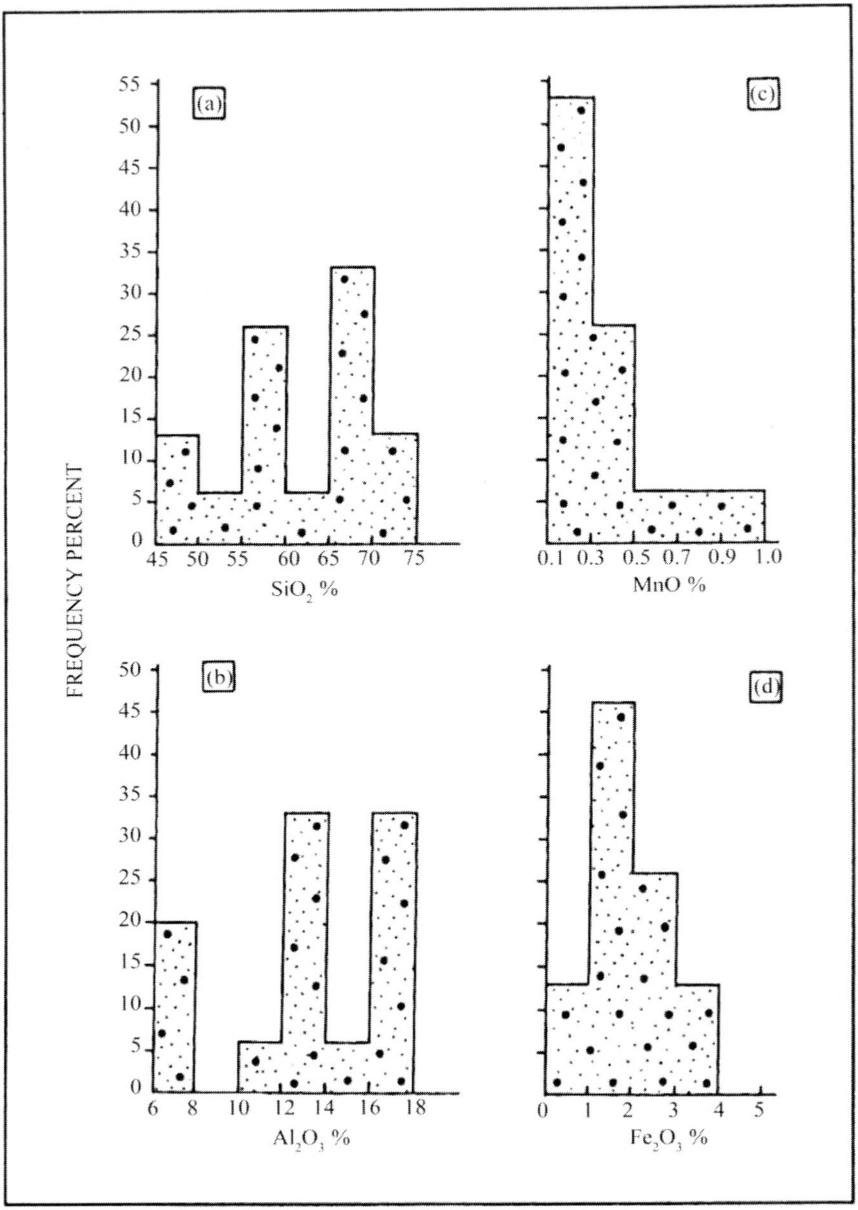

Figure 4. Histograms showing frequency percent distribution of SiO_2, MnO, Al_2O_3 and Fe_2O_3 in Khondalite and Charnockite Groups of Garividi, Garbham and Chipurupalle areas, Srikakulam-Visakhapatnam Belt (A.P.).

Table 4. Average major oxide (wt%) of the host rocks (Khondalite and charnokite) compared with other Khondalites and Charnockites.

	I	II	III	IV	V	VI	VII	VIII
SiO_2	60.86	63.90	68.36	60.75	60.08	58.04	67.10	69.90
Al_2O_3	13.01	13.19	14.08	14.49	12.38	17.30	14.10	15.10
TiO_2	0.77	0.56	0.73	0.36	0.65	1.65	0.80	0.41
Fe_2O_3	1.49	2.10	1.22	1.75	3.28	1.20	4.36	2.36
FeO	5.70	5.97	4.51	9.98	4.20	8.04	3.60	1.30
MnO	0.35	0.39	0.11	0.10	0.18	2.27	0.05	0.02
MgO	3.90	2.77	0.85	3.26	1.95	5.90	2.31	1.09
CaO	10.30	5.80	2.37	5.10	9.43	1.40	2.82	3.20
Na_2O	1.47	1.72	1.52	2.62	2.59	0.88	3.91	4.79
K_2O	1.66	3.03	5.93	1.20	3.06	3.24	2.23	1.49
P_2O_5	0.26	0.28	0.24	0.05	-	0.75	0.02	0.08
H_2O	0.55	0.46	0.44	0.51	1.80	0.15	0.47	0.16

4.3. Manganese (MnO)

The histogram (Figure 4(d)) represents frequency percent distribution, showing unimodal positively skewed nature of Fe_2O_3. The maxima is at 1 to 2 percent class and the typical mode also lies in the same class.

4.5. Ferrous Oxide (FeO)

The quantitative variation trend of FeO in the host rocks ranges from 1.20 percent to 13.11 percent and gives and average value of 5.79 percent (Table 3 and Table 4).

The Khondalites and Charnockites having higher concentration of FeO as compared to Fe_2O_3 (Table 3 and Table 4), [22] reported 1.12% to 8.04% from same area, [26] reported 9.98% from hypersthene-granulite of Narasimhunipeta, [11] reported 8.04% from Khondalites of Garividi and [23] reported 3.60% and 1.30% from Charnockites of Tamil Nadu and Karnataka (Table 3 and Table 4).

The histogram (Figure 5(a)) represents frequency percent distribution, showing polymodal nature of FeO. The maxima is at 3 to 5 percent class and typical mode also lies in the same class.

4.6. Phosphorus (P_2O_5)

The quantitative variation trend of P_2O_5 in the host rocks ranges from 0.10% to 0.51% and gives an average value of 0.27 percent (Table 3, and Table 4).

The concentration of P_2O_5 is abnormally on the higher side in the manganese ores as well as in the host rock of the study area, whereas [22] reported 0.18% to 1.01% from the various areas of Srikakulam district, [11] reported 0.75% from Khondalites of Garividi, and [24] reported 0.24% from feldspar garnetiferous granite-gneiss of peddapantupalli (Table 3 and Table 4).

The histogram (Figure 5(b)) represents frequency percent distribution, showing bimodal nature of P_2O_5. The maxima is at 0.3% to 0.4% class. The typical mode of P_2O_5 is also lies between 0.3 percent to 0.4 percent class.

4.7. Titania (TiO_2)

The quantitative variation trend of TiO_2 in the host rocks ranges 0.40 percent to 1.45 percent and gives an average value of 0.70 percent (Table 3).

[24] reported 0.73% from feldspar garnetiferous granite-gneiss of Peddapantupalli, Khondalite of Sri Lanka shows 0.65% TiO_2 [25] . [11] reported 1.65% from Khondalite of Garividi and [23] reported 0.80% and 0.41% from Charnockites of Tamil Nadu and Karnataka.

The histogram (Figure 5(c)) represents frequency percent distribution showing unimodal, positively skewed nature of TiO_2. The maxima is at 0.4 to 0.6 percent class and the typical mode also lies between 0.4 to 0.6 percent class.

4.8. Lime (CaO)

The quantitative variation trend of calcium oxide in the host rocks of the study area ranges from 1.25 percent to 25.61 and gives and average value of 8.80 percent (Table 3 and Table 4).

The CaO in the Khondalites and Charnockites of the study area shows a wide variation range from 1.25% to 25.61% (Table 3 and Table 4), Rao, K.V.R. (1955) reported 5.10% from hypersthene-granulite from the Narasimhunipeta and Krishnan (1934) reported 9.43% CaO in Khondalites of Sri Lanka (Table 3 and Table 4).

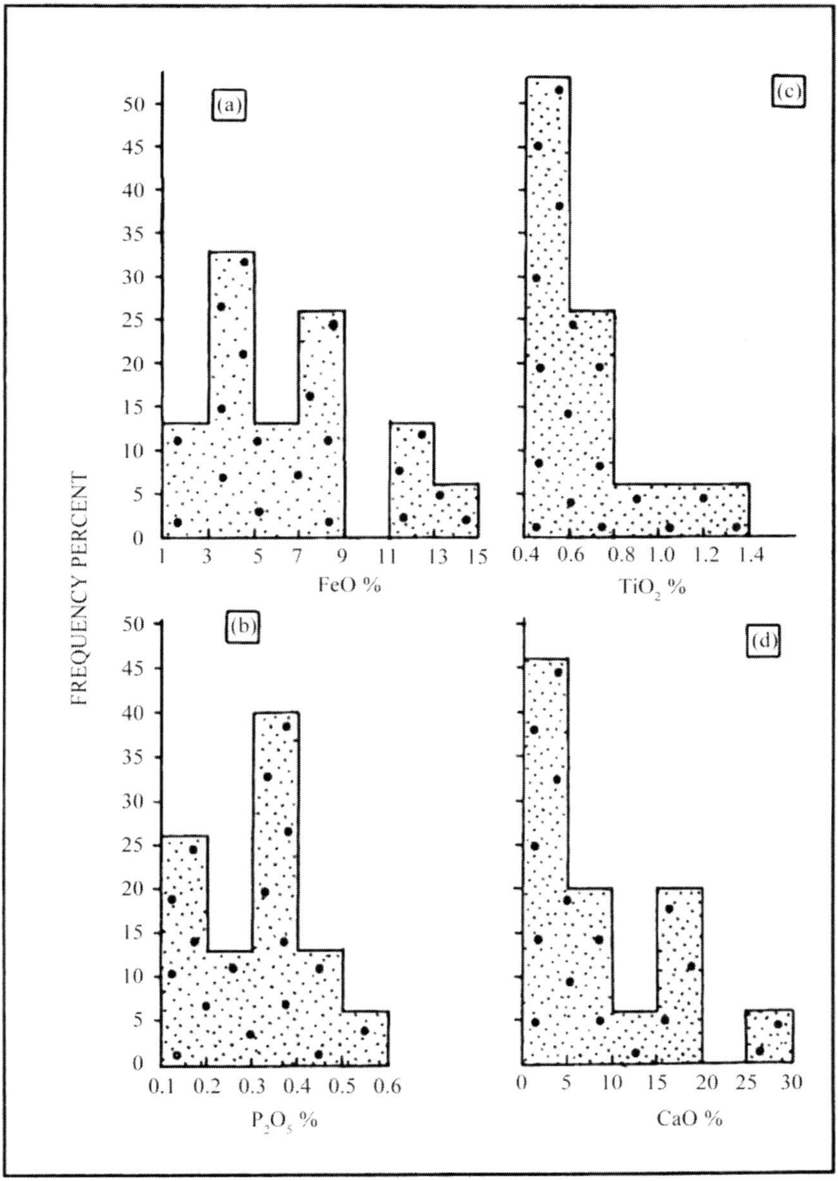

Figure 5. Histograms showing frequency percent Distribution of FeO, TiO_2, P_2O_5 and CaO in Khondalite and Charnockite Groups of Garividi, Garbham, and Chipurupalle areas, Srikakulam-Visakhapatnam Belt (A.P.).

The histogram (Figure 5(d)) represents frequency percent distribution, showing bimodal nature of CaO. The maxima is at 1 percent to 5 percent class. The typical mode lies between 1 percent to 5 percent class.

4.9. Magnesia (MgO)

The quantitative variation trend of MgO in the host rocks ranges from 0.86 percent to 7.90 percent and gives and average value of 3.52 percent (Table 3 and Table 4).

Rao, K.V.R. (1955) reported 3.26% from hypersthene-granulite of narasimhunipeta, [11] reported 5.90% from Khondalites of Garividi and [23] reported 2.31% from Charnockites of Tamil Nadu (Table 3 and Table 4).

The histogram (Figure 6(a)) represents frequency percent distribution, showing bimodal nature of MgO. The typical mode of MgO and maxima lie between 0 to 2 and 4 to 6 percent classes.

4.10. Soda (Na_2O)

The quantitative variation trend of Na_2O in the host rocks ranges from 0.34 percent to 2.84 percent and gives an average value of 1.55 percent (Table 3 and Table 4).

Rao, K.V.R. (1955) reported 2.62% from hypersthene granulite of Narasimhunipeta, 2.59% reported in Khondalites of Sri Lanka Krishnan (1934), and [23] reported 3.91% and 4.79% from Charnockites of Tamil Nadu and Karnataka (Table 4).

The histogram (Figure 6(b)) represents frequency percent distribution, showing polymodal nature of Na_2O. The maxima is at 1.5 percent to 2.0 percent class, and the typical mode also lies in the same class.

4.11. Potash (K_2O)

The quantitative variation trend of K_2O in the host rock ranges from 0.02 percent to 5.49 percent and gives an average value of 2.12 percent. [24] reported 5.93% from feldspar garnetiferous granite-gneiss of peddapantupalli, 3.06% reported from Khondalites of Sri Lanka [25] .

[11] reported 3.24% from Khondalites of Garividi, and [23] reported 2.23% from Charnockites of Tamil Nadu (Table 4).

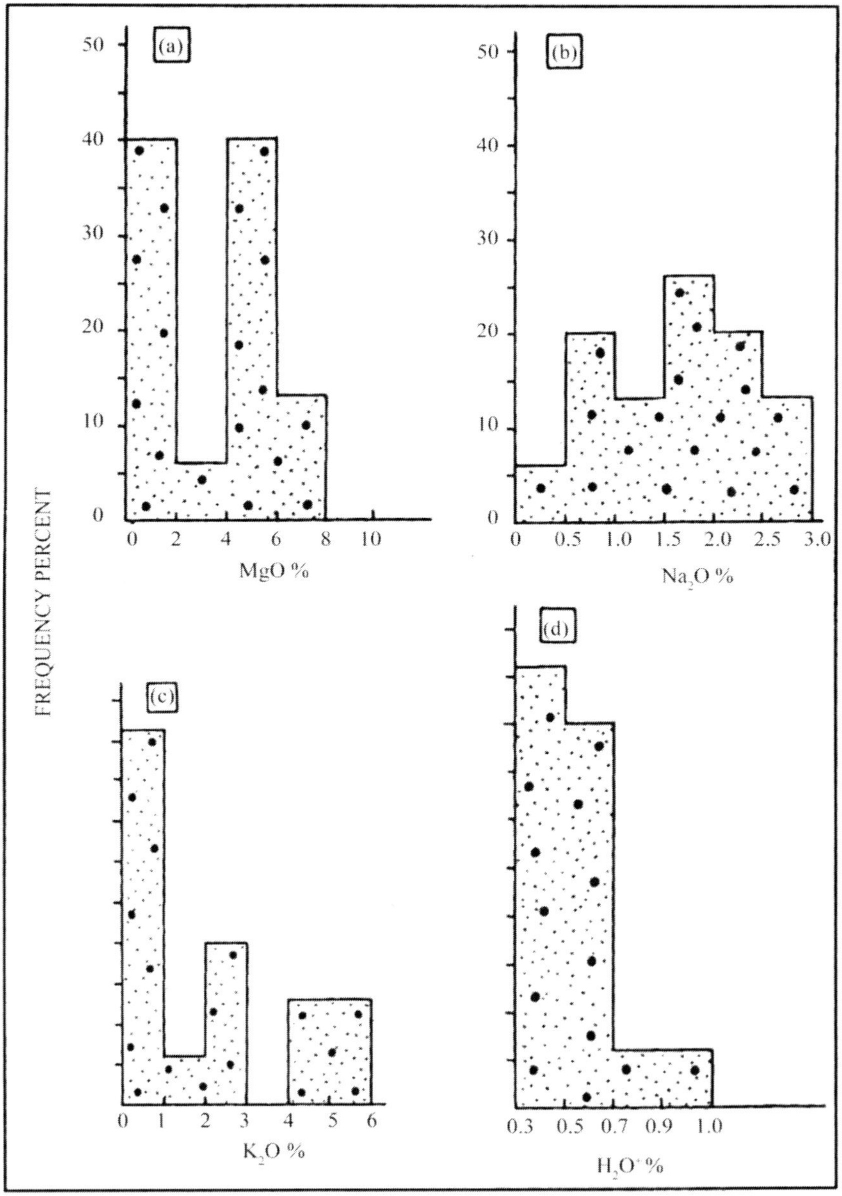

Figure 6. Histograms showing frequency percent distribution of MgO, K_2O, Na_2O and H_2O^+ in Khondalite and Charnockite Groups of Garividi, Garbham and Chipurupalle areas, Srikakulam-Visakhapatnam Belt (A.P.).

The histogram (Figure 6(c)) represents frequency percent distribution, showing polymodal nature of K_2O. The maxima is at 0 to 1 percent class and typical mode also lies in the same class.

The Khondalites and Charnockites have wide variation of potassium oxide. This higher amount of K_2O is due to the presence of potash feldspar and micaceous mineral in these rocks. The high potash contents in Khondalites and Charnockites are possibly due to the migmatisation, which took place by the injection of feldspathic (K-feldspar) fluid. The megacrystic K-feldspar in them crystallised from the injected fluid and also by the segregation of the neo-mobilised feldspathic components during migmatisation. According to [27] , the main supply of alkalies came probably from the intrusive igneous rocks, another probable source might be the quartz with its numerous fluid inclusion.

4.12. Water (H_2O^+)

The quantitative variation trend of H_2O^+ in the host rocks ranges from 0.30% to 0.96% and gives an average value of 0.52 percent.

[24] reported 0.44% from feldsdpar garnetiferous granite-gneiss of peddapantupalli, [26] reported 0.51% from hypersthene-granulite of Narasihunipeta, and [23] reported 0.47% from Charnockites of Tamil Nadu.

The histogram (Figure 6(d)) represents frequency percent distribution, showing unimodal positively skewed nature of H_2O^+. The maxima is at 0.3 to 0.5 percent and typical mode also lies in the same class.

5. MUTUAL RELATIONSHIP

An attempt has been made to determine the geochemical relationship of MnO with SiO, Al_2O_3, FeO, K_2O, Na_2O, CaO (Figures 7(a)-(f)), and Fe_2O_3 in Khondalite and Charnockite Groups of Garividi, Garbham and Chipurupalle areas, Srikakulam-Visakhapatnam Belt (A.P.). The mutual relationships of significant oxides are plotted in various diagrams and discussed as follows:

5.1. Silica (SiO_2)

The computed correlation coefficient of silica shows strongly antipathetic relationship with MnO (Figure 7(a)) and FeO (relationship of SiO_2 with Fe_2O_3 and FeO and Al_2O_3 with Fe_2O_3 in Khondalite and Charnockite Groups of Garividi, Garbham and Chipurupalle areas, Srikakulam-Visakhapatnam

Belt (A.P.)) and MgO and shows slightly negative relationship with TiO_2, Na_2O and H_2O^+ (Table 5).

These relationships reveal that with the increase of SiO_2 there is correspondingly decrease in MnO, FeO, TiO_2, CaO, MgO, Na_2O and H_2O^+ and vice versa. The negative relationship of silica with manganese and FeO probably indicates that manganese and FeO are the substitutes of silica. It appears that the negative relationship of MnO with silica is possibly due to the alteration of manganese bearing minerals into silicate minerals. The highly variable silica/alumina ratio (Table 6) and sympathetic relationship of silica with Al_2O_3, Fe_2O_3 and K_2O favours the coexistence of quartz and aluminosilicate minerals in the Khondalite and Charnockite rocks. The higher concentration of silica is due to the presence of quartz and other siliceous minerals a characteristic feature of the original pellitic sedimentary rocks. Furthermore the high concentration of silica in the rocks of study area is possibly due to the presence of scapolite.

Silica shows a sympathetic relationship with Al_2O_3, Fe_2O_3, P_2O_5 and K_2O ($r = 0.732$) (Table 6). The positive relationship of silica with these elements is possibly due to the presence of potash feldspar and other alumino-silicate minerals.

5.2. Alumina (Al_2O_3)

The alumina is negatively but insignificantly related with MnO (Figure 7(b)), FeO (Figure 7(c)), TiO_2, MgO, and H_2O^+ and strongly antipathetic with CaO (r = 0.79. As evident from the (Table 6), alumina shows positive relationship with SiO_2, Fe_2O_3 (r = 0.73, P_2O_5, Na_2O and K_2O). However, a strong positive relationship appears between Al_2O_3 & Fe_2O_3 ($r = 0.73$). By correlation coefficient it therefore appears that like other few oxides, e.g., FeO, TiO_2, MgO, H_2O^+, the alumina is also not significantly related with MnO. Further the positive relationship of alumina with silica, Fe_2O_3, Na_2O and K_2O indicates the presence of sillimanite, kyanite, andalusite, orthoclase, microcline. The high Al_2O_3/Fe (total iron) ratio (Table 6) in both Khondalite and Charnockite is identical to that found in alumino-silicate minerals.

The abnormally high Al_2O_3 content is some samples of host rocks may be due to the presence of white specks of kaoline, found in cleavages and cracks of garnet [11] . These specks appear to be the cause of these abnormally high values. The abundance of alumina and silica can be assigned to the argillaceous nature of the parent rocks.

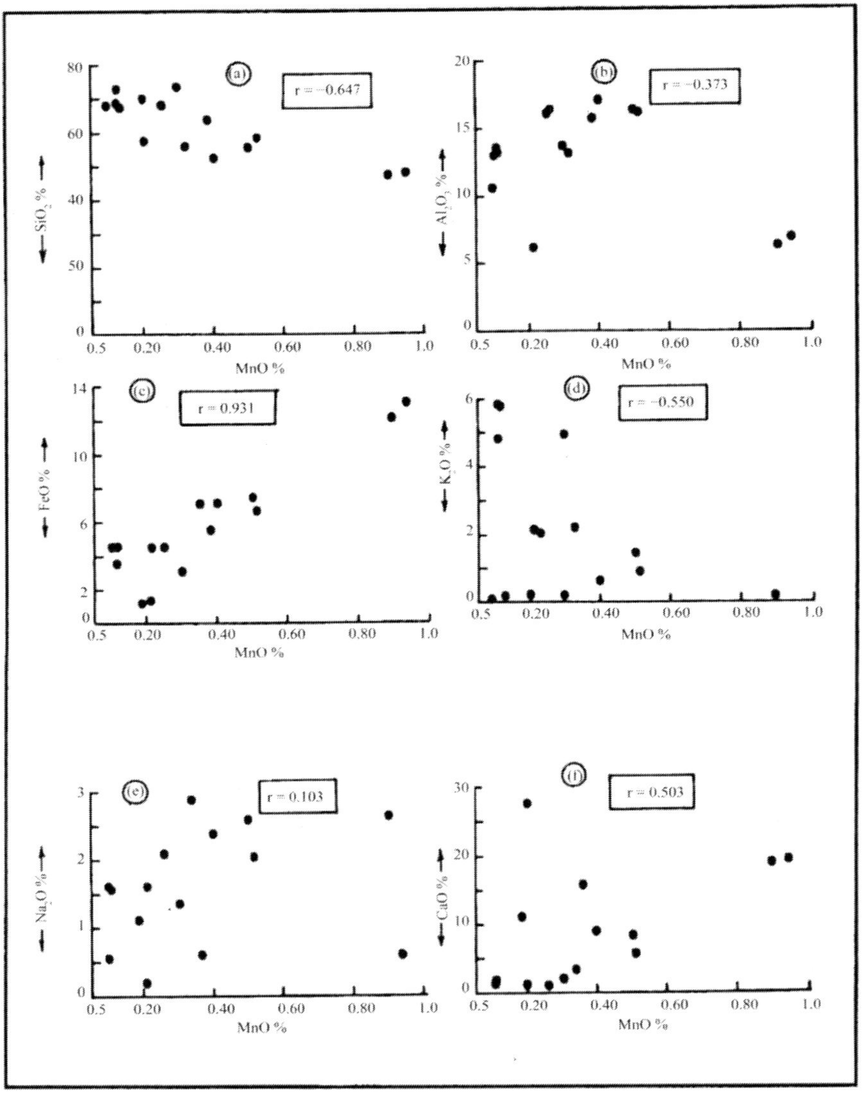

Figure 7. Relationship of MnO with SiO_2, Al_2O_3, FeO, K_2O, Na_2O and CaO in Khondalite and Charnockite Groups of Garividi, Garbham and Chipurupalle areas, Srikakulam-Visakhapatnam Belt (A.P.).

Table 5. Correlation coefficient between the various constituents of Khondalite and Charnockite Groups of Garividi, Garbham and Chipurupalle areas, Srikakulam-Visakhapatnam Belt (A.P.).

S.No.	SiO_2	Al_2O_3	MnO	Fe_2O_3	FeO	P_2O_5	TiO_2	CaO	MgO	Na_2O	K_2O	H_2O^+	Ni	Co	Pb	Zn	Cr	Sr	V	Li
SiO_2	1	0.3863	−0.0348	0.0348	−0.7988	0.2349	−0.4119	−0.7385	−0.9348	−0.1741	0.7329	−0.2199	0.1747	0.0957	0.6072	−0.2634	0.0899	0.0178	−0.2974	0.0779
Al_2O_3		1	−0.3746	0.5374	−0.2285	0.4139	−0.0116	−0.7976	−0.4863	0.3681	0.2813	−0.2005	0.3542	0.2514	0.0668	0.0683	0.239	−0.0308	−0.1491	−0.3063
MnO			1	0.1993	0.9319	−0.474	0.0923	0.5032	0.7941	0.1035	−0.5504	0.0573	−0.0163	0.2032	−0.249	0.0689	−0.2299	−0.219	0.1018	0.1688
Fe_2O_3				1	0.2476	0.0803	−0.3351	−0.4456	−0.2154	0.1929	−0.0402	−0.1686	0.3964	0.6076	0.2426	−0.3531	0.1611	−0.3805	−0.1415	0.091
FeO					1	−0.467	0.0407	0.3514	0.6977	0.0139	−0.4058	0.0484	−0.1092	0.1699	−0.2559	0.0429	−0.0546	−0.0091	0.1808	0.0044
P_2O_5						1	0.298	−0.2433	−0.1491	0.2804	0.0272	0.1629	0.2422	0.1302	−0.1324	0.0787	0.2954	−0.3134	0.3457	−0.3347
TiO_2							1	0.3417	0.3732	0.2686	−0.2434	−0.0778	−0.1786	−0.4448	−0.5717	0.4041	0.2491	0.0493	0.5239	−0.1731
CaO								1	0.7822	−0.1503	−0.7267	0.2387	−0.225	−0.3526	−0.5382	0.1953	−0.2272	0.0612	0.238	0.0786
MgO									1	0.1214	−0.7766	0.3612	−0.1259	−0.1526	−0.6402	0.3623	−0.2005	−0.0782	0.2634	−0.0911
Na_2O										1	−0.1265	−0.1089	0.0251	0.3904	−0.0253	0.0677	0.1261	−0.3141	−0.0978	−0.4564
K_2O											1	−0.1963	−0.1984	0.0819	0.6583	−0.2417	0.2086	0.0984	−0.0662	0.1648
H_2O+												1	−0.4218	−0.365	−0.2954	0.2541	0.1423	0.1041	−0.005	−0.0424
Ni													1	0.429	0.0073	−0.0018	−0.5629	−0.4311	−0.0747	0.2016
Co														1	0.5297	−0.347	−0.1564	−0.5655	−0.2778	0.0288
Pb															1	−0.6955	0.0855	−0.338	−0.2282	0.3454
Zn																1	−0.2832	0.3346	−0.2687	−0.125
Cr																	1	0.2135	0.3407	−0.4597
Sr																		1	−0.1375	−0.378
V																			1	−0.1025
Li																				1

The generally high concentration of alumina is possible due to the presence of alumina-silicate minerals, i.e., feldspar, micaceous minerals and sillimanite (in garnet-sillimanite gneisses). In feldspar Si^{4+} (radius 0.42 Å) is replaced by Al^{3+} (radius 0.50 Å) in Si-O tetrahedra [28] .

5.3. Manganese (MnO)

The MnO shows strong antipathetic relationship with SiO_2 ($r = -0.84$) and K_2O (Figure 7(d)), whereas it shows feebly negative relationship with Al_2O_3 and P_2O_5.

Table 6. Fe_2O_3/FeO, SiO_2/Al_2O_3, Na_2O/K_2O, CaO/MgO, MnO/P_2O_5, MnO/CaO, Al_2O_3/TiO_2, MnO/Fe (total iron) Al_2O_3/Fe (total iron) ratios of Khondalite and Charnockite Groups of Garividi, Garbham and Chipurupalle areas, Srikakulam-Visakha- patnam Belt (A.P.).

Sample Nos.	Fe_2O_3/FeO	SiO_2/Al_2O_3	Na_2O/K_2O	CaO/MgO	MnO/P_2O_5	MnO/CaO	Al_2O_3/TiO_2	MnO/Fe (Total Iron)	Al_2O_3/Fe (Total Iron)
1	0.26	5.29	0.255	2.75	0.5	0.050	17.93	0.020	2.284
2	0.318	5.22	0.322	2.22	1.2	0.049	33.62	0.031	3.539
3	0.730	4.17	0.157	1.45	0.7	0.168	35.63	0.026	2.104
4	0.099	7.60	33.5	2.51	0.94	0.047	10.7	0.065	0.445
5	0.147	4.18	2.91	3.91	3.6	0.022	16.9	0.043	1.625
6	0.366	6.25	7.5	2.98	0.42	0.014	18.86	0.109	6.67
7	0.372	9.22	8.0	5.59	0.58	0.008	5.74	0.105	3.175
8	0.305	2.99	3.87	1.52	0.88	0.045	18.35	0.042	1.853
9	0.197	3.43	1.80	1.45	1.42	0.068	11.26	0.056	1.823
10	0.331	3.68	2.21	1.47	1.57	0.076	22.25	0.058	1.795
11	0.360	4.09	1.29	1.02	1.18	0.118	35.59	0.048	2.005
12	0.284	5.17	0.094	1.22	0.23	0.051	18.50	0.021	2.445
13	0.518	5.36	0.281	2.16	3.0	0.125	21.86	0.062	2.733
14	0.732	4.14	1.03	1.45	0.74	0.208	34.16	0.032	2.062
15	0.189	6.44	83.5	2.87	6.42	0.045	12.36	0.062	0.514

As evident from Table 4, the MnO, in the Khondalite and Charnockites of the present area have strong positive relationship with FeO ($r = 0.93$) and MgO ($r = 0.79$) and slightly positive relationship with Fe_2O_3, TiO_2, Na_2O (Figure 7(e)), Cao (Figure 7(f)) and H_2O^+, MnO/CaO ratio is constant. The positive relationship of MnO with Fe_2O_3, FeO and the constant MnO/Fe (total iron) (Table 6) ratio indicate that iron and manganese were delivered in the same proportion. The constant MnO/Fe ratio suggests coprecipitation of Mn and Fe under fairly oxidizing condition. The small quantity of manganese determined from the country rocks has most probably been derived from the manganese silicate minerals like manganese bearing garnets, pyroxenes, and other Mn bearing minerals presents in the host rocks ([14] [29] - [31]) possibly in association of ferruginous magnesium rich calc-granulite and garnet sillimanite gneiss and hypersthene gneisses.

5.4. Iron Oxides

The relationship of Fe_2O_3 with TiO_2, CaO, MgO, K_2O and H_2O^+ is negative as Fe_2O_3 increase with the decrease of these elements or vice versa.

The antipathetic relationship of Fe_2O_3 with MgO in Khondalite and Charnockite rocks may be explained by the replacement of Fe^{3+} (0.67 Å) by

Mg^{2+} (0.66 Å) [32] . The higher FeO/Fe_2O_3values is Khondalite from the Eastern Ghat granulite belt of India indicate lower oxidation conditions in Early Proterozoic [33] . In Eastern Ghat granulite belt of India the variation in FeO/Fe_2O_3 ratio from one area to the other suggests that these sediments were deposited in different basins with different pH. Eh conditions of deposition [33] . The Khondalite and char- nockite rocks of the area under study, exhibit composition variation in iron (total iron), ranging from 1.64% to 14.41% and exhibit high MgO values. Random variation is also observed in K_2O, Na_2O and iron in ternary diagram showing high and random distribution of MgO in Khondalites (Δ) and Charnockites (•) of Garividi, Garb- ham and Chipurupalle areas, Srikakulam-Visakhapatnam Belt (A.P.) Fe_2O_3 (Total Iron), (Table 7 and Table 8).

The significant relationship between various other oxides are also taken into consideration for example relationships of Al_2O_3 with TiO_2 (Figure 8(a)), CaO with MgO (Figure 8(b)) and Fe_2O_3(Figure 8(c)) and FeO (Figure 8(d)).

5.5. Titania (TiO_2)

Titanium oxide shows positive correlation with MnO, FeO, P_2O_5, CaO, MgO and Na_2O whereas SiO_2, Al_2O_3 (Figure 8(a)), Fe_2O_3, K_2O and H_2O^+ are negatively correlated with TiO_2. The positive correlation of TiO_2 with CaO ($r = 0.34$), MgO ($r = 0.37$) and Na_2O ($r = 0.26$) clearly indicates that much of the TiO_2 contains in Khondalite and Charnockite rocks were possibly derived from the titanium bearing minerals like sphene.

5.6. Lime (CaO)

The graph and computed correlation coefficient of calcium show positive relationship with MnO in relationship of MnO with SiO_2, Al_2O_3, FeO, K_2O, Na_2O and CaO in Khondalite and Charnockite Groups of Garividi, Garbham and Chipurupalle areas, Srikakulam-Visakhapatnam Belt (A.P.), FeO, TiO_2 and H_2O^+ and strongly positive relationship with MgO ($r = 0.78$), (Figure 8(b)). CaO show antipathetic relationship with Fe_2O_3, P_2O_5 and Na_2O and strongly antipathetic relationship with SiO_2 ($r = 0.78$), Al_2O_3 ($r = -0.79$) and K_2O ($r = 0.726$). The ratio of CaO and MnO is constant. The Khondalite and Charnockite rocks of the study area having fairly high concentration of CaO (range from 1.25% to 25.61%). Furthermore diopside, wollastonite, anorthite, scapolite, piedmontite and calcite also contributed varying proportion of CaO to the Khondalite and Charnockite rocks. The lime content seems to be of migmatitic origin [38] . The relationship between various other oxides are also taken into consideration for example relationships of Fe_2O_3 (Figure 9(a)), SiO_2 with FeO (Figure 9(b) and Figure 9(c)) and in these host rocks of Garividi, Garbham and Chirpurupalle areas.

5.7. Phosphorus (P_2O_5)

The concentration of P_2O_5 is abnormally on the higher side in the manganese ores as well as in the host rocks of the study area. However, P_2O_5 shows insignificant negative relationship with MnO, FeO, CaO and MgO.

The ratio of MnO and P_2O_5 is variably low and high. Whereas P_2O_5 shows positive relationship with SiO_2, Al_2O_3, Fe_2O_3, TiO_2, Na_2O, K_2O and H_2O^+. The positive relationship of P_2O_5 with SiO_2, $Al_2O_3$and K_2O appears

Table 7. Recalculted to 100 wt% of bulk components (Fe_2O_3 + MgO), Na_2O and K_2O in Khondalite rocks of Garividi, Garbham and Chipurupalle areas, Srikakulam-Visakhapatnam Belt (A.P.).

Sample No	(Fe_2O_3 + MgO)	Na_2O	K_2O
1	46.9	10.8	42.2
2	42.7	13.9	43.3
3	77.9	3.0	19.0
4	97.0	2.9	0.08
5	92.9	5.2	1.8
6	81.0	16.7	2.2
7	78.4	19.1	2.3
8	82.9	13.5	3.4
9	77.8	14.2	7.9
10	81.7	12.5	5.6

Table 8. Recalculated to 100 wt% of bulk components (Fe_2O_3 + MgO), Na_2O and K_2O in Charnockite rocks of Garividi, Garbham and Chipurupalle areas, Srikakulam-Visakhapatnam Belt (A.P.).

Sample No	(Fe_2O_3 + MgO)	Na_2O	K_2O
1	68.4	17.7	13.7
2	53.2	4.0	42.6
3	48.1	11.4	40.4
4	67.4	16.5	16.0
5	92.6	7.2	0.08

Fe_2O_3 total iron.

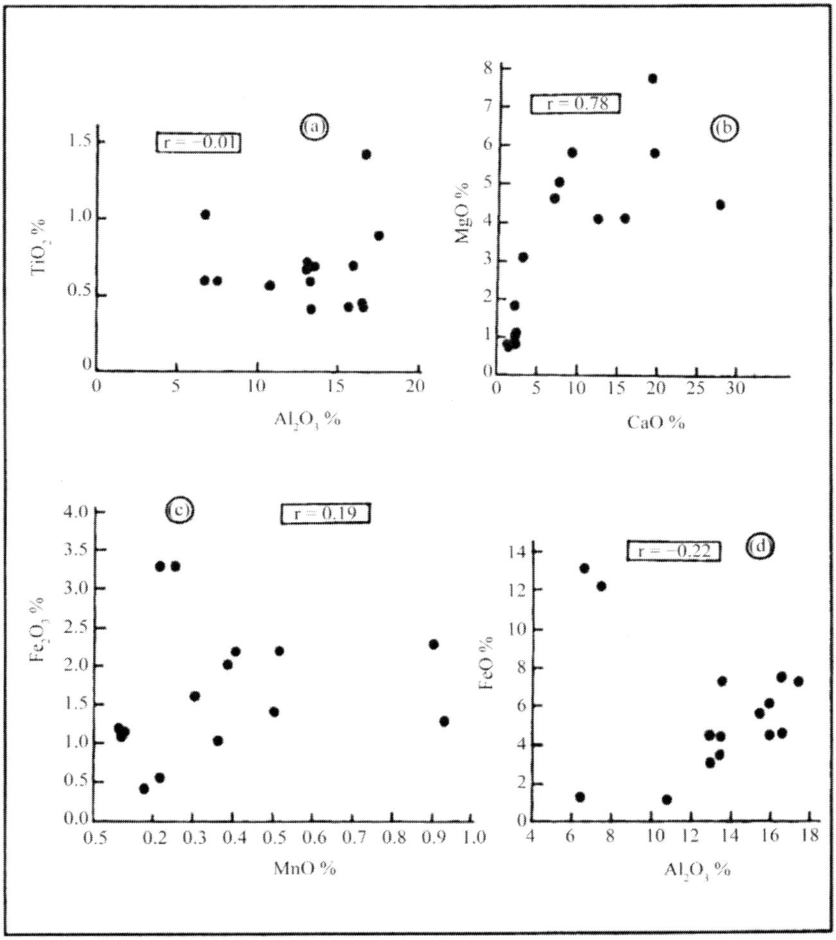

Figure 8. Relationship of Al_2O_3 with TiO_2 and FeO, and CaO with MgO and MnO with Fe_2O_3 in Khondalite and Charnockite Groups of Garividi, Garbham and Chipurupalle areas, Srikakulam-Visakhapatnam Belt (A.P.).

to be due to the presence of alumino-phosphate minerals probably leucophosphate formed by the reaction of phosphate rich solutions with clay minerals. The chemical reaction which indicates the formation of leucophosphate as given by [34] is as follows:

$$Al_2Si_2O_5(OH)_4 + 2HPO_4^{2-} \rightarrow 2AlPO_4 \cdot 2H_2O + 2SiO_2 + H_2O$$

$$2AlPO_4 \cdot 2H_2O + KOH \rightarrow KAl_2(PO_4)_2OH \cdot 2H_2O + 2H_2O$$

Further the positive relation of P_2O_5 with Fe_2O_3 clearly indicates the ferruginous nature of phosphate minerals. According in [1] [27] [31] [35] , the source of phosphorus in the original manganiferous rocks is apatite and spessartite which are the common constituent of the rocks associated with gondite and Kodurite types.

[36] reported up to 4.10 percent P_2O_5 in spessartite. The presence of apatite in the host rocks indicates that they were subjected to metamorphism in a more hydrous environment as also suggested by [37] .

The manganese silicate-carbonate, clay and silica. In few mines as in Avagudem the manganese carbonate was either deposited in the form of carbonate or formed through diagenesis of manganese oxide [39] .

5.8. Magnesia (MgO)

The present study reveals that MgO is positively correlated with FeO, CaO, MnO, TiO_2, Na_2O and H_2O^+ whe- reas it is negatively correlated with Al_2O_3, Fe_2O_3, P_2O_5, SiO_2 and K_2O (Figures 9(a)-(c)).

The Khondalites and Charnockites of the study area show higher concentration of MgO (ranges from 0.86% to 7.90%). [33] also reported higher concentration of MgO in Khondalites from the Eastern Ghat granulite belt of India.

These Khondalite with high MgO content can be compared with Late-Archaean-Early Proterozoic metapelites. In the area under study the Khondalite and Charnockite rocks exhibit compositional variations (Table 9 and Table 10) in MgO (ranges from 0.86% to 7.90%), K_2O (ranges from 0.02% to 5.94%) and Na_2O (ranges from 0.34% to 2.84%). The ternary diagram (Figure 10) shows wide scatter distribution of MgO, K_2O and Na_2O.

The concentration of magnesium oxide is possibly due to the presence of garnet, spinel, hornblende, magnesium rich pyroxenes and biotite in the rocks of the study area. All these minerals pointing to the argillaceous nature of the original sediments which gave rise to these rocks.

5.9. Soda (Na_2O)

In the Khondalite and Charnockite rocks soda shows positive relationship with Al_2O_3, MnO in relationship of MnO with SiO_2, Al_2O_3, FeO, K_2O, Na_2O and CaO in Khondalite and Charnockite Groups of Garividi, Garbham and Chipurupalle areas, Srikakulam-Visakhapatnam Belt (A.P.), Fe_2O_3, FeO, P_2O_5 and TiO_2.

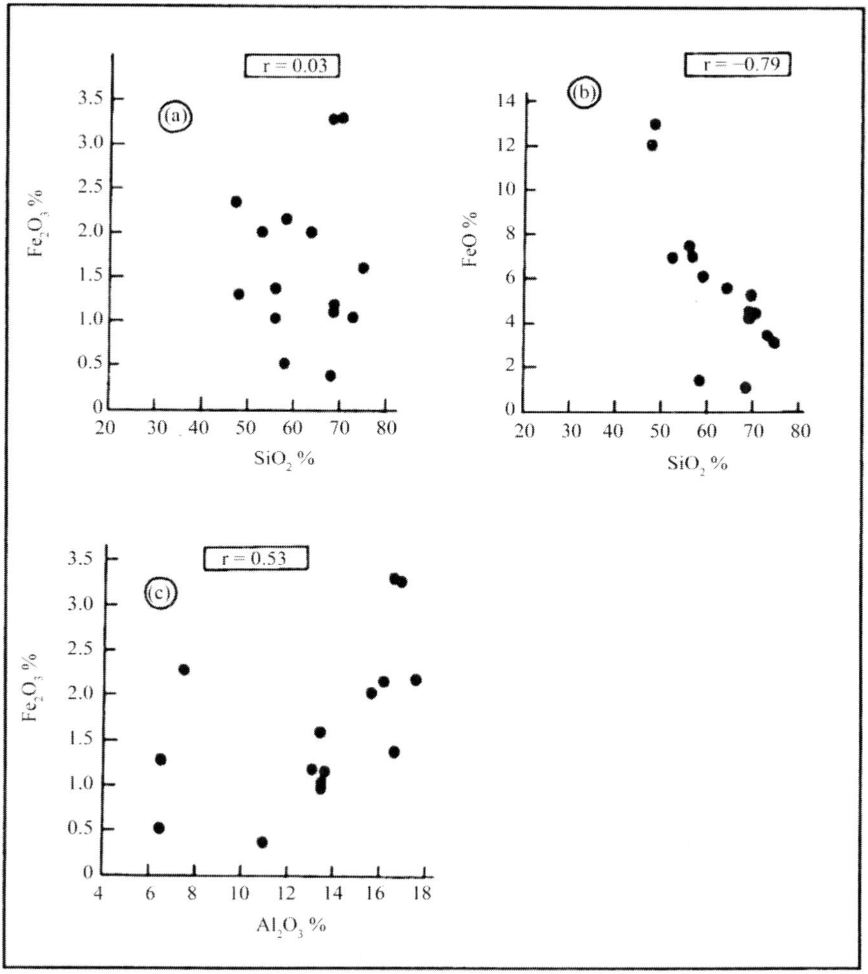

Figure 9. Relationship of SiO_2 with Fe_2O_3 and FeO and Al_2O_3 with Fe_2O_3in Khondalite and Charnockite Groups of Garividi, Garbham and Chipurupalle areas, Srikakulam-Visakha- patnam Belt (A.P.).

Table 9. Recalculated to 100 wt% of bulk components MgO, Na_2O and K_2O in Khondalite rocks of Garividi, Garbham and Chipurupalle areas, Srikakulam-Visakhapatnam Belt (A.P.).

Sample No	MgO	Na_2O	K_2O
1	10.3	18.2	71.3
2	14.2	20.9	64.8
3	25.5	10.1	64.2
4	92.9	7.8	.2
5	81.4	13.7	4.7
6	75.4	21.6	2.8
7	71.7	25.0	3.1
8	65.0	27.8	7.1
9	55.8	28.4	15.7
10	60.3	27.2	12.3

These rocks in few samples show exceptionally high Na_2O/K_2O ratio. The ternary diagram (Figure 11) shows wide scatter distribution of K_2O and Na_2O.The presence of soda is due to the presence of feldspar and also due to the presence of scapolite in the Khondalites of study area [11] . The main supply of alkalies probably came from the igneous intrusion which cut across the Khondalite and Charnockite rocks. The high soda concentration in these rocks suggests the existence of predominantly tonalite/trondhjemite rocks in the source area. A study of granitic/gneissic episodes of the South Indian shield [40] - [42] indicates that the Early Archaean.

Post-greenstone granitic and gneissic episodes of the stable craton are mostly tonalitic/trondhjemitic in nature and K-rich granites appeared only in Late Proterozoic which explains better the high soda content in Proterozoic Khondalites of the Eastern Ghats. Predominance of soda rich tonalites and trondhjemites in Early Archean crust is also reported from other shield areas [43] [44] . Wide distribution in Na_2O from area to area may indicate to different temperatures of Na-rich and mafic source material indifferent basins [33] .

5.10. Potash (K_2O)

K_2O shows insignificant negative relationship with CaO, MgO, MnO in relationship of MnO with SiO_2, Al_2O_3, FeO, K_2O, Na_2O and CaO in Khondalite and Charnockite Groups of Garividi, Garbham and Chipurupalle areas, Srikakulam-Visakhapatnam Belt, Fe_2O_3, FeO, TiO_2, Na_2O and H_2O^+. However, it shows strongly positive relationship with SiO_2 and slightly positive relationship with Al_2O_3 and P_2O_5 (Table 4). The dominance of K_2O over

Table 10. Recalculated to 100 wt% of bulk components MgO, Na_2O and K_2O in Charnockite rocks of Garividi, Garbham and Chipurupalle areas, Srikakulam-Visakhapatnam Belt (A.P.).

Sample No	MgO	Na_2O	K_2O
1	38.2	34.8	26.9
2	22.7	6.6	70.5
3	14.8	18.7	66.4
4	16.8	42.2	40.9
5	80.3	19.4	0.2

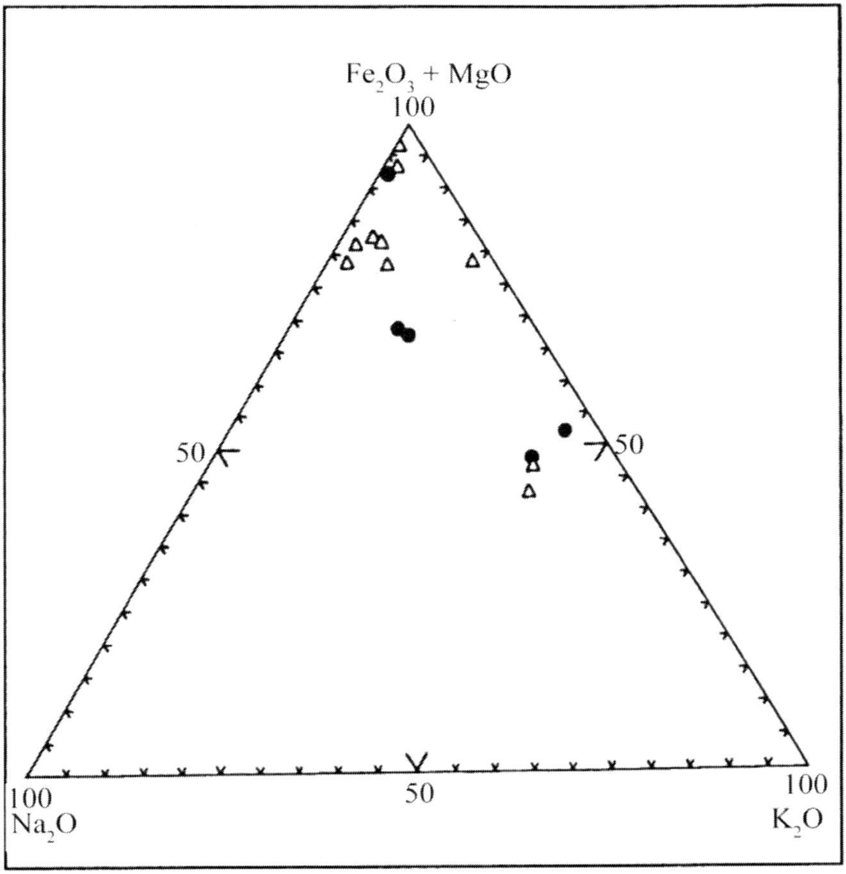

Figure 10. Ternary Diagram showing high and random distribution of MgO in Khondalites (Δ) and Charnockites (•) of Garividi, Garbham and Chipurupalle areas, Srikakulam-Vi- sakhapatnam Belt (A.P.) Fe_2O_3 (Total Iron).

Na_2O is characteristic of parametamorphic rocks [20] and these rocks have exceptionally high and variable Na_2O/K_2O ratio. Potassium and sodium also show close coherence with aluminium, which is possibly due to the presence of soda and potash feldspar in these rocks.

5.11. Water (H_2O^+)

Water shows positive relationship with MnO, FeO, P_2O_5, CaO and MgO and negative relationship with SiO_2, Al_2O_3, Fe_2O_3, TiO_2, Na_2O and K_2O.

6. CONCLUSIONS

The analysis of 15 host rocks samples reveals that:

¾ Silica increases with the increases of Al, phosphorous and potash and vice versa.

¾ Alumina increases with the increases of SiO_2, P_2O_5, Na_2O and K_2O and vice versa.

¾ Mn increases with the increases of Fe and vice versa.

¾ MnO also shows strong sympathetic relationship with MgO and Na_2O.

¾ Phosphorous decreases with increase of MnO and vice versa.

¾ Ti increases with the increases of Mn and vice versa.

¾ Lime increases with the increases of Mn and vice versa.

¾ Mg increases with the increases of MnO and vice versa.

¾ Soda increases with the increases of both Mn and Fe and vice versa.

¾ Potash decreases with the increases of both Mn and Fe and vice versa.

The geochemical behavior of the host rocks indicate that the evolution of the ores took place in two different stage.

In first stage, sediments were rich in Mn, deposited in a basin during typical condition of oxidation. The chemical alteration and metamorphism of these sediments gave rise to the formation of primary metamorphic

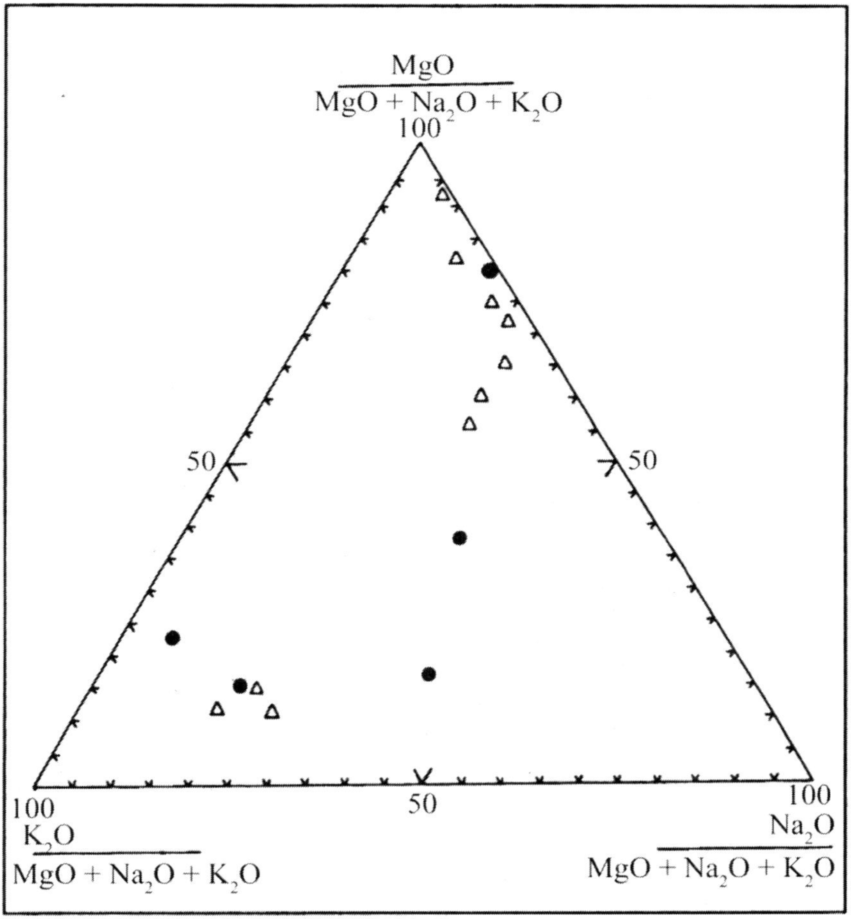

Figure 11. Ternary Diagram showing Wide Scatter of Na_2O and K_2O in Khondalites (Δ) and Charnockites (•) of Garivi- di, Garbham and Chipurupalle areas, Srikakulam-Visakha- patnam Belt (A.P.).

Mn minerals. In the later stages these metasediments suffered metasomatism due to intense igneous intrusions which caused hybridization at places.

In the second stage, the previously formed Mn ores and Mn-silicates of the country rocks like Mn-rich garnets and pyroxenes which played an important role in the formation of secondary Mn minerals viz. psilomelane-cry- ptomelane, hollandite, pyrolusite and wad etc. by the alteration. The meteoric water rich in CO_2 played an important role in solution, transportation, precipitation, concentration and finally redeposition of the secondary Mn ores.

The concentration and redeposition have largely taken place from colloidal state or solution in oxidizing environment. It is assumed that Mn and Fe are in colloidal state. The anions and cations of minor and trace elements are adsorbed on the surface of negatively charged $Mn(OH)_4$ and positively charged $Fe(OH)_3$ particles. The high K_2O/Na_2O, CaO/MgO and Co/Ni ratios in these ores are due to high adsorption capacities of K_2O, CaO and Co then Na_2O, MgO and Ni:

Presumably the Mn and Fe solutions first form in feebly acidic medium with at pH value around 7. However with an increase in alkalinity Mn^{2+} and Fe^{2+} they are converted to colloidal $Mn(OH)_4$ and $Fe(OH)_3$. The solubility of SiO_2 also increases with the increase pH (about 9), the Al_2O_3 dissolves into solution. The addition of $Al(OH)_3$ into solution causes drastic changes in character of $Mn(OH)_4$colloids which are precipitated as oxide ores. Originally the Mn deposits have been derived from sources on the bordering Cratons, which are also the probable provinces for the sediments in this part of the basin.

ACKNOWLEDGEMENTS

Farhat Nasim Siddiquie, Mohd Shaif The authors are grateful to Prof. M. Rahimullah, Chairman Department of Geology and Dean Faculty of Science, AMU Aligarh. To Director National Institute of Oceanography, Dona Paula, Goa for providing research facilities.

REFERENCES

1. Siddiquie, F.N. (2001) Source of Unusually High P_2O_5 in Manganese Ores of Srikakulam District (A.P.), India. Journal of Applied Geochemistry, 3, 78-84.
2. Siddiquie, F.N., Ali, S.A., Bhat, S.H. and Burhamuddin, Kh. (2015) Genetic Association of Trace Elements of with Special Emphasis on Cu, Ni, and Cr in the Manganese Ores of Vizianagaram District, (A.P.), India. Journal of Applied Geochemistry, 17, 66-77.
3. Acharya, D.C. and Nayak, B.K. (2002) Mineral Chemistry of Manganese Ores Associated with Precambrian Eastern Ghats Complex of Andhra Pradesh, Orissa, India. In: Siddiquie, F.N., Ed., Metallogenesis of Manganese Ores of Srikakulam-Visakhapatnam Belt, Icon Publication Ltd., New Delhi, 213 p.
4. Sarkar, S.N. (1968) Precambrian Stratigraphy and Geochronology of Peninsular India. Dhanbad Publishers, India, 33 p.
5. Roy, S. (1966) Syngenetic Manganese Formations of India. Jadavpur University Press, Calcutta, 219 p.
6. Mahnert, K.R. (1968) Migmatites and the Origin of Granitic Rocks. Els. Pub. Co., 393 p.

7. Rao, G.V. (1969) The Geology and Manganese Ore Deposits of Parts of Vizakhapatnam (Visakhampatnam) Manganese Belt, Srikakulam District, A.P., India. Bulletin of Geological Survey of India, Series A, No. 35, 129 p.
8. Fermor, L.L. (1909) Manganese Ore Deposits of India. Memoir of Geological Survey of India, 37, 272.
9. Middlemiss, C.S. (1914) General Report of Geological Survey of India. Geological Survey of India, 36, 102-105.
10. Rao, S.V.G.K. and Sarma, K.J. (1978) Report on Assessment of Manganese Ore in Srikakulam and Visakhapatnam Districts, A.P., India. Unpublished Report, Geological Survey of India.
11. Prabhakhar Rao, G. (1950) The Geology of Manganese Bearing Rocks of Garividi and Garbham, Vizagapatnam District, A.P., India. Quarterly Journal of the Geological, Mining and Metallurgical Society of India, 22, 25-42.
12. Venkataramaiah, K., Doshi, C.M. and Rao, S.V.G.K. (1978) A Report on the Assessment of the Manganese Ore Deposits in Srikakulam and Visakhapatnam District, A.P., India. Unpublished Report, Indian Bureau of Mines.
13. Siddiquie, F.N. (1982) A Geochemical Appraisal of the Manganese Ores of Garividi, Garbham and Chirpurupalle mines, Srikakulam District, A.P., India. Unpublished M.Phil. Thesis, Geology Department, Aligarh Muslim University, Aligarh, 92 p.
14. Siddiquie, F.N. (1986) Geology and Geochemistry of Manganese Ore Deposits of Srikakulam District (Andhra Pradesh), India. Unpublished Ph.D. Thesis, Geology Department, Aligarh Muslim University, Aligarh, 253 p.
15. Siddiquie, F.N. (2000) On the Occurrence of Lepocinclis from Khondalites of Srikakulam District (A.P.). Bulletin of the Indian Geologist's Association, 33, 63-68.
16. Siddiquie, F.N. (2003) Some Modification in the Stratigraphy of Manganese Bearing Formations, Srikakulam District (A.P.), India. Journal of Science and Technology, 11, 137-154.
17. Siddiquie, F.N. (2004) Metallogenesis of Manganese Ores of Srikakulam-Visakhapatnam Belt. Icon Publication Pvt. Ltd., New Delhi, 194.
18. Sharma, N.L., Ali, I. and Perraju, P. (1974) Geology of Parts of Srikakulam, Chipurupalli and Palakonda Taluks, Srikakulam District, A.P., India. Unpublished Report, Geol. Surv. India.
19. The Indian Express (Express Newsline) (2002) Geologist Discovers New Green Alga. Science Talk, 7.
20. Krishna Rao, J.S.R. (1954) The Geology of Chipurupalle Area, Vizagapatnam District, A.P., India, with Special Reference to the Origin of Manganese Ores. Quarterly Journal of the Geological, Mining and Metallurgical Society of India, 26, 36-44.
21. Siddiquie, F.N. and Bhat, S.H. (2010) Geochemistry of Trace Elements in Manganese Ores and Their Host Rocks in Vizianagaram Manganese Ores Belt (A.P.), India. Journal of Applied Geochemistry, 10, 38-72.
22. Janardhan, A.S., Newton, R.C. and Hansen, E.C. (1983) Transformation of Peninsular Gneiss to Charnockite in Southern Karnataka. Memoir 4, Geological Society of India. Proceedings of the Indo-U.S. Workshop, Hyderabad, 417-435.
23. Krishna Rao, J.S.R. (1975) Some Problems on the Manganese Ore Deposits of A.P., India. Acta Mieralogica-Petrogra- phica, 22, 133-141.
24. Gupta, B.P. (1973) Petrology of Leptynites and Garnetiferous Apatite Gneiss around Garbham, Srikakulam District, A.P., India. Journal of the Geological Society of India, 14, 113-123.
25. Krishnan, M.S. (1934) Khondalite from the Road between Banderwala and Nuwara, Eliya M.P. 1. Ceylon (Sri Lanka). Record of Geological Survey of India, 67, 395.
26. Raghava Rao, K.V. (1955) In: Rao, G.V., 1959. The Geology and Manganese Ore Deposits of Parts of Vizakhapatnam (Visakhapatnam) Manganese Belt, Srikakulam District, A.P. India. Bulletin of Geological Survey of India, Series A, No. 35, 1-129.
27. Mookherjee, A. (1961) Distribution of Minor Elements in the Gonditic Manganese Ore and Its Geochemical Significance. Economic Geology, 56, 723-729.

28. Rankama, K. and Sahama, Th.G. (1950) Geochemistry. University of Chicago Press, Chicago, 912 p.
29. Siddiquie, F.N. and Raza, M. (1990) Major and Trace Elements Variability in Manganese Ores of Srikakulam District (A.P.). Indian Geologist's Association, Chandigarh, Vol. 23, 99-109.
30. Siddiquie, F.N. and Raza, M. (1990) Trace Elements in the Manganese Ores of Srikakulam Distt. (A.P.) India (Abstract). National Seminar on Precambrian Geology, Madras, 61.
31. Siddiquie, F.N. and Raza, M. (2008) Geochemical Environment and Metallogenesis of Manganese Ores of Vizianagaram Manganese Belt Andhra Pradesh, India. Journal of the Indian Association of Sedimentologist, 27, 1-16.
32. Drever, J.I. (1971) Magnesium Iron Replacement in Clay Minerals in Anoxic Marine Sediments. Science, 172, 1334- 1336.
33. Divakara Rao, V. (1983) Khondalites from the Eastern Ghat Granulite Belt Geochemistry and Origin. Geophysical Research Bulletin, 20, 233-242.
34. Nriagu, J.O. (1972) Stability of Vivianite and Iron-Pair Formation of the System $Fe_3(PO_4)_2$-H_3PO_4-H_2O. Geochimica et Cosmochimica Acta, 36, 459-470.
35. Krishna Rao, J.S.R. and Venkataramaraju, D. (1966) A Study of Apatites from Srikakulam District, A.P., India. Quarterly Journal of the Geological, Mining and Metallurgical Society of India, 38, 177-181.
36. Mason, B. and Berggren, T. (1941) A Phosphate-Bearing Spessartite Garnet from Wodgina, Western Australia. Geologiska Föreningen i Stockholm Förhandlingar, 63, 413-418.
37. Mitchell, L., Faust, G.T., Hendricks, S.B. and Reynolds, D.S. (1943) The Mineralogy and Genesis of Hydroxyl Apatite. American Mineralogist, 28, 356.
38. Bhattacharyya, C. (1972) Granitisation in Relation to Evaluation of the Charnockite Series from the Eastern Ghat Hills, Srikakulam District, A.P., India. Neues Jahrbuch für Mineralogie Monatshefte, No. 5, 220-240.
39. Bostrom, K. (1967) The Problem of Excess Manganese in Pelagic Sediments. In: Ableson, P.H., Ed., Researches in Geochemistry, Vol. 2, John Wiley and Sons, New York, 420-452.
40. Naqvi, S.M., Divakara Rao, V. and Narain, H. (1978) The Primitive Crust: Evidence from the Indian Shield. Precambrian Research, 6, 323-345.
41. Ali, S. and Devakara Rao, V. (1980) Geochemistry and Origin of Acidic Rocks of Shimoga, Karnataka, India. Journal of Gelogical Society India, 21, 91-96.
42. Divakara Rao, V. and Rama Rao, P. (1982) Granitic Activity and Crustal Growth in India Shield. Precambrian Research, 16, 257-271.
43. Sheraton, J.W. (1980) Geochemistry of Precambrian Metapelites from East Antarctica: Secular and Metamorphic Variations. Bureau of Mineral Resources Journal of Australia Geology and Geophysics, 5, 279-288.
44. Glikson, A.Y. (1979) Early Precambrian Tonalite-Trondhjemite.

CHAPTER 5

Petrology, Geochemistry and Tectonomagmatic Setting of Neshveh Intrusion (NW Saveh)

Reza Keshavarzi[1], Dariush Esmaili[1], Mehdi Rezaei Kahkhaei[1], Mir Ali Asghar Mokhtari[2*], Reza Jabari[3]

[1]*School of Geology, Faculty of Science, University of Tehran, Tehran, Iran*

[2]*Geology Group, Faculty of Science, University of Zanjan, Zanjan, Iran*

[3]*Geology group, Faculty of Science, Islamic Azad University of Science and Research of Tehran, Tehran, Iran*

ABSTRACT

Neshveh intrusion is located in the NW of Saveh City and is a part of Orumieh-Dokhtar magmatic arc. This intrusion consists of quartz monzodiorite, granodiorite and granite that have intruded into the Eocene volcano-sedimentary rocks. This intrusion is high-K calk-alkaline and metaluminous and is classified as I-type granitoids. Field investigations along with petrographic and geochemical studies indicate that all phases of Neshveh intrusion are derived from a common magma source as a result of mineral differentiation. Different phases of this intrusion have low Mg#, Ni, Cr, Co and V which are indicative for higher evolution of magma during the magma ascent and before complete crystallization. All phases of Neshveh granitoid are characterized by LREE-rich patterns with high LREE/HREE ratio and negative Eu anomalies. Similarity of the mentioned patterns suggests a comagmatic source for these rocks and demonstrates the role of magmatic differentiation in their evolution. There are negative anomalies in the Nb and Ti along with positive anomalies of Rb, Ba, K and Pb on the spider diagrams. These anomalies are indicative for a subduction setting for magma source of these rocks. Geochemical studies indicate that the Neshveh granitoid is formed in a volcanic arc and active continental margin. In this base, it is assumed that this intrusion is formed as a result of Neo-Tethys oceanic lithosphere subduction beneath the Central Iran zone which is replaced in the Orumieh-Dokhtar magmatic arc.

Keywords: Petrology, Geochemistry, Intrusion, Granitoid, Neshveh, Saveh

1. INTRODUCTION

Neshveh intrusion is located in the north west of Saveh and is a small part of the Orumieh-Dokhtar magmatic arc in the Alpine-Himalayan orogenic belt. Volcanic rocks are so common in the Orumieh-Dokhtar magmatic arc. These rocks have varieties in composition and tectonic setting which are varying from acidic to basic and continental to shallow marine environments. There are some intrusions also within the volcanic rocks. Acidic-intermediate volcanic and intrusive rocks are widespread in contrast to basic rocks that are probably indicative for the effect of continental crust in their formation.

Several studies had been carried out in the Saveh region, such as studies of [1] and [2] which were focused on the petrography and Petrochemistry of igneous rocks in the south west of Saveh, volcanic and plutonic rocks in the south of studied area. According to [3] , volcanic rocks around the Neshveh village consist of three parts: the oldest volcanic rocks include rhyolite, dacite and some andesite; pyroclastic rocks and lava flows along with intercalation of marl, sandstone and limestone with Lutetian age; and finally andesites and latites with porphyritic textures in the upper part with Priabonian age. The Eocene volcanic flows are widespread in the studied area, which are the result of hybrid crustal-mantle source. The hybrid source of these rocks is attributed to the ascending of basic magma with mantlic source and melting of the continental crust.

Neshveh intrusion located between 35°10' to 35°14' latitudes and 50°10' to 50°14' longitude at the 40 km north west of Saveh. This intrusion was intruded into the volcanic-sedimentary rocks of Eocene (Figure 1). This intrusion didn't investigate for detail petrology and geochemistry, yet. Also, no reliable analyses of trace elements have been published from this intrusion. This research is a detailed study on the petrology and geochemistry of major and trace elements and tectonomagmatic setting of the Neshveh intrusion.

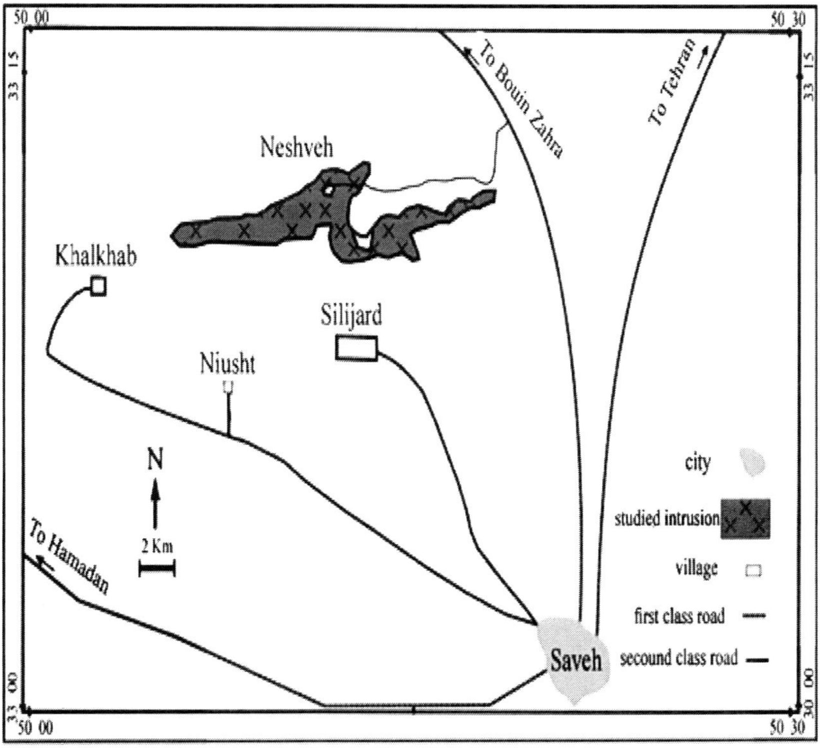

Figure 1. Geographic location of the Neshveh intrusion in the northwest Saveh.

2. GEOLOGY

Neshveh region which is located in the north west of Saveh is a small part of the Orumieh-Dokhtar magmatic arc within the Central Iran zone. All rocks crops out in this region have Cenozoic age and older rocks don't exposed. Based on Saveh geological map 1:100,000 [4] , the main rocks in this area consist of sedimentary, volcanic and intrusive rocks have Paleocene-Eocene age and later (Figure 2). These rocks composed of conglomerate equal to Fajan formation, volcanic-sedimentary rocks of Middle-Upper Eocene consist of intermediateacidic tuffs and lavas, agglomerate and intermediate-basic lavas along with the tuffaceous and sandy limestone and finally Upper Eocene ignimbrite, Lower Red Formation (conglomerate, sandstone, siltstone and red-brown marl), Qom Formation (limestone, marly limestone and marl), Upper Red Formation (conglomerate with red sandstone intercalation), Mio-Pliocene marl, conglomerate equal to Hezar Dareh Formation and quaternary alluvial deposits.

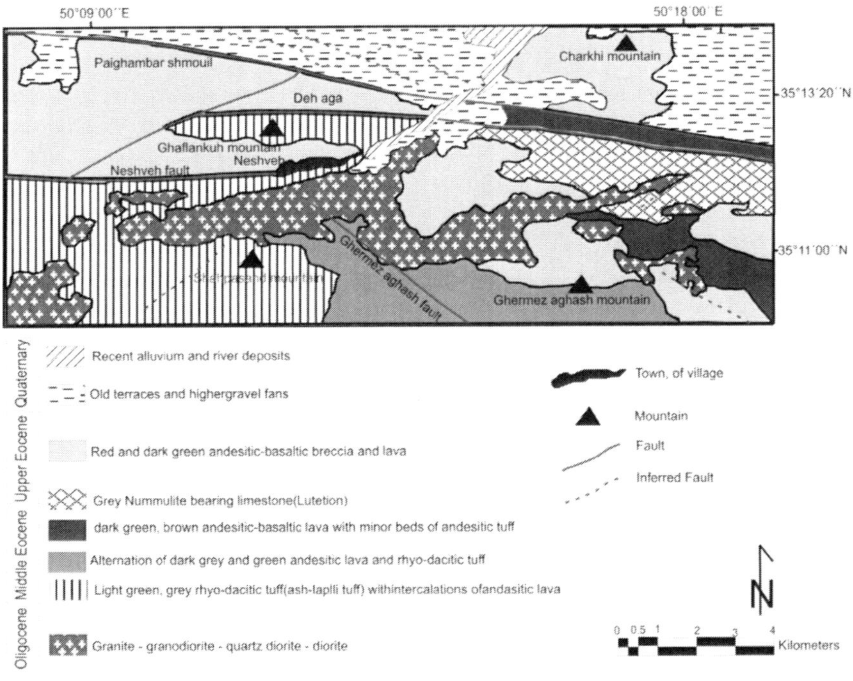

Figure 2. The geological map of studied area based on 1:100,000 geological map of Saveh [4] .

Several granitoid intrusions were intruded into the Eocene volcanic-sedimentary rocks. Dating of some of these intrusions by [1] by K-Ar method on the amphibole, biotite, K-feldspar and whole rock indicate 28 - 42 Ma (Table 1). The Silijard granite, Kalkhab tonalite-diorite, Neshveh granitoid (this study) and Niousht micro-granite are the most important intrusions in the northwest of Saveh.

Table 1. The result of dating analysis on the some intrusions around the Saveh [1] .

Biotites	37 - 41 Ma
Amphiboles	31 - 38 Ma
Alkali Feldspars	35 - 42 Ma
Whole rocks	28 - 39 Ma

In addition to mentioned intrusives, some sub-volcanic domes also exist in different parts of the northwest Saveh that have acidic-intermediate composition. These domes intruded into the Upper Eocene-Oligocene intrusives. In this base, these sub-volcanic intrusions attributed to the Post Oligocene.

There are some acidic and basic dykes in this area. Basic dykes with andesite-andesitic basalt composition are widespread in comparison to acidic dike with dacite-andesite composition. These dikes cut the Eocene volcanic rocks and Oligocene intrusives. In this base, their intrusions have been attributed to the post Oligocene.

3. RESEARCH METHODS

Our researches in this study consist of two parts: field and laboratory studies. Field studies include identifying the different phases of intrusion, relationship between them and host rock and finally sampling of different phases for laboratory studies. Laboratory investigations include preparing of 90 thin section and petrographic studies, analyzing of 9 samples by ICP-AES and ICP-MS for major, trace and rare earth elements at the ALS Chemex (Table 2). Finally, these analyses were processed by using Excel, Minpet, GCDKit and Igpet programs.

4. PETROGRAPHY

Neshveh granitic intrusion with an area about 22 km^2, consist of intrusive rocks in the mountains of Shahpasand and Zarou, north of Jafarabad Village, north of Aftabroo Qeshlaq village and south of Neshveh village. This intrusion crops out in the east-west trend and intruded into the Upper

Table 2. The result of samples analysis from the studied area by ICP-AES and ICP-MS methods (major elements in % and trace elements in ppm).

Rock Type	Andesitic Basalt	Granodiorite					Granite	Qz Monzodiorite	
Sample No.	SN05	SN15	SN17	SN31	SN48	SN44	SN52	SN10	SN11
SiO_2	49	62.7	62.5	62.3	64	65.1	69.4	56.5	59.6
TiO_2	1.12	0.57	0.51	0.59	0.53	0.49	0.32	0.73	0.71
Al_2O_3	14.65	16.15	15.45	15.75	15.65	14.8	13.8	17.1	16.7
Fe_2O_3	13.4	6.03	5.07	5.98	5.17	3.89	3.05	8.66	7.66
MnO	0.22	0.15	0.13	0.1	0.09	0.05	0.07	0.14	0.14
MgO	5.09	1.78	2.05	2.58	1.98	1.52	1.1	3.26	2.48
CaO	7.89	4.81	3.38	4.15	3.8	4.84	2.27	7.16	5.89
Na_2O	4.28	3.49	3.64	3.36	3.41	3.47	3.06	3.23	3.59
K_2O	1.14	2.73	3.46	2.95	3.26	2.9	4.19	1.85	2.31
P_2O_5	0.15	0.19	0.15	0.19	0.16	0.15	0.09	0.23	0.2
Total	96.94	98.6	96.34	97.95	98.05	97.21	97.35	98.86	99.28
V	506	128	89	118	87	76	47	241	194
Cr	15.9	10	10	10	10	10	10	20	10
Ni	19	8	5	6	5	5	5	14	8
Co	15	12.8	8	12	10.2	5	4	23.8	17.4
Cu	19	47	22	59	17	5	8	53	48
Zn	98	76	80	40	51	23	44	51	75
Ga	17.6	16.2	15	15.3	15.3	14.4	13.2	17.5	17.7
Sn	8	1	1	1	1	2	1	2	1
W	1	2	2	2	2	1	1	1	2
Ba	354	675	701	618	665	759	808	465	548
Sr	271	449	347	381	352	353	248	503	424
Rb	29.5	71.4	74.2	63.5	78.2	38.1	90.8	48.7	60.9
Nb	1.5	6.6	7.9	6.5	8.2	7.4	8.7	4.8	6.6
Y	22	20.4	21.6	17.5	19.7	18.4	16.2	20.9	23
Zr	81	121	134	113	145	145	124	83	119
Cs	0.71	1.22	1.49	1.19	1.74	0.4	1.47	1.95	2.34
Hf	2.3	3.6	3.8	3.1	4.1	4.1	3.8	2.4	3.5
Ta	0.1	0.5	0.5	0.5	0.5	0.5	0.7	0.3	0.5
Th	0.35	5.55	5.89	4.98	6.18	5.91	8.42	3.17	5.11
U	1.87	1.43	1.3	1.05	0.97	1.56	1.78	0.71	1.23
La	23.2	24.2	16.8	13.1	13.1	16	19.7	14	15.3
Ce	39.5	45.7	32.6	25.6	26.6	31.6	34.9	28.4	30.8
Pr	3.93	5.24	4.07	3.3	3.53	3.85	3.9	3.64	3.85

Nd	13.2	19.3	16.2	13.3	14.2	14.9	14.2	14.9	15.4
Sm	2.9	3.76	3.7	3.1	3.43	3.21	2.84	3.38	3.53
Eu	1.53	0.95	0.91	0.98	0.9	0.88	0.6	0.99	0.94
Gd	3.77	3.91	4.15	3.4	3.69	3.65	2.99	3.67	3.87
Tb	0.62	0.56	0.64	0.54	0.59	0.55	0.46	0.59	0.62
Dy	4.14	3.48	3.88	3.22	3.56	3.26	2.8	3.73	4
Ho	0.83	0.71	0.83	0.66	0.74	0.69	0.6	0.77	0.8
Er	2.53	2.12	2.41	1.99	2.28	2.12	1.89	2.18	2.45
Tm	0.36	0.32	0.37	0.29	0.34	0.32	0.29	0.32	0.34
Yb	2.37	2.14	2.44	1.93	2.29	2.16	2.06	2.24	2.39
Lu	0.39	0.36	0.37	0.29	0.36	0.33	0.34	0.33	0.4
Mo	2	2	2	2	2	2	2	2	2
Pb	0.5	13	21	10	15	6	17	7	11
Tl	0.5	0.5	0.5	0.5	0.5	0.5	0.5	0.5	0.5

Eocene volcanic-sedimentary rocks (Figure 3). There are chloritic and argillic alterations present at the contact of intrusion with host rocks. Major faults include Kooshk Nosrat, Neshveh and QermezAghaj with NW-SE trend cut the Neshveh granitic intrusion and caused to crushing and fracturing the intrusion (Figure 3). Moreover, this intrusion cut by a series of andesiticbasaltic dikes. Some xenoliths of the Eocene volcanic rocks present at the marginal parts of the intrusion, which their dimension and frequency decrease to the interior part. In addition, there are some mafic micro-granular enclaves in different parts of the Neshveh intrusion which are characterized by rounded shapes. These enclaves mineralogically composed of the monzodiorite and diorite.

Based on Petrographic studies of 90 samples from different parts of the Neshveh intrusion, it was subdivided into two compositional zones: quartz monzogabbro in the West and quartz monzodiorite, granodiorite and granite in the East (Figure 4). The contacts are gradational but the rock types are distinct and easily recognized both in outcrop and in aerial photos and satellite images. Compositionally, dikes include andesite-basaltic andesite. Samples Neshveh intrusion has medium to coarse grained granular texture which is locally accompanied by porphyroidic texture with plagioclase phenocrysts, micro-granular and graphic texture.

Figure 3. (a) View from the Neshveh intrusion and its contact with Eocene volcanic-sedimentary rocks (view to the south); (b) Close view from the fracturing in the Neshveh intrusion as a result of tectonic.

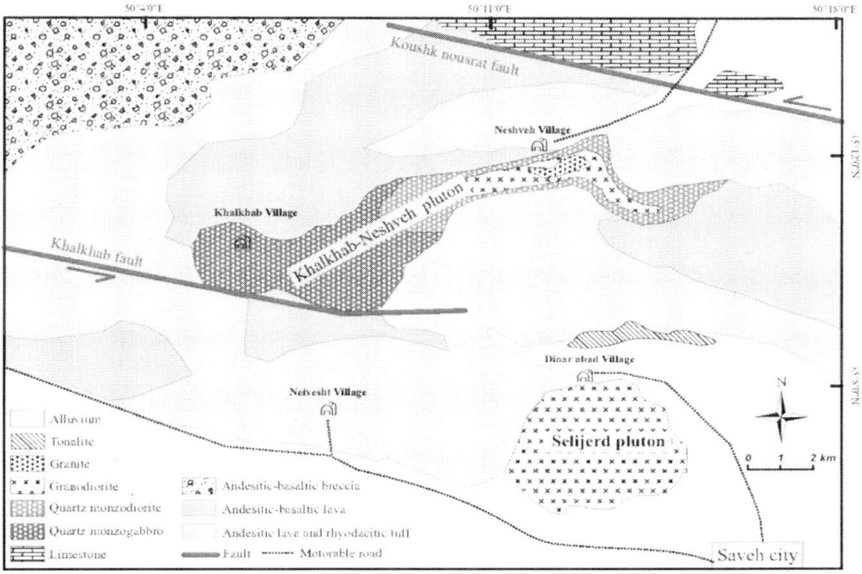

Figure 4. Simplified geological map of NW Saveh intrusions (Based on satellite data and the geological map of Saveh in 1:100,000 scale; Galamgash and Fonudi, 1998).

The quartz monzogabbro occupies about 50% of the pluton and is medium to coarse-grained with various textures; some samples show inter-granular and poiklitic textures, while others show hetero-granular texture. The inter-granular texture is constituted by grains of clinopyroxene which occupy the angular interstices between plagioclase crystals. The quartz monzogabbro consists dominantly of plagioclase (51.5 - 55.6 modal %), clinopyroxene (20.2% - 25.7%), K-feldspar (9% - 14.1%), quartz (7.9% - 11.7%), and subordinate opaque minerals (2.3% - 3.5%) and apatite (0.6%).

The quartz monzodiorite surrounds the granodiorite and granite. To the west, quartz monzodiorite is transitional to more mafic rocks, the quartz monzogabbro while to the east it forms the margin of the pluton. The quartz monzodiorite is generally medium-grained and characterized by equi-granular texture. It consists dominantly of plagioclase (41.1% - 52.4 modal %), K-feldspar (9% - 19%), quartz (10.9% - 18.4%), hornblende (6.9% - 16.3%), subordinates clinopyroxene (0% - 12.1%), biotite (2.5% - 5%) and opaque minerals (2.3% - 3%) (Table 1). Accessory minerals such as titanite and apatite are rare.

About 20% of the Neshveh intrusion is constituted by granodiorite, emplaced within the quartz monzodiorite hosted by volcanic rocks. These rocks are mainly included plagioclase (40% - 45%), quartz (approximately 20%), alkali feldspar (15% - 20%), hornblende (~10%) and biotite (~5%). Granites are the next order and generally emplaced at the eastern part of the intrusion. These rocks are predominantly of medium-grained granular texture, but porphyroidic texture with plagioclase phenocrysts present also. Granites composed of plagioclase (35% - 40%), quartz (~25%), alkali feldspar (~30%), hornblende (~5%) and biotite (~1%).

In the all type of rocks at the studied area, plagioclase present as euhedral and subhedral crystals and are of andesine and oligoclase type. Some of them show zoning (Figure 5). These minerals underwent low grade alte ration to sericite and sometimes epidote and calcite (Figure 5). Alkali feldspars (orthoclase) indicate argillic alteration in some samples. At the granitic samples, perthite texture was formed as a result of sub-solidus processes. Quartz presents as fine to medium-grained xenomorphic crystals between the other minerals. Graphic texture was formed as a result of quartz and orthoclase in the granites and occasionally in granodiorites (Figure 5). In some samples, quartz shows wavy extinction indicating tectonic stresses on the intrusion. Mafic minerals

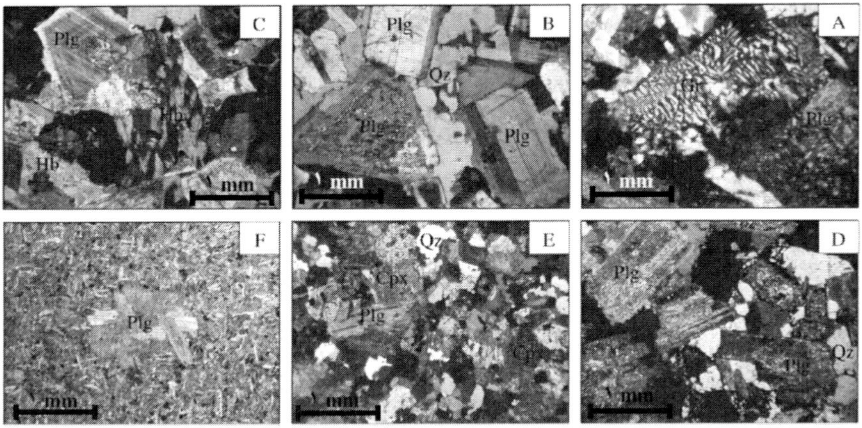

Figure 5. (A) Graphic texture in the granites and granodiorites (XPL light); (B) Plagioclase crystal that underwent epidote, sericite and chlorite alteration from the center (XPL light); (C) A plagioclase with zoning (XPL light); (D) Sericitized plagioclase crystals (XPL light); (E) Clinopyroxene bearing microgranular texture in the quartz-monzodiorite (XPL light); (F) Porphyritic texture includes plagioclase phenocrysts within the plagioclase, amphibole and pyroxene bearing fain-grained matrix (XPL light). (Gr: graphic, Plg: plagioclase, Qz: quartz, Hb: hornblende, Cpx: clinopyroxene).

in the Neshveh intrusion are mainly hornblende and biotite is less. Hornblende present as euhedral and subhedral crystals at the all phases of Neshveh intrusion (Figure 5). Some crystals of hornblende, altered to actinolite which sometimes accompanied by biotite and iron hydroxides. Biotite present at the all phases of this intrusion which its content increase at the granites. Clinopyroxene is observed with low frequency in the quartz-monzodiorite which mainly has been replaced by actinolite. The rests of the clinopyroxene are found as islands form within the actinolite (Figure 5). Sphene, opaque, zircon and apatite are the minor minerals in the Neshveh intrusion and secondary minerals include sericite, epidote, actinolite, chlorite and clay minerals.

As mentioned before, there are some medium-basic dykes within the Neshveh intrusion that compositionally are andesite and basaltic andesite. They have porphyry, microlitic porphyry, inter-granular and intersertal textures phenocrysts of plagioclase, clinopyroxene and hornblende in a fine-grained to microlitic matrix (Figure 5). Some plagioclase crystals show zoning and classified as andesine-labradorite. Some plagioclase crystals altered to sericite and some clinopyroxene crystals show alteration to chlorite. Fine-grained matrix underwent alteration to chlorite.

5. GEOCHEMISTRY

For studying of geochemical characteristics of Neshveh intrusion, 8 samples from plutonic rock and one sample from basaltic andesite dikes were analyzed at the ALS Chemex laboratories for major, trace and rare earth elements by ICP-MS and ICP-AES analysis method. The results of this analysis are presented in Table 1.

Based on K_2O + Na_2O versus SiO_2 diagram [5] and [6] and the AFM triangular diagram [5] , all rocks units of the Neshveh intrusion are calc-alkaline character (Figure 6). At the SiO_2 against K_2O diagram [7] , they are classified as high potassium calc-alkaline rocks (Figure 7). Based on calcic-alkali index [8] , all samples Neshveh intrusion have Calc-alkali nature (Figure 7). On the ASI versus A/NK diagram [9] , all samples are classified as metaluminous rocks (Figure 7). K_2O/Na_2O ratio is more than one in the studied rocks.

Granitic rocks have been classified as M, I, S, A and H-type by some researchers [e.g., [10] -[15] . Several evidences suggest that the Neshveh intrusion is I-type and possibly Andean type: 1-Frequency of hornblende, magnetite, biotite and titanite in theses rocks and absence of muscovite, garnet, cordierite, andalusite and sillimanite. 2-SiO_2 content varies between 56.5% - 69.4% in these rocks. 3-Presence of mafic micro-granular enclaves

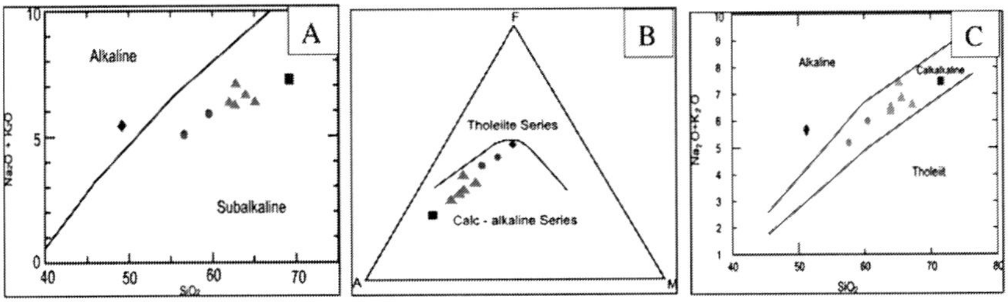

Figure 6. Location of the samples from studied area on the: (A) SiO_2versus Na_2O+K_2O diagram [5] , (B) AFM diagram [5] , and (C) SiO_2versus Na_2O+K_2O diagram [6] .

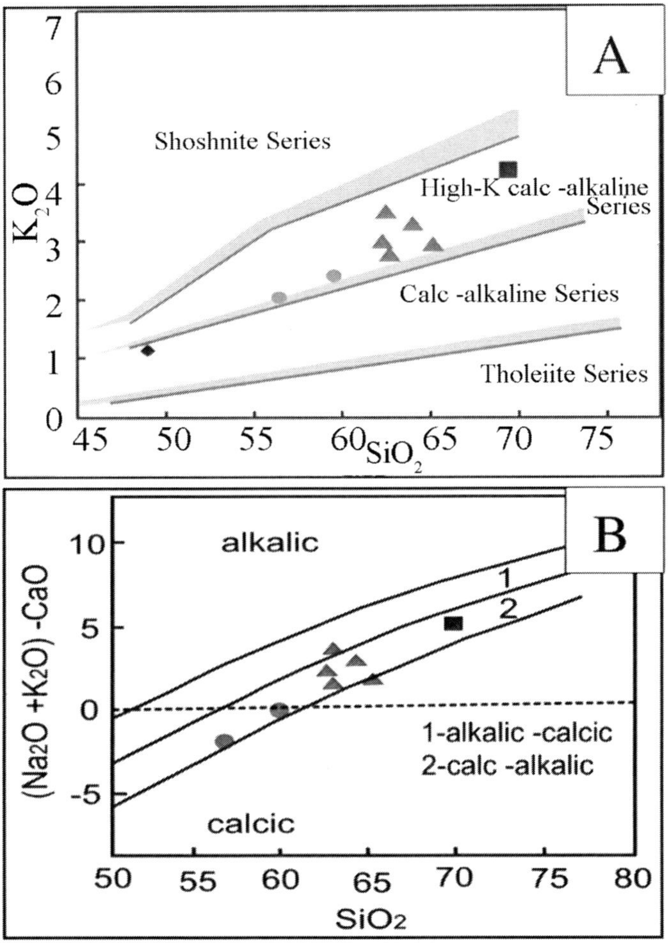

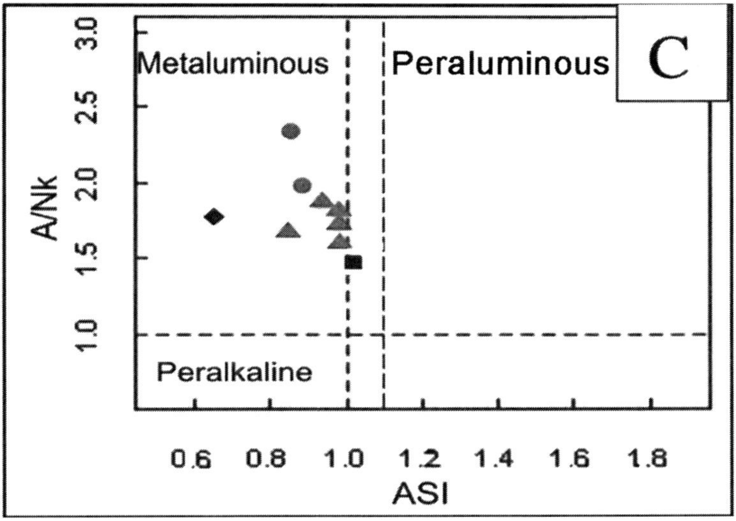

Figure 7. Location of the samples from studied area on the: (A) SiO_2 versus K_2O diagram [7] ; (B) SiO_2 versus (Na_2O+K_2O)-CaO diagram [8] , and (C) ASI versus A/NK diagram [9] .

in different parts of the intrusion and the absence of micaceous xenoliths [16] . 4-Absence of normative corundum in the samples and metaluminous nature of them. 5-Absence of pegmatites. 6-Decreasing P_2O_5 versus SiO_2 [17] (Figure 8). 7-Based on SiO_2 vs. K_2O [15] and ACNK versus SiO_2 diagrams [18] , the Neshveh intrusion classified as I-type granitoids (Figure 8).

[19] classified granitoids into 7 main groups. Based on this classification, Neshveh intrusion is classified as amphibole-bearing calc-alkaline granitoids (ACG).

On the variation diagrams, with increasing SiO_2, contents of Al_2O_3, P_2O_5, CaO, TiO_2, MgO, MnO and Fe_2O_3 are decreased and K_2O and Na_2O demonstrate increasing trend. These trends may reflect the crystal fractionation process in the evolution of the Neshveh intrusion. The Nd/Sr vs. La/Nb diagram [20] , is useful indicator for fractionation of plagioclase in the magma. Based on Figure 8, crystallization and fractionation of plagioclase led to increasing of the Nd/Sr ratio in the magma while the La/Nb ratio will be constant.

[21] exhibited a method to recognition of the three main processes of partial melting, fractional crystallization and magma mixing in the magmas. In this method, an incompatible element is used against a consistent

element. According to this method, the consistent element put in the horizontal axis and the incompatible element is in the vertical axis. If the trend of samples is a smooth and straight and almost horizontal line, fractional crystallization is the probable process in the magma evolution. If the trend is a smooth and straight and almost vertical line, partial melting is the effective process in the evolution of magma. In the cases that the trends of samples show a concave curved, this trend indicating that the magma mixing is the most probable process in the evolution of magma. In the V vs. Rb and Ba diagrams (Figure 9), despite the very small scattering, samples of the Neshveh intrusion has relatively horizontal trend indicating that the main mechanism for evolution of the Neshveh intrusion was the fractional crystallization.

Based on geochemical studies, Neshveh granitoid can be classified as sodic granite ($Na_2O > K_2O$) and has low values of Mg#, Ni, Cr, Co and V. Lower contents of these elements indicative for evolution of magma during ascending and before complete crystallization [22] .

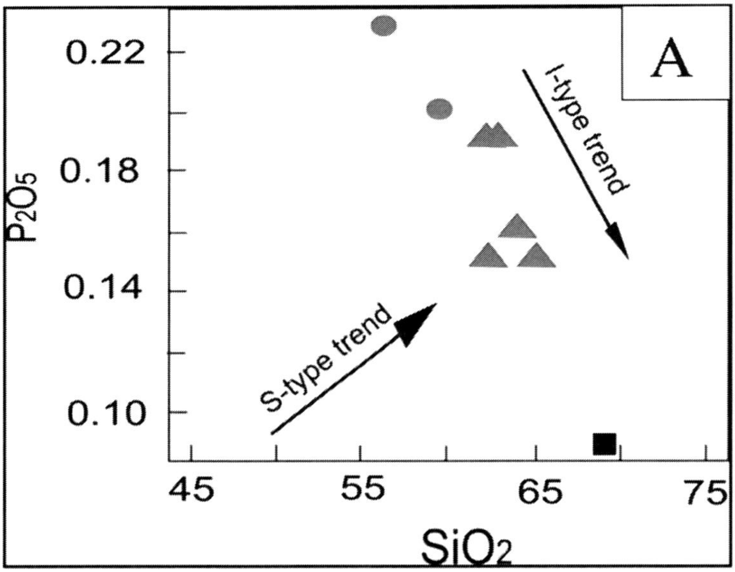

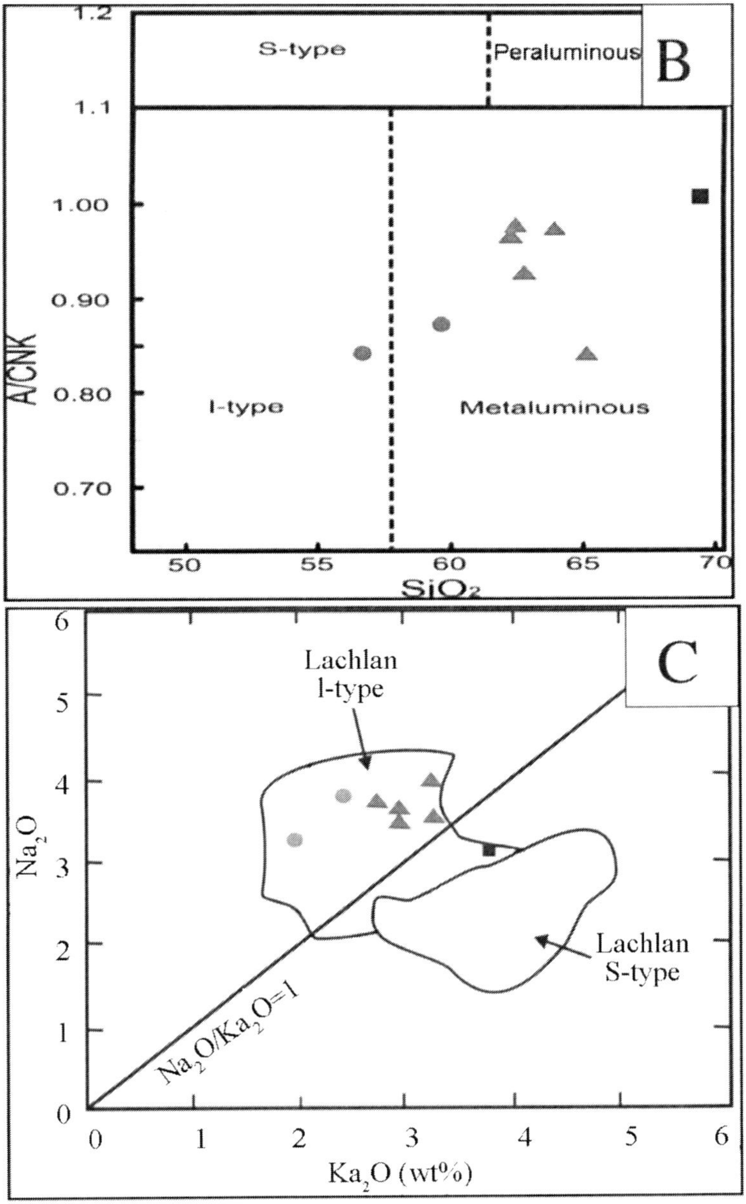

Figure 8. (A) Decreasing trend of P_2O_5 versus SiO_2 [17] . Location of the samples from studied area on the: (B) SiO_2 versus A/CNK [18] , and (C)SiO_2 versus Na_2O [15] .

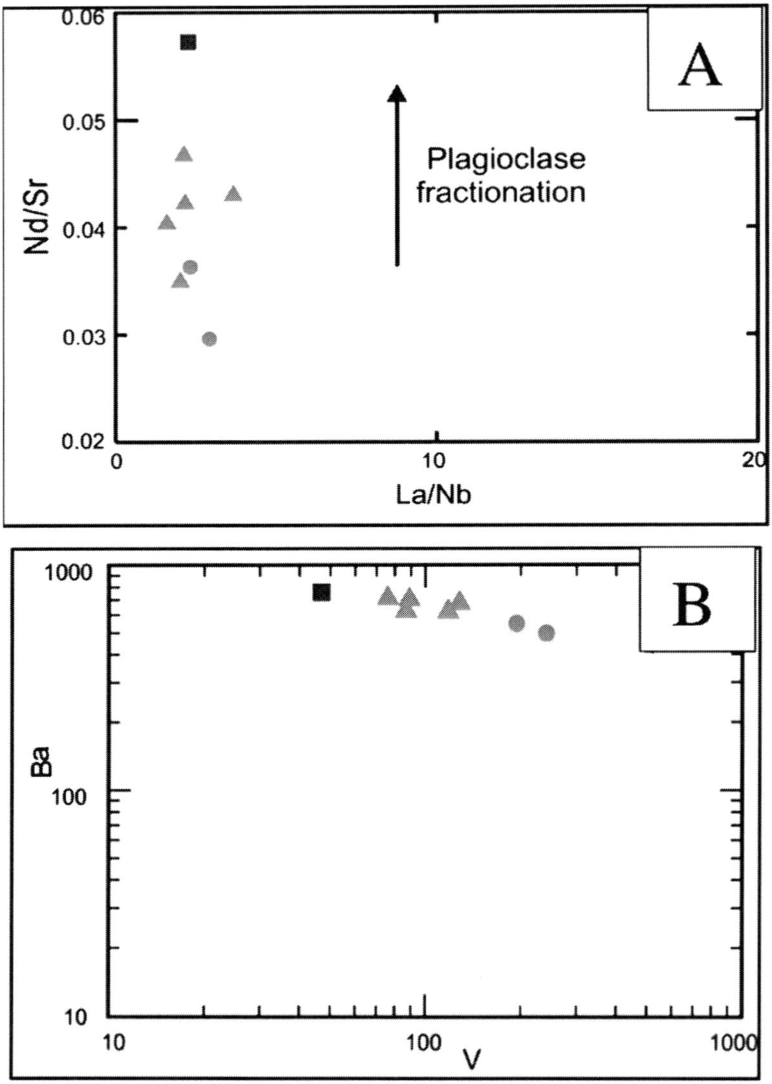
A
0.06
0.05
0.04
0.03
0.02
Nd/Sr
Plagioclase
fractionation
0
10
20
La/Nb
B
1000
100
10
Ba
10
100
1000
V

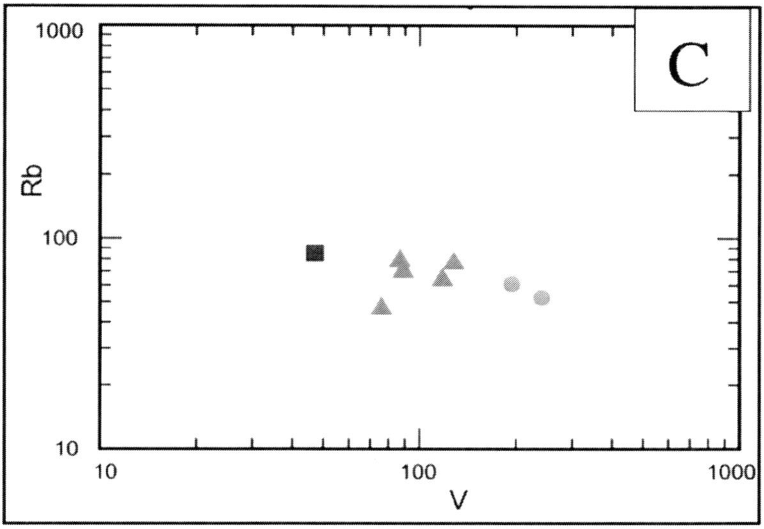

Figure 9. (A) Increasing trend in the Nd/Sr versus La/Nb which is indicative for plagioclase fractionation; (B) and (C) Decreasing trend of Rb and Ba versus V which is indicative for fractional crystallization on the evolution of Neshveh intrusion.

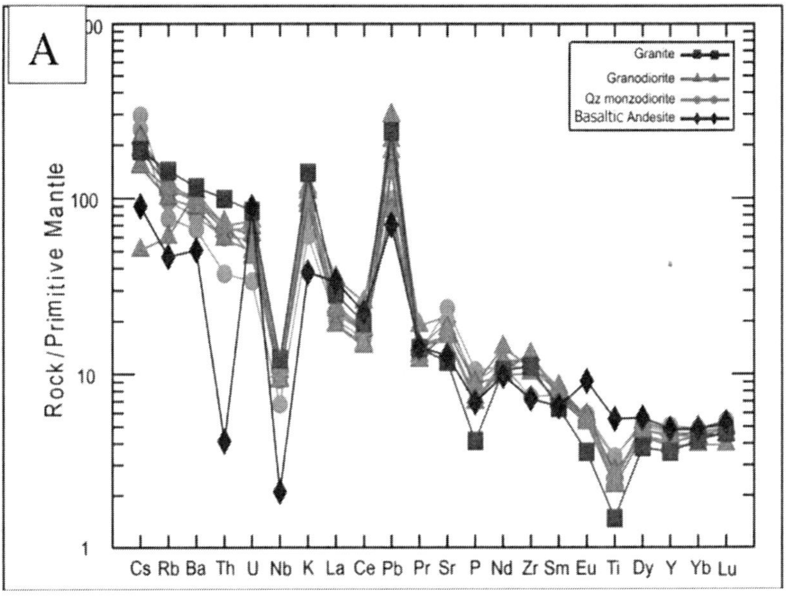

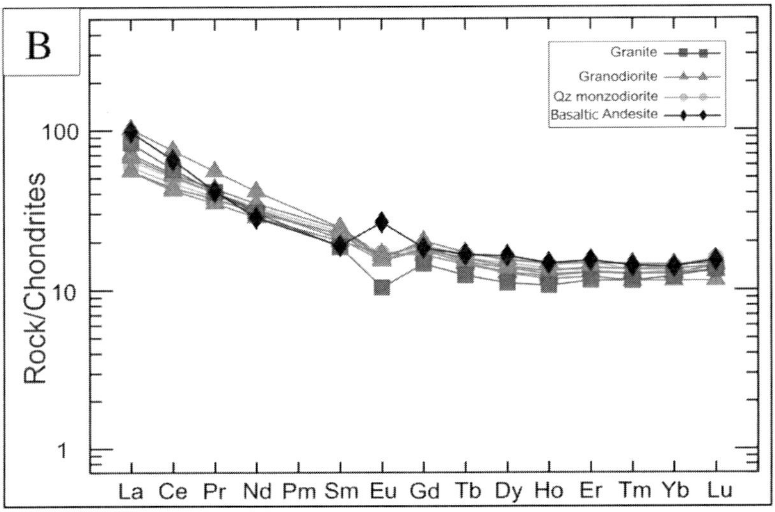

Figure 10. (A) Primitive mantle [23] normalized spider diagram for the studied rocks; (B) Chondrite normalized [23] REE patterns for the studied rocks.

The primitive mantle normalized spider diagrams for different units of Neshveh intrusion demonstrate negative anomalies of HFSEs including Nb, Ti and P, along with positive anomalies in LILEs including Rb, Ba, Th, K and Pb (Figure 10). The negative anomalies of Ti and Nb are controlled by titanium bearing minerals such as titanite, ilmenite, rutile, garnet and some amphiboles. With increasing pressure, the solubility of titanium bearing minerals reduces in aqueous fluids [24] . Therefore, theses minerals that are rich in HFSEs will be stable during partial melting process at depths greater than 30 km that is led to negative anomalies in the melt [24] . The negative anomaly in P indicative for low content of apatite in the studied rocks. As mentioned before, in the I-type granitoids, phosphorus act as a consistent element and fractionation this element in the early stages of magmacrystallization, led to negative anomalies in the crystalized rocks. The sample from dikes has negative Th anomalies (Figure 9). Based on mineralogical and geochemical composition of dykes; it seems that their parental magma source is the mantle. Since the Th content in the mantle is higher than the crust, this negative anomaly indicating that this magma rise up rapidly and underwent very low crustal contamination.

Fluids and melts derived from the subducted oceanic slab, led to the metasomatism of upper mantle wedge and thereupon, negative anomalies of Nb and Ta [25] -[27] . Positive anomalies in Pb, K, Rb and Ba are attributed to metasomatism of mantle wedge by fluids derived from the subducted slab or/and contamination with continental crust [28] [29] .

The Chondrite-normalized REE patterns in the different phases of Neshveh intrusion show a LREE rich pattern with a high LREE/HREE ration (Figure 10). All phases of Neshveh intrusion indicate negative anomalies Eu while sample from the dikes show a positive anomaly in Eu (Figure 10). Negative anomalies in Eu are related to the feldspar fractionation during magma crystallization or remaining of feldspar at the source. If these negative anomalies are associated with negative anomalies in Sr, fractional crystallization of plagioclase is responsible for these anomalies. While, the negative anomalies in Eu are associated with negative anomalies in Ba, K-feldspar fractionation has the main role in this respect [25] . In the studied rocks, negative anomalies of Eu occur along with Sr, which refers to the plagioclase fractionation. Positive anomaly in Eu from the dike sample is indicative for the absence of plagioclase fractionation during the magma crystallization and high concentration of this mineral in these rocks.

6. TECTONOMAGMATIC SETTING

Besides the geochemical characteristics of rare elements such as spider diagrams, tectonomagmatic setting discrimination diagrams also demonstrate a subduction setting relation for the Neshveh granitoid.

Based on the [30] diagram that is offered with considering, two cationic parameters include R2 = 6Ca + 2Mg + Al and R1 = 4Si − 11(Na + K) − 2(Fe + Ti), samples of Neshveh intrusion put into the pre-plate collision setting (Figure 11). On the [31] triangular diagrams that are drowned based on immobile elements include Ta and Hf and mobile element of Rb, samples of studied intrusion put into the volcanic arc (VA) setting (Figure 11)

Based on the Rb versus Y + Nb and Rb versus Ta + Yb diagrams [32] samples from the Neshveh intrusion put in the volcanic arc granite (VAG) setting (Figure 12). On the Th/Hf versus Ta/Hf, Th/Ta versus Yb and Ta/Yb versus Th/Ta diagrams [33] , all samples from the Neshveh intrusion put into the active continental margins are (Figure 13). The sample from the dikes is situated in the within plate setting (Figure 13).

Based on the AFM diagram [34] , the studied rocks located in the continental arc granites (CAG) setting (Figure 14), indicating that these granitoids originated from the subduction of an oceanic crust under a continental crust at the active continental margins. The parental magma was medium to high K calc-alkaline.

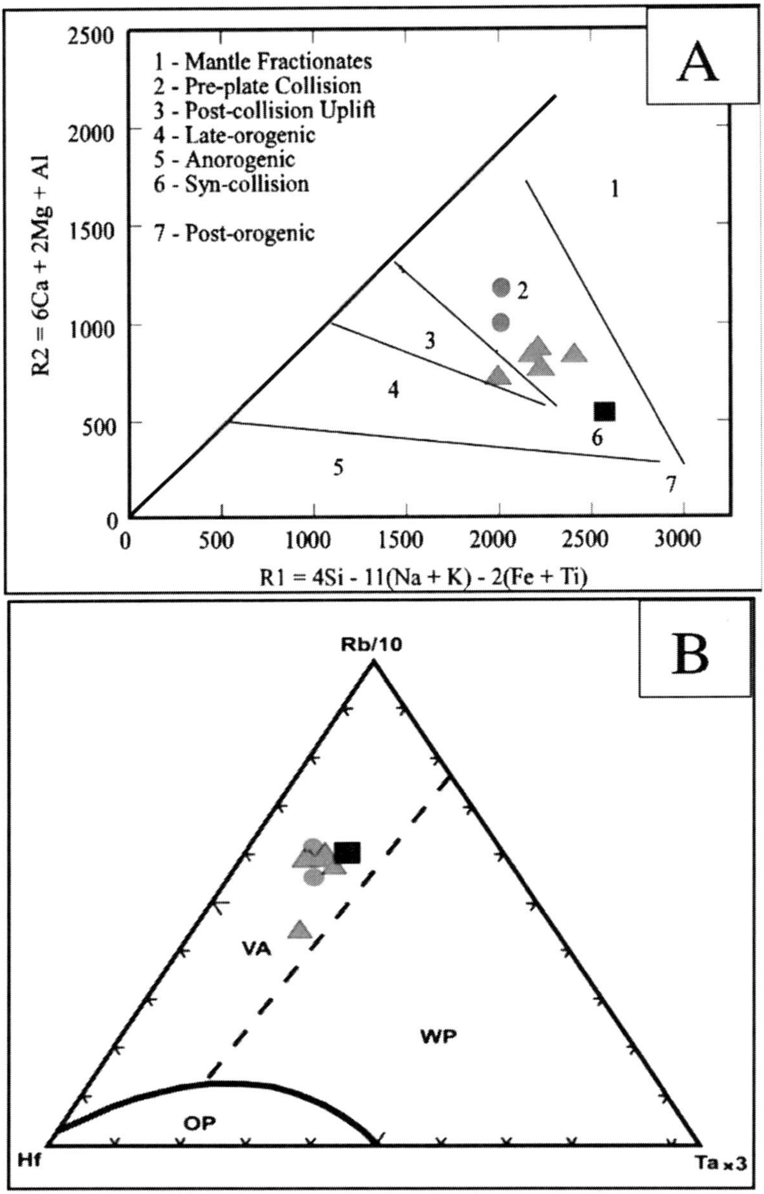
A
2500
2000
1500
1000
500
0
R2 = 6Ca + 2Mg + Al
1 - Mantle Fractionates
2 - Pre-plate Collision
3 - Post-collision Uplift
4 - Late-orogenic
5 - Anorogenic
6 - Syn-collision
7 - Post-orogenic
1
2
3
4
5
6
7
0
500
1000
1500
2000
2500
3000
R1 = 4Si - 11(Na + K) - 2(Fe + Ti)
B
Rb/10
VA
WP
OP
Hf
Ta x3

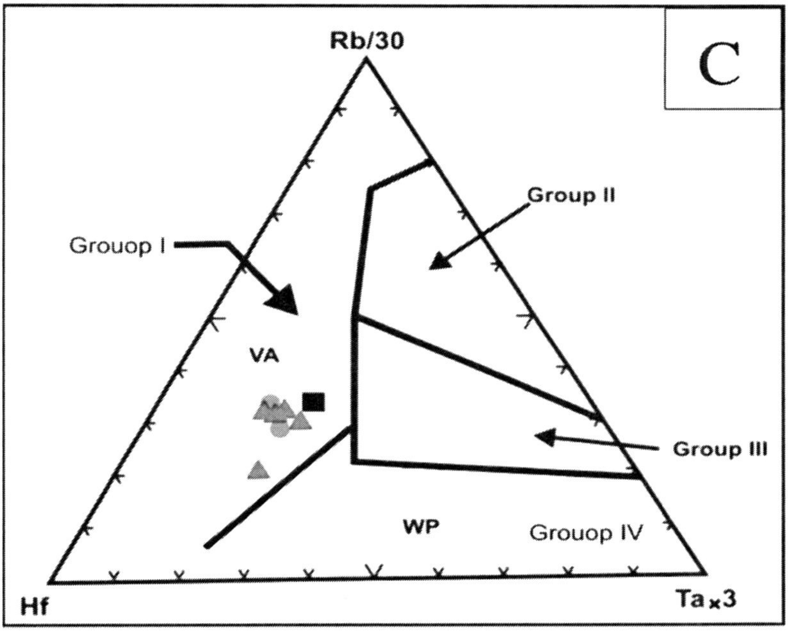

Figure 11. (A)Location of the Neshveh intrusion samples on the two cationic parameters of R1 and R2 [29] . (B) and (C) Location of the Neshveh intrusion samples on the Hf-Ta-Rb triangular diagrams [30] .

The Th/Yb versus Ta/Yb diagram [35] was used to recognition the source magma for Neshveh intrusion. According to this diagram, the studied samples are located within the calc-alkaline Andean granitoids (Figure 15). Moreover, the variation trend of samples on this diagram indicates the evolution of magma by fractional crystallization.

As mentioned in the previous parts, based on [19] classification, Neshveh intrusion can be classified as ACG granitoids. According to him, ACG granitoids are hybrid type (mantle + crust) and are related with subduction settings. In this base, it is supposed a subduction setting to Neshveh intrusion.

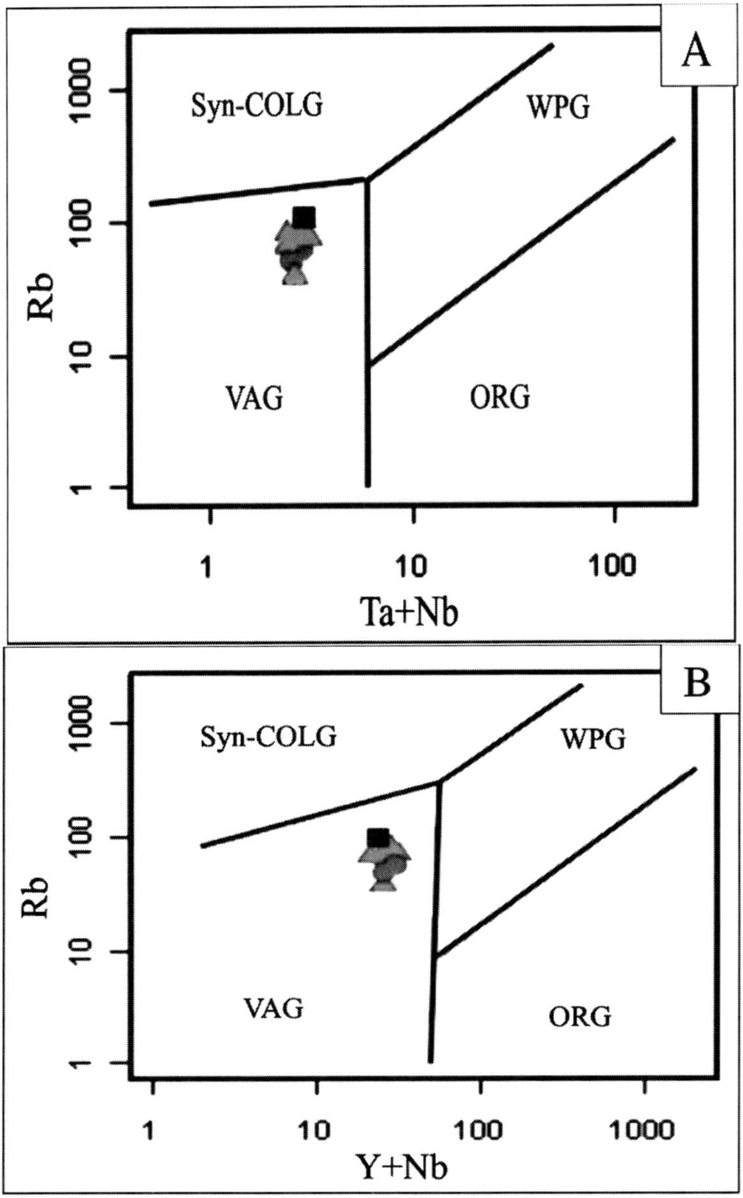

Figure 12. Location of the Studied samples on the: (A) Rb versus Ta+Yb diagram [31] ; and (B) Rb versus Y+Nb diagram [31] .

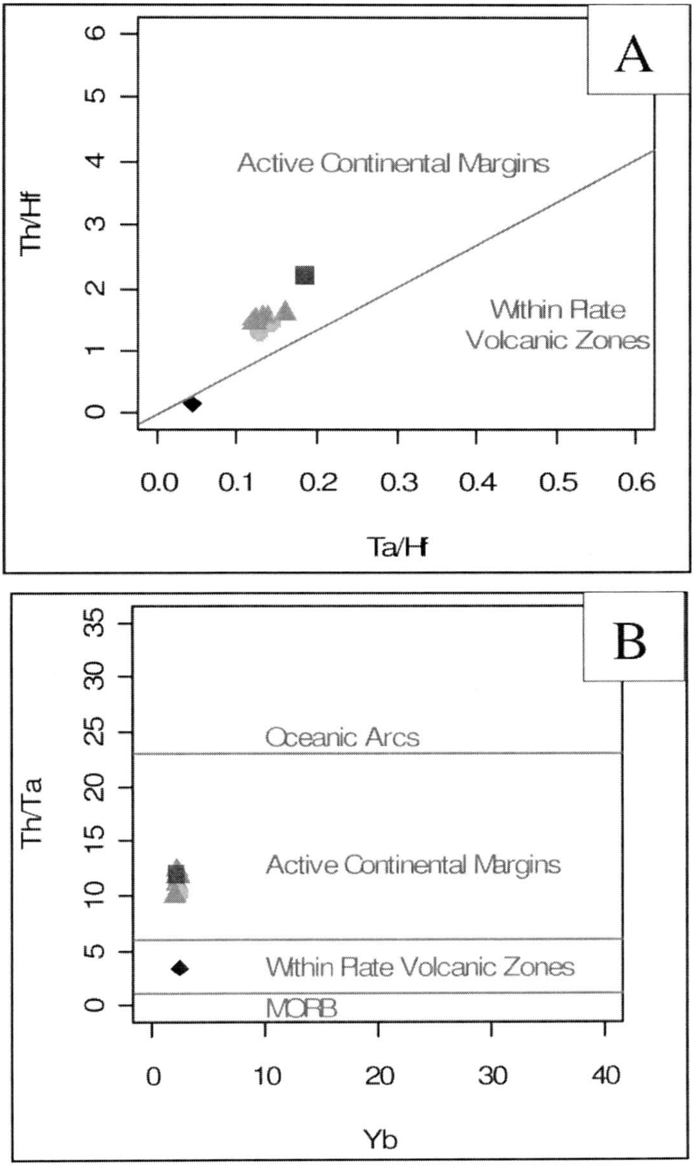
A
Th/Hf
0
1
2
3
4
5
6
Active Continental Margins
Within Plate
Volcanic Zones
0.0
0.1
0.2
0.3
0.4
0.5
0.6
Ta/Hf
B
Th/Ta
0
5
10
15
20
25
30
35
Oceanic Arcs
Active Continental Margins
Within Plate Volcanic Zones
MORB
0
10
20
30
40
Yb

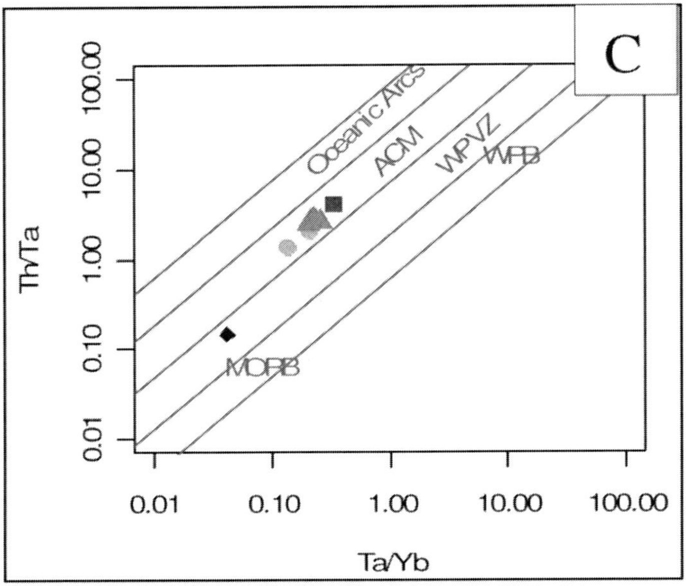

Figure 13. Location of the Studied samples on the: (A) Th/Hf versus Ta/Hf; (B) Yb versus Th/Ta; and (C) Th/Ta versus Ta/Yb diagrams [32]
.

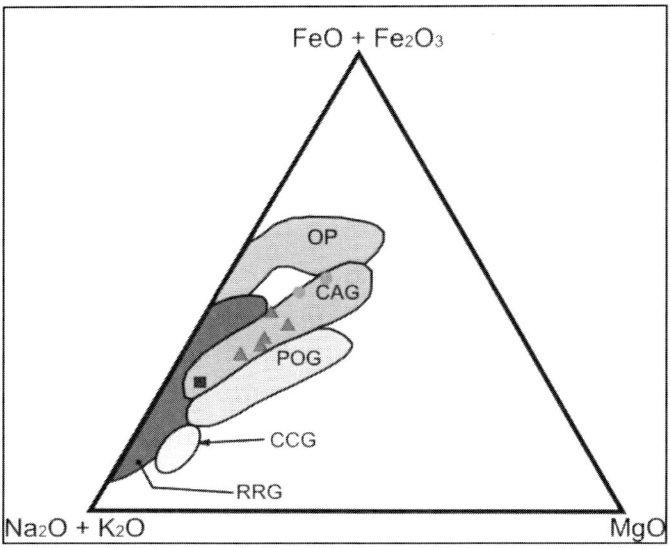

Figure 14. Location of the studied samples on the AFM diagram [33]
.

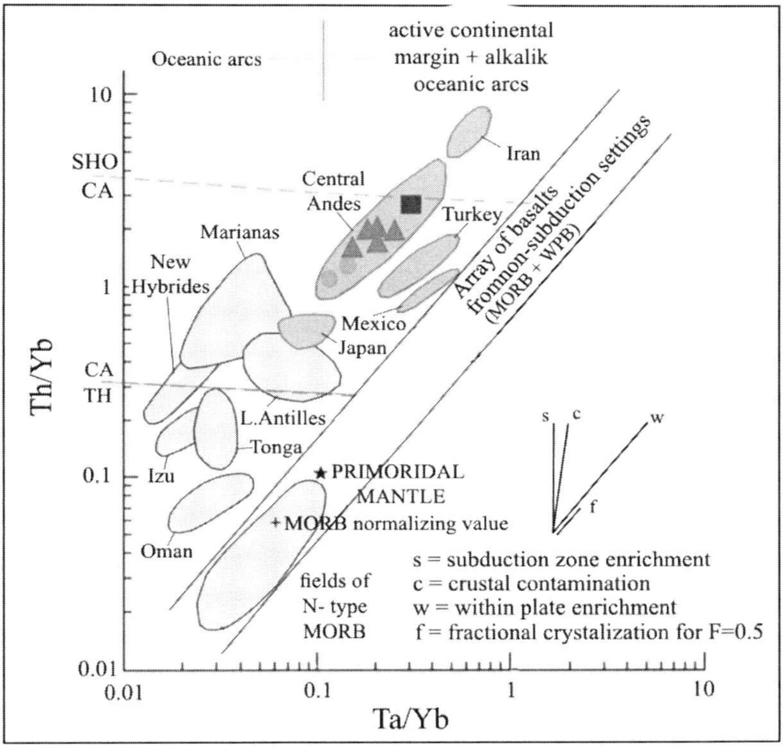

Figure 15. Location of the Neshveh intrusion sample on the Th/ Yb versus Ta/Yb diagram [34] .

7. CONCLUSION

Field studies, petrographic and geochemical investigations indicate that the Neshveh granitoid consists of the quartz-monzodiorite, granodiorite and granite. Field relations and geochemical studies demonstrate that all of these rocks are related to a single parent magma which underwent fractional crystallization. The Neshveh granitic intrusion has high-K calc-alkaline nature. This intrusion is metaluminous and classified as I-type granitoids. Negative anomalies of Nb and Ti along with positive anomalies of Rb, Ba, K, and Pb in the spider diagrams are indicative for the formation of this intrusion in relation with subduction setting. In terms of geodynamic setting, this intrusion is classified as volcanic arc granites (VAG) and active continental margin granites (CAG) which have been formed as a result of

Neotethys oceanic crust subduction beneath the Central Iran continental crust.

ACKNOWLEDGEMENTS

Reza Keshavarzi, Dariush Esmaili, Mehdi Rezaei Kahkhaei, Mir Ali Asghar Mokhtari, Reza Jabari (northwest Saveh). Our gratitude is expressed to their helps.

REFERENCES

1. Caillat, C., Dehlavi, P. and Martel, J.B. (1978) Geologie de la region de Saveh (Iran). Contribution a l'etude du volcanism et du plutonism tertiares de la zone de I Iran central. PhD Thesis, Grenoble, 325 Pages.
2. Helmi, F. (1991) Petrology and geochemistry of igneous rocks in Niousht area (NE Saveh). M.Sc. Thesis, University of Tehran, Iran (in Persian).
3. Moinvaziri, H. (1999) Introduction to the Magmatism of Iran. Tarbiat Moallem University, Tehran (in Persian).
4. Galamgash, J. and Fonudi, M. (1998) Explanatory Text of Saveh. Geological Map 1:100,000. Geological Survey of Iran, Tehran.
5. Irvine, T.N. and Baragar, W.R.A. (1971) A Guide to Chemical Classification of the Common Volcanic Rock. Canadian Journal of Earth Sciences, 8, 523-548.
6. Kuno, H. (1968) Differentiation of Basalt Magmas. In: Hess, H.H. and Poldervaart, A., Eds., Basalts: The Poldervaart Treatise on Rocks of Basaltic Composition, Interscience Publishers, New York, 623-688.
7. Peccerillo, A. and Taylor, S.R. (1976) Geochemistry of Eocene Calc Alkaline Volcanic Rocks from Kastamonu Area, Northern Turkey. Contribution to Mineralogy and Petrology, 58, 63-81.
8. Frost, B.F., Barnes, G.G., Collins, W.J., Arculus, R.J., Ellis, D.J. and Frost, C.D. (2001) A Geological Classification for Granitic Rocks. Journal of Petrology, 42, 2033-2048.
9. Shand, S.J. (1943) Eruptive Rocks. 2nd Edition, John Wiley, New York.
10. Chappell, B.W. and White, A.J.R. (1974) Two Contrasting Granite Types. Pacific Geology, 8, 173-174.
11. Collins, W.J., Beams, S.D., White, A.J.R. and Chappell, B.W. (1980) Nature and Origin of A-Type Granites with Particular Reference to South Eastern Australia. Contribution to Mineralogy and Petrology, 80, 189-200.
12. Whalen, J.B., Currie, K.L. and Chappell, B.W. (1987) A-Type Granites: Geochemical Characteristics, Discrimination and Petrogenesis. Contribution to Mineralogy and Petrology, 95, 407-419.
13. Castro, A., Monero-Venatas, I. and De La Rosa, J.D. (1991) H-Type (Hybrid) Granitoids: A Proposed Revision of the Granite Type Classification and Nomenclature. Earth Science Reviews, 31, 237-253.
14. Pitcher, W.S. (1993) The Nature and Origin of Granite. Blackie Academic and Professional Edition. Shopman and Hall, London.
15. Chappell, B.W. and White, A.J.R. (2001) Two Contrasting Granite Types. 25 Years Later. Australian Journal of Earth Sciences, 48, 489-499.

16. Didier, J., Duthou, J.L. and Lameyre, J. (1982) Mantle and Crustal Granites: Genetic Classification of Orogenic Granites and the Nature of Their Enclaves. Journal of Volcanology and Geothermal Research, 14, 125-132.
17. Chappell, B.W. and White, A.J.R. (1992) I- and S-Type Granites in the Lachlan Fold Belt. Transactions of the Royal Society of Edinburgh: Earth Sciences, 83, 1-26.
18. Chappell, B.W. and White, A.J.R. (1983) Granitoid Types and Their Distribution in the Lachlan Fold Belt, Southeastern Australia. Geological Society of America Memoirs, 159, 21-34.
19. Barbarin, B. (1999) A Review of the Relationships between Granitoid Types, Their Origins and Their Geodynamic Environments. Lithos, 46, 605-626.
20. Bourdon, E., Eissen, J.P., Monzier, M., Rovin, C., Martin, H., Cotten, J. and Hall, M.L. (2002) Adakite-Like Lavas from Antisana Volcano (Ecuador): Evidence for Slab Melt Metasomatism beneath the Andean Northern Volcanic Zone. Journal of Petrology, 43, 199-217.
21. Cocherie, A. (1986) Systematic Use of Trace Element Distribution on Patterns in Log-Log Diagrams for Plutonic Suites. Geochimica et Cosmoshimica Acta, 50, 2517-2522.
22. Woodhead, J.D. and Johnson, R.W. (1993) Isotopic and Trace Element Profile across the New Britain Island Arc, Papua New Guines. Contributions to Mineralogy and Petrology, 113, 479-491.
23. Sun, S.S. and McDonough, W.F. (1989) Chemical and Isotopic Systematics of Oceanic Basalts: Implications for Mantle Composition and Processes. In: Saunders, A.D. and Norry, M.J., Eds., Magmatism in the Ocean Basins, Vol. 42, Geological Society Special Publication, London, 313-345.
24. Glenn, A.G. (2004) The Influence of Melt Structure on Trace Element Partitioning near the Peridotite Solidus. Contributions to Mineralogy and Petrology, 147, 511-527.
25. Wilson, M. (2007) Igneous Petrogenesis. Chapman and Hall, London.
26. Sajona, F.G., Maury, R.C., Bellon, H., Cotton, J. and Defant, M. (1996) High Field Strength Elements Enrichment of Pliocene-Pleistocene Island Arc Basalts Zamboanga Peninsula, Western Mindanao (Philippines). Journal of Petrology, 37, 693-726.
27. Chappell, B.W. (1999) Aluminium Saturation in Iand S-Type Granites and the Characterization of Fractionated Haplogranites. Lithos, 46, 535-551.
28. Kamber, B.S., Ewart, A., Collerson, K.D., Bruce, M.C. and McDonald, G.D. (2002) Fluid-Mobile Trace Element Constraints on the Role of Slab Melting and Implications for Archean Crustal Growth Models. Contributions to Mineralogy and Petrology, 144, 38-56.
29. Atherton, M.P. and Ghani, A.A. (2002) Slab Breakoff: A Model for Caledonian, Late Granite Syncollisional Magmatism in the Orthotectonic (Metamorphic) Zone of Scotland and Donegal, Ireland. Lithos, 62, 65-85.
30. Batchelor, R.A. and Bowden, P. (1985) Petrogenetic Interpretation of Granitoid Rock Series Using Multicationic Parameters. Chemical Geology, 48, 43-55.
31. Harris, N.B.W., Pearce, J.A. and Tindle, A.G. (1986) Geochemical Characteristics of Collision Zone Magmatism. Collision Tectonic, Geological Society of American Bulletin, Special Publication, 19, 67-81.
32. Pearce, J.A. (1996) Sources and Settings of Granitic Rocks. Episodes, 19, 120-125.
33. Gorton, M.P. and Schandl, E.S. (2002) From Continents to Island Arc: A Geochemical Index of Tectonic Setting for Arc-Related and within Plate Felsic to Intermediate Volcanic Rocks. Canadian Mineralogist, 38, 1065-1073.
34. Bowden, P., Batchelor, R.A., Chappelle, B.W., Didier, J. and Lameyer, J. (1984) Petrological, Geochemical and Source Criteria for the Classification of Granitic Rocks: A Discussion. Physics of the Earth and Planetary Interiors, 35, 1-11.

CHAPTER 6

Na-Montmorillonite Edge Structure and Surface Complexes: An Atomistic Perspective

Aric G. Newton [1,2], Jin-Yong Lee [1,2] and Kideok D. Kwon [1,2,*]

[1]*Department of Geology, Kangwon National University, Chuncheon 24341, Korea*

[2]*Critical Zone Frontier Research Laboratory (CFRL), Kangwon National University, Chuncheon 24341, Korea*

ABSTRACT

The edges of montmorillonite (MMT) react strongly with metals and organic matter, but the atomic structure of the edge and its surface complexes are not unambiguous since the experimental isolation of the edge is challenging. In this study, we introduce an atomistic model of a Na MMT edge that is suitable for classical molecular dynamics (MD) simulations, in particular for the B edge, a representative edge surface of 2:1 phyllosilicates. Our model possesses the surface groups identified through density functional theory (DFT) geometry optimizations performed with variation in the structural charge deficit and Mg substitution sites. The edge structure of the classical MD simulations agreed well with previous DFT-based MD simulation results. Our MD simulations revealed an extensive H-bond network stabilizing the Na-MMT edge surface, which required an extensive simulation trajectory. Some Na counter ions formed inner-sphere complexes at two edge sites. The stronger edge site coincided with the exposed vacancy in the dioctahedral sheet; a weaker site was associated with the cleaved hexagonal cavity of the tetrahedral sheet. The six-coordinate Na complexes were not directly associated with the Mg edge site. Our simulations have demonstrated the heterogeneous surface structures, the distribution of edge surface groups, and the reactivity of the MMT edge.

Keywords: clay edge; mineral surface; Na-montmorillonite; nanoporous minerals; molecular dynamics; density functional theory; computational geochemistry

1. INTRODUCTION

Smectite minerals, common products of the weathering and low-temperature hydrothermal alteration of primary minerals, are omnipresent in the environment with an outsized role in the fate and transport of soil solution solutes. This significant role is due to the highly specific surface area that is a consequence of the nanometer thickness (~1.0 nm) of these layered 2:1 phyllosilicate minerals. Montmorillonite (MMT), a subgroup of smectite minerals, is comprised of an Al sheet between two Si sheets arranged into tetrahedral-octahedral-tetrahedral (TOT) layers. In MMT, more than 50% of the permanent negative charge (0.2–0.6 per formula unit) originates from substitutions of divalent cations (e.g., Mg^{2+} or Fe^{2+}) for Al^{3+} in the octahedral sheet [1]. Counter ions balance this permanent charge in the interlayer nanopores. The hydration of these interlayer cations and osmotic pressure results in the swelling of the smectite layers. The permanent negative charge imparts significant cation exchange capacity to the MMT layers. These physicochemical properties of the MMT interlayer have been used to great advantage in engineered barriers [2,3,4], nanocomposite materials [5,6,7], and pharmaceutical and biomedical applications [8].

The edges of 2:1 phyllosilicate minerals are no less important to their reactivity, despite being a much smaller fraction of the total surface area. The edges of these clay nanosheets are a source of variable charge in soils and an important sorption site for cations, anions, and any sorbates too large to cross the boundary from the neighboring mesopore into the interlayer nanopore, such as soil organic matter [1,9]. A comprehensive understanding of the clay edge is challenging as this small fraction of the total surface area is difficult to isolate experimentally and the edge face is highly disordered. The choice of a non- or low-swelling clay mineral combined with the careful manipulation of the experimental pH and ionic strength has been used in sorption studies to discriminate between clay edge and interlayer sorption [10,11,12,13]. A surface complexation model of sorption that treated both the permanent structural and variable charge has proved successful in modeling sorption to MMT by employing a general description of the edge sorption sites [14]. This 2 Site Protolysis Non Electrostatic Surface Complexation and Cation Exchange (2 SPNESC/CE) model discriminates between strong and weak energy sites at the MMT edges. Extended X-ray absorption fine structure (EXAFS) spectroscopy has been used to isolate a strong edge sorption site in the plane of the octahedral sheet at low-loadings but was unable to resolve the weak reactive edge sites at higher sorbate loadings [15]. These experimental protocols and surface complex models can provide a spatially-averaged perspective of edge surface sorption based on the reporting unit queried experimentally. However, these experimental methods cannot provide sufficient resolution for a true atomic-scale description of individual features of the edge surface. Increasingly, atomistic simulations are employed to address this limitation of experimental methods as they afford a detailed atomistic description of the clay edge.

An accurate atomic structure of the 2:1 edge face was first proposed by White and Zelazny [16], who used crystal growth theory [17] to identify three periodic bond chains (PBCs) that are parallel to the edges of 2:1 dioctahedral phyllosilicates. Two of these PBCs are symmetrically equivalent, and one is unique. These PBCs are composed of linked tetrahedron-octahedron-tetrahedron (TOT) units. White and Zelazny [16] chose to refer to the PBCs according to the center of symmetry across which the tetrahedra are attached to the octahedron. We preserve this nomenclature and refer to the symmetrically equivalent PBCs as the AC edge and the unique PBC as the B edge. These edge structures have been the initial configuration for atomistic simulations of the 2:1 clay edge at both the quantum mechanical (QM) and molecular mechanical (MM) levels of theory. Many QM simulations have been used for a detailed characterization of the 2:1 edge structures, surface energies, acid-base reactivity, and cationic surface complexes [18,19,20,21,22,23,24,25,26,27]. These simulations are the most theoretically rigorous available, but the theoretical rigor comes with high computational costs that limit the model size and the duration of any

QM-based molecular dynamics (MD) simulations. Constraints on model size often result in structures that do not sufficiently isolate the surface groups or clay edges from one another. By trading some theoretical rigor for computational efficiency, MM simulations can employ larger models of the clay edge that isolate the edges and explore extended time-scales. The increased computational efficiency is achieved through the simplification of interatomic interactions to a set of pair-wise potentials, referred to as a force field, that are parameterized by using experimental or QM results. With a thoughtful simulation design, the advantages of individual QM and MM simulations can be leveraged for a comprehensive understanding of the 2:1 clay edges.

MM simulations of clay mineral edges, however, have been hampered by the lack of validated force fields for the mineral edges. While the development of reliable force fields for clay mineral edges is an ongoing research topic in the computational chemistry community, a recent classical molecular dynamics (MD) simulation study [28] demonstrated that the edge structures can be described within a reasonable accuracy by using the Clayff force field [29], a force field widely used in the research of interlayer or basal surfaces of clay minerals. The classical MD simulations not only reproduced the main edge structrual features obtained by QM-based MD simulations but also offered evidence for defect structures at the pyrophyllite AC edge, denoted as Edelman Favejee (EF) [30] defects by the authors who first proposed a similar structure for 2:1 smectites. Defects may play an important role in surface reactivity, but defects at the clay mineral edges have not been discussed yet. Our atomistic thermodynamics study based on density functional theory (DFT) [31] tested the thermodynamics stability of the EF defects and showed that the defects on the pyrophyllite AC edge are feasible at water chemical potentials and temperatures at which the pyrophyllite B edge would exist. Recently, our classical MD simulations [32] using Clayff explored the effects of layer charge arising from isomorphic substitutions in the octahedral layer on the AC edge of MMT by employing DFT models with different layer charges. The simulations demonstrated the correlation between local charge deficits in the octahedral layer and coordination of Al and Mg at the AC edge. The presence of local charge deficits increased the disorder of the edge Al polyhedra relative to the unsubstituted MMT AC edges, including five- and four-fold coordination with O and EF defects at the edge.

In the present study, we extend our previous simulations to the B edge of Na-MMT. We develop a model of the edge informed by the energetics, edge surface groups, and partial charges from DFT geometry-optimized MMT edge structures suitable for classical MD simulations. Based on this model, our classical MD simulations provide a detailed characterization of the edge structure, the interfacial hydrogen bond network, and edge sorption sites as influenced by isomorphic substitutions in the octahedral sheet.

2. MATERIALS AND METHODS

2.1. Molecular Dynamics Simulations

The atomistic model of the Na-montmorillonite (Na-MMT) edge was derived from a previous model of the pyrophyllite edge, with the details of that model described elsewhere [28]. The B edge surface model was created from a 4 × 4 × 2 ($a \times b \times c$) expansion of the atomic coordinates of the pyrophyllite-1*M* unit cell [33]. The octahedral sheet of this 2:1 phyllosilicate unit cell is trans-vacant (i.e., the two cis-sites, where the anionic positions are shared along an octahedral edge, are occupied; Figure 1). The B edge structure is created by cleavage of the monoclinic cell along the (010) plane. A hydrated 60 Å pore was created perpendicular to the B edge faces; the opposing edge interfaces are separated across the bulk by approximately 30 Å. The simulation cell c dimension was expanded to 30.6 Å, and the two MMT layers were distributed equally in the c direction to accommodate a hydrated bi-layer in the two interlayer spaces. The initial monoclinic simulation cell geometry was a = 20.84, b = 94.29 Å, c = 30.64 Å, α = 90.00°, β = 100.46°, and γ = 90.00°.

The unsaturated bonds of the edge surface are healed through the chemi- and physi-sorption of an integer number of water molecules to maintain the surface at the point of zero net proton charge (p.z.n.p.c.). The two unsaturated surface Si per TOT unit were healed with one chemisorbed water molecule (i.e., ≡SiO–H + ≡Si–OH). The unsaturated octahedron of each TOT unit is healed with one physisorbed water molecule. Unlike the DFT-based molecular dynamics (DFT-MD), wherein energetically unfavorable initial distributions of physi- and chemi-sorbed water molecules may result in a re-distribution of protons amongst the surface oxygen [20,24,34], classical mechanical simulations must proceed from surface structures chosen a priori; the initial surface configuration was created based on our DFT geometry optimization results. The force fields used in classical mechanical simulation assign an atom type to each atom in the molecular structure based on the chemical element of the atom and its coordination environment. We have chosen atom type assignments for the B edge surface groups such that no new atom types or revisions of partial charges reported for bulk clay minerals [29] are necesssary. Surface coordinated –OH groups are assigned parameters consistent with the partial charges, short-range potentials, and harmonic bonds of the hydroxyl oxygen and hydrogen atom types; the atoms of surface coordinated water molecules are identical to the bulk water molecules.

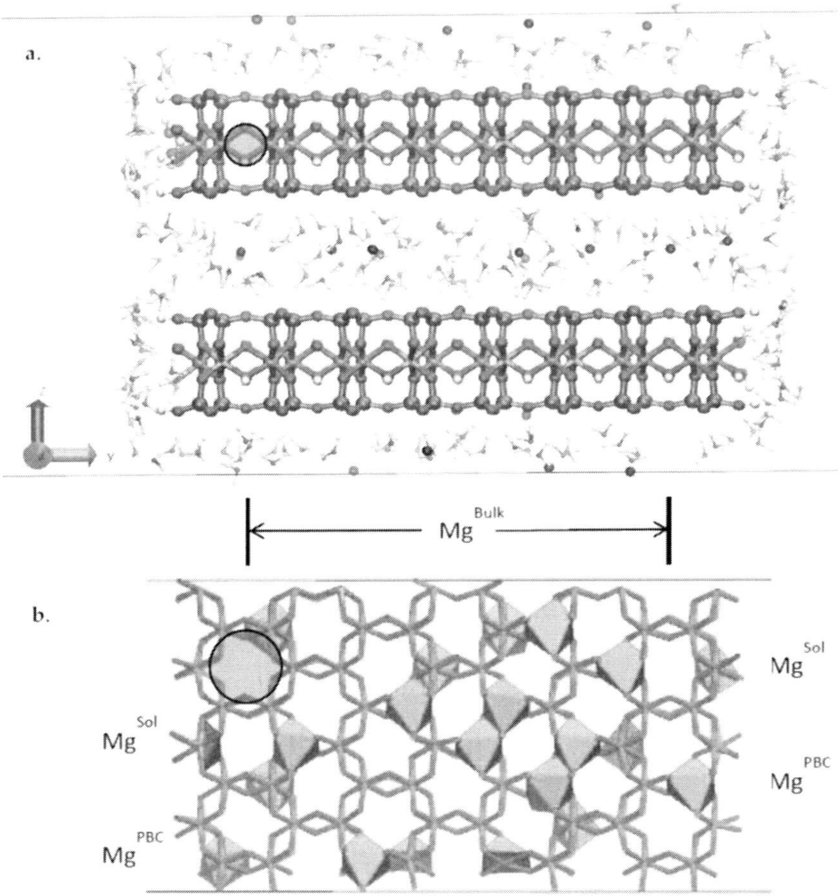

Figure 1. (**a**) Profile of Na-montmorillonite (Na–MMT) model after the equilibration of the aqueous phase and just prior to the release of the montmorillonite (MMT) atoms from their initial positions. Al = magenta; Mg = green; Si = ochre; O = red; H = white; Na = blue; (**b**) Plan view of the octahedral sheets from the Na–MMT structure showing the locations of the random Mg substitutions (green polyhedra). The Mg substitutions in the lower sheet can be distinguished from those in the upper sheet by the stick representations of the octahedral Al–O bonds imposed above the green polyhedra. Mg^{Sol} = solvent accessible Mg; Mg^{PBC} = solvent inaccessible Mg substitutions in the linking PBC positions; Mg^{Bulk} = Mg

substitutions in the bulk of the MMT octahedral sheets. Trans-vacant octahedral sites noted with black shaded circles.

Isomorphic substitutions of Al^{3+} by Mg^{2+} in the octahedral layer were assigned at random and were subject to the two constraints described previously [32]. The random assignment of substitutions and the first constraint that prohibits the presence of $Mg–(OH)_2–Mg$ pairs in the octahedral sheet are based on ^1H magic angle spinning nuclear magnetic resonance (MAS NMR) and FT-IR spectra from MMTs that show no contributions from $Mg–(OH)_2–Mg$ moieties and QM simulations that demonstrated that Mg^{2+} substitutions are randomly distributed within the octahedral layer [35,36,37]. The second constraint, informed by the properties of synthetic MMTs [38,39,40,41], limits the substitution ratio of the edge PBC (i.e., Mg edge substitutions: total number of octahedral edge sites in the PBC) to less than 0.20, which is the maximum substitution ratio for a synthetic MMT with only octahedral substitutions. The structural characterization of synthetic MMTs with divalent cation substitution ratios greater than 0.20 revealed octahedral sheet occupancies greater than 2.00, which is indicative of an MMT structure with some trioctahedral character, and isomorphic substitutions in the tetrahedral sheet. The octahedral substitution ratio (i.e., the number of substitutions per total octahedral sites) is one half the layer charge, x, in number of moles of excess electron charge per the M-MMT chemical formula: $M_x[Al_{(2-x)}Mg_x]Si_4O_{10}(OH)_2$, where M is the interlayer counter ion and the term in brackets represents the octahedral sheet cations. Any assignments that violated either of the two constraints were ignored, and a new random assignment was made. The two edge interfaces created are comprised of the edge PBCs of two MMT layers. Each interface possesses an edge PBC with a solvent accessible substitution (Mg^{Sol}) and a second edge PBC with a solvent inaccessible Mg (Mg^{PBC}) substitution. The remaining Mg substitutions were distributed in the MMT bulk (Mg^{bulk}). An equal number of random substitutions were assigned to each layer [N(sub/layer) = 13], and an equal number of Na ions were assigned to each interlayer space to balance this permanent structural charge. This atomistic model of Na-MMT had a layer charge of 0.40, which is consistent with the low- to medium-charged MMT [42] (cation exchange capacity (CEC) ~110 $cmol_c \cdot kg^{-1}$). The chemical formula for the MMT without sorbed water molecules was $(Mg_{26}Al_{102})Si_{256}O_{640}(OH)_{128}$. The interlayer and 60 Å macropore were hydrated with a total of 1320 water molecules (Figure 1).

The interatomic potentials of the Clayff force field [29] were used for the Coulomb and short-range interactions as well as the harmonic bonds of the hydroxyl groups in the atomistic model of Na–MMT. Adjustments were made to the partial Coulombic charges of the octahedral Mg atoms (q_{mgo} = 1.3598 e) and the bridging O atoms with double substitution (q_{obss} = −1.3116 e) to maintain the charge neutrality of the atomistic model. Other than these two minor adjustments, all assigned atom types in the MMT structure are those species available in the Clayff force field. The aqueous phase was described by the extended simple point charge (SPC/E) water model with harmonic bond stretching and angle bending terms [43,44]. Long-range electrostatic interactions and the attractive contribution to the van der Waals interaction were evaluated using the Ewald summation [45], with the repulsive contribution of the van der Waals interaction truncated at a distance of 10.0 Å. All Ewald summations were determined to an accuracy of 0.0001 kcal/mol.

All MD simulations were performed with the LAMMPS software package [46], installed on the supercomputers at the National Energy Research Scientific Computing Center (NERSC; Berkeley, CA, USA). The simulation algorithm has been described in detail elsewhere [32]. In brief, the solution phase was initiated with a series of short MD simulations in the microcanonical (NVE, T = 298.15 K) ensemble, where the number of atoms (N), simulation cell volume (V), and energy (E) were held constant, with the temperature (T) controlled with a direct temperature-rescaling thermostat and the SHAKE algorithm [47] for water molecule bonds and angles. The timestep increased in a progression of Δt = [0.001, 0.01, 0.2, 0.3, 0.4, 0.5 fs] for 40,000 steps each before the solution phase linear velocity was zeroed, the atoms of the MMT phase were released from their rigid positions, and the thermodynamic ensemble was changed to isobaric-isothermal [NPT; (P = 1.0 atm, T = 298.15 K)], wherein pressure (P) and temperature were held constant with a chain of three Nose-Hoover thermostats. The entire system was then equilibrated for an additional 200,000 steps with Δt = 0.25 and 0.5 fs. The approximately 200 ps of equilibration were followed by a 5.0 ns production run in the NPT ensemble. The internal routines of the LAMMPS code were used to calculate the metal-oxygen radial distribution functions (RDFs) and coordination numbers (CNs). The charactersitic properties of the RDFs (i.e., metal-oxygen distances, r_{MeO}; the first-shell minimum of the RDF, ρ; and the coordination numbers, CNs) were calculated from 10 time-averaged blocks of 100 ps duration. The hydrogen bonds (H-bonds) at the NaMMT edge-water interface were defined by the donor-acceptor separation distance and the donor-hydrogen···acceptor angle criteria developed by Kumar, *et al.* [48] for the SPC/E water model ($d_{H\text{-}bond} \leq 3.2$ Å; $\angle$H–D ··· A ≤ 40°) and rendered and quantified with the open-source program, Visual Molecular Dynamics (VMD) [49]. The volumetric density maps were calculated for a 0.05 Å grid resolution and a uniform atom weight of 1.0 using the VMD VolMap plugin.

2.2. Density Functional Theory (DFT) Geometry Optimizations

All DFT computations were performed with CASTEP, a planewave DFT code [50], using ultrasoft pseudopotentials [51] under the generalized gradient approximations [52]. The planewave basis set was expanded to include kinetic energy of up to 600 eV. A 6 × 4 × 4 grid was used for the *k*-point sampling using a Monkhorst-Pack grid [53] for the pyrophyllite unit cell, and one *k*-point was used for supercells, which represent surface models. With the chosen kinetic-energy cutoff and *k*-point grid, the atomic forces converged to <<0.01 eV/Å. The geometry optimization permitted the coordinates of all ions to relax. Geometry optimizations were performed until the residual atomic force was ≤0.03 eV/Å and the root-mean-square stress of the bulk structure was ≤0.02 GPa.

A periodic slab model of the B edge surface for DFT calculations (Figure 2) was created based on a 2 × 2 × 1 supercell of a geometry optimized pyrophyllite-1*M* unit cell from Bleam, et al. [33]. This slab model has two symmetric edge surfaces at the left and right of Figure 2. The B edge structure was cleaved from the supercell in the same manner as the molecular mechanics simulations. The generic chemical formula for the surface slab models without sorbed water molecules was $[Al_{(16-n)}Mg_n]Si_{32}O_{80}(OH)_{16}$, where n = 2 or 3 in our DFT simulations, corresponding to a layer charge of 0.25 or 0.375. The slab model was comprised of four PBCs (~18 Å thickness). The vacuum space perpendicular to the edge surface slab model was ~20 Å. The supercell dimensions for the DFT edge surface model were 10.45 × 37.53 × 10.12 Å ($a \times b \times c$) with each edge surface ~1.0 nm^2. A distance greater than 17.5 Å separated the periodic images of the surface models.

Isomorphic substitutions of Mg for Al were introduced into the PBC of each edge surface and, in some models, the bulk of the octahedral sheet. The effect of the substitution location in the edge PBC was investigated by considering two different models. In the first model, the isomorphic substitution occupied a solvent accessible octahedron in the edge PBC (Figure 2a, Mg^{Sol}); the isomorphic substitution of the second model occupied a linking octahedral position in the PBC (Figure 2b, Mg^{PBC}). A single isomorphic substitution in each edge PBC results in a MMT with a layer charge, x_1 = 0.25. The layer charge of the models was increased to x_2 = 0.375 through the introduction of an additional isomorphic

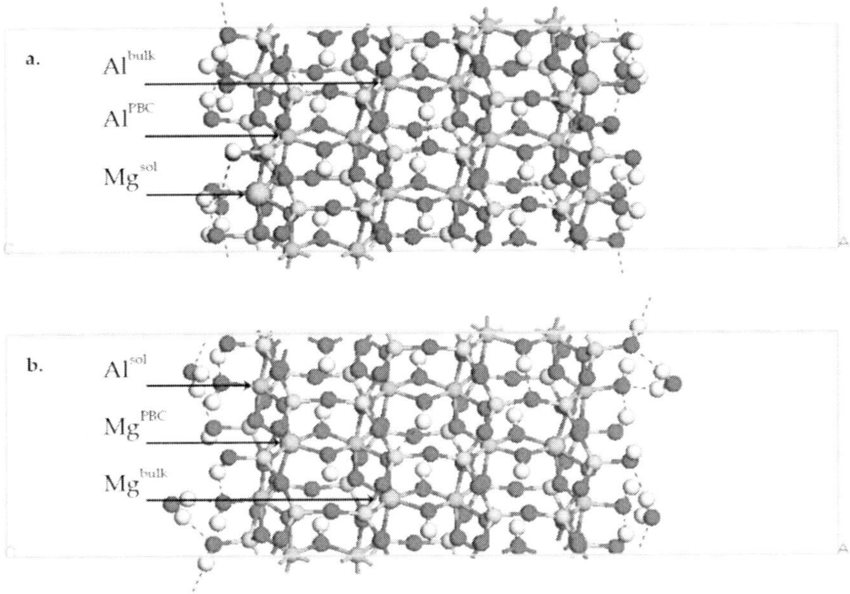

Figure 2. Density functional theory (DFT) geometry-optimized Na–MMT B edge structure at (**a**) a low layer charge (x_1 = 0.25) and (**b**) a higher layer charge (x_2 = 0.375). The bulk, linking periodic bond chain (PBC), and solvent accessible atom positions in the octahedral sheet are defined. The atom legend is as in Figure 1. The hydrogen bonds are shown as dashed blue lines.

substitution into the bulk of the octahedral sheet (Figure 2b, Mg^{Bulk}). The inclusion of the Mg^{Bulk} atom did not change the substitution ratio of the edge PBCs; the ratio remained 0.25. Although in excess of our defined constraint, this edge substitution ratio is the lowest possible for the current DFT model. The small size of these atomistic models imposes limits on the substitution ratios and layer charges that can be considered. Thus, we have chosen models that most closely agree with those of the classical mechanical simulations described above and are within the range of the octahedral substitution ratios of synthetic MMT. The isomorphic substitution of Mg for Al results in a net negative charge in the supercell representing the surface. In this study, the charge deficit was compensated by application of a uniform positive background [54].

Unsaturated bonds at the B edge were healed via the dissociative chemisorption and physisorption reactions with water molecules to create an edge surface at the p.z.n.p.c. The Si edge tetrahedra each possess one silanol (≡SiOH) group. Upon cleavage of the B edge surface, each solvent accessible octahedron is in five-fold coordination with oxygen and possesses one solvent accessible, singly coordinated hydroxyl group. Saturation of these surface octahedra requires the physisorption of a water molecule (or other neutral species) to maintain conditions at the p.z.n.p.c. A total of four H_2O molecules per edge surface were used to saturate the solvent-accessible TOT units.

When all edge surface octahedra are equivalent, the assignment of one water molecule and one hydroxyl group to each edge octahedron is a reasonable initial structure [i.e., $\equiv Al(OH)(OH_2)$]. When the surface octahedra include isomorphic substitutions, however, this homogeneous assignment is perhaps insufficient. The observed spontaneous transfer of protons on the pyrophyllite B edge surface during DFT-MD simulations [20,24,34]; the effects of isomorphic substitution on the p$K_a$of edge surfaces groups [25]; and the considerable difference between the first hydrolysis constants of Mg and Al in solution (pK_{a1} = 11.4 v. 5.0, respectively [55]) all suggest that the investigation of alternative initial surface proton distributions is prudent. Liu, et al. [24] used constrained DFT-MD to calculate the free-energy profile for H_2O dissociation from the phyllosilicate edges and identified variations in the surface functional groups when Mg is substituted for solvent accessible Al sites. Our simulations are designed to examine explicitly the effects of local charge deficits and the location of isomorphic substitutions on the edge structure. For the most thorough investigation of these effects, we also considered an alternate initial proton distribution at the p.z.n.p.c., where the surface octahedron is coordinated to two hydroxyls or two water molecules [i.e., $\equiv Al(OH)_2 + \equiv Mg(OH_2)_2$ and $\equiv Mg(OH)_2 + \equiv Al(OH_2)_2$].

3. RESULTS

3.1. DFT Calculation Results

The geometry optimized structures and relative energies demonstrate a correlation between the distribution of the active surface groups, as identified by Liu, et al. [24], and the magnitude of the local charge deficit (Table 1). At lower substitution densities (Figure 3a,c), the edge cations of the octahedral sheet are more likely to be in octahedral coordination with a hydrolyzed Al^{sol} [$\equiv Al(OH)_2$] adjacent to a Mg^{sol} coordinated to two water molecules [$\equiv Mg(OH_2)_2$]. This arrangement of surface groups at a low layer charge can be rationalized by the large difference in the first hydrolysis constant (pK_a) of aqueous Al^{3+} and Mg^{2+}. At high local charge densities, the isomorphic substitutions are less likely to be solvent accessible, and the edge Al favor trigonal bipyramidal coordination

(Figure 3b,d). These findings are important for the atom type assignments in the classical mechanical model of the Na–MMT B edge.

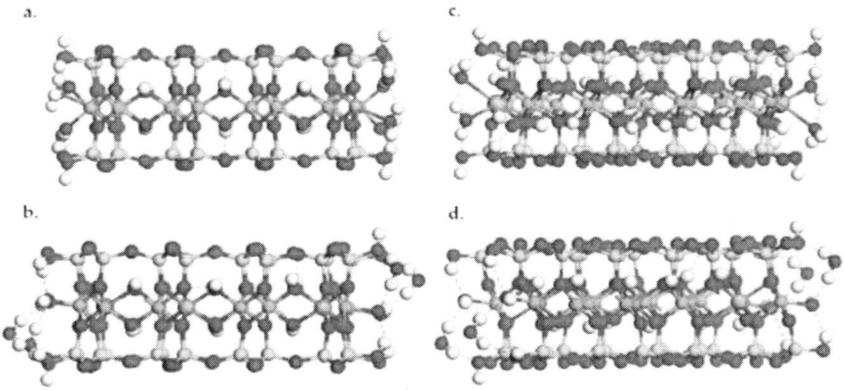

Figure 3. Na–MMT models geometry-optimized by DFT with layer charges (**a**) $x_1 = 0.25$ and (**b**) $x_2 = 0.375$; oblique views are provided for the layer charges (**c**) $x_1 = 0.25$ and (**d**) $x_2 = 0.375$. Both the Mg^{sol} and Al^{sol} of the low layer charge model are in six-fold coordination. The Al^{sol} of the high layer charge model are in trigonal bipyramidal coordination with the surface hydroxyl stabilized through the H-bonds of water bridges. The asymmetric Mg–OH_2 bonds and H-bond patterns are evident in the oblique views of the interfaces. The atom legend is the same as in Figure 1; hydrogen bonds are shown as dashed blue lines.

Table 1. Relative total energies (E_{rel}) for the low- and high-layer charge (x_1 and x_2) Na–MMT B edges with different proton distributions on the surface octahedra at the p.z.n.p.c.

Octahedral Surface Groups	E_{rel} (kJ/mol)	
	$x_1 = 0.25$	$x_2 = 0.375$
≡Al–(OH)(OH_2) + ≡Al–(OH)(OH_2) *	0	0
≡Mg–(OH)(OH_2) + ≡Al–(OH)(OH_2)	−56.9	158.1
≡Mg–$(OH_2)_2$ + ≡Al–$(OH)_2$	−98.4	85.8
≡Mg–$(OH)_2$ + ≡Al–$(OH_2)_2$	79.1 †	n/r

* model with solvent-inaccessible Mg (MgPBC), whose total energy is the reference for the relative energy calculation;
† Initial octahedral surface groups are not stable. A spontaneous proton transfer occurred during geometry optimization. n/r: not reported.

At a higher layer charge ($x_2 = 0.375$), the most favorable location for the Mg substitution was the solvent inaccessible position of the edge PBC (Table 1, Mg^{PBC}). The Al^{sol} of this structure was in five-fold coordination with oxygen. This coordination is evident upon visual inspection of Figure 3b and from the large Al^{sol}-water oxygen (Al^{sol}–Ow) distance that places the water oxygen outside of the first Al–O shell (Table 2). The contraction of the average Al^{sol}–O distance relative to the Al^{bulk}–O distance also confirms the reduced number of oxygens coordinated to the Al^{sol} atoms. The exposed surface hydroxyl coordinated to Al^{sol} is stabilized via an intrasurface H-bond with the nearest silanol group and the H-bonds of a water molecule that forms a bridge to the second silanol group of the surface TOT structure.

Table 2. Interatomic distances for the lowest-energy DFT models of MMT B edges with layer charges of x_1 = 0.25 or x_2 = 0.375. The bulk octahedral cation-oxygen distances (Al^{bulk}– or Mg^{bulk}–O) are included for comparison. The water and hydroxyl O atoms are denoted as Ow and Oh, respectively.

α-β Pair	Interatomic Distance (Å)	
	$x_1 = 0.25$	$x_2 = 0.375$
Al^{sol}–O	1.94	1.85
Al^{sol}–Ow	-	3.74
Al^{sol}–Oh	1.85	1.80
Al^{pbc}–O	1.92	1.91
Mg^{sol}–O	2.12	-
Mg^{sol}–Ow	2.32	-
Mg^{pbc}–O	-	2.07
Al^{bulk}–O	1.93	1.92
Mg^{bulk}–O	-	2.07

At a lower layer charge, however, the Mg^{sol} substitution is preferable (Table 1: x_1 = 0.25); the ≡Mg–$(OH_2)_2$ and ≡$(Al–OH)_2$edge structures are the most energetically favorable. In this lowest energy structure, the Mg^{sol} is in six-fold coordination with oxygen, including two water molecules (Figure 3a). The Mg-water oxygen [Mg–Ow] distances were longer than the average Mg^{sol}–O distance (Table 2). The adjacent Al^{sol} is also in octahedral coordination with oxygen, including two surface hydroxyl groups. The Al-hydroxyl oxygen [Al^{sol}–Oh] distances at the surface are contracted relative to the Al–Oh distances in the bulk structure [1.85 v. 1.90 Å, respectively]. The Mulliken charges [56] of these two surface hydroxyl groups in single coordination with the Al^{sol} and adjacent to the Mg^{Sol} substitution were approximately equal (−0.64 e) and slightly greater than both the charges of the hydroxyl of the edge silanol groups and the bulk structure (−0.60 e).

The initial proton distribution, wherein the Mg^{sol} was hydrolyzed [≡Mg–$(OH)_2$] and the Al^{sol} was coordinated to two water molecules [≡Al–$(OH_2)_2$], was unstable at the low layer charge. A spontaneous proton transfer from the edge Al to the edge Mg octahedra occurred during the geometry optimization. The Mg^{sol} octahedra were coordinated to one hydroxyl and a water molecule, and the Al^{sol} was in five-fold coordination with a hydroxyl group in the final structure. At the higher layer charge, a spontaneous proton transfer occurred. However, the geometry optimization of this initial proton distribution did not result in a structure that satisfied the convergence criteria.

The minimum energy edge structures predicted by our DFT optimization concur with those edge structures obtained via DFT-based molecular dynamics simulations of the B edge with an octahedral substitution [24]. These previous simulations employed a model that possessed a layer charge of 0.25, was half the size of our model, and considered only a Mg^{sol}substitution. The active surface groups in this model were an octahedral Mg [≡Mg–$(OH_2)_2$], two different octahedral Al [≡Al–$(OH)(OH_2)$; ≡Al–$(OH)_2$], and a trigonal bipyramidal Al [≡Al^V–OH].

3.2. Classical MD Simultion Results

All of the relevant surface groups identified via the DFT geometry optimization and DFT-MD simulations of the Na–MMT B edge are recreated in our classical mechanical model. The surface groups associated with both low- and high-local charge deficits could be accommodated in the classical mechanical model due to the greater length-scale of the model. The greater areal extent of the edge surface in the classical model created a more heterogeneous distribution of isomorphic substitutions and permitted the creation of a model with the representative surface groups. The atom type assignment for edges where the Mg substitution is solvent inaccessible is straightforward as the edge surface groups are all coordinated to solvent accessible Al atoms. The initial assignment of surface groups for the Al^{sol} is one singly coordinated hydroxyl and one water molecule (i.e., ≡Al–$(OH)(OH_2)$). When the Mg substitution is in the solvent accessible position, our atom type assignment is informed by the DFT results. In bulk, the Clayff hydroxyl oxygen atoms associated with isomorphic substitutions have a different partial charge than unsubstituted hydroxyl oxygens ($q(ohs)$ = −1.0808 e versus $q(oh)$ = −0.95 e, respectively). Cleavage of the bulk mineral phase can expose these substituted hydroxyl groups (*ohs*) at the edge (i.e., ≡Mg–$(OH)_2$). The DFT geometry optimizations demonstrated that this arrangement of surface hydroxyls was unstable, and protons were spontaneously transferred to lower the relative energy of the structure (Table 1). As the Mulliken charge of the hydroxyl groups coordinated to the hydrolyzed Al^{sol} was greater than the other surface and bulk hydroxyls, we have assigned the two substituted hydroxyl groups to the hydrolyzed Al^{sol} of the classical mechanical model and saturated the Mg^{sol} with two water molecules (i.e., ≡Al–$(OH)_2$ and ≡Mg–$(OH_2)_2$). This choice of atom type assignments maintains the edge surface at the p.z.n.p.c. and the partial charge neutrality of the classical mechanical model.

The Al^{sol}–O distance calculated by our classical mechanics MD simulation (Table 3: 1.87 Å) is intermediate between the distances for the low- and high-layer charge DFT models. This result is expected given that the larger classical MD model possesses Al^{sol} groups that are in six- and five-fold coordination with O (Table 3: CN(Al^{sol}–O) = 5.72; Figure 4). The mix of octahedral and trigonal bipyramidal coordination results from the free exchange of the coordinated water molecule with the bulk, which is thermodynamically favorable [24]. On average, slightly more than half of the ≡Al–(OH)(OH_2) surface groups are in octahedral coordination at any time throughout the MD simulation. The Al^{sol}–O distances of the DFT models that bracket the MD model distances reflect Al^{sol} that are in either octahedral or trigonal bipyramidal coordination. The Al^{sol}–O distance is contracted relative to the bulk Al–O distance (Al^{bulk}–O). This contraction is attributed to the presence of trigonal bipyramidal Al at the edge. The Al^{sol}–Oh distance on the Na–MMT B edge agrees well with the corresponding distance in the DFT model at low layer charge (x_1 = 0.25) and previously reported B edge Al–OH distance of pyrophyllite [28]. The Al^{sol}–Ow distance of the classical mechanical model of the Na–MMT edge cannot be compared with the DFT models as the Al^{sol} is hydrolyzed at a low layer charge (i.e., ≡Al–$(OH)_2$) or in five-fold coordination at a higher layer charge in the smaller DFT models. Comparison of the Al^{sol}–Ow distance of the Na–MMT B edge with the corresponding distance on the pyrophyllite B edge reveals that it is contracted somewhat (r_{peak}(Al–Ow) = 2.06 and 2.25 Å, respectively). The Al in the linking position in the edge PBC (Al^{pbc}) has a first oxygen shell that is indistinguishable from the Al^{bulk}.

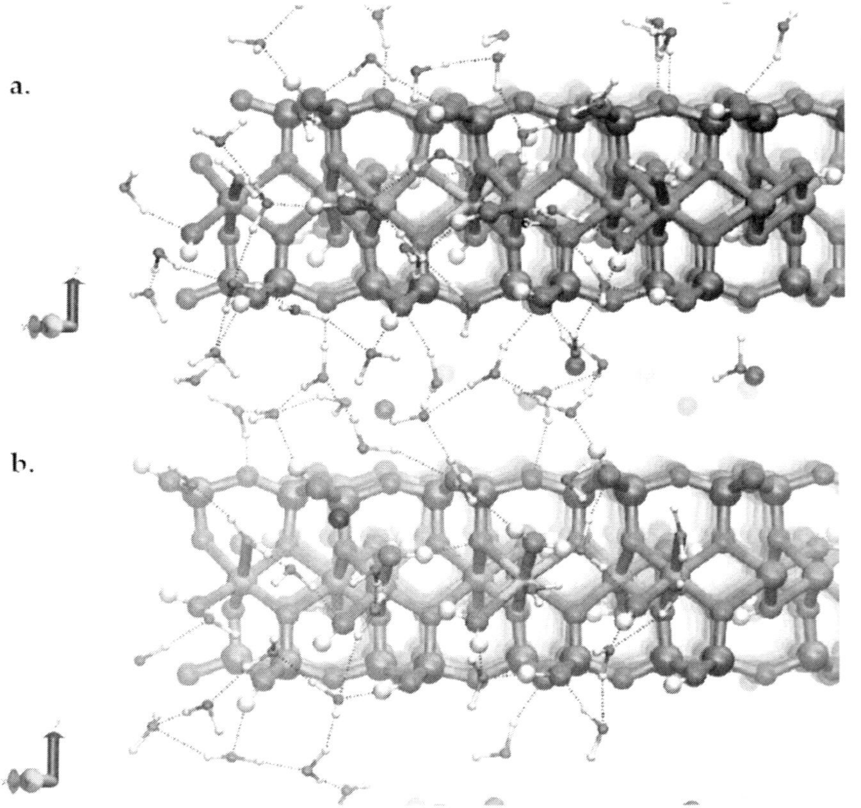

Figure 4. Na–MMT edge structures taken from the MD simulation trajectory at t = 5.0 ns with substitution in the (**a**) Mg^{pbc} and (**b**) Mg^{sol} positions. Both of the Al^{sol} that share an edge with the Mg^{pbc} are in trigonal bipyramidal coordination with oxygen. The Mg^{sol} and adjacent Al^{sol} are both in octahedral coordination with oxygen. The atom legend is the same as in Figure 1; Hydrogen bonds are shown as dashed blue lines.

The classical MD model reproduced the solvent accessible Mg-oxygen distance (Table 3: $r_{peak}(Mg^{sol}–O)$ = 2.12 Å) of the DFT model at the low layer charge. The MD simulations underestimated the Mg^{sol}–Ow distance when compared to the DFT results ($r_{peak}(Mg^{sol}–Ow)$ = 2.20 versus 2.32 Å, respectively). A closer examination of Mg^{sol}–Ow distances from the DFT geometry-optimization reveals two distinct clusters of distances at 2.17 and 2.53 Å. With this additional perspective, the first peak of the Mg^{sol}–Ow RDF reproduces the first cluster distance, and the first minimum of the RDF encompasses the distance of the second cluster [Table 3: r_{peak}(Mg^{sol}–O) = 2.52 Å]. The water ligands associated with Mg^{sol} were not exchanged with the bulk phase during the timescale of our classical MD trajectory, a result consistent with the slow exchange of first shell water molecules from hydrated Mg ions (average residence time: 1–2 μs [57,58]). The Mg^{pbc}–O pair distribution is indistinguishable from the Mg–O pairs in bulk.

Table 3. Cation-oxygen (α-β) pair distribution functions of the octahedral sheet of the Na–MMT model from classical molecular dynamics (MD) simulations. The pair distribution functions are summarized by the distance to the first peak (r_{peak}), coordination number (CN), the first minimum (ρ) of the cation–O pair distribution for the solvent accessible (sol), and the solvent-inaccessible (pbc) positions of the edge PBC and bulk positions in the octahedral sheet. The standard deviation for each property (+/−) is provided. Ow and Oh represent the O of water and hydroxyl, respectively.

α-β Pair	r_{peak} (Å)	+/−	CN	+/−	ρ (Å)	+/−
Al^{sol}–O	1.87	0.011	5.72	0.037	2.81	0.067
Al^{sol}–Ow	2.06	0.022	0.57	0.038	2.74	0.101
Al^{sol}–Oh	1.85	0.008	1.14	0.000	2.11	0.022
Al^{pbc}–O	1.95	0.010	6.00	0.001	2.67	0.036
Mg^{sol}–O	2.12	0.017	5.99	0.005	2.52	0.030
Mg^{sol}–Ow	2.20	0.019	1.99	0.004	2.52	0.028
Mg^{pbc}–O	2.11	0.012	6.00	0.003	2.44	0.027
Al^{bulk}–O	1.93	0.010	6.00	0.000	2.71	0.028
Mg^{bulk}–O	2.11	0.005	6.00	0.000	2.46	0.025

3.3. H-Bond Network at the Na-Montmorillonite B Edge

Our MD simulations showed that an extensive H-bond network exists at the edge-water interface and likely contributes to the stability of the disordered surface structures. The hydroxyl groups of the trigonal-bipyramidal Al that share an edge with the solvent-inaccessible Mg substitution (Mg^{pbc}) are H-bonded to water molecules that bridge the surface and form H-bonds with the adjacent octahedral Al (Figure 4a). Similarly, water bridges stabilize the two water molecules coordinated with the solvent accessible Mg. These water bridges accept H-bonds from the Mg coordinated water molecules and accept or donate H-bonds to the neighboring aluminol or silanol groups (Figure 4b). The hydroxyl groups of the surface Al accept and donate H-bonds to water bridges to the adjacent surface groups (i.e., ≡Al–(OH_2) and ≡Mg(OH_2)$_2$) as well as to the silanol groups. The surface hydroxyl groups occasionally form intrasurface H-bonds directly by donating a H-bond to neighboring silanol or the O that bridges the tetrahedral and octahedral sheets (i.e., ≡(Al or Mg)–O–Si≡). These surface H-bonds described by the classical mechanics model are qualitatively similar to those observed in the DFT geometry-optimized structure (Figure 3).

The observed surface H-bond network described above was quantified for the final 1.0 ns of the MD trajectory (Table 4). The ≡Mg–(OH_2)$_2$ groups almost exclusively behaved as H-bond donors in our model. The ≡Al–(OH)$_2$ surface groups accepted twice as many H-bond as were donated. Our findings are a marked departure with previous results from a DFT-based MD study by Liu, et al. [24] that reported that the water ligands of ≡Mg–(OH_2)$_2$ behave as very weak H-bond donors and that ≡Al–(OH)$_2$ accepts one H-bond from solvating waters but typically does not donate H-bonds to water due to the orientation of the hydroxyl groups. The Al–O–H angle frequency distribution from our MD simulations possesses a dominant peak at 150° with a lesser peak near 110° (Figure 5). The lesser peak is due to one of the four Al–O–H angles in the model that is a H-bond donor in an intrasurface H-bond (Figure 6b: upper Al–O–H group). This one hydroxyl group does not donate H-bonds to the water and its orientation is consistent with that reported in the DFT–MD simulations. The majority of these hydroxyls, however, are more likely oriented toward the solvent phase. In the DFT-MD simulations, the reported absence of H-bond donors from these ≡Al–(OH)$_2$ surface groups may be attributed to limitations in the physical model size and the duration of the simulation trajectory (45 ps).

The H-bonds of the ≡Al–(OH_2)(OH) surface group are reported separately for the OH and OH_2 moieties (Table 4). The hydroxyl group functions as both a H-bond donor and acceptor in approximately the same proportion as the surface Al coordinated to two hydroxyl groups (i.e., ≡Al–(OH)$_2$). The coordinated water molecule behaves almost exclusively as a H-bond donor similar to the water molecules coordinated to the surface Mg. The interfacial H-bonding of the ≡Al–(OH)(OH_2) on the B edge agrees quantitatively with the sum of H-bond donors and acceptors for this group reported in the previous DFT-based MD of the pyrophyllite B edge [20], which is equivalent to the pyrophyllite (010) edge in a monoclinic unit cell. We also examined the possible influence of an octahedral Mg substitution on the H-bonds of the silanol surface groups (Table 4). The hydrolyzed surface Al and the surface Mg coordinated to two water molecules did not affect the silanol H-bond pattern when compared to the silanol groups that were un-

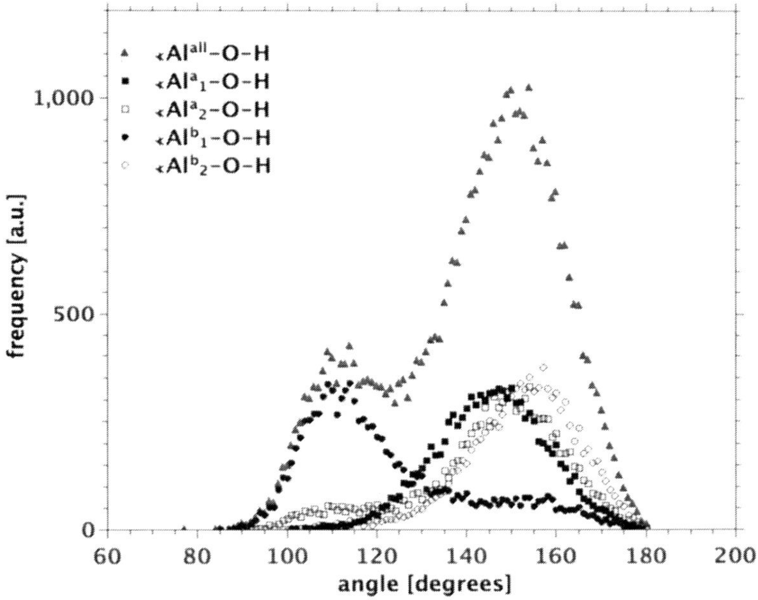

Figure 5. Frequency distribution for the ≡Al–O–H angles of the hydrolyzed Al adjacent to the Mg^{sol} sites during the final 1.0 ns of the 5.0 ns MD trajectory. The individual and cumulative frequency distributions are provided for each of the four relevant ≡Al–O–H angles. The Al superscripts correspond to the edge complexes in Figure 6; the subscript corresponds to the upper (1) or lower (2) hydroxyl coordinated to the edge Al.

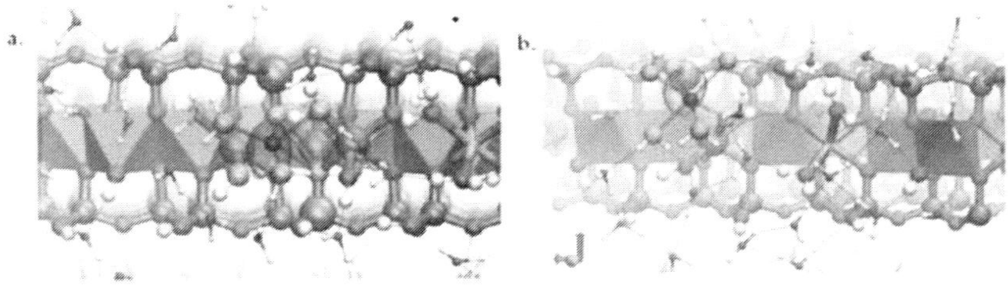

Figure 6. Na surface complexes (blue sphere) on the MMT B edge surface that are associated with (**a**) the octahedral sheet and (**b**) the tetrahedral sheet. The atom legend is the same as in Figure 4, with the additions of Al as octahedra; the first shell oxygen atoms being shown as orange; and the first shell Al, Mg, and Si as enlarged spheres. The half-maximum isosurface from the Na ion volumetric density maps are shown as transparent, blue, ovoid surfaces.

Table 4. Hydrogen bonding to the edge surface groups. The average number of total H-bonds, *N*(H-bonds), is provided as well as the average number of H-bonds donated (D) and accepted (A) by the surface group. The H-bonds to the silanol surface groups are summarized for silanol groups in TOT units with ≡Al–(OH)(OH_2) (i.e., ≡Si–OH), in TOT units with ≡Al–$(OH)_2$ (i.e., ≡Si^{Al}–OH), and in TOT units with ≡Mg–$(OH_2)_2$ (i.e., ≡Si^{Mg}–OH). The average number of H-bonds and standard deviation (+/−) was calculated from the final 1.0 ns of the 5.0 ns trajectory.

Surface Group	H-Bond	*N*(H-Bonds)	+/−
≡Mg–$(OH_2)_2$	Total	1.54	0.287
-	D	1.53	0.281
-	A	0.01	0.050
≡Al–$(OH)_2$	Total	1.46	0.329
-	D	0.47	0.244
-	A	0.99	0.186
≡Al(OH_2)–(OH)	Total	1.75	0.198
-	D	0.65	0.132
-	A	1.10	0.140
≡Al(OH)–(OH_2)	Total	1.24	0.233
-	D	1.24	0.231
-	A	0.01	0.043
≡Si–OH	Total	1.31	0.132
-	D	0.78	0.081
-	A	0.53	0.106
≡Si^{Al}–OH	Total	1.18	0.308
-	D	0.79	0.200
-	A	0.39	0.230
≡Si^{Mg}–OH	Total	1.50	0.375
-	D	0.86	0.173
-	A	0.64	0.297

influenced by the isomorphic substitution [compare N(H-bond) for ≡Si^{Al}–OH and ≡Si^{Mg}–OH versus ≡Si–OH).

3.4. Na-Montmorillonite B Edge Adsorption Sites

During the course of the 5.0 ns MD trajectory, ~10% of the Na counter ions in the interlayer crossed the edge plane and migrated to the bulk mesopore. The volumetric density maps for these Na ions for the interval $4.0 < t < 5.0$ ns revealed the formation of inner-sphere Na complexes at two distinct sites: Na was associated with a site in the plane of the octahedral sheet (Figure 6a) and the cleaved hexagonal cavity of the tetrahedral sheets (Figure 6b). The Na RDFs further showed that these

Na edge complexes formed near the substituted Mg site but are not directly associated with the Mg^{sol} site (Table 5). The r_{peak}(Na–Mg) distance was greater than 6 Å (6.5 and 7.9 Å). Sodium formed surface complexes with Al octahedra or Si tetrahedra but not with Mg octahedra.

Table 5. Pair distribution functions for the Na edge complexes in Figure 6a,b. The total Na–O RDF is further factored into the constituent O atom types: Ow, water oxygen; Oh, hydroxyl oxygen; and Ob, bridging oxygen. The a/b values for distances and CNs in the table correspond to the complexes in Figure 6a,b, respectively.

α-β Pair	r_{peak}, Å	CN	ρ, Å
	a/b	a/b	a/b
Na–O	2.34/2.36	6.17/5.89	3.02/3.01
Na–Ow	2.37/2.37	1.98/3.99	2.83/3.12
Na–Oh	2.28/2.32	2.01/2.27	2.62/3.26
Na–Ob	2.49/5.26	1.97/14.37	3.18/6.23
Na–Al	3.27/3.53	2.99/1.01	3.91/3.99
Na–Mg	7.94/6.54	0.99/1.00	8.34/7.20
Na–Si	3.46/3.47	1.96/1.12	4.10/3.97

One can infer from the volumetric density and RDF analysis that the octahedral sheet sites would be the stronger of the two types of sorptive sites identified in our simulations. While all edge Na complexes were in octahedral coordination, the first shell of the complex associated with the hexagonal cavity (tetrahedral sheet) was somewhat more relaxed than the complexes associated with octahedral sheets as evidenced by the greater extent of the first shell (i.e., ρ = 3.3 v. 2.6 Å). The Na surface complex at the hexagonal cavity had four water oxygens, as compared to two in the first shell of the complexes in the plane of the octahedral sheet. The Na complexes in the plane of the octahedral sheet each had two bridging oxygen atoms in their first O shell (i.e., CN(Na–Ob) $\approx$ 2.0); the bridging oxygen atoms were absent from the first O shell of the Na complex at the hexagonal vacancy as the average Na–Ob distance was much greater than the average Na–O distance.

The Na edge complexes in the plane of the octahedral sheet have three octahedral nearest-neighbors as well as the two Si tetrahedra of the shared bridging O atoms. These Na ions clearly occupy the dioctahedral vacancy of the B edge and form; what can more generally be described as multinuclear, tetradentate, inner-sphere complexes. The Na edge complex associated with the tetrahedral sheet was coordinated with the hydrolyzed Al and the hydroxyl of the Si at the corner of the cleaved hexagonal cavity. This Na edge surface complex can be described as a binuclear, bidentate, inner-sphere complex. It is worth noting that, due to the interconversion of the edge Al between five- and six-fold O coordination, the arrangement of hydroxyl groups at the edge is not static. The tangible result of this is that the nearest dioctahedral vacancy may not always possess the necessary coaxial arrangement of hydroxyl groups to form an inner-sphere complex. This necessary surface hydroxyl configuration was absent in the instance where we observed a binuclear, bidentate complex; a water molecule occupies the coaxial position of the Al across the dioctahedral vacancy.

The crystallographic description of the first site at the dioctahedral vacancy on the B edge (Figure 6a) is consistent with previous DFT simulations of the B edge complexes for Cd^{2+} [59] and UO_2^{2+} [22] and the interpretation of spectra from polarized X-ray adsorption fine structure spectroscopy (P-EXAFS) from Ni and Zn sorption on MMT [60,61]. This site is attributed to the strong site in the 2SPNE SC/CE model of surface complexation on 2:1 phyllosilicates [14]. The second sorption site associated with the cleaved hexagonal cavity (Figure 6b) is consistent with the second of two most probable sites identified for UO_2^{2+} sorption but differed from the Cd^{2+} surface complexes. The surface charge condition was not equivalent in any of the simulations; the simulations of Cd^{2+} complexes had no structural charge and one deprotonated hydroxyl group per sorbate, while the UO_2^{2+} simulations had structural charge balanced with a counter ion and two deprotonated hydroxyl groups per sorbate, and our simulations possessed only structural charge at the edge. The Cd^{2+}DFT-MD simulations sampled two low-energy sites of assumed configurations; a mononuclear, bidentate site associated with the edge Al (i.e., $\equiv Al-(OH)_2-Cd^{2+}$) and a mononuclear, monodentate silanol site (i.e., $\equiv Si-OH-Cd^{2+}$). In fact, our MD trajectory showed that at early times the Na ion was in a mononuclear, bidentate octahedral complex with the $\equiv Al-(OH)_2$surface groups (i.e., $\equiv Al-(OH)_2-Na^+$) but later occupied the dioctahedral vacancy. This well illustrates the advantage of classical MD simulations in obtaining sufficient structural data, which could be critical in the interpretation of metal surface speciation.

4. DISCUSSION

The results of our present atomistic simulations provide new insights when compared to previous simulations of the Na–MMT AC edge [32]. The DFT geometry optimizations of the AC and B edges both demonstrated a relationship between the local charge deficits and structural disorder of the Na–MMT edge. The relative stability of octahedral substitutions in the edge PBC, however, was different for the two stoichiometrically-equivalent edge structures. The AC edge favored substitutions in the solvent-inaccessible positions (Mg^{pbc}) at both low and high layer-charges [32]. In contrast, on the B edge, the favorable site changed from Mg^{sol} at low a layer-charge to Mg^{pbc} at a high layer-charge. An increase in layer charge reversed the location of the lowest relative energy substitution in the B edge chain and altered the coordination of edge Al of the octahedral sheets. Trigonal bipyramidal Al structures were predicted at both the AC and B edges of Na–MMT, but an Edelman-Favejee [30] defect co-occurs in the tetrahedral sheet with five-fold Al structures on the AC edge [32]. The Al^{sol} and Mg^{sol} of the AC edge were stable as octahedral and square-pyramidal structures due to the free exchange of the coordinated water molecule with the bulk. In contrast, no stable square-pyramidal Al structures were observed on the B edge, being the edge Al transition between octahedral and trigonal-bipyramidal polyhedra. In addition, the water molecules coordinated with Mg^{sol} on the B edge were not freely exchanged with the bulk on the timescale of our MD simulations, a result consistent with previous constrained DFT-based MD that predicted that this exchange was thermodynamically unfavorable [24]. The present simulations, when considered with previous simulations of the Na–MMT AC edge, demonstrate that a relationship between disorder at the 2:1 phyllosilicate edges and excess local charge exists but also identify aspects of the disordered structures that are unique to the B edge.

Our exploration of the B edge sorption sites under the influence of octahedral substitutions was conducted in the absence of variable charge (i.e., pH = p.z.n.p.c.). This pH condition imposes some limitations on our findings; however, the results should be considered in light of the calculated pK_as of the B edge surface groups. The calculated pK_as for the B edge are ~7 for silanol groups and ~8 for aluminol but are subject to relatively large uncertainties that result in the overlap of the two values [25,26,27]. On the AC edge, the pK_a for the amphoteric Al site (i.e., ≡Al–OH_2) is 5.5 [26]. The AC edge also possesses a lower surface energy than the B edge and would be a greater fraction of the total edge surface area [19,28,31]. This difference in surface acidities and fraction of total surface area indicate that the reactive site density of the AC edge will be greater than that of the B edge. Thus, our findings with regard to the B edge surface complexes are applicable under even mildly acidic conditions.

Our current classical MD simulations and DFT geometry optimizations provide a unique perspective on the structure of the montmorillonite edge. We have used DFT geometry optimizations to affirm the identity of the relevant surface groups in the octahedral sheet and the relationship of these groups to the local charge deficit. These relevant surface groups can be successfully reproduced in classical MD simulations using the Clayff force field without modification. When combined with previous simulation of the AC edge [32], the Clayff force field can be used for future simulations to explore the properties of the dominant crystallographic edges, nanoparticles, and transport across the macro- to nano-pore boundary. A limited number of Na counter ions freely crossed this boundary during the course of our simulations and proved useful as probes in identifying two distinct edge sites associated with isomorphic substitutions at the edge. The identification of two sorptive sites at the edge surface and the influence of octahedral substitutions on the sorbed complexes should be useful in constraining the number and type of sites in surface complexation models of MMT. These results will also guide the interpretation of experimental spectra and any future atomistic simulations to determine the free energies for sorption at the B edge. The present simulations have focused on the role of permanent structural charge originating from the octahedral sheet of 2:1 aluminosilicates with atomistic models whose layer charges were constrained by synthetic MMTs. The layer charge of 2:1 phyllosilicates can be considerably greater if the isomorphic substitutions originate from or occur in combination with the tetrahedral sheets [62]. Previous DFT simulations of the 2:1 phyllosilicate edge have considered the effects of isolated isomorphic substitutions in the tetrahedral sheets on properties of the edge [24,34]. The effects of tetrahedral substitutions on the edge structure of high-layer charge minerals (e.g., vermiculite, illite, mica, etc.) and unconstrained classical MD simulations of these 2:1 edges are deserving topics of future classical mechanical simulations.

ACKNOWLEDGMENTS

This research was supported by the Basic Science Research Program through the National Research Foundation of Korea (NRF), funded by the Ministry of Science, ICT, and Future Planning (NRF-2015R1A4A1041105). Kideok D. Kwon was supported by the NRF Ministry of Education (2016R1D1A1B03931919). A portion of the computations were performed using resources of the National Energy Research Scientific Computing Center, which is supported by the Office of Science of the U.S. Department of Energy under Contract No. DE-AC02-05CH11231.

AUTHOR CONTRIBUTIONS

Aric G. Newton performed the molecular dynamics (MD) simulations; Kideok D. Kwon performed the density functional theory (DFT) geometry optimizations; and Aric G. Newton, Jin-Yong Lee, and Kideok D. Kwon wrote the manuscript.

CONFLICTS OF INTEREST

The authors declare no conflict of interest. The funding sponsors had no role in the design of the study; in the collection, analyses, or interpretation of data; in the writing of the manuscript; or in the decision to publish the results.

REFERENCES

1. Sposito, G. The Chemistry of Soils, 2nd ed.; Oxford University Press: Oxford, UK; New York, NY, USA, 2008.
2. Miller, A.W.; Wang, Y. Radionuclide interaction with clays in dilute and heavily compacted systems: A critical review. Environ. Sci. Technol. 2012, 46, 1981–1994.
3. Pusch, R.; Karnland, O. Physico chemical stability of smectite clays. Eng. Geol. 1996, 41, 73–85.
4. Tournassat, C.; Bourg, I.C.; Steefel, C.I.; Bergaya, F. Natural and engineered clay barriers. In Developments in Clay Science 6; Elsevier: Amsterdam, The Netherlands, 2015; Volume 6, pp. 5–432.
5. Boulet, P.; Coveney, P.V.; Stackhouse, S. Simulation of hydrated Li+-, Na+- and K+-montmorillonite/polymer nanocomposites using large-scale molecular dynamics. Chem. Phys. Lett. 2004, 389, 261–267.
6. Liu, X.D.; Lu, X.C.; Wang, R.C.; Zhou, H.Q.; Xu, S.J. Interlayer structure and dynamics of alkylammonium-intercalated smectites with and without water: A molecular dynamics study. Clays Clay Miner. 2007, 55, 554–564.
7. Yu, Y.R.; Yang, X.N. Molecular simulation of swelling and interlayer structure for organoclay in supercritical CO2. Phys. Chem. Chem. Phys. 2011, 13, 282–290.
8. Kim, M.H.; Choi, G.; Elzatahry, A.; Vinu, A.; Choy, Y.B.; Choy, J.H. Review of clay-drug hybrid materials for biomedical applications: Administration routes. Clays Clay Miner. 2016, 64, 115–130.
9. Lal, R. Soil carbon sequestration impacts on global climate change and food security. Science 2004, 304, 1623–1627.
10. Brouwer, E.; Baeyens, B.; Maes, A.; Cremers, A. Cesium and rubidium ion equilibria in illite clay. J. Phys. Chem. 1983, 87, 1213–1219.
11. Elzinga, E.J.; Sparks, D.L. Nickel sorption mechanisms in a pyrophyllite-montmorillonite mixture. J. Colloid Interface Sci. 1999, 213, 506–512.
12. Morton, J.D.; Semrau, J.D.; Hayes, K.F. An X-ray absorption spectroscopy study of the structure and reversibility of copper adsorbed to montmorillonite clay. Geochim. Cosmochim. Acta 2001, 65, 2709–2722.

13. Zachara, J.M.; Smith, S.C.; Liu, C.X.; McKinley, J.P.; Serne, R.J.; Gassman, P.L. Sorption of Cs+ to micaceous subsurface sediments from the hanford site, USA. Geochim. Cosmochim. Acta 2002, 66, 193–211.
14. Bradbury, M.H.; Baeyens, B. A mechanistic description of Ni and Zn sorption on Na-montmorillonite 2. Modelling. J. Contam. Hydrol. 1997, 27, 223–248.
15. Dähn, R.; Baeyens, B.; Bradbury, M.H. Investigation of the different binding edge sites for Zn on montmorillonite using P-EXAFS—The strong/weak site concept in the 2SPNE SC/CE sorption model. Geochim. Cosmochim. Acta 2011, 75, 5154–5168.
16. White, G.N.; Zelazny, L.W. Analysis and implications of the edge structure of dioctahedral phyllosilicates. Clays Clay Miner. 1988, 36, 141–146.
17. Hartman, P. Crystal Growth: An Introduction; North-Holland Publishing Co./American Elsevier: Amsterdam, The Netherlands; New York, NY, USA, 1973.
18. Bickmore, B.R.; Rosso, K.M.; Nagy, K.L.; Cygan, R.T.; Tadanier, C.J. Ab initio determination of edge surface structures for dioctahedral 2:1 phyllosilicates: Implications for acid-base reactivity. Clays Clay Miner. 2003, 51, 359–371.
19. Churakov, S.V. Ab initio study of sorption on pyrophyllite: Structure and acidity of the edge sites. J. Phys. Chem. B2006, 110, 4135–4146.
20. Churakov, S.V. Structure and dynamics of the water films confined between edges of pyrophyllite: A first principle study. Geochim. Cosmochim. Acta 2007, 71, 1130–1144.
21. Kremleva, A.; Kruger, S. Comparative computational study of Np(V) and U(VI) adsorption on (110) edge surfaces of montmorillonite. Clays Clay Miner. 2016, 64, 438–451.
22. Kremleva, A.; Kruger, S.; Rosch, N. Uranyl adsorption at solvated edge surfaces of 2:1 smectites. A density functional study. Phys. Chem. Chem. Phys. 2015, 17, 13757–13768.
23. Kremleva, A.; Krüger, S.; Rösch, N. Toward a reliable energetics of adsorption at solvated mineral surfaces: A computational study of uranyl(VI) on 2:1 clay minerals. J. Phys. Chem. C 2016, 120, 324–335.
24. Liu, X.; Lu, X.; Meijer, E.J.; Wang, R.; Zhou, H. Atomic-scale structures of interfaces between phyllosilicate edges and water. Geochim. Cosmochim. Acta 2012, 81, 56–68.
25. Liu, X.; Lu, X.; Sprik, M.; Cheng, J.; Meijer, E.J.; Wang, R. Acidity of edge surface sites of montmorillonite and kaolinite. Geochim. Cosmochim. Acta 2013, 117, 180–190.
26. Liu, X.D.; Cheng, J.; Sprik, M.; Lu, X.C.; Wang, R.C. Surface acidity of 2:1-type dioctahedral clay minerals from first principles molecular dynamics simulations. Geochim. Cosmochim. Acta 2014, 140, 410–417.
27. Tazi, S.; Rotenberg, B.; Salanne, M.; Sprik, M.; Sulpizi, M. Absolute acidity of clay edge sites from ab-initio simulations. Geochim. Cosmochim. Acta 2012, 94, 1–11.
28. Newton, A.G.; Sposito, G. Molecular dynamics simulations of pyrophyllite edge surfaces: Structure, surface energies, and solvent accessibility. Clays Clay Miner. 2015, 63, 277–289.
29. Cygan, R.T.; Liang, J.J.; Kalinichev, A.G. Molecular models of hydroxide, oxyhydroxide, and clay phases and the development of a general force field. J. Phys. Chem. B 2004, 108, 1255–1266.
30. Edelman, C.H.; Favejee, J.C.L. On the crystal structure of montmorillonite and halloysite. Z. Kristallogr. 1940, 102, 417–431.
31. Kwon, K.D.; Newton, A.G. Structure and stability of pyrophyllite edge surfaces: Effect of temperature and water chemical potential. Geochim. Cosmochim. Acta 2016, 190, 100–114.
32. Newton, A.; Kwon, K.; Cheong, D.-K. Edge structure of montmorillonite from atomistic simulations. Minerals 2016, 6, 25.
33. Bleam, W.F.; Welhouse, G.J.; Janowiak, M.A. The surface coulomb energy and proton coulomb potentials of pyrophyllite (010), (110), (100), and (130) edges. Clays Clay Miner. 1993, 41, 305–316.

34. Suter, J.L.; Kabalan, L.; Khader, M.; Coveney, P.V. Ab initio molecular dynamics study of the interlayer and micropore structure of aqueous montmorillonite clays. Geochim. Cosmochim. Acta 2015, 169, 17–29.
35. Cadars, S.; Guegan, R.; Garaga, M.N.; Bourrat, X.; Le Forestier, L.; Fayon, F.; Huynh, T.V.; Allier, T.; Nour, Z.; Massiot, D. New insights into the molecular structures, compositions, and cation distributions in synthetic and natural montmorillonite clays. Chem. Mater. 2012, 24, 4376–4389.
36. Sainz-Diaz, C.I.; Timon, V.; Botella, V.; Artacho, E.; Hernandez-Laguna, A. Quantum mechanical calculations of dioctahedral 2:1 phyllosilicates: Effect of octahedral cation distributions in pyrophyllite, illite, and smectite. Am. Mineral. 2002, 87, 958–965.
37. Sainz-Díaz, C.I.; Cuadros, J.; Hernández-Laguna, A. Analysis of cation distribution in the octahedral sheet of dioctahedral 2:1 phyllosilicates by using inverse monte carlo methods. Phys. Chem. Miner. 2001, 28, 445–454.
38. Lantenois, S.; Champallier, R.; Bény, J.M.; Muller, F. Hydrothermal synthesis and characterization of dioctahedral smectites: A montmorillonites series. Appl. Clay Sci. 2008, 38, 165–178.
39. Reinholdt, M.; Brendlé, J.; Tuilier, M.-H.; Kaliaguine, S.; Ambroise, E. Hydrothermal synthesis and characterization of Ni-Al montmorillonite-like phyllosilicates. Nanomaterials 2013, 3, 48–69.
40. Reinholdt, M.; Miehe-Brendle, J.; Delmotte, L.; Le Dred, R.; Tuilier, M.H. Synthesis and characterization of montmorillonite-type phyllosilicates in a fluoride medium. Clay Min. 2005, 40, 177–190.
41. Reinholdt, M.; Miehe-Brendle, J.; Delmotte, L.; Tuilier, M.H.; le Dred, R.; Cortes, R.; Flank, A.M. Fluorine route synthesis of montmorillonites containing Mg or Zn and characterization by XRD, thermal analysis, MAS-NMR, and EXAFS spectroscopy. Eur. J. Inorg. Chem. 2001, 2831–2841.
42. Emmerich, K.; Wolters, F.; Kahr, G.; Lagaly, G. Clay profiling: The classification of montmorillonites. Clays Clay Miner.2009, 57, 104–114.
43. Berendsen, H.J.C.; Grigera, J.R.; Straatsma, T.P. The missing term in effective pair potentials. J. Phys. Chem. 1987, 91, 6269–6271.
44. Lopez-Lemus, J.; Chapela, G.A.; Alejandre, J. Effect of flexibility on surface tension and coexisting densities of water. J. Chem. Phys. 2008, 128, 174703.
45. Karasawa, N.; Goddard, W.A. Acceleration of convergence for lattice sums. J. Phys. Chem. 1989, 93, 7320–7327.
46. Plimpton, S.J. Fast parallel algorithms for short-range molecular dynamics. J. Comput. Phys. 1995, 117, 1–19.
47. Ryckaert, J.-P.; Ciccotti, G.; Berendsen, H.J. Numerical integration of the cartesian equations of motion of a system with constraints: Molecular dynamics of n-alkanes. J. Comput. Phys. 1977, 23, 327–341.
48. Kumar, R.; Schmidt, J.R.; Skinner, J.L. Hydrogen bonding definitions and dynamics in liquid water. J. Chem. Phys.2007, 126, 204107.
49. Humphrey, W.; Dalke, A.; Schulten, K. VMD: Visual molecular dynamics. J. Mol. Graph. 1996, 14, 33–38.
50. Clark, S.J.; Segall, M.D.; Pickard, C.J.; Hasnip, P.J.; Probert, M.J.; Refson, K.; Payne, M.C. First principles methods using CASTEP. Z. Kristallogr. 2005, 220, 567–570.
51. Vanderbilt, D. Soft self-consistent pseudopotentials in a generalized eigenvalue formalism. Phys. Rev. B Condens. Matter Mater. Phys. 1990, 41, 7892–7895.
52. Perdew, J.P.; Burke, K.; Ernzerhof, M. Generalized gradient approximation made simple. Phys. Rev. Lett. 1996, 77, 3865–3868.
53. Monkhorst, H.J.; Pack, J.D. Special points for brillouin-zone integrations. Phys. Rev. B 1976, 13, 5188–5192.
54. Leslie, M.; Gillan, M.J. The energy and elastic dipole tensor of defects in ionic-crystals calculated by the supercell method. J. Phys. C Solid State Phys. 1985, 18, 973–982.
55. Stumm, W.; Morgan, J.J. Aquatic Chemistry: Chemical Equilibria and Rates in Natural Waters, 3rd ed.; Wiley: New York, NY, USA, 1996.

56. Mulliken, R.S.; Roothaan, C.C.J. Electronic population analysis on LCAO–MO molecular wave functions. I. J. Chem. Phys. 1955, 23, 1833–1840.
57. Bleuzen, A.; Pittet, P.A.; Helm, L.; Merbach, A.E. Water exchange on magnesium(II) in aqueous solution: A variable temperature and pressure 17O NMR study. Magn. Reson. Chem. 1997, 35, 765–773.
58. Neely, J.; Connick, R. Rate of water exchange from hydrated magnesium ion. J. Am. Chem. Soc. 1970, 92, 3476–3478.
59. Zhang, C.; Liu, X.; Lu, X.; Meijer, E.J.; Wang, K.; He, M.; Wang, R. Cadmium(II) complexes adsorbed on clay edge surfaces: Insight from first principles molecular dynamics simulation. Clays Clay Miner. 2016, 64, 337–347.
60. Churakov, S.V.; Dähn, R. Zinc adsorption on clays inferred from atomistic simulations and EXAFS spectroscopy. Environ. Sci. Technol. 2012, 46, 5713–5719.
61. Dähn, R.; Scheidegger, A.M.; Manceau, A.; Schlegel, M.L.; Baeyens, B.; Bradbury, M.H.; Chateigner, D. Structural evidence for the sorption of Ni(II) atoms on the edges of montmorillonite clay minerals: A polarized X-ray absorption fine structure study. Geochim. Cosmochim. Acta 2003, 67, 1–15.
62. Bailey, S.W. Structures of layer silicates. In Crystal Structures of Clay Minerals and Their X-ray Indentification; Brindley, G.W., Brown, G., Eds.; Mineralogical Society: London, UK, 1980; Volume 5, pp. 1–124.

CHAPTER 7

Geochemistry of Muro Banded Iron-Formation, Central Nigeria

John Adeyinka Adekoya*, Chukwuemeka Tony Okonkwo, Mohammed Olatoye Adepoju

Department of Applied Geology, Federal University of Technology, Akure, Nigeria

ABSTRACT

Muro Banded Iron-Formation occurs in the Proterozoic Toto Schist Belt, central Nigeria. It consists preponderantly of oxide facies and minor carbonate facies. The oxide facies is made up of alternating bands of quartz (metachert) with those of hematite + magnetite + martite ± goethite, chlorite, pyrrhotite and garnet. The carbonate facies consists of quartz (metachert) + siderite ± goethite. In the oxide facies the total iron content (Fe_2O_3t) ranges from 33.95% to 48.08% and the SiO_2 content from 50.33% to 64.50%. In the case of the carbonate facies, the Fe_2O_3t content varies from 15.42% to 20.66% and SiO_2content from 66.84 to 72.86%. The Al_2O_3 content is generally low ranging from 0.1% to 0.54% in the oxide facies, and 0.24% to 0.31% in the carbonate facies. Chemically, the Muro Iron-Formation is similar to the Lake Superior-type iron-formations in terms of the distribution of the major and trace elements. This taken together with similarities in lithological associations indicates its deposition in similar environments i.e. shallow intra-continental or restricted/barred marine basin. The very low Al_2O_3 contents indicate minor clastic dilution of the original chemical precipitates.

Keywords: Muro Iron-Formation; Oxide Facies; Carbonate Facies; Geochemistry; Central Nigeria

1. INTRODUCTION

Muro Banded Iron-Formation (BIF) is one of the banded iron-formations (BIFs) associated with the Proterozoic metasedimentary "schist belts" in northern Nigeria (**Figure 1**). The BIF and associated metasediments form a range of elongate but rugged Muro Hills stretching for 13 km in a NE-SW direction. Two facies of the BIF, namely, oxide and carbonate facies [1] have been recognised with the former predominating and the latter being poorly developed and rather sporadic [2].

Since the discovery of the iron deposit by the Geological Survey of Nigeria in 1981, only reconnaissance studies have been done on its geological and structural setting as well as its geochemistry [3-6]. The present investigation is aimed at studying in some detail the geochemical characteristics of the iron-formation. Bulk chemical analysis of the iron-bearing rocks was carried out, and major and trace element contents of the BIF were compared with those of other Nigerian and Lake Superior-type and Algoma-type BIFs.

2. GEOLOGICAL SETTING

2.1. Regional Geology

The Muro Banded Iron-Formation occurs in the Toto Schist Belt which is part of the Proterozoic supracrustal assemblage in the Nigerian basement complex [7]. The Nigerian basement is the southern part of Neoproterozoic Pan-African mobile belt situated east of the West African craton. Most of the supracrustals forming the schist belts in Nigeria are exposed predominantly west of longitude 8°. The age of the metasediments is uncertain as both Paleoproterozoic and Neoproterozoic ages have been obtained from them. In general, they bear imprints of polyphase deformation (D1 - D4) and metamorphism [8]. They consist essentially of pelitic and semi-pelitic schists and phyllites interlayered with minor to substantial quartzites, metaconglomerate, iron-formation, marble and amphibolites of tholeiitic affinity. The main deformation phase (D2) produced major folds such as the Muro antiform in the study area.

Both the basement and its supracrustal cover (the Schist Belts) have been intruded by a suite of Pan-African granites (Older Granites) with minor contact metamorphic and local migmatisation effects in the host rocks during the Pan African orogeny (ca 600 ± 150 Ma).

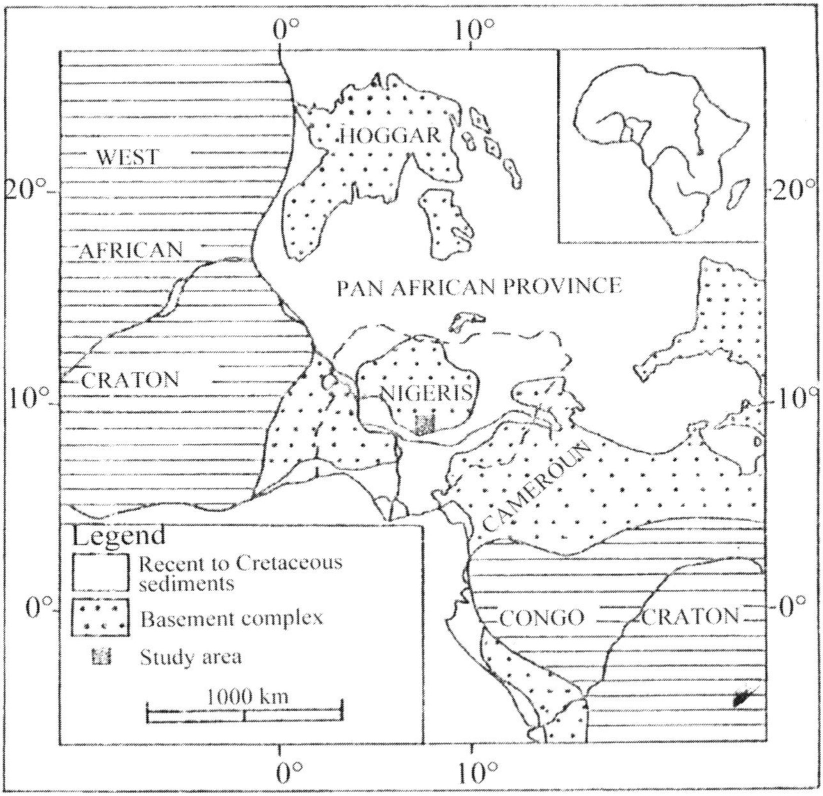

Figure 1. Index map showing the location of the Muro district in the Pan-African province in relation to the West African and the Congo cratons.

2.2. Local Geology of Muro Area

Four major lithological units which are interbanded have been mapped in the Muro area (**Figure 2**). These include marble, banded iron-formation, quartzites, and metasediments of pelitic to semi-pelitic composition. The various lithologies are folded into an overturned antiform trending NE-SW [4]. The marble occupies both the innermost core (M1) and flanks (M2) of the fold.

Successively following the former are bands of pelitic and semi-pelitic metasediments, quartzites and banded iron-formation. West of the Muro Hills, the metasediments are unconformably overlain by the Cretaceous Nupe Sandstone.

Bands of iron-formation (35 - 100 m wide/thick) occur between the outer zone marble and the quartzites. They outcrop in elongate ridges and irregular hills which take the arcuate form of the Muro Antiform (**Figure 2**) and can be traced for 20 km. Two generations of minor folds are exhibited in the iron-formation. These include tight, isoclinal, overturned folds having attenuated limbs with thickened or sharp crests and open folds with nearly horizontal axes. Both oxide and carbonate facies [1,9] occur but the former predominates with the latter being poorly developed. Occasionally bands of jasper are associated with the iron-formation.

The oxide facies iron-formation is a dark grey, finegrained, crystalline rock which displays banding on both centimetric and millimetric scales (i.e. mesoand microbanding). The banding consists of alternating light-coloured quartz-rich and darker-coloured iron oxide-rich layers. The mesobands range in width from about 1 cm to 2 cm (**Figure 3**) and the microbands from <0.5 - 2 mm (**Figure 4**).

The carbonate facies is a very hard, fine grained, massive rock made up of irregular brown siderite-rich and metachert-rich portions (**Figure 5**). It commonly occurs as rubble or vuggy boulders as found at west of Kuttiko in the southern area of the Muro Hills. Banding which is characteristic of the oxide facies is either non-existent or rare in this rock.

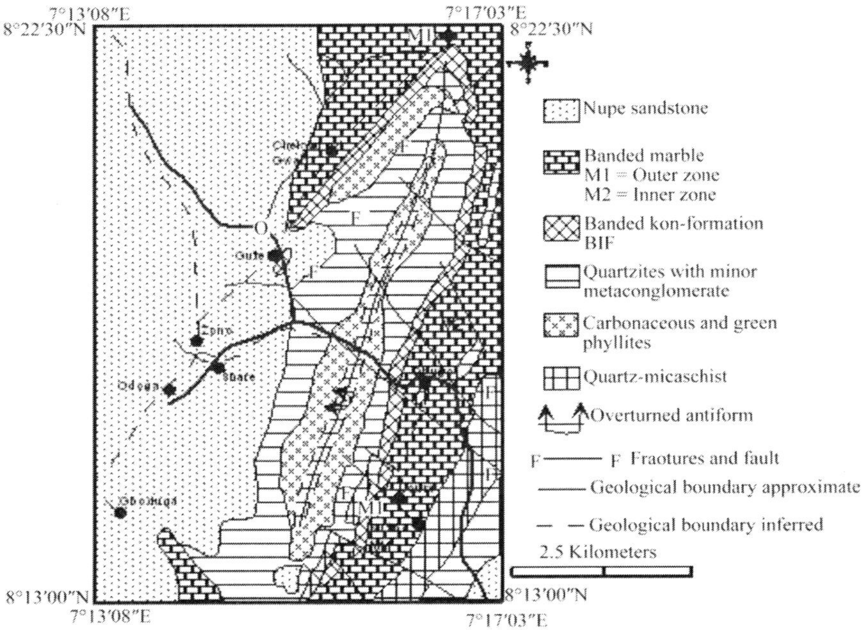

Figure 2. Geological map of Muro district, central Nigeria.

Figure 3. Photograph of Muro iron-formation oxide facies displaying dark iron-rich mesobands interbanded with lightcoloured quartz-rich ones.

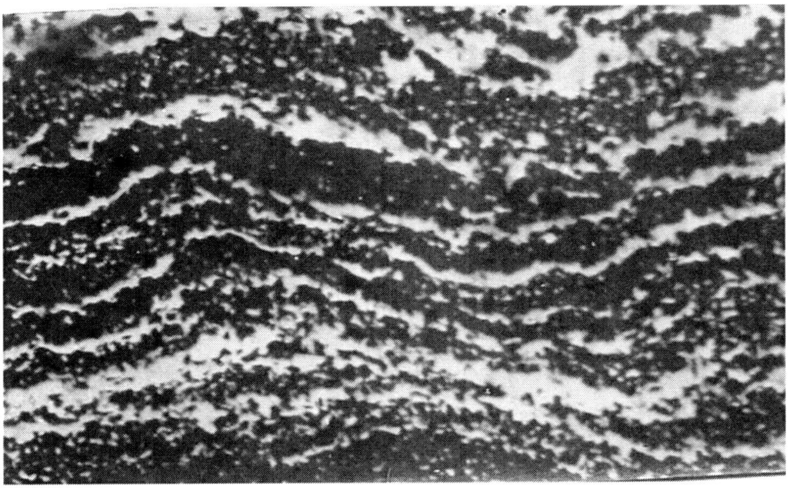

Figure 4. Photomicrograph of polished section of the oxide facies of the Muro iron-formation displaying grey, hematite-rich microbands alternating with dark, quartz-rich ones. Reflected light; Magnification ×20.

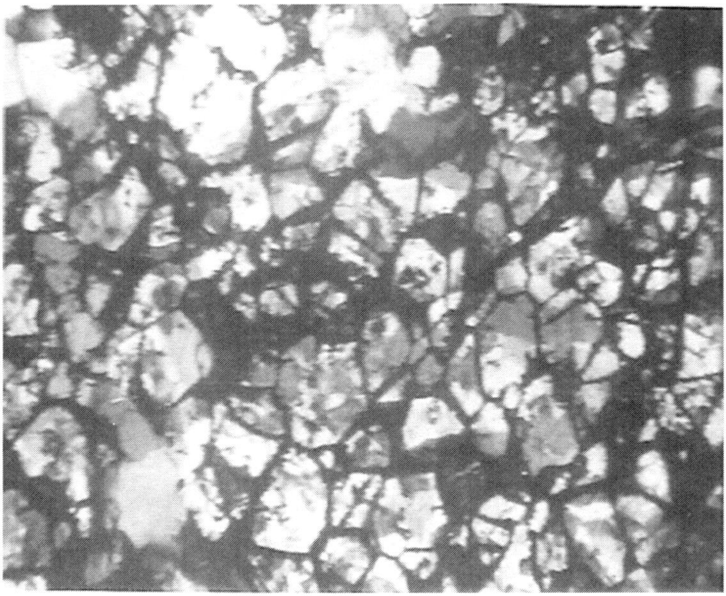

Figure 5. Photomicrograph of carbonate facies iron-formation showing siderite grains oxidized to black iron oxides at grain boundaries. Transmitted light; crossed polars; Magnification ×25.

3. ANALYTICAL METHODS

Representative samples of the two facies of the BIF were selected and analysed for both major and trace elements using a Phillips PW 1450 automatic sequential X-ray fluorescence (XRF) in the Department of Geological Sciences, McGill University, Montreal, Canada. Analyses were done on the fused beads for SiO_2, TiO_2, Al_2O_3, Fe_2O_3, MnO, MgO, CaO, Na_2O K_2O, P_2O_5, Ba, Cr, Ni, V and Zn. Other trace elements—Nb, Pb, Rb, Th, Sr, Cu, Co, U, Y and Zr—were analysed on pressed rock powder pellets by XRF spectrometry. The loss on ignition (LOI) of each rock powder was calculated after igniting to a constant weight.

4. RESULTS AND DISCUSSION

4.1. Major Element Geochemistry

Geochemical analyses (**Table 1**) have revealed the essentially bimodal composition of the iron-formation with silica (SiO_2) and total iron (as Fe_2O_3t) constituting over 85% of its components irrespective of the facies. Other components except the volatiles (LOI) occur in very small amounts. An antithetic relationship exists between the SiO_2 and total Fe_2O_3, which is reflected in the strong negative correlation between the two components (Figure 6) and in the empirical inverse distribution of quartz and iron minerals in the iron-formation. In this respect Muro iron-formation is similar to iron-formations in other parts of the world [10-12]. A plot of the Muro samples in the compositional diagram after [10] shows that they lie in the Precambrian BIF field (**Figure 7**), thus indicating that the Muro BIF is geochemically similar to other Precambrian iron-formations.

Silica (SiO_2) content in the oxide facies ranges from 50.33% - 64.50% with an average of 56.53%. Higher values obtained from the carbonate facies samples could be due to highly irregular distribution of the metachert and siderite and the difficulty in obtaining representative samples on the field. However, the generally high silica contents in all the BIF samples are attributable largely to the high metachert or quartz constituents of the iron-formation.

TiO_2 content (0.01% - 0.05%) is very low with essentially uniform distribution as most samples are in the range of 0.01% - 0.02 % (**Table 1**). Alumina content is low, ranging from 0.01% - 0.54% with an average of 0.28% for either the oxide or carbonate facies. In spite of the low concentration of alumina, it exhibits a weak positive correlation with TiO_2 (**Figure 8**). This is probably inherited from the common association of Ti and Al in clayey materials derived from residual weathering products. Both Al and Ti are hydrolysate elements [13] which are transported together by rivers to the basin of deposition [14]. The presence of clay material in the iron-formation of which alumina is an index indicates an initial clastic dilution in the basin of deposition.

Total iron as Fe_2O_3 ranges from 33.15% to 48.08%, with an average of 41.69%, in the oxide facies. The iron content of the carbonate facies is extremely variable and its total iron content (Fe_2O_3t) ranges from 3.68% - 20.66%. The relatively low values of iron in this facies are attributable to the same factors that accounted for its very high SiO_2 values as explained above. Chemical analyses (**Table 1**) have revealed high $Fe_2O_3$and low FeO values in the oxide facies in consonance with the preponderant hematite and subordinate magnetite in the rock. Similarly, the carbonate facies contains relatively high FeO and low Fe_2O_3 components, thus reflecting the presence of siderite as the main mineral in that facies.

Table 1. Major element concentrations of the oxide facies (1 - 13) and the carbonate facies (14 - 18) of the muro banded iron-formation.

Element	1	2	3	4	5	6	7	8	9	10	11	12	13	14	15	16	17	18
	Oxide Facies													Carbonate Facies				
SiO_2	59.41	53	53.01	56.23	63.27	50.33	63.64	58.73	55.54	54.06	51.05	62.18	64.5	93.87	70.47	66.54	72.86	68.39
TiO_2	0.01	0.05	0.02	0.02	0.01	0.01	0.05	0.02	0.04	0.02	0.01	0.01	0.01	0.01	0.01	0.04	0.01	0.01
Al_2O_3	0.18	0.21	0.37	0.41	0.09	0.1	0.54	0.19	0.45	0.41	0.01	0.2	0.49	0.27	0.24	0.25	0.28	0.31
Fe_2O_3	ND	36.58	ND	ND	ND	47.72	33.38	35.94	36.56	36.37	46.95	36.27	32.41	3.68	1.58	3.72	1.49	2.13
FeO	ND	8.47	ND	ND	ND	0.23	0.19	4.87	7.85	8.04	1.02	0.63	0.67	ND	14.32	15.13	12.54	16.68
Fe_2O_3t	40.04	45.99	45.9	42.62	35.7	47.98	33.59	41.35	45.28	45.3	48.08	36.97	3.15	3.68	17.49	20.53	15.42	20.66
MnO	0.08	0.05	0.08	0.03	0.04	0.06	0.12	0.06	0.05	0.03	0.04	0.07	0.1	1.06	2.75			
MgO	0.01	0.06	0.05	0.01	0.01	0.01	0.01	0.01	0.01	0.02	0.01	0.01	0.01	0.02	0.01	0.02	0.03	0.05
CaO	0.02	0.01	0.01	0.04	0.01	0.1	0.01	0.05	0.03	0.01	0.01	0.01	0.01	0.06	0.08	0.11	0.07	0.06
Na_2O	0.01	0.01	0.01	0.01	0.01	0.01	0.01	0.01	0.01	0.01	0.01	0.01	0.01	0.01	0.01	0.01	0.01	0.01
K_2O	0.01	0.07	0.03	0.07	0.01	0.01	0.01	0.01	0.04	0.02	0.01	0.01	0.01	0.03	0.05	0.08	0.03	0.04
P_2O_5	0.023	0.05	0.046	0.009	0.035	0.15	0.042	0.03	0.01	0.03	0.13	0.04	0.05	0.14	0.16	0.27	0.18	0.2
LOI	0.13	0.57	0.51	0.01	0.72	1.45	2.07	0.52	0.1	0.72	0.77	0.83	2.14	1.25	11.21	10.97	11	10.06
Total	99.89	99.12	100.03	99.44	99.87	100.14	100.04	100.41	100.67	99.73	99.98	100.24	100.51	100.39	100.88	100.22	100.41	100.26
Total Fe	28	32.17	32.1	29.81	24.97	33.56	23.49	28.92	31.67	31.68	33.63	25.86	23.18	2.57	12.23	14.36	10.78	14.45
SiO_2/Fe_2O_3	1.5	1.2	1.2	1.3	1.8	1.1	1.9	1.42	1.23	1.19	1.06	1.68	1.95	25.5	4.03	3.24	4.72	3.31
Al_2O_3/SiO_2	0.0030	0.0040	0.0069	0.0073	0.0014	0.0020	0.0085	0.0032	0.0081	0.0075	0.00019	0.0032	0.0076	0.0029	0.0034	0.0038	0.0038	0.0045

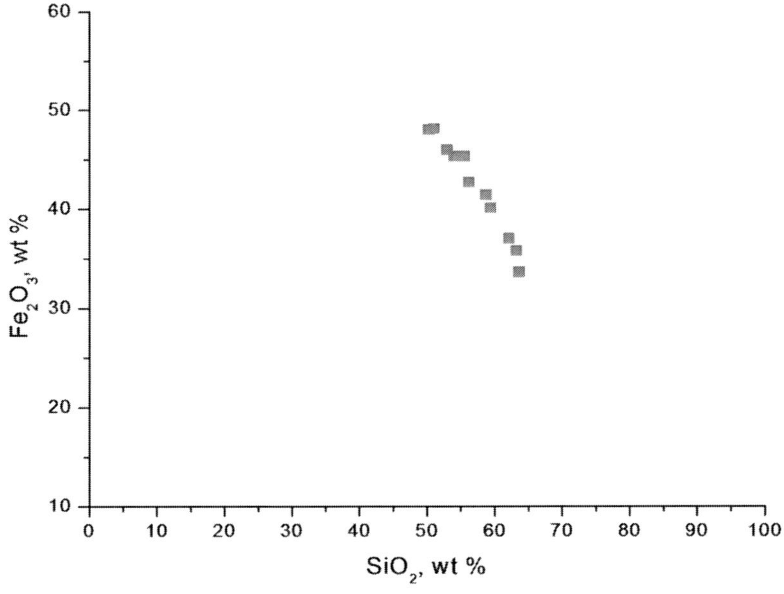

Figure 6. Plot of SiO_2 against Fe_2O_3 for the Muro oxide facies showing a negative correlation between silicon and iron.

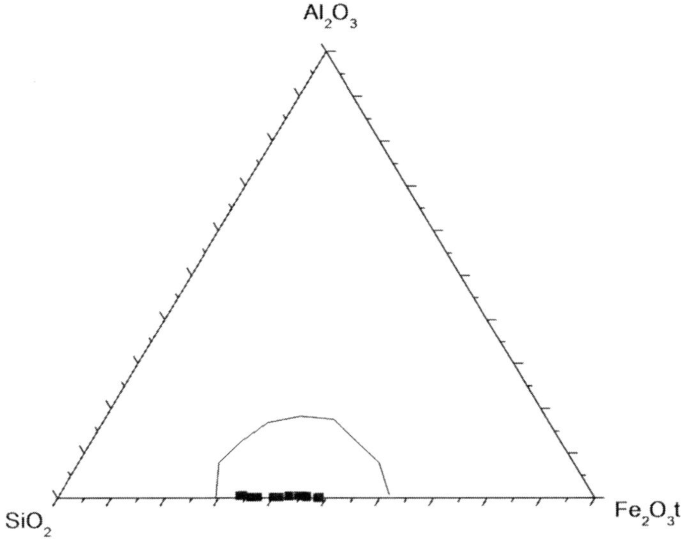

Figure 7. Al_2O_3-SiO_2-Fe_2O_3t triangular diagram showing the field of Precambrian Banded iron-formations (BIF) after Govett (1966) and the plots of the Muro oxide facies samples correlation between the two elements.

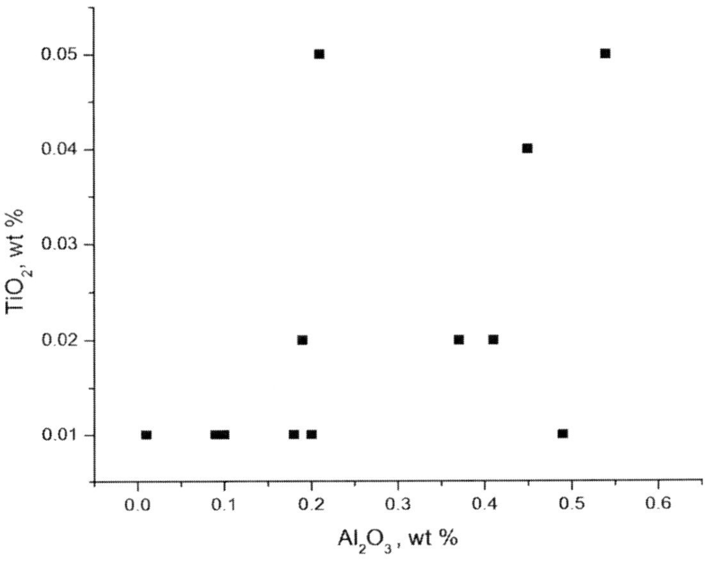

Figure 8. Plot of Al_2O_3 against TiO_2 for the Muro oxide facies showing some positive correlation albeit with some scatter.

Manganese content is low (mean 0.06%) and is distributed somewhat uniformly in the oxide facies but it is much higher (mean 2.23%) in the carbonate facies.

Very low concentrations of magnesium and calcium are present in the iron-formation irrespective of its facies. MgO content ranges from 0.01% - 0.06% and CaO content from 0.01% - 0.11%. The low concentrations of the two elements especially when combined with the low volatile contents (LOI) of the oxide facies indicate paucity of carbonate and silicate minerals in the facies [15], which is also confirmed from petrographic studies.

The alkali metal contents (Na_2O and K_2O) are very low with most samples having values of 0.01% especially in the case of sodium. This also reflects the rarity of silicate minerals in the iron deposit.

Phosphorous content (P_2O_5) is low varying from 0.01% - 0.27% in the iron-formation. Its distribution tends to be uniform in the oxide facies which also displays much lower concentration (average 0.05%) than the carbonate facies (average 0.19%).

Compared with the data from [16] and [12] the mean contents of Fe_2O_3, TiO_2, MnO, MgO, CaO, Na_2O, K_2O, P_2O_5 and Al_2O_3 in the Muro oxide facies are similar to those of the Lake Superior and the Algoma oxide facies (**Figure 9**, **Table 2**). However, the Muro oxide facies has a slightly higher SiO_2 content relative to the others (**Figure 9**).

4.2. Trace Element Geochemistry

The trace elements presented in **Table 3** are expressed in averages and ranges in **Table 4** for comparison with the average trace element contents of other iron-formations. The low values of these elements in the Muro Iron-Formation are similar to their values in Precambrian ironformations in other parts of the world [12,15,17-19].

Its oxide facies (**Table 3**, analyses 1 - 13) appears to be more impoverished in trace elements than the carbonate facies (analyses 14 - 18) which contains relatively high concentrations of Ba, Cu, Ni, Mn, Co, and V compared with even the Lake Superior-type and Algoma-type BIFs (**Table 4**). Occurring in particularly low concentrations (lower than the detection limits of the analytical instrument) are Ni, Rb, Y and Zn in the oxide facies and Rb and Zn in the carbonate facies (**Table 3**).

In general, the level of trace element concentrations in the Muro oxide facies is less than those of its Lake Superior-type and Algoma-type BIFs (**Table 4**, **Figure 10**).

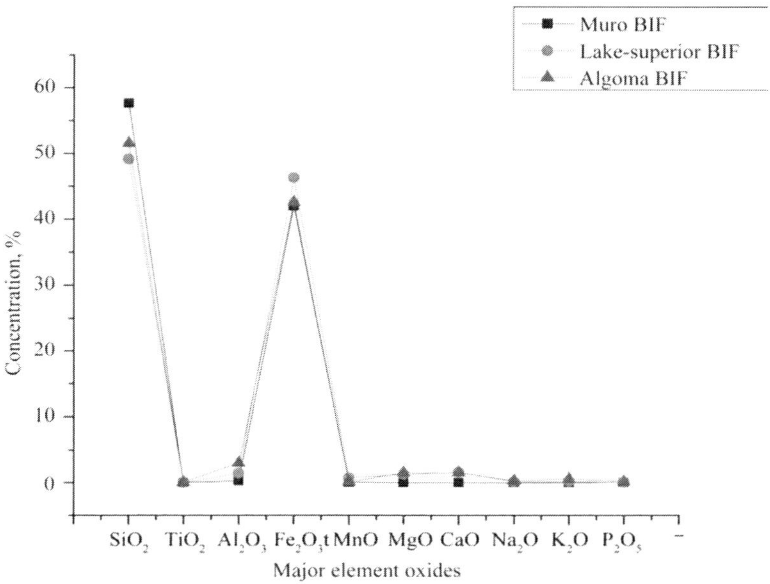

Figure 9. Spider diagram comparing the major element composition of Muro BIF with those of Lake Superior-type and Algoma-type BIFs.

Table 2. Major element summary statistics (range, mean and number of samples, n) of Muro iron-formation and some other iron-formations from different parts of the world.

	Muro Iron-Formation n = 13		Yilgarn Block Australia n = 35		Brockman Iron-Formation n = 10		Biwabik Iron-Formation n = 7		Lake Superior-Type Mean		Algoma-Type Mean	
	Range	Mean	Range	Mean	Range	Mean	Range	Mean	Oxide	Silicate	Oxide	Silicate
SiO_2	51.00 - 64.96	57.66	21.53 - 68.28	49.07	43.44 - 60.21	48.2	46.93 - 53.96	50.62	49.13	63.38	51.58	48.62
TiO_2	0.01 - 0.05	0.02	0.00 - 0.18		0.01 - 0.07	0.03	0.22 - 0.13	0.06	0.02	0.2	0.09	0.2
Al_2O_3	0.09 - 0.55	0.28	0.01 - 3.51	0.7	0.13 - 1.22	0.47	0.33 - 2.28	1.13	1.45	2.59	3.06	7.96
Fe_2O_3	34.07 - 48.62	38.29	1.98 - 37.55	18.98	17.99 - 35.91	29.09	12.00 - 26.25	20.28	36.84	9.35	27.43	16.63
FeO	0.20 - 8.60	3.58	14.37 - 66.61	23.65	13.32 - 19.64	16.49	17.44 - 26.65	21.43	8.54	17.51	13.26	19.05
Fe_2O_3t	38.71 - 48.62	42.02	ND		ND	47.41	ND	44.09	46.33	28.81	42.62	37.8
MnO	0.03 - 0.12	0.06	0.04 - 2.76	0.55	0.00 - 0.27	0.08	0.41 - 1.02	0.72	0.76	0.37	0.14	0.77
MgO	0.01 - 0.06	0.01	1.00 - 9.63	3.46	1.70 - 3.70	2.39	2.59 - 4.08	3.17	1.29	2.93	1.56	3.04
CaO	0.01 - 0.10	0.02	0.32 - 7.61	2.68	0.05 - 3.80	1.81	1.02 - 3.17	1.98	1.64	2.58	1.54	3.04
Na_2O	0.01 - 0.01	0.01	0.02 - 0.57	0.11	0.03 - 1.33	0.49	0.02 - 0.09	0.06	0.12	0.21	0.32	0.05
K_2O	0.01 - 0.07	0.02	0.00 - 0.61		0.11 - 2.85	0.71	0.07 - 0.31	0.17	0.15	0.68	0.59	0.43
P_2O_5	0.01 - 0.15	0.05	0.02 - 0.45	0.16	0.22 - 0.49	0.22	0.05 - 0.17	0.09	0.06	0.11	0.21	0.44
S	ND	ND	ND	0 - 5.20	0 - 0.06	0.03	ND		0.02	0.09	0.30	1.95
-O=S	ND	ND	ND	ND	ND	0.09	ND		ND	ND	ND	ND
Total		100		100		100		99.71	100	100	100	100

Table 3. Trace element concentrations of the oxide facies (1 - 13) and the carbonate facies (14 - 18) of the muro banded iron-formation.

Element	1	2	3	4	5	6	7	8	9	10	11	12	13	14	15	16	17	18
	Oxide Facies													Carbonate Facies				
V	10	14	21	10	12	10	17	10	19	23	10	14	22	112	98	102	80	110
Cr_2O_3	15 (10)	15 (10)	28 (19)	15 (10)	15 (10)	15 (10)	15 (10)	15 (10)	15 (10)	15 (10)	15 (10)	15 (10)	30 (21)	33 (23)	35 (24)	28 (19)	18 (12)	32 (22)
Co	ND	194	ND	ND	ND	189	142	170	154	195	140	180	130	154	148	150	130	128
Ni	10	10	10	10	10	10	10	10	10	10	10	10	10	10	10	22	26	17
Cu	14	10	10	10	264	32	10	10	10	170	28	18	10	357	208	299	345	313
Zn	10	10	10	10	10	10	10	10	10	10	10	10	10	10	10	10	10	10
Nb	5	5	5	7	6	5	7	5	5	6	7	5	6	8	8	7	8	6
Zr	44	39	41	39	38	38	40	38	40	39	42	36	38	43	40	42	44	43
Y	5	5	5	5	5	5	5	5	5	5	5	5	5	5	45	5	5	5
Sr	19	20	17	20	18	17	19	18	16	17	19	20	19	36	34	35	38	37
Rb	5	5	5	5	5	5	5	5	5	5	5	5	5	5	5	5	5	5
Pb	5	16	6	10	6	5	8	5	16	12	10	5	6	5	5	6	8	5
Th	5	12	5	6	5	5	5	5	5	6	10	5	5	5	5	5	5	5
U	10	12	8	6	9	12	7	8	6	9	10	8	7	9	9	10	8	6

Table 4. Trace element summary statistics (range, mean and number of samples, n) of Muro Iron-Formation and mean of some other iron-formations from different parts of the world.

	Muro Oxide Facies N = 13		Muro Carbonate Facies		Lake Superior Oxide Facies	Lake Superior Silicate Facies	Algoma Oxide Facies	Algoma Silicate Facies	Orissa Oxide Facies	Minas Gerais Itabirite
	Range	Mean	Range	Mean	Mean	Mean	Mean	Mean	Mean	Mean
Ba	<9 - 99	41	614 - 1144	924	180	170	170	70	10	179
Co	142 - 194	166	128 - 154	142	27	27	38	36	<10	69
Cr	<10 - 19	10	12 - 24	20	122	102	78	38	30	28.5
Cu	<10 - 264	40	208 - 357	304	10	37	96	99	10	22
Mn	232 - 264	465	8200 - 23900	17300	4600	3400	1400	7300	120	1785
Nb	<5 - 7	5	6 - 8	7	ND	ND	ND	ND	ND	ND
Ni	<10	<10	17 - 26	22	32	46	83	151	15	20.5
Pb	<5 - 16	8	<5 - 8	5	ND	ND	ND	ND	ND	ND
Rb	<5	<5	<5	<5	ND	ND	ND	ND	ND	ND
Sr	16 - 20	18	34 - 37	36	42	30	98	23	15	ND
Th	<5 - 12	5	<5 - 5	5	ND	ND	ND	ND	ND	ND
Ti	60 - 300	120	<60 - 240	60	160	1900	860	1910	40	216.6
U	6 - 12	9	6 - 10	8	ND	ND	ND	ND	ND	ND
V	10 - 23	12	80 - 112	100	30	124	97	54	30	35
Y	<5	<5	<5 - 45	8	41	43	54	84	ND	ND
Zn	<10	<10	<10	<10	2	ND	33	34	ND	ND
Zr	36 - 41	39	40 - 43	42	56	165	84	267	10	17.3

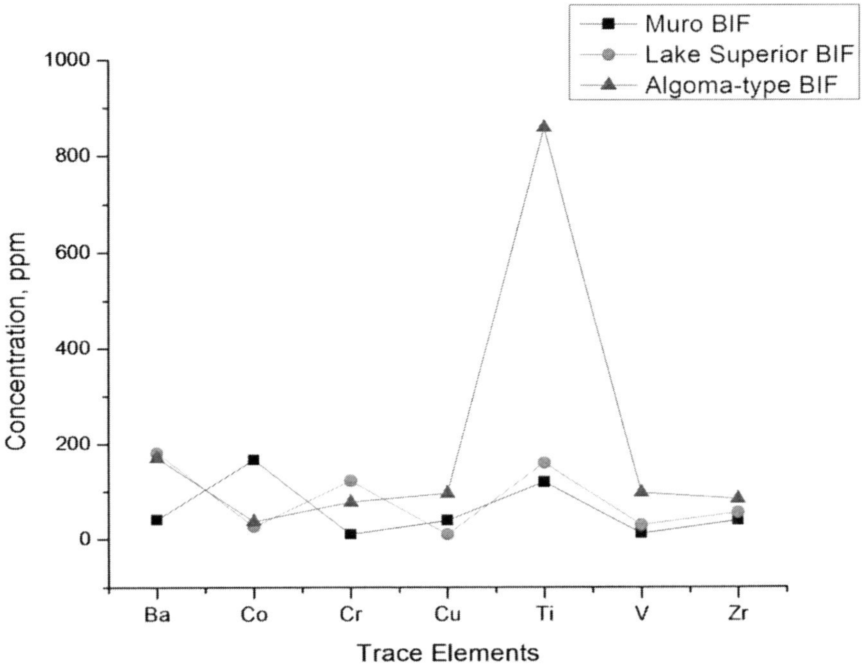

Figure 10. Spider diagram comparing the trace element composition of Muro BIF and those of Lake Superior-type and Algoma-type BIFs.

These differences in the trace element concentrations may be attributed to the level of detrital material (clay constituents) in the original iron-formation. The Al_2O_3 content, which is a measure of the detrital dilution of iron-formations [20], is less in the Muro oxide facies (0.28%) than in the Lake Superior oxide facies (3.06%) (**Table 2**). The geochemistry of the various trace elements is summarised as follows.

Chromium and Niobium: Chromium concentration is low being in the range of 10 - 19 ppm in the oxide facies and 12 - 24 ppm in the carbonate facies. Compared with those of the Lake Superior-type and the Algoma-type iron-formations (**Table 4**) the Cr concentration is particularly low but it is closer in value to those of the Minas Gerais itabirite and the Orissa BIF. Such a low Cr concentration has been reported in Swedish iron ores and BIF deposited as oxidates in sedimentary environments [17,21]. Niobium concentration is low (5 - 8 ppm) but its distrbution tends to be uniform in both the oxide and carbonate facies. Because this element has not been determined in most iron-formations it is impossible to compare its level of concentration in the Muro oxide facies with those of other iron-formations.

Vanadium: Vanadium distribution pattern in the Muro iron-formation is facies dependent. The vanadium content of the oxide facies is low (average 12 ppm) while that of the carbonate facies is comparatively higher (average 100 ppm). A similar relationship exists in the Lake Superior-type BIF (**Table 4**).

Yttrium: Yttrium concentration is very low. If one sample that gave an anomalous value of 45 ppm is ignored, all the remaining samples contain little (less than 5 ppm) or no yttrium at all.

Zirconium, Uranium and Thorium: Zirconium content is relatively low and its distribution is somewhat uniform in both the oxide and carbonate facies with its concentration ranging from 36 - 43 ppm. The average Zr contents of the Muro oxide-facies (39 ppm) and carbonate-facies (42 ppm) are less than those of their corresponding facies in the Lake Superior-type and Algomatype iron-formations. Since the distribution of this element is attributed to provenance rather than to sorting of sediments [17], its low level of concentration therefore indicates that the iron-formation is derived from low Zr source materials. This possibility excludes igneous derivation of the iron-formation but favours a sedimentary origin.

Uranium: Uranium concentration is very low (ranging from 6 - 12 ppm) with a uniform distribution pattern as both oxide and carbonate facies display similar average U contents (8 and 9 ppm, respectively).

Thorium: Thorium occurs in generally very low concentrations (5 - 12 ppm) with most samples being below the detection limit (5 ppm).

Cobalt and Nickel: Cobalt content of the Muro ironformation is high, varying from 128 - 194 ppm, and it contrasts greatly with low contents of the Lake Superiortype, Algoma-type and Orissa BIF as well as the Minas Gerais itabirite (**Table 4**). In the Muro iron-formation the cobalt concentration of the oxide facies (142 - 194 ppm) is higher than that of the carbonate facies (128 - 154 ppm). Nickel concentration is generally very low in the Muro iron-formation, ranging from <10 - 26 ppm. It is particularly low in the oxide facies (<10 ppm).

Copper, Lead and Zinc: The Muro iron-formation is relatively enriched in copper but is impoverished in lead and zinc. Cu distribution tends to be facies dependent with lower concentration in the oxide facies (average 40 ppm) compared with its higher content in the carbonate facies (average 304 ppm). A similar relationship has been found in the Isua BIF in Greenland [22].

Barium, Strontium and Rubidium: Both barium and strontium exhibit a similar distribution pattern. The concentrations of both elements are relatively high in the carbonate facies and are comparatively much less in the oxide facies (**Table 3**). The sympathetic behaviour of the two elements is attributable to the fact that Sr associates with Ba and Ca in nature [17]. Sr content is generally low in the Muro iron-formation being in the range of 16 - 37 ppm. The average Sr contents of both the oxide and carbonate facies compare favourably with those of the Lake Superior-type BIF but are appreciably less than those of the Algoma-type BIF (**Table 3**). On the other hand, the Ba content of the carbonate facies is very high (614 - 1144 ppm) with an average of 924 ppm while those of the oxide facies is not only low but highly variable (9 - 93 ppm). Rubidium is strongly depleted in the iron-formation, its value being below the detection limit (5 ppm) of the analytical equipment Generally, the trace element concentrations of the Muro oxide facies is similar to those of the Lake Superior oxide facies except for the much lower contents of Ba and Cr, and much higher contents of Co in the Muro BIF (**Figure 10**). On the other hand, the trace element contents of the Muro oxide facies differ from those of the Algoma oxide facies (**Figure 10**). The Muro BIF has significantly lower contents of Ba, Cr, Cu, Ti, V and Zr compared to the Algoma-type BIF.

5. MINERAL CHEMISTRY

Electron microprobe analysis of hematite in the oxide facies of Muro Iron Formation is presented while that of the carbonate mineral in the carbonate facies is presented. The hematite contains only minor amounts of MnO, Al_2O_3 and SiO_2, while the carbonate is dominantly rhodochrosite with minor amounts of magnesia and siderite. Both analyses reveal the minor importance of manganese in the iron formation [cf 23].

6. SUMMARY AND CONCLUSIONS

The major and trace element distribution patterns of the Muro BIF are essentially similar to those of the Lake Superior-type BIFs as described by [12,15,17]. The observed trace element distribution and abundance patterns in the iron-formation also lend support to the shallow water sedimentary deposition of the iron-formation. Lower concentrations of Cr, Ti, Ni and Zr, (which have a strong igneous affinity) in the Muro Iron-Formation than in the Algoma analogue indicate a sedimentary origin of the iron-formation. Paucity of trace elements especially of Ba, Co, Ca, Mn, Ni and Pb in the iron-formation suggests shallow water deposition [15,17]. According to [17] very low concentrations of V, Cr, Zr, Ni and Cu in BIF indicate low temperature precipitation under sedimentary conditions. The Muro BIF contains particularly low contents of these elements (see **Table**

3), thus supporting the low temperature deposition of the iron-formation in a sedimentary environment.

The weak positive correlation of TiO_2 with Al_2O_3 as well as their very low concentrations in the Muro BIF suggests only very minor terrigeneous contribution to its source materials (silica and iron) [24,25]. This observation is supported by the scarcity of silicate minerals in the mineral assemblage. Silica was deposited as chert as revealed by the occurrence of jasper and metachert in the iron-formation. The presence in the jasper of siliceous spheroidal structures which are ascribed to biogenic origin [26] suggests some biological contribution to the silica content of the BIF.

Muro Iron-Formation, which is made up of predominant oxide facies and subordinate carbonate facies, is of the Lake Superior-type. This is demonstrated by its similarity in both lithological association and geochemistry to the Lake Superior-type iron-formations. These similarities suggest common depositional environments and source materials.

7. ACKNOWLEDGEMENTS

John Adeyinka Adekoya, Chukwuemeka Tony Okonkwo, Mohammed Olatoye Adepoju We wish to express our profound gratitude to Professor Bob Martins of the Department of Geological Sciences, McGill University, Montreal, Canada for permitting the analysis of the samples to be done in the geochemical laboratories of his department. We are also grateful to colleagues in the Nigerian Geological Survey Agency and the Obafemi Awolowo University, Ile-Ife, Nigeria for their logistic support. The study was largely funded by the Federal University of Technology, Akure through the University Research Grant number URG 87-88/03.

REFERENCES

1. H. L. James, "Sedimentary Facies of Iron-Formations," Economic Geology, Vol. 49, No. 3, 1954, pp. 235-293.
2. J. A. Adekoya, "The Geology of Banded Iron-Formations in the Precambrian Basement Complex of Northern Nigeria," Ph.D. Dissertation, University of Ibadan, Ibadan, 1991.
3. J. A. Adekoya and B. O. Oladeji, "Geological and Structural Setting of the Muro Hills Banded Iron-Formation, Plateau State, Nigeria," Journal of Mining and Geology, Vol. 22, 1985, p. 161.
4. E. O. G. Muotoh, P. O. Oluyide, A. U. Okoro and O. Mogbo, "The Muro Hills Banded Iron-Formation," In: P. O. Oluyide, W. C. Mbonu, A. E. Ogezi, I. G. Egbuniwe, A. C. Ajibade and A. C. Umeji, Eds., Precambrian Geology of Nigeria, Geological Survey of Nigeria, Kaduna, 1988, pp. 219-227.

5. L. O. Anike, A. C. Umeji and A. C. Onyeagocha, "Geology and Geochemistry of the Muro Banded Iron-Formation, S. W. Plateau State, Nigeria," Journal of Mining and Geology, Vol. 17, 1990, pp. 21-26.
6. L. O. Anike, A. C. Umeji and I. P. Orajaka, "Geology of Precambrian Banded Iron Formation from Muro Hill, Nigeria," Economic Geology, Vol. 88, No. 5, 1993, pp. 1237-1241.
7. M. A. Rahaman, "Recent Advances in the Study of the Basement Complex of Nigeria," In: P. O. Oluyide, W. C. Mbonu, A. E. Ogezi, I. G. Egbuniwe, A. C. Ajibade and A. C. Umeji, Eds., Precambrian Geology of Nigeria, Geological Survey of Nigeria, Kaduna, 1988, pp. 11-43.
8. A. C. Ajibade and J. B. Wright, "Structural Relationships in the Schist Belts of Northwestern Nigeria," In: P. O. Oluyide, W. C. Mbonu, A. E. Ogezi, I. G. Egbuniwe, A. C. Ajibade and A. C. Umeji, Eds., Precambrian Geology of Nigeria, Geological Survey of Nigeria, Kaduna, 1988, pp. 103-110.
9. H. L. James, "Precambrian Iron-Formations: Nature, Origin, and Mineralogic Evolution from Sedimentation to Metamorphism," In: K. H. Wolf and C. V. Chilingarian, Eds., Developments in Sedimentology, Vol. 47, 1992, pp. 543-589.
10. G. J. S. Govett, "Origin of Banded Iron-Formation," Geological Society of America Bulletin, Vol. 77, No. 11, 1966, pp. 1191-1212.
11. M. N. Gole, "Archean Banded Iron-Formations, Yilgarn Block, Western Australia," Economic Geology, Vol. 76, No. 7, 1981, pp. 1954-1974.
12. G. A. Gross, "Geochemistry of Iron-Formation in Canada," In: J.-J. Chauvel, C. Yugi, E. M. El-Shazly, G. A. Gross, K. Laajoki, M. S. Markov, K. L. Rai, V. A. Stulchikov and S. S. Augustithis, Eds., Ancient Banded iron Formations (Regional Representations), Theophrastus, Athens, 1990, pp. 3-26.
13. B. Mason, "Principles of Geochemistry," 3rd Edition, John Wiley & Sons, New York, 1966.
14. H. Lepp and S. S. Goldich, "Origin of the Precambrian Iron-Formation," Economic Geology, Vol. 59, No. 6, 1964, pp. 1025-1060.
15. T. Majumder, K. L. Chakraborty and A. Bhattacharya, "Geochemistry of Banded Iron-Formation of Orissa, India," Mineralium Deposita, Vol. 17, No. 1, 1982, pp. 107- 118.
16. G. A. Gross and C. R. Macleod, "A Preliminary Assessment of the Chemical Composition of Iron-Formations in Canada," Canadian Miner-Alogist, Vol. 18, 1980, pp. 223 -229.
17. K. L. Rai and P. R. Paul, "Geochemistry of Banded IronFormation, Iron Ores and Associated Lithologies from Jamda-Koira Valley of Bihar, India," In: J.-J. Chauvel, C. Yugi, E. M. El-Shazly, G. A. Gross, K. Laajoki, M. S. Markov, K. L. Rai, V. A. Stulchikov and S. S. Augustithis, Eds., Ancient Banded iron Formations (Regional Representations), Theophrastus, Athens, 1990, pp. 311- 326.
18. A. F. Trendall and J. G. Blockley, "The Iron-Formations of the Hamersley Group, Western Australia, with Special Reference to the Associated Crocidolite," Geological Survey of Western Australia, Perth, 1990.
19. H. L. James, "Data of Geochemistry," 6th Edition, US Geological Survey, Professional Paper 440-W, 1966.
20. R. L. Stanton, "Ore Petrology," McGraw Hill Book Co., New York, 1972.
21. S. Landergren, "On the Geochemistry of Sweedish Iron Ore and Associated Rocks," Sveriges Geologika Undersokning, Vol. 42, No. 5, 1948, pp .1-182.
22. P. W. U. Appel, "Geochemistry of the Early Archean Isua Iron-Formation, West Greenland," In: P. W. U. Appel and G. L. La Berge, Eds., Precambrian Iron-Formations, Theophrastus, Athens, 1987, pp. 31-68.
23. A. Mucke, "The Nigerian Manganese-Rich Iron-Formations and Their Host Rocks—From Sedimentation to Metamorphism," Journal of African Earth Sciences, Vol. 41, No. 5, 2005, pp. 407-436.
24. C. A. Spier, S. M. B. de Oliveira, A. N. Sial and F. J. Rios, "Geochemistry and Genesis of the Banded Iron Formations of the Caue Formation, Quadrilatero Ferrifero, Minas Gerais, Brazil," Precambrian Research, Vol. 152, No. 3-4, 2007, pp.170-206.

25. E. Pecoits, M. K. Gingras, M. E. Barley, A. Kappler, N. R. Posth and K. O. Konhauser, "Petrography and Geochemistry of of the Dales Gorge Banded Iron Formation: Paragenetic Sequence, Source, and Implications for Paleo- Ocean Chemistry," Precambrian Research, Vol. 172, No. 1-2, 2009, pp. 163-187.
26. J. A. Adekoya, "Relict Sedimentary Features in Metamorphosed Proterozoic Banded Iron-Formations, Northern Nigeria," Nigerian Journal of Science, Vol. 36, No. 2, 2002, pp. 129-138.

CHAPTER 8

Natural Gas Geochemistry in the Offshore Nile Delta, Egypt

Mohamed Abdel-Aziz Younes[1]

[1] *Geology Department (Moharrem Beck), Faculty of Science, Alexandria University, Alexandria, Egypt*

ABSTRACT

The offshore Nile Delta basin is considered as one of the most promising province in Egypt which has an excellent potential gas and condensate reserves for future exploration. This study aimed to characterize the origin of natural gas and to determine whether it is biogenic, thermogenic or mixed processes which are considered the dominant generation pathways. Regional geochemical studies were conducted on 20 test gas samples produced from the Miocene and Plio-Pleistocene reservoirs of Abu Qir, N. Abu Qir, Temsah, Wakkar and Port Fouad fields representing the western and the eastern Nile Delta province. The geochemical analyses revealed that the main constituents of the natural gases are methane (96.2%–99.37%) with minor contribution from ethane and propane. Methane carbon isotopic composition ranges between _65.6‰ and _40.3 ‰ PDB displaying a strong indication for gas mixing of thermogenic and early microbial methane. The gas chromatography_mass spectrometry carried out on the produced natural gases exhibits high oleanane index that ranges between 19% and 42%, the medium concentration of moretane index between 11% and 16% and the absence of gammacerane index indicating that the natural gases were derived from siliciclastic source rocks containing type III kerogen of terrestrial origin and higher land plants input of Tertiary age. The calculated maturity parameters of the studied natural gas proportions based on various sterane isomerisation distributions, i.e. C29αßß/(αßß+ααα)_and C29ααα20S/(S+R) reached 0.6 and 0.5 respectively indicating a medium stage of thermal maturation equivalent to the main peak of oil generation window (0.85 Ro%). The sterane isomerization ratios may reflect the rapid rate of subsidence and sedimentation in the Nile Delta and appear to have been generated during the early stage of source rock maturation.

Keywords: biogenic, thermogenic, Nile Delta, biomarkers, Egypt

1. INTRODUCTION

1.1. Location of The Study Area

The Nile Delta and the offshore Mediterranean Sea basin are hydrocarbon rich provinces that have generated natural gas and condensate from siliciclastic reservoirs ranging in age from Plio_Pleistocene to Miocene. Currently, these areas are the most active exploration and development province in Egypt. The proven huge reserves of the Pliocene gas discoveries made in the last decade firmly established the Pliocene sequence as a primary hydrocarbon potential target for the exploration activities in the Nile Delta and offshore Mediterranean region.

The study area lies in the offshore Mediterranean Sea to the north of Alexandria and Port Said cities within the northern eastern Nile Delta between longitudes 30° 10′- and 32° 54′-E and latitude 31° 25′- and 32° 00′- N. This area includes four giant natural gas fields namely; Abu Qir Field located to the north of Alexandria city and Temsah, Wakkar and Port Fouad fields located 65 kilometers to the north of Port Said in the offshore Mediterranean Sea, about 30 km from the Egyptian coast (Figure 1).

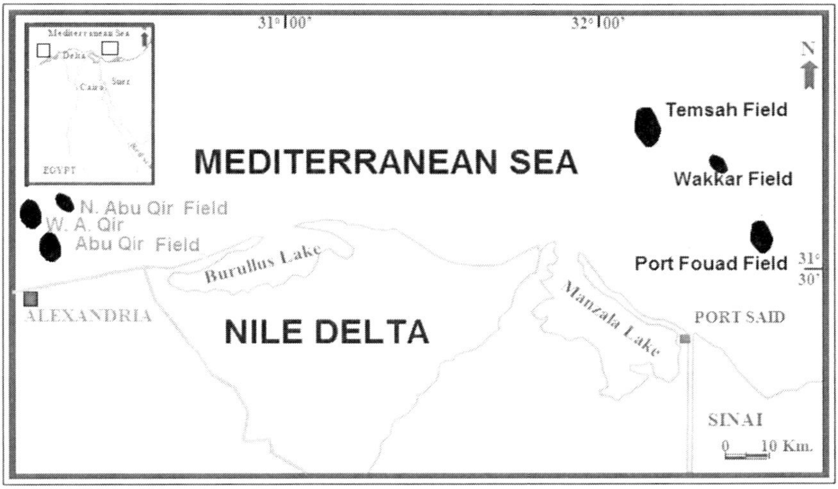

FIGURE 1. Location map of the study gas fields, offshore Nile Delta, Egypt.

1.2. Geologic Setting and Lithostratigraphy

The present Nile Delta lies on the north_external margin of the African plate. Its structural evolution is largely contemporaneous with the break–up of the African plate margin consequent to the opening of the Red Sea and the northward transition of the Arabian Peninsula [1, 2]. The northern part of the Nile Delta is characterized by a series of major tectonic features with different orientations. Some of them are active during the deposition of Tertiary section while other flatten with depth, become horizontal and dies on top of the Rosetta Formation affecting only through the Kafr El-Sheikh Formation.

The Nile Delta is characterized by asymmetric fold, overthrust faults and diapers. Its dates back to the Syrian Arc system having an arcuate trend from northeast to southwest through the northern part of the Nile Delta to the Western Desert of Egypt (Figure 2).

The regional subsurface stratigraphic succession of the Nile Delta has been represented in Figure 3. This succession is discussed by many authors [3-11].

The lithostratigraphy of the Nile Delta is represented by Miocene sediments:

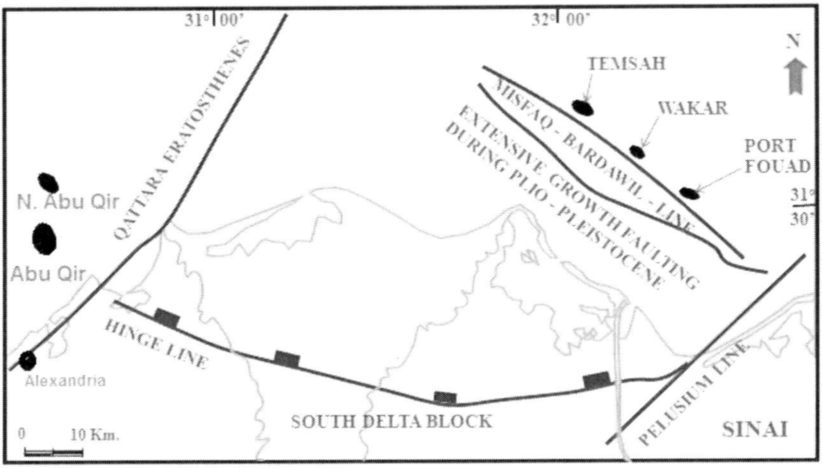

FIGURE 2. Regional structural setting, north Nile Delta from Mosconi et al. [38] and Hemdan et al. [39].

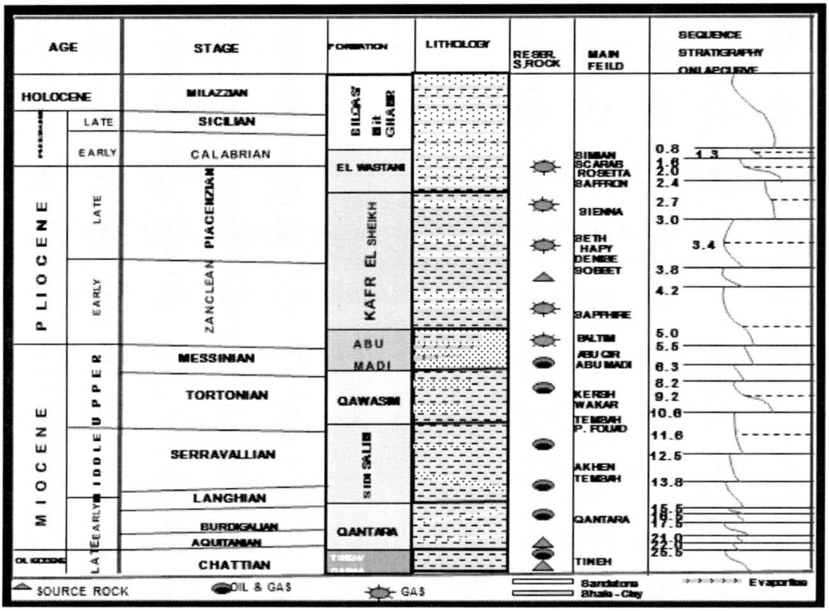

FIGURE 3. Lithostratigraphic succession of Northern Nile Delta, Egypt modified from Kamel et al. [8].

Moghra, Sidi Salem and Abu Madi formations (from bottom to top). In addition, the Pliocene–Pleistocene sequences are represented mainly by shale or clay with sandstone interbeds. The Miocene sequence is summarized as follows:

1.2.1. Moghra Formation (Early Miocene)

Moghra Formation consists of marine to fluvio–marine deposits (shale and sandstone) of presumed deltaic origin, although it is associated with marine carbonate intervals. This formation is unconformably overlain by Sidi Salem Formation.

1.2.2. Sidi Salem Formation (Middle Miocene)

The lower limit of this formation is not known in the central part of the Nile Delta where it overlies the Moghra Formation to the west and south. Sidi Salem Formation is composed mainly of mudstone with interbeds of marls, sandstones and siltstones. This formation corresponds to a lower neritic slope environment. Sidi Salem Formation is unconformably overlained by Abu Madi Formation.

1.2.3. Abu Madi Formation (Late Miocene)

Abu Madi Formation is composed of interbedded sandstones and mudstones. This formation was deposited as an incised valley fill at the end of the Late Miocene. The depositional environment of Abu Madi Formation changes from fluvial to lagoonal with marine influence increasing upward.

The Plio- Pleistocene sequences are represented by the following formations.

1.2.4. Kafr El Sheikh Formation (Early–Middle Pliocene)

Kafr El Sheikh Formation consists of mudstone sequence with thin limestone and sandstone interbeds. This formation extends all over the Delta area. In the Nile Valley it is accumulated at outer neritic to upper bathyal water depths, following a major lower Pliocene transgression resulted in the flooding of the former Eonile canyon and transforming it into a long, narrow gulf far south Aswan [12]. Rizzini et al. [4] suggested that-the Kafr El Sheikh Formation is accumulated as neritic mudstones on the present onshore delta- and in a basinal setting offshore. They suggested that the sands incorporated into this formation are essentially storm sands. This formation is conformably overlain by El Wastani Formation.

1.2.5. El Wastani Formation (Late Pliocene)

El Wastani Formation consists of thick sand beds interbedded with thin clay beds which become thinner toward the top of the formation. This formation is transitional between the shelf facies of the underlying Kafr El Sheikh Formation and the coastal and continental sands of the overlying Mit Ghamr Formation. It shows large forests due to the progradation and could also contain deltaic deposits.

1.2.6. Mit Ghamr Formation (Late Pliocene)

Mit Ghamr Formation consists of thick layers of sand and pebbles which show clay interbeddings of limited thickness. The depositional environment of this formation is probably shallow marine to fluvial. It is a typical fill-up basin with coastal sands, coquina beds, clay and peat. It conformably underlies the Bilqas Formation.

1.2.7. Bilqas Formation (Holocene)

Bilqas Formation is the top sedimentary cover of the Nile Delta province. It consists of sand interbedded with clay rich in pelecypod, gastropod and ostracod fragments. The clays contain many fragments of vegetal matter and peat deposits. The deposition occurred most likely in lagoons and brackish swamp, interrupted by beach sands.

During the last two decades, the Nile Delta and the offshore Mediterranean Sea have been intensively investigated by the petroleum industry. Till now over 36 TCF of gas have been discovered since drilling began in 1966 [13]. The geologic history of the Nile Delta created multiple source, reservoir and seal combinations. Main source rocks are believed to occur in Late Mesozoic and Oligocene to Miocene sedimentary succession.

2. SAMPLING AND ANALYTICAL TECHNIQUES

A total of twenty gas samples were obtained through drill stem testing or modular dynamic testing from 20 wells located in the offshore area of the eastern and western Nile Delta fields namely, Abu Qir and North Abu Qir from the western Nile Delta while Temsah, Wakkar and Port Fouad representing the eastern Nile Delta-province and the different reservoir stratigraphic ages namely Miocene and Plio-Pleistocene.

The isotopic composition of twenty gas samples was determined in BGR (Bundesanstalt für Geowissnschaften und Rohstoffe) lab in Hannover. Gas chromatograph (GC Varian CP-3800) was equipped with a Flame Ionization Detector and a Thermal Conductivity Detector for separation of gas components (C1–C5) using three capillary columns (First, Silica PLOT 30 m×0.32 mm, Second, CP-Sil 5CB 30 m x 0.32 mm). Permanent gases (atmospheric gas components) were separated using three packed columns (Molsieve-12× 1.5 m ×1/8 in., Hayesep Q 0.5 m ×1/8 in. nickel, and a Hayesep T 0.5 m ×1/8 in. nickel). All six columns are placed in one oven, heated to 50 °C at the start of the analysis. After 10 min, temperature was increased at 10 °C/ min up to 180 °C, which is held for another 10 min. Finally the oven temperature was reduced to become 50 °C.

Measurements of $\delta\ ^{13}C$ and δ D on the hydrocarbon gases were conducted with a Thermo Finnigan Delta plus XL mass spectrometer. Gas components were separated on a gas chromatograph and injected into the mass spectrometer. Isotope values are reported in the δ notation in per mil (‰) relative to the common PDB and SMOW standards.

Data of analysis including gas chromatography, gas chromatography-mass spectrometry were conducted through Stratochem and Corex Laboratories (New Maadi, Cairo).

3. RESULTS AND DISCUSSIONS

3.1. Bulk Geochemical Characteristics of Natural Gas

The abundant composition of the natural gases is dominated by methane ranging between 92.72% and 99.96%. The natural gas wetness denoted by the formula C2+/∑(C1-C5 ×100 and cumulative contents of the higher hydrocarbons from ethane up to pentane range from 0.05% to 6.76%. According to Schoell [14] the natural-gas may be classified as wet if it exhibits a wetness index greater than 5%. The natural gas samples examined from the western Nile Delta can therefore be classified as wet gas, while the major number of gas samples from the eastern Nile Delta is classified as dry gas except the samples produced from the two wells PSE-1and DEN-1. Wet gas is thermogenic in origin but associated with condensate in most cases while the dry gas may have either a microbial or late mature thermal origin [14] or may be related to that microbial alteration occurring within the reservoirs through the preferential removal of the C3+ components [15].

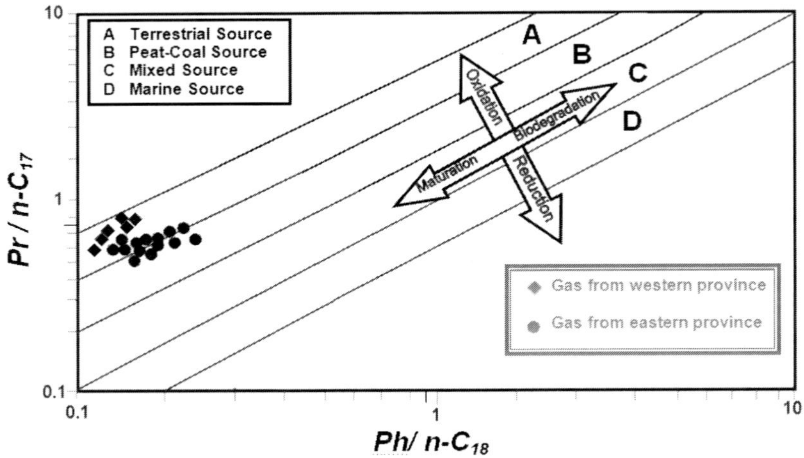

FIGURE 4. Relationship between isoprenoids Pr/n-C_{17} versus Ph/n-C_{18} showing source and depositional environments of the natural gas samples from the two provinces of the Nile Delta from Shanmugam[**17**]. All the study gas samples are located within the terrestrial to peat coal source depositional environments.

The predominance of pristane over phytane (Pr/Ph) ratio was used as an indicator of source rock depositional environments [16]. The ratio of Pr/Ph was ranged between 2 and 3 for the natural gas produced from the eastern province indicating fluviomarine and coastal swamp environment. The same ratio was found to be in the range between 6.4 and 9.4 which indicates peat swamp depositional environment. Figure 4 summarizes the achieved results concerning the nature of the source rock depositional environments is strongly supported by the plot of the isoprenoid ratios Pr/n-C_{17} versus Ph/n-C_{18} [17]. It can be seen from this figure that both the natural gases produced from the eastern and western provinces of the offshore Nile Delta and Mediterranean Sea are located within the region of terrestrial to peat coal source depositional environments. The high odd even carbon preference index (CPI > 1) for the studied natural gas samples from the two provinces reflect the generation from source facies dominating terrigenous and wax-rich components [18].

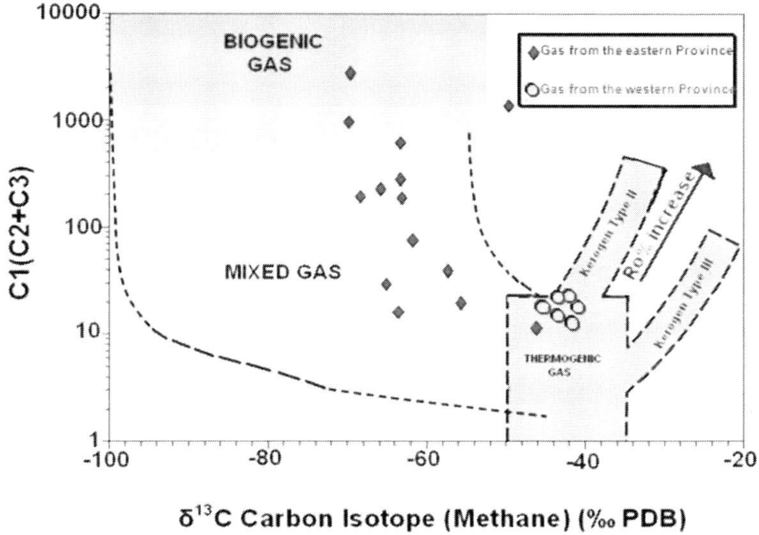

FIGURE 5. Genetic characterization of natural gases using carbon isotopes and C1 / C2 + C3 ratios modified after Bernard et al.[29] and Faber and Stahl [30].

3.2. Source and Maturation Dependent Biomarker Distributions

The results of organic geochemical analyses of twenty gas samples representing the western Nile Delta province namely, Abu Qir and North Abu Qir fields and the eastern Nile Delta province, namely Temsah, Wakkar and Port Fouad fields. The results of gas chromatogram mass fragmentogram of the study gas samples show a predominance of C_{29} stereoisomers > 50% over C_{27} regular steranes between 20% and 27% suggestings a significant input of terrestrial organic matter [19, 20]. Also, the achieved result is strongly supported by the relative abundance of diasterane index that ranges between 55% and 69% of the study gas samples.

All the study gas samples that exhibit higher oleanane index ranging from 19% and 42% strongly support the enrichment of angiosperm higher land plants input to the siliciclastic source kerogen which is thought to be derived from Late Cretaceous to Tertiary age [21, 22]. The medium concentration of moretane index in all the study samples ranges between

11% and 16% and the absence of gammacerane index strongly supports a terrestrial input [23, 24].

Biomarker maturity parameters, include the sterane isomerization ratios-C29 ααα20S/(S+R), and C29 αßß/(αßß+ααα) which according to Seifert and Moldowan [25] are genetically related to the effect of thermal maturity processes. The average sterane isomerization ratios C_{29}ααα 20*S*/(*S*+*R*) and C_{29}αßß/(αßß+ααα) of natural gas are 0.6 and 0.5 respectively indicating a medium stage of thermal maturation equivalent to the main peak of oil generation window (0.85 Ro%) [18]. The medium sterane isomerization ratios may reflect the rapid rates of subsidence and sedimentation in the Nile Delta and appear to have been generated during the early stage of source rock maturation from type III kerogen [26].

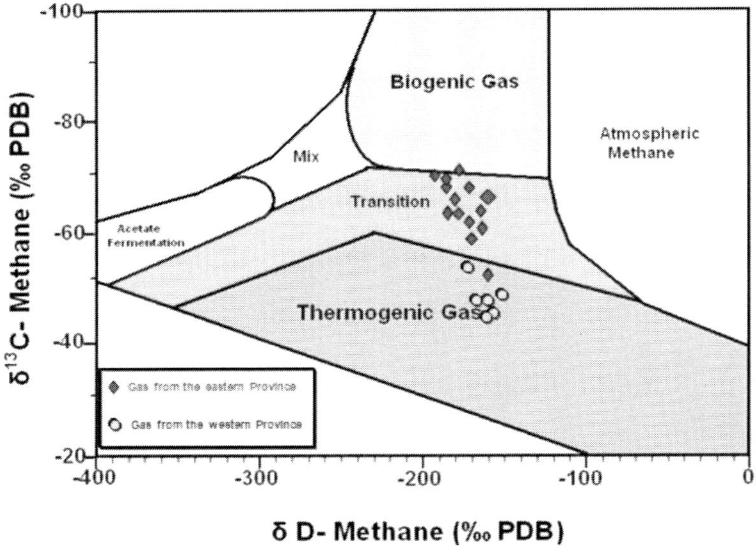

FIGURE 6. Carbon and Hydrogen isotope ratios of natural gases. Genetic fields according to Schoell[27, 31] and Whiticar et al. [32].

3.3. Relation Between Isotopic Composition and Origin of Natural Gas

The isotopic composition of natural gas depends on the type and maturity of source rock [27, 28]. Stable isotope geochemistry helps to differentiate the original gas derived from a single source rock. The methane carbon isotope compositions range between -65.6 ‰ and -41.1‰. Methane hydrogen isotope data ranges from -193 ‰ to -149‰. The isotopically lightest δ ^{13}C sample was obtained from the SEM-1 well, while the heaviest δ ^{13}C value was received from AQ-16 well. Many of the δ ^{13}C values have intermediate positions between these end members, which are consistent with what is commonly interpreted as mixed microbial and thermogenic gas [14, 28]. A general increase in the δ ^{13}C values with depth from Miocene to Plio-Pleistocene reservoir is consistent with the increasing thermal maturity and the decreasing relative concentration of biogenic gas.

The wet gas (C2+) components also display a significant variability in their isotopic compositions. The δ ^{13}C of ethane (δ $^{13}C2$) ranges from -37.5‰ to -27.9‰ but that of propane (δ $^{13}C3$) ranges from -32.2‰ to -11.1‰.

Natural gases become isotopically heavier and contain relatively more methane with increasing thermal maturity. Thermogenic methane is generally enriched in δ ^{13}C compared with microbial methane [27].

Figure 5 represents the Bernard plot in which the molecular ratio C1/ (C2 + C3) is plotted versus methane's stable carbon isotope ratio according to Bernard et al.[29] and Faber and Stahl [30]. The relationship between these two parameters allows distinguishing microbially generated methane from thermogenic hydrocarbon gases. The figure provides a strong indication for gas mixing of thermogenic and early microbial methane.

The δ ^{13}C methane versus δ D methane plot (Figure 6) represents a scheme for the carbon and hydrogen isotope ratios of methane and their genetic implications. Although this diagnostic plot was initially established for determining the thermogenic origin of methane [31] it is useful for the discrimination of microbial methanogenesis, i.e. carbon dioxide (CO_2) reduction versus acetate fermentation [32]. The δ ^{13}C values between -65.6 ‰ to -41.1‰ and δ D values between -193 ‰ and -149‰ obtained for the methane of the samples from this study classify methane as mixtures between thermogenic and biogenic origin. Figure 7 gives a good matching with the published regional gas data. The relatively heavy hydrogen isotopic compositions (Figure 7) suggest that the microbial gas proportions are probably generated through CO_2 reduction [31, 32].

In the offshore Nile Delta and Mediterranean Sea, the Oligocene to Early Miocene sedimentary sequence is considered as the primary source of natural gas and condensate characterized as mixed type II/III kerogen [8, 26].

The isotopic composition of the individual gas components is a function of thermal maturity of the generated source rock kerogen. Stahl [33] proposed an empirical relationship between methane carbon isotopic composition and vitrinite reflectance. Berner et al. [34] and Berner and Faber [35] illustrated a model between the carbon isotopic compositions of methane,

ethane, and propane and source rock maturities represented by vitrinite reflectance measurements. In this study we used the isotope maturity models developed by Berner and Faber [35] to estimate the type and maturity of the precursor (source) material (Figures 7and 8). The geochemical characteristics of natural gas from the Nile Delta indicating a derivation from source rocks rich in terrestrial precursors [8, 36, 37]. Consequently, the admixture of light microbial methane to the thermogenic hydrocarbon gas takes place during secondary processes, such as migration, microbial oxidation, or mixing. These processes usually affect the gas properties within the reservoir [28, 40].

Isotopic compositions of ethane and propane from the west and east Nile Delta samples proved their mixing origin and derivation from terrestrial organic matter with a source rock maturity between 0.1% and 2.0% Vitrinite Reflectance measurements.

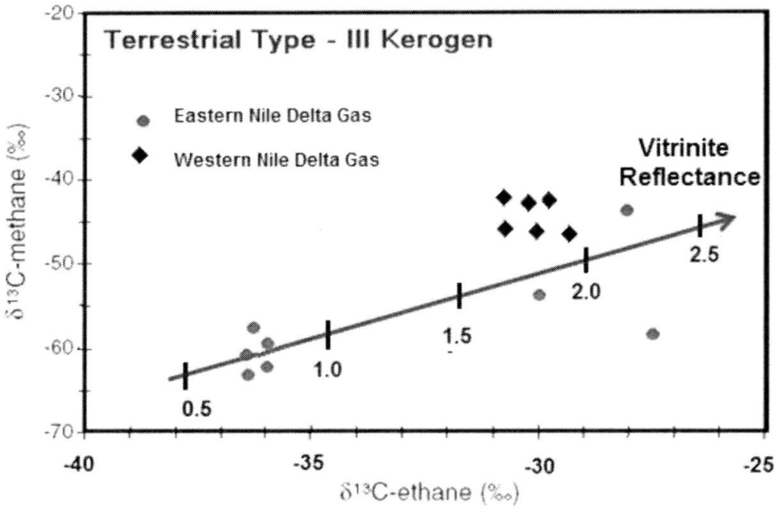

FIGURE 7. Isotopic composition of methane and ethane in natural gases of the western and eastern Nile Delta. The maturity line is according to Berner and Faber [**35**]. The terrestrial organic matters indicate the maturity of the source rock in vitrinite reflectance measurements

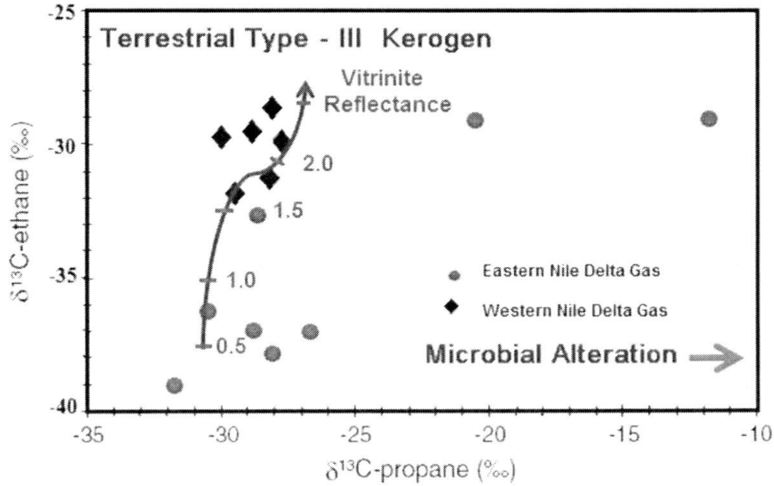

FIGURE 8. Isotopic composition of ethane and propane in natural gases of the western and eastern Nile Delta. The maturity line is according to Berner and Faber [35].

4. SUMMARY AND CONCLUSIONS

The molecular and isotopic composition of twenty natural test gas samples produced from the Miocene and Plio-Pleistocene reservoirs of Abu Qir, N. Abu Qir, Temsah, Wakkar and Port Fouad fields representing the western and the eastern Nile Delta province were used to examine the generation and alteration of natural gas and condensate accumulations. The geochemical analyses revealed that the main constituents of the natural gases are methane 96.2–99.37% with minor contribution from ethane and propane. Methane carbon isotopic composition ranges from -65.6‰ to -40.3‰ PDB displaying a strong indication for gas mixing of thermogenic and early microbial methane. The gas chromatography_mass spectrometry carried out to the produced natural gases exhibit high oleanane index that ranges between (19% and 42%), a medium concentration of moretane index between 11% and 16% and the absence of gammacerane index and predominance of C_{29} stereoisomers > 50% over C_{27} regular steranes (between 20% and 27%) and a high diasterane index that ranges between 55% and 69%. The above mentioned results pointed out that the natural gases were derived from siliciclastic source rocks containing type-III kerogen of terrestrial origin and higher land plants input from Late Cretaceous to Tertiary age. The calculated maturity parameters based on various sterane isomerisation ratios, i.e.$C_{29}\alpha\beta\beta/(\alpha\beta\beta+\alpha\alpha\alpha)$and $C_{29}\alpha\alpha\alpha 20S/(S+R)$ reached to 0.6 and 0.5 respectively indicating a medium stage of thermal maturation equivalent to the main peak of oil generation window (0.85 Ro%). The medium sterane isomerization ratios may reflect the rapid rates of subsidence and sedimentation in the Nile Delta and appears to have been generated during the early stage of source rock maturation.

5. ACKNOWLEDGEMENTS

The author wish to express the deepest gratitude to the management of the Western Desert Operating Petroleum Company (Wepco) and Belayim Petroleum Company (Petrobel) for providing test gas and condensate samples to accomplish this work. Sincere thanks and gratitudes are also due to the Egyptian General Petroleum Corporation (EGPC) for granting the permission to publish this paper.

REFERENES

1. Said R., (1962) The Geology of Egypt: Elsevier, Amsterdam, pp. 337
2. Said R., (1981) The geology and evolution of the River Nile: Springer, New York, N.Y., 151 p.
3. Ross, D. A., and Uchupi E., (1977) Structure and sedimentary history of southern Mediterranean Sea-Nile cone area: Am. Ass. Petrol. Geol. Bull., Vol. 61, p. 872–902.
4. Rizzini, A., Vezzani, F., Cococcetta V., and Milad, G., (1978) Stratigraphy and Sedimentation of A Neogene-Quaternary section in the Nile Delta area, (A.R.E.). Mar. Geol., vol. 27, p.327–348.
5. Zaghloul Z. M., EL Shahat A., and Hegah, O., (1980) Mineralogy of the Tertiary–Quaternary subsurface sediments, West Nile Delta area. Egypt. Jour. of Geology, vol. 24, p.177–188.
6. Helmy, M., and Fouad, O., (1994) Prospectivity and play assessment of Abu Qir area, Nile Delta, Egypt. EGPC, 12th Petroleum Exploration and Production Conference, Cairo, vol. 1, 17, pp. 276–292.
7. Sarhan, M., Talaat, M., Barsoum, K., Bertello, F., and Nobili, M., 1996. The Pliocene play in the Mediterranean offshore: Structural setting and growth faults controlled hydrocarbon accumulations in the Nile Delta basin. A comparison with the Niger Delta basin. EGPC, 13th Petroleum Exploration and Production Conference, vol. 1,7, p. 121–139.
8. Kamel, H., Eita, T. and Sarhan, M., (1998) Nile Delta hydrocarbon potentiality. EGPC, 14th Petroleum Exploration and Production Conference, vol. 2, pp. 485–503.
9. El Barkoky, A., and Helal, M., (2002) Some Neogene stratigraphic aspects of the Nile Delta. Mediterranean offshore conference. MOC, 2002.

10. Dewidar Kh. and Frihy, O. (2007) Pre- and post-beach response to engineering hard structures using Landsat time-series at the northwestern part of the Nile Delta, Egypt : Journal of Coast Conservation vol. 11: pp. 133–142.
11. Dewidar Kh. and Frihy, O. (2010) Automated techniques for quantification of beach change rates using Landsat series along the North-eastern Nile Delta, Egypt: Oceanograpic and Mar. Sci., Vol. 1(2) pp. 028-039.
12. Chumakov, I. S., (1967) Pliocene and Pleistocene deposits of the Nile Vally in Nubia and Upper Egypt. Doki. Akad. Nauk. SSSR., p. 5–170.
13. Dolson, J.C., Boucher, P.J., Dodd, T., Ismail, J., (2002) Petroleum potential of an emerging giant gas province, Nile delta and Mediterranean Sea o. Egypt. Oil and Gas Journal, 32–37.
14. Schoell, M., (1980) The hydrogen and carbon isotopic composition of methane from natural gases of various origins. Geochimica et Cosmochimica Acta v.44, pp. 649–662.
15. James, A.T., Burns, B.J., (1984) Microbial alteration of subsurface natural gas accumulations. Am. Ass. Petr. Geol. Bulletin v. 68, pp.957–960.
16. Lijambach, G.W., (1975) On the origin of petroleum. Proceeding of the 9th World Petroleum Congress, Applied Science Publisher, London. 2, 357-369.
17. Shanmugam, G. (1985) Significance of coniferous rain forests and related oil, Gippsland Basin, Australia. Am. Ass. Pet. Geol. Bulletin, 69, 1241-1254.
18. Peters, K. E. Snedden, J. W., Sulaeman, A., Sarg, J. F. and Enrico, R. J. (2000) A new geochemical sequence stratigraphic model for the Mahakam delta and Makassar slope, Kalimantan, Indonesia: Am. Ass. Pet. Geol. Bulletin, 84, 12-44.
19. Huang, W.Y. and Meinschein, W.G., (1979) Sterols and ecological indicators: Geochim. et Cosmochim. Acta, v. 43, p.739-745.
20. Huang, H. P. (2000) The nature and origin of petroleum in the Chaiwopu Sub-Basin (Jungar Basin) N. W. China. Journal of Petroleum Geology, v. 23, pp. 193-220.
21. Ekweozor, C. M., Okogun, J. I., Ekong, D.E.U. and Maxwell, J. R. (1979) Preliminary organic geochemical studies of samples from the Niger Delta (Nigeria). Analyses of crude oils for triterpanes. Chemical Geology, 27, 11-29
22. Moldowan, J.M., Dahl, J., Huizinga, B. and Fago, F., (1994) The molecular fossil record of oleanane and its relation to angiosperms. Science 265, 768-771.
23. Connan, J., Bouroullec, J., Dessort, T. D. and Albrecht, P. (1986) The microbial input in carbonate anhydrite facies of sabkha environment from Guatemala; A molecular approach In : Leythaeuser, D. and Rullkotter, J. (Eds.) Advances in Organic Geochemistry, 1985. Oxford-Pergamon, 20-50.
24. Mann, A. L., Goodwin, N. S. and Lowes, S. (1987) Geochemical characteristics of lacustrine source rocks: A combined palynological - molecular study of Tertiary sequence from offshore China. In Proced. Indonesian Petroleum Association, 16th Annual Convention, Jakarta, 241-258.
25. Seifert, W.K. and Moldowan, J. M. (1981) Paleoreconstruction by biological markers. Geoch. et Cosmoch. Acta, 45, 783-795.
26. Sharaf, L.M. (2003) Source rock evaluation and geochemistry of condensates and natural gases, offshore Nile Delta, Egypt. Journal of Petroleum Geology, V.26(2), pp. 189-209.
27. Schoell, M., (1984) Stable isotopes in petroleum research. In: Brooks, J.B., Welte, D. (Eds.), Advances in Petroleum Geochemistry, vol. 1. Academic Press, London, pp. 215–245.
28. Whiticar, M.J., (1994) Correlation of natural gases with their sources. In: Magoon, L.B., Dow, W.G. (Eds.), The Petroleum System From Source to Trap, vol. 60. Am. Ass. Pet. Geol. Memoir, pp. 261–283.
29. Bernard, B., Brooks, J.M. and Sackett, W.M., (1978) Light hydrocarbons in Recent Texas continental shelf and slope sediments. Journal of Geophysical Research vol. 83, 4053–4061.
30. Faber, E. and Stahl, W., (1984) Geochemical surface exploration for hydrocarbons in the North Sea. Am. Ass. Pet. Geol. Bulletin vol.68, pp.363–386.

31. Schoell, M., (1983) Genetic characterization of natural gases. Am. Ass. Pet. Geol. Bulletin vol.67, 2225–2238.
32. Whiticar, M.J., Faber, E., Schoell, M., (1986) Biogenic methane formation in marine and fresh water environments: CO2- reduction vs. acetate fermentation – Isotope evidence. Geochim. et Cosmochim. Acta 50, 693–709.
33. Stahl, W.J., (1977) Carbon and nitrogen isotopes in hydrocarbon research and exploration. Chemical Geology 20, 121–149.
34. Berner, U., Faber, E., Scheeder, G., Panten, D., (1995) Primary cracking of algal and land plant kerogens: kinetic models of isotope variations in methane, ethane and propane. Chemical Geology 126, 233–245.
35. Berner, U. and Faber, E., (1996) Empirical carbon isotope/maturity relationships for gases from algal kerogens and terrigenous organic matter, based on dry, open-system pyrolysis. Oganic Geochemistry 24, 947–955.
36. Wever, H.E., (2000) Petroleum and source rock characterization based on C7 star plot results: examples from Egypt. American Association of Petroleum Geologists Bulletin 84, 1041–1054.
37. Vandre´, C., Cramer, B., Gerling, P., Winsemann, J., (2004) The geochemistry of natural gas and condensates from the western o.shore Nile delta, Egypt. In: Proceedings of the DGMK Conference, Celle, Germany, pp. 371–380.
38. Mosconi, A., Rebora, A., Venturino, G., Bocc, P., and Khalil, M. H., 1996. Egypt – Nile Delta and North Sinai Cenozoic tectonic evolutionary model. EGPC, 13th Petrol. Exp. Prod. Conf., Cairo, vol. 2, p. 203–223.
39. Hemdan, K., El Alfy, M., Enani, N., Barrasi, M., and Monir, M., (2002) Structural Complexity of Pliocene and its Impact on Trapping Mechanism, N. Port Said Concession, Egypt. Mediterranean offshore conference. (MOC, 2002).
40. Claudius, V., Bernhard, C., Gerling, P.and Winsemann, J. (2007). Natural gas formation in the western Nile delta (Eastern Mediterranean): Thermogenic versus microbial. Organic Geochemistry v.38, pp. 523–539.

CHAPTER 9

Geochemistry and Petrology of Basic Volcanic Rocks of Jabal Al Haruj Al-Aswad, Libya

N. M. Al-Hafdh[1*], Abd El-Salam S. El-Shaafi[2]

[1]*Department of Earth Science, Faculty of Sciences, Mosul University, Mosul, Iraq*

[2]*Department of Earth Science, Faculty of Sciences, Benghazi University, Benghazi, Libya*

ABSTRACT

The Al Haruj Intra-continental Volcanic Province is the largest part of the extensive volcanic activity in Libya which is considered to be a typical within plate basalts. The volcano-tectonics evolution of this province, as well as its origin, are still widely disputed. According to K-Ar dating previously studied, the volcanic activity started in the Late Miocene and lasted until at least the Late Pleistocene. The field may still be volcanically active. The mafic rocks of Jabal Al Haruj have been classified into six major phases or groups. These phases have been differentiated using Landsat images together with aerial photographs of different scales as well as field observations. The topographic forms of the earliest phase are highly eroded while the forms of the latest phase are usually fresh and very well preserved as regards primary features. Mafic lavas of this field consist of alkali basalts to olivine tholeiites (transitional basalt) which contain olivine as essential constituent together with clinopyroxene, plagioclase and glass. The basalt exhibits intergranular, intersertal, ophitic and subophitic relations. Amygdaloidal and glomerporphyritic textures are also observed. The basaltic rocks of different ages and from different localities are petrographically rather similar. Phenocrysts of olivine probably the result from slow cooling in crustal magma chambers prior to eruptions, suggesting that magmas ascended slowly through the crust. 109 sam- ples have been carefully collected from various phases, some of these samples have been chosen for major and trace elements analyses, using XRF in order to determine the characteristics of the mantle source and investigate crustal interaction. The major and trace elements revealed a slightly significant chemical diversity among the phases and within each phase. The normative classification of most of these rocks shows close agreement with their modal classification. A vague correlation between MgO and most major oxides in the studied samples suggests different degrees of partial melting rather than fractional crystallization. A characteristic feature of the studied volcanic rocks is the relatively constant ratios of certain incompatible trace elements (Nb/Zr, Rb/Zr), which provides strong evidence of a common source. In addition, the rocks display similar patterns of the peaks and troughs; this strongly suggests that they have a common parent and common subsequent processes. The compatible transitional metals Ni (81 - 193 ppm) and Cr contents (238 - 361 ppm) and relatively low Mg# (Mg/(Mg + Fet)) (52 - 62) give an indication that the studied basaltic rocks have slightly to moderately fractionated olivine and/or spinel. The magmatism of this volcanic field seems to be related to reactivation of pre-existing structures during the passive rifting of the Sirt Basin that most likely produced in response to convergence between European and African plates since Jurassic until Holocene times.

Keywords: Jabal Al Haruj Al Aswad Area, Basaltic Rocks, Geochemistry, Petrology, Petrogenesis, Libya

1. INTRODUCTION

The Al Haruj (Jabal Al Haruj al Aswad and al Abiyad) area is located in central part of Libya, and is considered one of the main occurrences of young extrusive rocks. It is the largest of four volcanic provinces in Libya; the other three occurrences are in Gharyan province in north-western Tripoli, Jabal as Sawda between Gharyan and Jabal Al Haruj, and Jabal Nuqy farther south, close to the Tibesti massif at the Chadian-Libyan border (Figure 1).

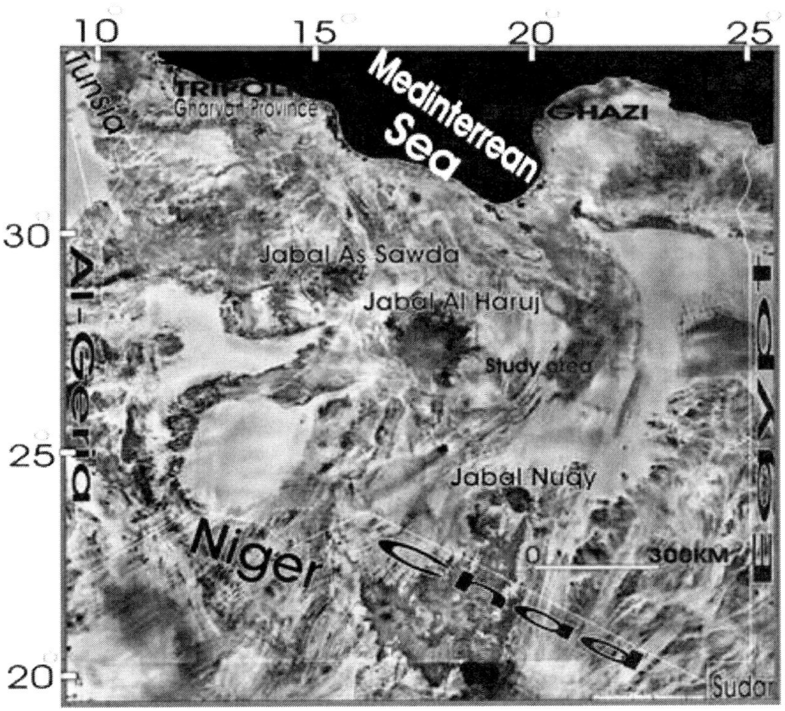

Figure 1. Landsat image showing the main occurrences of Tertiary volcanic provinces in Libya, in a trend NNW to SSE.

All these occurrences represent the main Tertiary-Recent volcanics in Libya, and are arranged with older volcanics in the NNW to the youngest in the SSE. Moreover, the smallest sizes of these volcanics are encountered in the extreme south-east of Libya, in Jabal Awayant, Busrewil and Wadsworth [1] . In general, these volcanics are composed essentially of voluminous alkaline basalts and less extensive transitional and theoliitic basalts; more differentiated rocks are less common, Busrewil and Suwesi [2] .

Al Haruj al Abiyad falls within the Al Haruj basalt field; it is located south-west of the main body and is the youngest region of the field, which is called Al Haruj Aswad. The extensive lava plateaus cover an area of at least 16,000 km^2 in Al Haruj al Aswad, preserved as variably eroded lava fields. The entire Al Haruj has a volcanic area of 45,000 km^2; thus, it is significantly larger than the Tibesti volcanic region, Nemeth et al. [3] . Most of the volume of the lava involved in the building of the lava plateau seems to have been extruded in an extremely fluid condition through thousands of fissures, which are not visible in the field, Busrewil and Wads- worth [4] .

The study area represents the north-eastern part of Jabal Al Haruj al Aswad and is limited by latitudes 27°00'00"N to 27°50'00"N and longitudes 18°00'00"E to 18°39'58.5"E (Figure 2). The landscape of the study area is characterized by a hilly morphology with frequent steep lava fronts, pressure ridges and blocky lava fields, with transverse intermittent desert sands dotted by balta (kipukas), but in other parts the relief is more or less flat.

Field geological mapping was carried out using Landsat TM multispectral satellite imagery with a scale of 1: 100,000, which can be fitted with any kind of coordinate system with an accuracy of at least 20 m, using a few calibration points that were measured during the fieldwork by GPS (Figure 2). This image formed the base of the geological map completed with altitude contours, locality names and other relevant data, Peregi et al. [5] .

As the study area is remote, rather inaccessible and extremely difficult terrain, the fieldwork was carried out with the Industrial Research Centre (IRC), Tripoli as part of the geological mapping project of SE Libya in March 2006. During fieldwork, systematic field traverses were carried out by researchers equipped with GPS units, digital cameras and so on.

During the field trip, all basaltic exposures were investigated. About 109 samples were carefully collected from the flows for each phase (Figure 3). Of these, 42 samples were prepared for thin sections, and 29 samples were chosen for the major and trace element analysis using the XRF (Philips 2440) technique carried out by the Central Laboratories of the Egyptian Geological Survey, Cairo.

The drainage patterns developed on the surfaces of the basaltic lava flows show systematic changes with the increasing age of basaltic lava flows, Busrewil and Suwesi [2] . The Landsat TM image shows the drainage pattern frequently increases in density as the basaltic rock become progressively older or decreases as the flow become younger, as in Phase 5, El-Shaafi [6] . The area investigated is cut by many wadis flowing from west and south-west towards the east and south-east, most of them starting from the central part of Al Haruj Al Aswad, outside the study area.

In the study area, the most common type of drainage pattern is dendritic, which is a well-integrated pattern formed by the main wadis (Wadi Bu al Hiddan, Wadi Al Baqar and Wadi Umm daladil), with tributaries branching and rebranching in several directions without systematic arrangement, Busrewil and Suwesi [2] .

The core of southern Libya was formed by Precambrian basement massifs represented by Jabal Awayant, the Tibesti massif and the Tihemboka Arch, which are parts of the Nile craton, East Saharan craton and Pan-African remobilized belt, Hallett [7] .

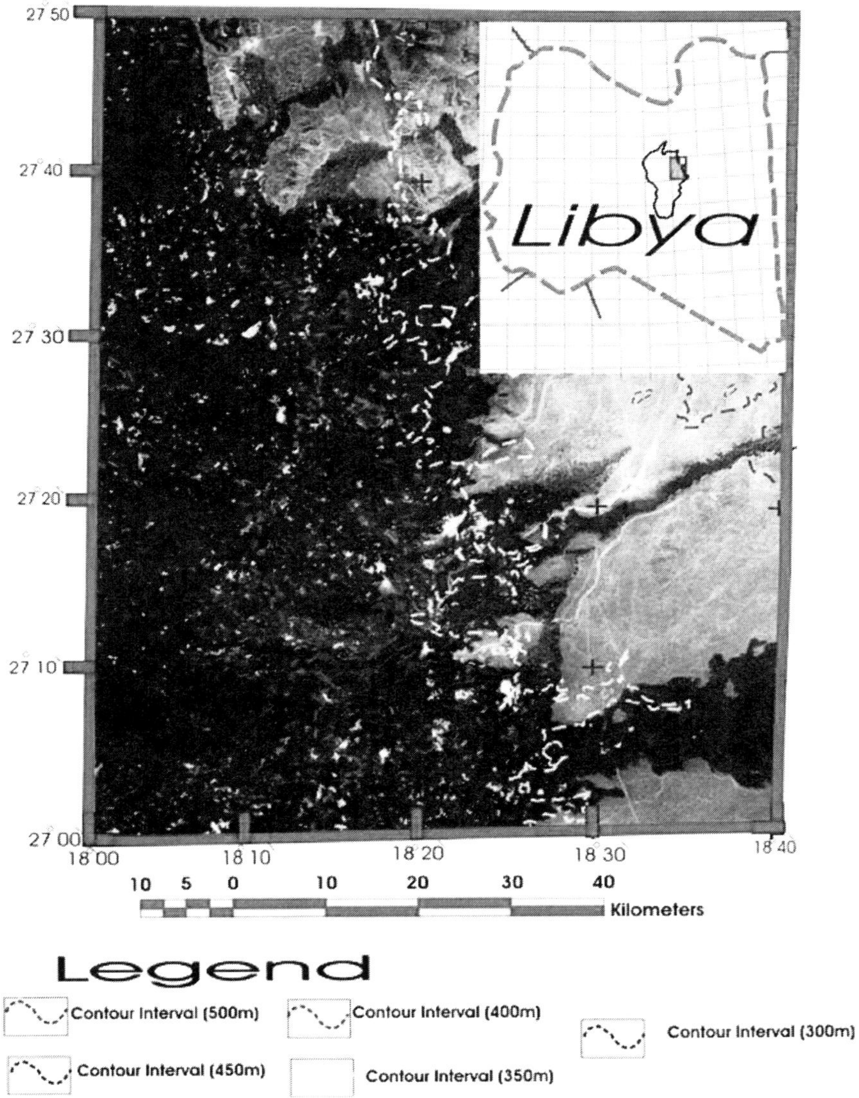

Figure 2. Landsat image of the north-eastern part of Al Haruj al Aswad, mosaiced and overlain by topographic contour lines to depict the spatial distribution of major volcanic phases.

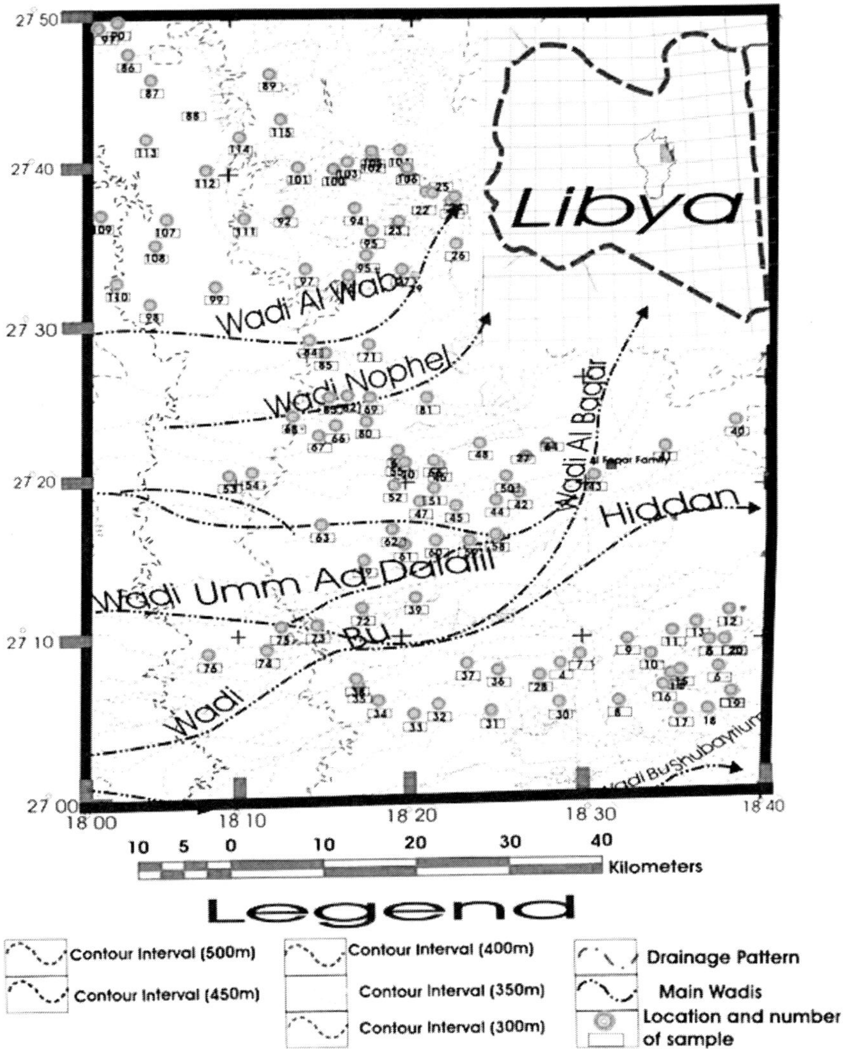

Figure 3. Topographic contour map showing the main wadis and their drainage patterns, as well as the localities of the samples.

During the Pan-African orogeny, the rocks within these terrains were uplifted and accompanied by intense deformation, shearing and metamorphism. The main phase of the pan-African orogeny was completed about 540 Ma, followed by post-orogenic magmatism until about 440 Ma, interpreted by Klitzsch [8] as the early manifestation of the Caledonian orogeny.

This event was accompanied by the formation of NNW-SSE trending extensional structures more than a thousand km long such as the Kalanshio trough, south Haruj uplift, Dor el Qussah trough and the Tripoli-Tibesti uplift, among others. These directly affected the geological history of Al Haruj al Aswad and al Abiyad, Peregi et al. [5] .

Toward, the Haruj uplift, the Palaeozoic geological development is much less clear than in the Dor el Qussah trough, as this area is covered by the volcanic pile of Al Haruj al Aswad. Following the Pan-African orogeny, Lower Paleozoic sediments were deposited and probably overlain by the Middle-Upper Devonian and Carboniferous sedimentary formations, Hallett [7] .

After the Hercynian orogeny (Figure 4) almost the entire Kalanshio trough became part of Sirt Arch and most of the Palaeozoic sequence was destroyed by erosion, Peregi et al. [5] .

The first data on the radiometric age lava of Jabal Al Haruj was given by Pesce [10] , who considered it to be Late Pliocene based on K/Ar determination.

Ade Hall et al. [11] published the first results of a detailed K/Ar whole rock radiometric age determination survey combined with a palaeomagnetic measurement of the Al Haruj volcanic area. The measurements give an interval from 6 to 0.4 Ma for the formation of the volcanic rocks, Less et al. [12] , with the majority of the flows being younger than 2.2 Ma. The modern study of age dating by Peregi et al. [5] was based on K/Ar analyses of 20 basalt samples carried out at Atomki in Hungary.

It shows that the volcanic activities continued from 5.27 Ma to about 0.1 Ma, which means that they can be assigned to the Pliocene and Pleistocene epochs, taking the limit of the Miocene at 5.3 Ma according to the Global Standard Stratigraphic Age (Table 1).

From this review it is clear that the study area has not been thoroughly studied so far, El-Shaafi [6] , except for the work of Klitzsch [13] , which deals with the regional geological mapping of Jabal Al Haruj (Al Abiyad and Al Aswad), showing the distribution of six major phases and one minor phase.

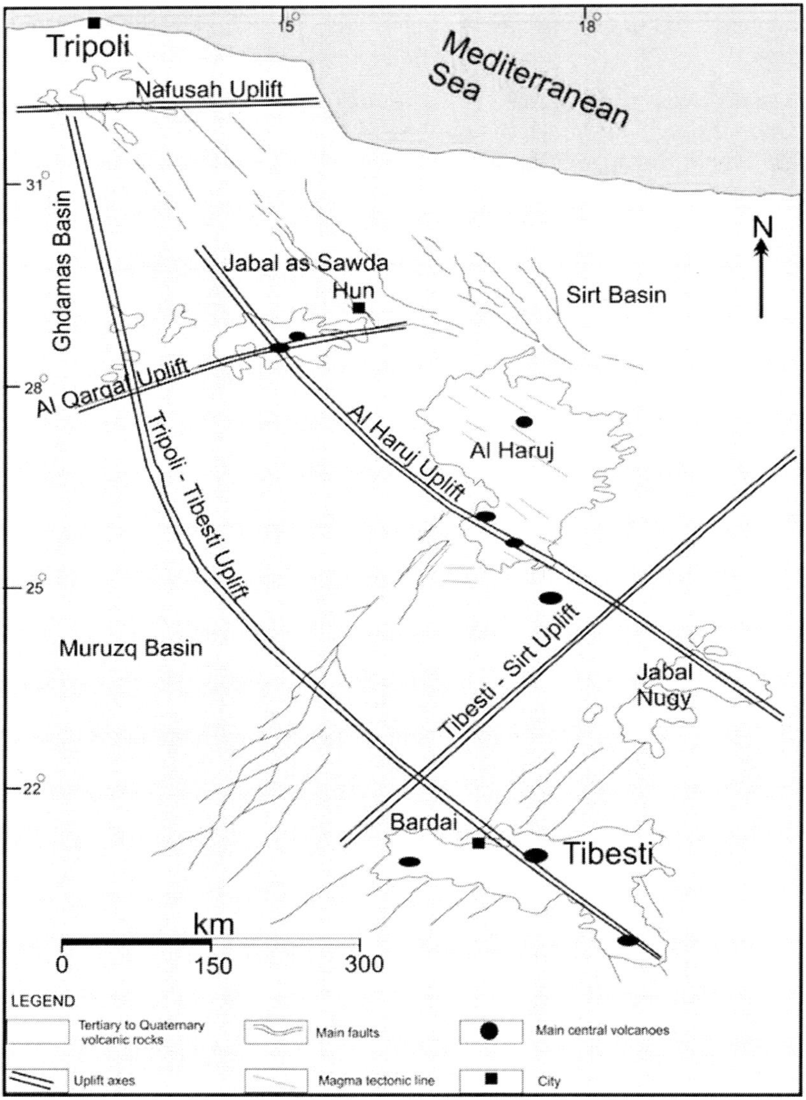

Figure 4. The volcanic distribution and relation between NW trending (Caledonian) and NE trending (Hercynian) structural elements and the volcanic eruptions, after Woller and Feduick [9] .

Table 1. Age of the volcanic phases according to radiometric age determinations and paleomagnetic measurements, and their geological interpretation, Peregi et al. [5] .

Volcanic Phases	Age in Ma	Presumed geological age
6	0.10 - 0.30	Late to Middle Pleistocene
5	0.30 - 0.50	Middle Pleistocene
4	0.50 - 1.22	Early to Middle Pleistocene
3	1.22 - 1.77	Early Pleistocene
2	1.77 - 2.24	Late Pliocene
1	2.73 - 5.27	Early to late Pliocene

On the other hand, Farahat et al. [14] mentioned that the Libyan low volcanicity rift is probably related to the large Afro-Arabian rift system that extends from Turkey to Mozambique. This rift system is thought to have developed episodically rather than continuously. Age dating has suggested that the eastern rift in Kenya has evolved in three major pulses as follows: Eocene (44 - 38 Ma), Middle Miocene (16 - 11 Ma) and Pliocene to Pleistocene (5 - 0 Ma). These ages are generally contemporaneous with the ages of the Libyan volcanics; for example, Jabal Al Haruj was contemporaneous with the third pulse and the Gharyan volcanic province with the three pulses.

The linear trend of the Libyan volcanics and their decrease in age towards the SSE is probably caused by intraplate extensional stresses. Continental rifting is normally associated with asthenospheric upwelling and high heat flow, but there is some controversy as to the mechanism involved, Farahat et al. [14] . It is difficult to decide whether the volcanic activity in Jabal Al Haruj was related directly to an uprising mantle plume (active rifting) or to the extension of the lithosphere (passive rifting), Peregi et al. [5] . The present work deals mainly with the different exposures in the study area. The exposed strata range in age from the Lower Miocene to Upper Pleistocene. Stratigraphically, the basalt flows in the study area disconformably overlie the older sequence of sedimentary rocks. The investigated volcanic rocks are overlay Maradah Formation and have a varied morphology, implying that they are produced episodically in consistency with available age data of 5.3 to 0.10 Ma from the Upper Miocene to Upper Pleistocene, Peregi et al. [5] .

The study area is not mountain range but rather a vast plateau. The commonest type of volcanic occurrence is represented by basalt lava flows, which shaped the present morphology. Generally, the central eruption of these lava flows are most markedly shown in the central part of Al Haruj, while elsewhere it is on its margins. In contrast, and despite the study area being located at the margin of Al Haruj, it is difficult to locate these features.

The Jabal Al Haruj Al Aswad volcanic province is characterized by the prevalence of lava of an effusive form rather than an intrusive form. Although some of the lava may have come from central volcanoes, the great volume of lava and the remarkable uniformity in thickness of individual lava sheets over a wide area indicate that the lava must have extruded from numerous fissures, Farahat et al. [14] . In the area investigated, plateau basalts are the most widespread; they are composed of several lava flows, whose thickness does not exceed several metres. Five major volcanic phases of different field appearances (including weathering surface, morphology, and primary structures) have been documented in the investigated area in the north-eastern part of Al Haruj Al Aswad province (Figure 5).

2. PETROGRAPHY

The petrographic characteristics of the Tertiary volcanic rocks of Jabal Al Haruj (al Aswad and al Abiyad) have been studied by many authors, among them Busrewil and Suwesi [2] , Less et al. [12] , Peregi et al. [5] , Vesely [15] and Woller [16] .

In the present study, 109 samples from the investigated area were described, and on the basis of preliminary petrographic descriptions 42 samples were selected for detailed petrographic study. Modal analyses of the various phases of basaltic rocks were carried out using an electronic point counter (Model E). The results of the modal analyses are presented in Table 2.

Microscopically, the rock samples are variably vesicular, amygdaloidal and almost medium grained and subordinately fine grained. Some of the samples are characterized by the presence of variable amounts of glass in the groundmass, but most of them are holocrystalline with an intergranular to intersertal texture.

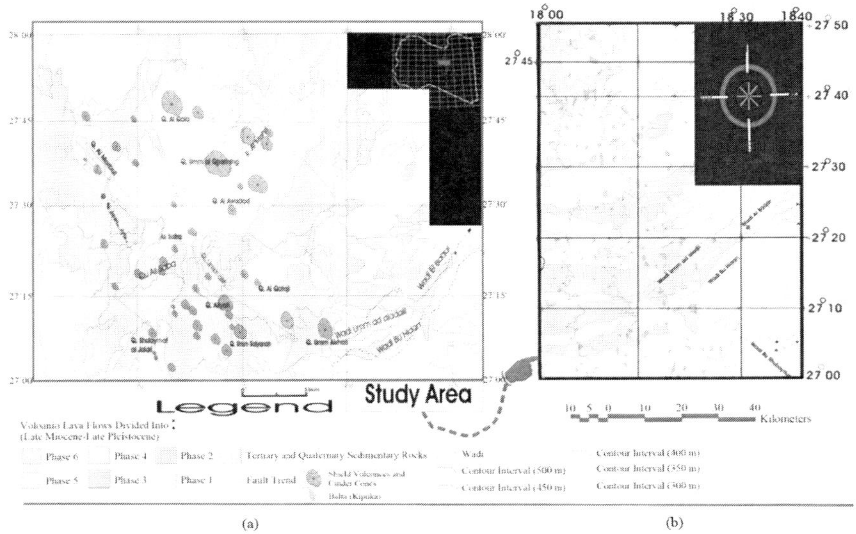

Figure 5. Geological map showing (a) The central part of al Haruj al Aswad, studied by Busrewill and Suwesi [2] and the north-eastern part of al Haruj al Aswad (study area) investigated in the present work; (b) Magnified geological map of the study area.

The phenocrysts consist predominately of olivine and a subordinate amount of clinopyroxene, while plagioclase is not found. Olivine phenocrysts form about 7% to 21% by volume of the rock and occur as subhedral and sometimes as euhedral crystals without zoning; therefore, they indicate the presence of enough time to equilibrate with the melting at mantle depth.

Olivine phenocrysts sometimes occur as clots displaying a glomerporphyritic texture. Pyroxene phenocrysts were found in some samples as elongated and prismatic crystals. The matrix consists mostly of medium grained plagioclase laths, olivine, clinopyroxene (augite) and opaque minerals (probably ilmenite).

Plagioclase laths are the most abundant, representing 32% to 48% of the matrix in most of the studied samples. They are both pilotaxitically and randomly arranged, and sometimes they exhibit a trachytic arrangement around olivine phenocrysts. Cryptocrystals fill the interstices between plagioclase laths. Some samples showed vesicles filled with carbonate secondary minerals.

The different textures and mineralogical composition of basaltic rocks of each phase are described in more detail in El-Shaafi [6] . A detailed petrographic description of Phase 1 follows.

Phase 1

This phase represents the earliest volcanic activity in the area. The lava flow of this phase is highly weathered, but although weathered in the field, the plagioclase and pyroxene (augite) are nearly fresh under the microscope.

Olivine phenocrysts are partially altered to iddingsite, which has stained the rocks a red colour. The most common texture of the rocks of this phase is an intersertal to intergranular texture (Plate 1). The intergranular texture is one in which the dimensions of the plagioclase and pyroxene are similar and the glass is relatively minor, which grades into intersertal when interstitial glass is a significant component.

Table 2. Model analyses of representative basaltic rocks from north-eastern part of Al Haruj al Aswad.

Phase Number	Sample Number	Olivine %	Clino-Pyroxene %	Plagioclase %	Opaque %	Glass %	Carbonate %	Texture Type
1	17	18.57	21.76	37.28	12.96	5.69	3.74	Porphyritic Intergranular
1	54	21.55	24.84	39.14	9.09	2.21	3.17	Porphyritic Intergranular
2	23	13.77	19.55	41.40	8.11	8.20	8.97	Porphyritic Intergranular
2	95	12.29	15.01	37.41	7.28	18.30	9.71	Intersertal, Amygdaloidal
3	35	15.87	24.11	45.31	8.21	3.99	2.51	Porphyritic Intergranular
3	51	16.21	20.01	43.72	6.25	10.41	3.41	Intergranular to Intersertal
3	68	16.45	19.92	48.46	4.29	7.48	3.40	Intergranular to Intersertal
4	30	18.01	19.56	46.09	5.02	10.71	0.61	Intergranular to Intersertal
4	69	19.10	21.14	47.16	4.77	3.68	4.15	Intergranular to Intersertal
4	108	6.71	26.88	46.43	2.56	13.86	3.56	Intergranular to Intersertal
5	86	7.83	12.99	32.92	0.78	42.76	2.72	Intersertal to Hyalo Ophitic

The rocks are composed mainly of plagioclase, clinopyroxene and olivine. Opaque minerals that might be ilmenite are present as accessory minerals. Reddish brown iddingsite and carbonate minerals are secondary products. Volcanic glass is also observed. The detailed petrography of Phase 1 as follows:

A?Plagioclase is the most abundant, occurring in the form of laths of lamellar twinned crystals, ranging in size from 0.40 to 0.95 mm in length and from 0.10 to 0.20 mm in width, and occurring in the form of irregular grains. The interference colour of plagioclase laths of the first order is grey to straw-yellow with broad twin lamellae. Generally, the relief of plagioclase is quite high, so this is a calcium-rich plagioclase, Mackenzie and Adams [17] .

Plagioclase crystals contain opaque crystals as inclusions, El-Shaafi [6] . There is more than one generation of plagioclase: the first generation produced small plagioclase crystals which occur as inclusions within clinopyroxene; the second produced lath-shaped crystals with lamellar twinning. Optical measurements indicate that the plagioclase compositions are generally in the range of An 61% to An 72% (labradorite-bytownite), which is characterized by broad twin lamellae under cross-polarized light. Laths of plagioclase are commonly seen to flow around phenocrysts and form a trachytic texture, which may have resulted from the flow of the magma before solidification, El-Shaafi [6] .

B?Olivine is the third of the abundant minerals which occur in the euhedral to subhedral form, or as rounded crystals. The olivine crystals range in size from 0.40 to 1.2 mm in length and from 0.40 to 0.70 mm in width, although rounded crystals averaging 0.38 mm diameter are recorded. It is recognized in the examined samples by its colourless, high relief and high birefringence, and it is commonly traversed by randomly oriented cracks.

Olivine phenocrysts are partially or completely converted into a strongly coloured reddish brown along their cracks and boundaries, known as iddingsite, El-Shaafi [6] . Multiple grains as clusters of adhering phenocrysts of olivine were observed and formed a glomerporphyritic texture, El-Shaafi [6] . Skeletal crystals of olivine were observed, where the edges and corners grow more rapidly than the faces, El-Shaafi [6] . More than one generation of olivine were observed: the first generation were the phenocrysts, which appear as highly cracked crystals, whereas the second generation produced small scattered crystals associated with clinopyroxene (augite) occupying the spaces between the plagioclase.

C?Clinopyroxene occurs in a subhedral to anhedral form, and as prismatic crystals varying from 0.35 mm to 0.85 mm in length and from 0.10 to 0.60 mm in width, which have a maximum extinction angle of 42° to 51°. They are identified as clinopyroxene (augite). Augite crystals are colourless to pale brown, and brown in some specimens. Rarely, the crystals show simple twining. The crystals are generally fresh, but may be slightly altered along the cracks and cleavages. Some of them contain inclusions of plagioclase and opaque minerals (most probably ilmenite).

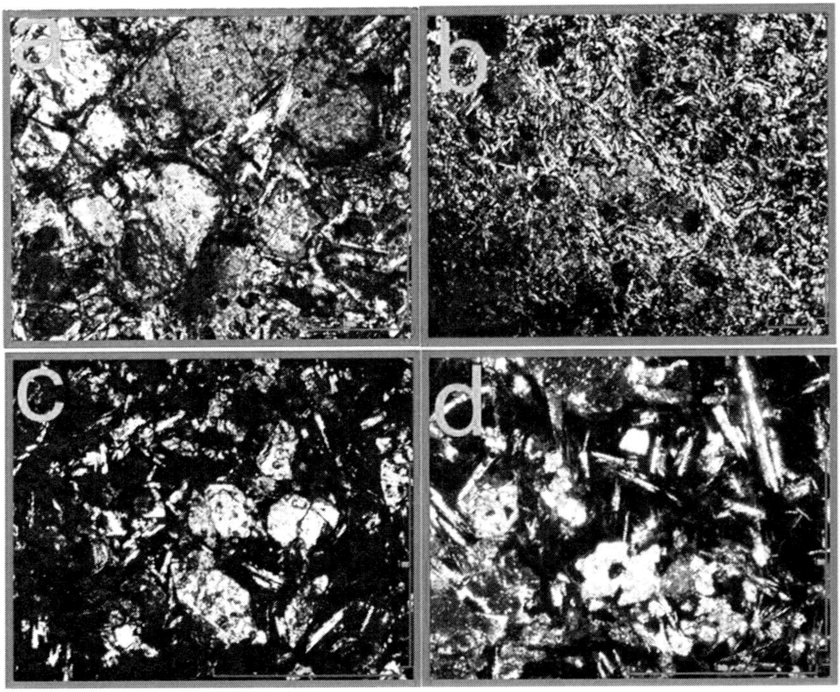

Plate 1. Microphotographs of olivine basalt of phase one, showing fine-grained rock with microphenocrysts of olivine (highest interference colour) and clinopyroxene (augite). The groundmass is composed of olivine, laths shaped of plagioclase and clinopyroxene (augite), with roughly rectangular crystals of opaque minerals. A higher magnification view reveals the olivine crystals and laths of plagioclase in more detail. Sometimes clinopyroxene crystals occur as a large poikilitic plate, enclosing several plagioclase and olivine crystals and forming an ophitic texture.

A subophitic texture is well developed in the rock, where plagioclase laths are partly enclosed with clinopyroxene, whereas the carbonate (most probably calcite) iddingsite occurs and secondary constituents.

D?Opaque minerals are present in considerable amounts, being in the form of discrete crystals that exhibit various forms. They are generally of a smaller size than associated silicate minerals, where they range in length from 0.05 mm to 0.20 mm and from 0.035 mm 0.10 mm in breadth, El-Shaafi [6] . They are the primary constituents, being crystallized straight from the melt.

E?Volcanic glass is present in some of the examined samples, but in small amounts between the plagioclase laths, manifesting an intersertal texture.

F?Carbonate occurs plentifully in some samples in the interstitial spaces and it fills irregular vesicles and forms an amygdaloidal texture, El-Shaafi [6] . Carbonate minerals have higher birefringence, so that when viewed in cross-polarized light (XPL), they do not show interference colours; instead, they show delicate pastel shades of colours because of the extreme birefringence, which reaches up to the sixth order of interference colour.

G?Iddingsite and bowlingite. In certain samples of the olivine basalts of Phase 1, the olivine phenocrysts are partially or completely altered into a strongly coloured yellowish or reddish brown mineral called iddingsite, El-Shaafi [6] . Bowlingite may be similar to iddingsite, but bowlingite is more strongly coloured than iddingsite and is rich in red colour, Hatch et al. [18] .

3. GEOCHEMISTRY

The geochemistry of Late Miocene-Pleistocene basaltic rocks of Al Haruj area has previously been treated by investigators. Busrewil and Wadsworth [1] , made the first attempt to undertake a preliminary study of chemical data on Al Haruj volcanics. These authors indicated that the rocks of Al Haruj belong to the alkali olivine basalt association, ranging in composition from basalts to hawaiites. Vesely [15] revealed that the rocks there are com- posed of alkaline to tholeiitic basalts. Busrewil and Suwesi [2] suggested the clustering of Al Haruj volcanics into four trends of differentiation, which may be described as follows: alkali basalts of the block lava series, alkali basalts of the main lava series, subalkali basalts and dolerites.

Peregi, et al. [5] studied the geochemistry of basaltic rocks from Al Haruj al Abiyad; they concluded that they were formed in an intracontinental (within plate) tectonic setting. They classified them into three main mafic groups based on their chemical composition: basanites, alkali basalts and olivine tholeiites. Moreover, they originated from different degrees of partial melting of a similar mantle source. The basanite magmas could have been formed by 0.5% to 1% partial melting, whereas the primary magmas of the alkaline basalts could have been generated by 2% to 6% partial melting, Peregi, et al. [5] .

The present study is an attempt to identify the geochemical characteristics of the Late Miocene?Pleistocene basaltic rocks of the NE part of Al Haruj al Aswad, evaluate the chemistry of these rocks within the framework of established magma types and determine the tectonic environment of their eruption. A major step in the development of evaluating the studied basaltic rocks involves choosing a method that enables rapid and accurate analysis to be conducted of major and trace elements present at low concentrations in silicate materials.

Twenty-nine samples were chemically analysed using X-ray fluorescence (Philips 2404) at the Central Laboratories of the Geological Survey of Egypt to determine the major and trace elements of the chemically analysed samples (Figure 6). Details of the analytical procedure and precision can be found in Appendix 1 of El-Shaafi [6] . The geochemical data of the 29 samples and the CIPW norms are presented in Table 3.

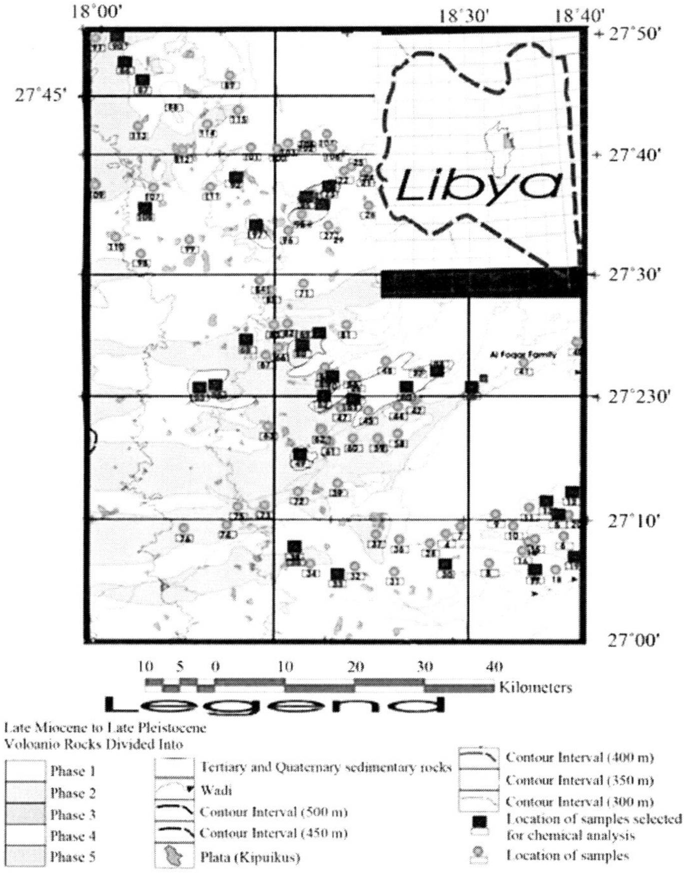

Figure 6. Geological map of the investigated area showing the sample localities of all collected samples during field inspection and location of samples selected for chemical analysis.

Table 3. Representative analyses of the major elements (wt%), trace elements (in ppm) and (CIPW) norms of the studied basaltic rocks. CIPW normative components are calculated for the volatile free data normalized to 100%. Fe_2O_3 and FeO were estimated based on ratio of Fe_2O_3/FeO = 0.2.

(a)

Representative analyses of the major elements (wt%) of Jabal al Haruj al Aswad																
	Intrusion	SiO_2	TiO_2	Al_2O_3	Fe_2O_3	FeO	MnO	MgO	CaO	Na_2O	K_2O	P_2O_5	LOI	SO_3	Cl	Sum	Total Fe
DN19	Phase 1.1	49.85	1.79	15.81	1.65332	8.2666	0.15	6.49	9.26	3.65	0.98	0.25	0.01	0.14	0.13	98.43	10.84
DN49	Phase 1.1	48.23	1.55	15.54	1.68688	8.4344	0.16	6.38	9.92	3.38	0.74	0.22	1.17	0.85	0.16	98.421	11.06
DN50	Phase 1.1	48.13	1.62	15.77	1.62282	8.1141	0.16	5.54	10.02	3.6	0.73	0.19	1.39	1.57	0.24	98.697	10.64
DN52	Phase 1.1	47.9	1.87	15.63	1.53	8.35	0.15	5.2	12.33	3.45	0.96	0.32	1.21	0.31	0.11	99.32	10.13
DN53	Phase 1.1	47.63	1.73	15.11	1.67467	8.3734	0.16	6.56	10.17	3.39	0.97	0.31	0.77	1.69	0.18	98.718	10.98
DN54	Phase 1.1	50.06	1.64	16.33	1.57554	7.8777	0.16	6.02	9.37	3.7	0.91	0.23	0.55	0.1	0.2	98.723	10.33
DN97	Phase 1.1	50.18	1.64	16.41	1.67162	8.3581	0.16	5.72	9.36	3.96	0.68	0.19	0.01	0.18	0.16	98.68	10.96
DN64	Phase 1.1	46.93	1.56	15.27	1.67162	8.3581	0.16	5.7	11.82	3.24	0.69	0.28	1.03	1.81	0.1	98.622	10.96
DN5	Phase 1.1	48.45	1.64	15.08	1.68993	8.4496	0.16	6.7	10.4	3.22	0.77	0.23	0.94	0.69	0.26	98.68	11.08
DN12	Phase 1.1	50.85	1.83	15.67	1.62892	8.1446	0.15	5.59	9.21	3.65	0.79	0.14	0.4	0.41	0.23	98.694	10.68
DN13	Phase 1.1	50.76	1.7	16.35	1.58164	7.9082	0.15	5.56	9.46	3.82	0.95	0.2	0.01	0.12	0.16	98.73	10.37
DN17	Phase 1.1	47.66	1.61	14.44	1.62282	8.1141	0.15	6.65	10.9	3	0.54	0.15	1.59	1.9	0.2	98.527	10.64
DN23	Phase n.2	48.25	1.88	15.81	1.61062	8.0531	0.16	5.26	10.46	3.55	0.81	0.31	1.78	0.7	0.17	98.804	10.56
DN80	Phase n.2	47.89	1.68	14.63	1.63197	8.1598	0.15	6.95	10.44	3.29	0.78	0.22	1.83	0.91	0.15	98.712	10.7
DN94	Phase n.2	49.95	1.71	17.5	1.52063	7.6031	0.15	4.75	10.56	4	0.54	0.27	0.19	0.09	0.05	98.884	9.97
DN95	Phase n.2	50.06	1.87	15.74	1.60451	8.0226	0.16	6.05	9.54	3.6	0.86	0.23	0.39	0.53	0.12	98.777	10.52
DN35	Phase n.3	48.75	2.07	15.22	1.7189	8.5945	0.16	6.32	9.9	4	0.96	0.37	0.19	0.4	0.1	98.753	11.27
DN43	Phase n.3	48.3	1.57	15.81	1.54961	7.748	0.15	5.66	10.38	3.53	0.77	0.19	1.26	1.67	0.18	98.768	10.16
DN51	Phase n.3	48.48	1.8	14.88	1.61977	8.0988	0.15	7.51	9.89	3	0.94	0.24	0.64	1.26	0.21	98.719	10.62
DN55	Phase n.3	50.12	1.83	16.62	1.61062	8.0531	0.16	4.71	10.34	3.85	0.74	0.29	0.15	0.14	0.09	98.704	10.56
DN68	Phase n.3	49.51	1.87	15.16	1.63502	8.1751	0.17	7.66	9.23	3.45	0.89	0.34	0.12	0.39	0.07	98.67	10.72
DN30	Phase n.4	51.63	1.82	16.54	1.45047	7.2523	0.15	5.36	8.97	4.03	1.06	0.3	0.01	0.05	0.13	98.753	9.51
DN33	Phase n.4	49.45	1.62	14.80	1.63654	8.1827	0.15	7.00	10.23	2.96	0.56	0.14	1.01	0.79	0.18	98.709	10.73
DN69	Phase n.4	49.65	1.68	16.25	1.59689	7.9844	0.16	5.82	10.24	3.75	0.83	0.26	0.13	0.16	0.18	98.691	10.47
DN92	Phase n.4	51.60	1.81	15.65	1.61977	8.0988	0.15	6.17	8.59	3.34	0.76	0.14	0.00	0.18	0.18	98.289	10.62
DN108	Phase n.4	48.44	1.87	14.75	1.7189	8.5945	0.17	7.15	10.42	3.29	0.80	0.27	0.84	0.12	0.16	98.593	11.27
DN86	Phase n.5	48.87	1.76	14.91	1.81499	9.075	0.17	7.35	10.09	2.99	0.68	0.15	0.11	0.49	0.20	98.66	11.90
DN87	Phase n.5	48.77	2.06	14.07	1.91261	9.563	0.17	6.96	10.01	3.22	0.86	0.21	0.14	0.46	0.13	97.946	12.54
DN90	Phase n.5	48.66	1.67	15.32	1.80737	9.0368	0.18	7.14	10.05	3.11	0.61	0.19	0.18	0.59	0.16	98.704	11.85

(b)

CIPW norms of the studied Jabal al Haruj al Aswad basaltic rocks. CIPW normative components are calculated for the volatile free data normalized to 100% Fe_2O_3 and

	Intrusion	Q	Or	Ab	An	Ne	Di	Hy	Ol	Mt	Il	Ap
DN19	Phase 1.1		5.91	30.89	23.86	0	17.1	2.03	12.72	2.44	3.46	0.58
DN49	Phase 1.1		4.55	28.6	25.045	0	19.3	0.45	13.8	2.54	3.06	0.53
DN50	Phase 1.1		4.49	29.04	24.714	0	20.4	0	11.8	2.46	3.23	0.46
DN52	Phase 1.1		5.79	20.94	24.326	4.87	29.3	0	7.6	2.28	3.63	0.74
DN53	Phase 1.1		5.79	24.45	23.147	1.85	21.6	0	13.41	2.52	3.42	0.74
DN54	Phase 1.1		5.5	31.31	25.262	0	16.4	3.2	10.98	2.32	3.19	0.56
DN97	Phase 1.1		4.08	33.51	24.992	0	16.9	1.155	12.28	2.46	3.17	0.44
DN64	Phase 1.1		4.25	21.79	25.084	3.22	27.3	0	10.01	2.51	3.1	0.67
DN5	Phase 1.1		4.73	27.25	24.419	0	21.7	0.84	13.04	2.54	3.21	0.56
DN12	Phase 1.1		4.79	30.89	24.04	0	17.4	10.84	4.4	2.41	3.55	0.32
DN13	Phase 1.1		5.73	32.32	24.659	0	17.5	3.3	9.53	2.33	3.29	0.46
DN17	Phase 1.1		3.37	25.39	24.34	0	24.7	3.29	10.11	2.48	3.23	0.37
DN23	Phase n.2		4.96	28.56	24.811	0.81	21.4	0	10.39	2.41	3.72	0.74
DN80	Phase n.2		4.85	26.25	22.847	0.88	23.5	0	13.17	2.45	3.34	0.53
DN94	Phase n.2		3.19	32.36	28.2	1.03	18.6	0	10.23	2.36	3.27	0.63
DN95	Phase n.2		5.2	30.46	24.248	0	18	0	8.68	2.36	3.63	0.56
DN35	Phase n.3		5.79	27.68	20.739	3.42	21.7	0	12.37	2.52	4.01	0.88
DN43	Phase n.3		4.79	28.12	25.019	1	21.6	0	11.17	2.33	3.11	0.46
DN51	Phase n.3		5.73	25.39	24.359	0	19.5	0	11.54	2.39	3.35	0.58
DN55	Phase n.3		4.43	32.58	25.882	0	19.8	0	9.31	2.35	3.53	0.67
DN68	Phase n.3		6.26	29.19	23.25	0	17	4.29	13.05	2.42	3.63	0.72
DN30	Phase n.4		6.38	34.1	23.91	0	15.5	6.03	6.91	2.13	3.51	0.7
DN33	Phase n.4		3.43	25.05	25.442	0	20.6	0	4.89	2.45	3.17	0.32
DN69	Phase n.4		5.02	30.06	25.055	0	20.1	0	11.68	2.35	3.25	0.6
DN92	Phase n.4	0.65	4.61	28.26	25.465	0	13.6	20.07	0	2.39	3.51	0.32
DN108	Phase n.4		4.85	26.97	23.116	0.5	22.5	0	14	2.54	3.65	0.65
DN86	Phase n.5		4.08	25.3	25.253	0	19.9	7.18	10.67	2.68	3.42	0.35
DN87	Phase n.5		5.2	27.25	21.397	0	22.6	2.53	12.67	2.83	4.01	0.49
DN90	Phase n.5		5.2	26.32	26.04	0	22.6	2.53	12.67	2.83	4.01	0.49

(c)

Representative analyses of trace elements (ppm) of Jabal al Haruj al Aswad basaltic rocks

	Intrusion	Sc	V	Cr	Co	Ni	Cu	Zn	Rb	Sr	Y	Zr	Nb	Mo
DN19	Phase 1.1		195.7	337.5	56.8	190.5	69.2	70.6	4.9	332.8	<1	86.8	10.1	1.6
DN49	Phase 1.1		227.4	319.9	52.9	120.4	48.0	70.2	5.2	179.2	<1	72.9	10.0	2.4
DN50	Phase 1.1		203.6	295.6	51.7	134.8	47.1	63.5	6.7	179.9	<1	70.4	9.7	2.1
DN52	Phase 1.1		207.5	281.4	50.7	116.9	54.4	59.6	4.2	184.0	<1	66.1	9.2	2.0
DN53	Phase 1.1		199.1	277.3	58.6	117.6	34.4	61.6	4.4	174.5	<1	64.0	7.8	2.3
DN54	Phase 1.1		201.6	281.1	53.9	160.5	73.0	63.8	6.3	197.3	<1	74.4	8.7	1.7
DN97	Phase 1.1		205.7	292.1	47.2	186.6	68.1	61.0	6.4	220.1	<1	81.0	11.0	2.7
DN64	Phase 1.1		206.9	322.1	51.1	184.9	53.4	62.6	5.5	306.2	<1	88.1	10.7	2.3
DN5	Phase 1.1		211	289.0	59.6	150.4	47.2	66.3	5.8	198.9	<1	76.5	11.1	2.5
DN12	Phase 1.1		196.5	273.4	54.6	141.5	46.0	66.0	5.1	238.5	<1	79.8	10.8	2.4
DN13	Phase 1.1		196.8	287.4	51.2	134.1	41.3	62.3	5.3	221.6	<1	71.0	9.5	1.7
DN17	Phase 1.1		204.7	325.3	58.0	124.6	47.0	63.8	5.4	318.8	<1	84.8	9.1	2.8
DN23	Phase n.2		193.2	267.8	53.5	118.6	41.5	52.5	5.2	240.5	<1	80.5	11.2	2.0
DN80	Phase n.2		210.3	305.3	54.2	136.9	68.5	70.0	5.6	206.1	<1	83.2	11.9	1.5
DN94	Phase n.2		219.2	292.7	51.7	141.6	45.5	59.7	5.4	266.2	<1	82.8	10.7	1.8
DN95	Phase n.2		196.7	238.7	42.6	81.9	52.9	58.7	3.4	176.9	<1	59.5	7.2	1.8
DN35	Phase n.3		197.3	339.1	53.5	193.1	55.0	66.8	5.6	306.4	<1	84.2	8.6	1.6
DN43	Phase n.3		208.3	329.4	47.1	114.0	60.2	62.2	4.9	405.6	<1	99.3	12.4	1.5
DN51	Phase n.3		186.1	265.0	45.2	124.2	48.1	56.9	4.6	247.0	<1	71.0	9.0	1.9
DN55	Phase n.3		201.2	282.1	43.8	137.3	50.5	60.0	5.6	235.9	<1	80.4	10.4	2.0
DN68	Phase n.3		228.1	316.0	61.2	153.8	47.7	65.0	5.2	388.3	<1	101.4	12.4	1.9
DN30	Phase n.4		217.1	268.5	48.8	108.3	32.6	62.5	7.1	186.6	<1	76.6	9.8	1.9
DN33	Phase n.4		211.9	304.1	53.4	114.7	44.5	74.7	6.0	183.1	<1	68.9	8.3	1.5
DN69	Phase n.4		211.2	283.3	46.5	103.0	35.2	58.9	5.8	187.9	<1	71.5	10.4	2.1
DN92	Phase n.4		237.1	358.4	56.9	133.6	47.9	70.7	6.3	176.9	<1	75.4	9.8	2.6
DN108	Phase n.4		215.0	345.2	56.7	172.8	49.4	71.7	4.7	202.9	<1	75.1	10.7	1.6
DN86	Phase n.5		243.2	327.6	57.3	139.1	52.4	68.8	6.1	200.1	<1	75.1	9.6	1.0
DN87	Phase n.5		260.0	361.3	60.1	165.3	59.1	70.8	7.0	236.6	<1	91.6	13.9	1.7
DN90	Phase n.5		229.1	300.1	57.1	130.6	54.3	76.8	6.1	178.5	<1	71.0	9.2	2.4

(d)

Representative analyses of trace elements (ppm) continued									
	Intrusion	Sn	Ba	La	Yb	Hf	Ta	Pb	Th
DN19	Phase 1.1	3.2	333.4	16.5	4.4	3.7	3.7	3.9	<1
DN49	Phase 1.1	5.9	176.6	6.0	4.1	2.8	4.2	3.9	<1
DN50	Phase 1.1	2.2	177.5	17.4	4.2	3.7	3.7	4.7	2.3
DN52	Phase 1.1	2	191.6	13.5	3.9	1.8	4.0	4.8	4.4
DN53	Phase 1.1	2.4	166.4	15.0	4.1	3.4	4.0	4.1	<1
DN54	Phase 1.1	2.4	241.3	11.4	4.2	3.0	3.5	7.7	<1
DN97	Phase 1.1	3.2	194.7	13.7	4.3	3.4	4.0	<1	<1
DN64	Phase 1.1	2.4	616.0	7.1	4.1	2.2	3.6	4.8	3.0
DN5	Phase 1.1	0.6	203.6	19.0	4.2	3.4	5	6.4	4.0
DN12	Phase 1.1	3.7	930.6	15.3	4.2	4.6	3.6	4.3	2.0
DN13	Phase 1.1	2.9	157.7	17.9	4.2	3.5	3.8	4.4	<1
DN17	Phase 1.1	2.5	214.2	15.5	4.1	2.2	3.4	4.5	1.0
DN23	Phase n.2	1.9	194.3	10.2	4	2.8	3.6	2.8	2.2
DN80	Phase n.2	3.4	189.8	12.3	4	3.1	3.6	5.5	0.8
DN94	Phase n.2	2.0	1070	13.4	4	4.9	2.8	2.2	2.1
DN95	Phase n.2	4.3	164.1	7.2	3.9	1.4	3.2	4.9	2.8
DN35	Phase n.3	2.4	1062	15.7	4.1	4.4	3.3	6.3	1.0
DN43	Phase n.3	1.5	764.0	15.4	4	3.4	4.1	3.8	2.3
DN51	Phase n.3	2.5	310.3	10.5	3.9	3.5	3.2	6.4	<1
DN55	Phase n.3	2.2	321.3	12.3	4	2.3	3.1	5.0	2.1
DN68	Phase n.3	3.8	663.0	17.0	4.1	3.8	3.3	3.7	<1
DN30	Phase n.4	3.7	219.8	12.6	3.9	3.5	3.6	5.5	3.0
DN33	Phase n.4	4.1	352.8	16.4	4.1	1.5	4.6	4.9	<1
DN69	Phase n.4	3.5	199.4	15.6	4.1	3.3	4	6.7	1.4
DN92	Phase n.4	2.1	176.8	19.5	4.1	4	4	4.6	<1
DN108	Phase n.4	2.5	292.0	11.3	4.2	4.1	3.9	5.9	1.0
DN86	Phase n.5	1.1	180.9	14.7	4.2	3.4	4	4.3	2.0
DN87	Phase n.5	2.4	182.0	29.9	4.2	3.3	3.4	6.4	2.5
DN90	Phase n.5	2.8	173.0	22.9	4.1	2.5	3.5	6.1	2.0

The chemical composition ranges of the investigated samples as a whole are relatively narrow within each individual basaltic phase and also within the different phases that existed there. The chemical ranges of the studied samples are somewhat restricted.

According to the CIPW norm, the analysed rocks are characterized by a high normative olivine content, with at most a low content of normative nepheline. Hence, these rocks can be classified as olivine basalt. The occurrence of normative hypersthene leads to rock being classified as normal olivine transitional basalt. SiO_2 ranges from 49.5% to 52.7% together with Mg# (Mg/(Mg + Fet)) values of 52 to 62 indicate that the studied basaltic rocks are slightly to moderately evolved (less primitive). Thus, they are not considered as the primary melt of the primitive mantle. The studied basaltic rocks are actually basalt on the basis of their silica content.

The petrographical criteria obtained from the studied basaltic rocks are not strictly conclusive in their classification. However, the modal contents cannot be accurately determined in most cases in volcanic rocks because of the fine grain size and presence of glassy materials. Therefore, some chemical criteria should be used to determine the basalt type of the studied rocks. The chemical classifications use parameters such as the existence of oxides, functions of oxides and ratios. Various suggestions have been proposed in this domain by different authors.

The volcanic rocks may be subdivided into two members of the major magma series, alkalic and subalkalic; however, although the boundary between them is marked as a solid line, it is actually transitional.

From the various literature dealing with petrochemical classification, it is clear that four types of basaltic rocks can be distinguished: tholeiitic basalts, transitional basalts, alkali basalts and high alumina basalts or calc-alkaline basalts, Cox et al. [19] . The concept of transitional basalts between the tholeiitic and alkali basalt type has been clarified by many authors, e.g. Middlemost, [20] . For the chemical classification of the basaltic rocks under consideration variation diagrams are employed using major elements, which are introduced in the following.

According to Middlemost [21] the investigated samples are transitional basalts of either a soda or potash series. This can be accomplished by an empirical boundary line on a soda vs potash diagram. The studied transitional basalts were found to belong to the soda series (Figure 7).

In addition to the classifications based on major elements, certain variation diagrams have been suggested by Winchester and Floyd [23] to classify the basaltic rocks using five minor and trace elements (Ti, P, Zr, YNb) to produce a set of binary diagrams that discriminate between tholeiitic and alkali basalts (Figure 8).

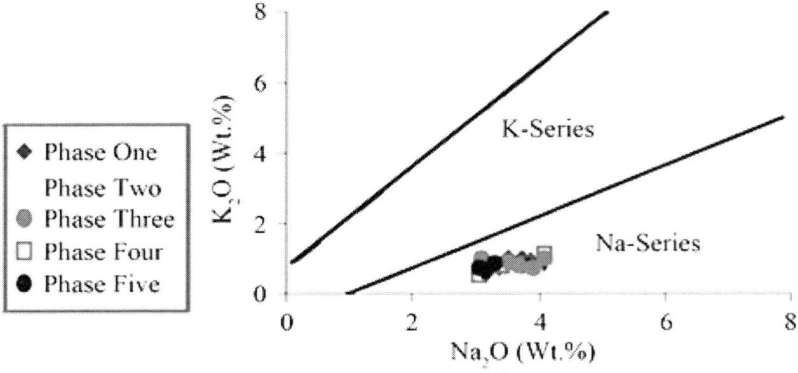

Figure 7. K_2O versus Na_2O (wt%) diagram, showing the subdivision of the transitional magma series into the Na sub-series of the samples in question, after Winter [22] .

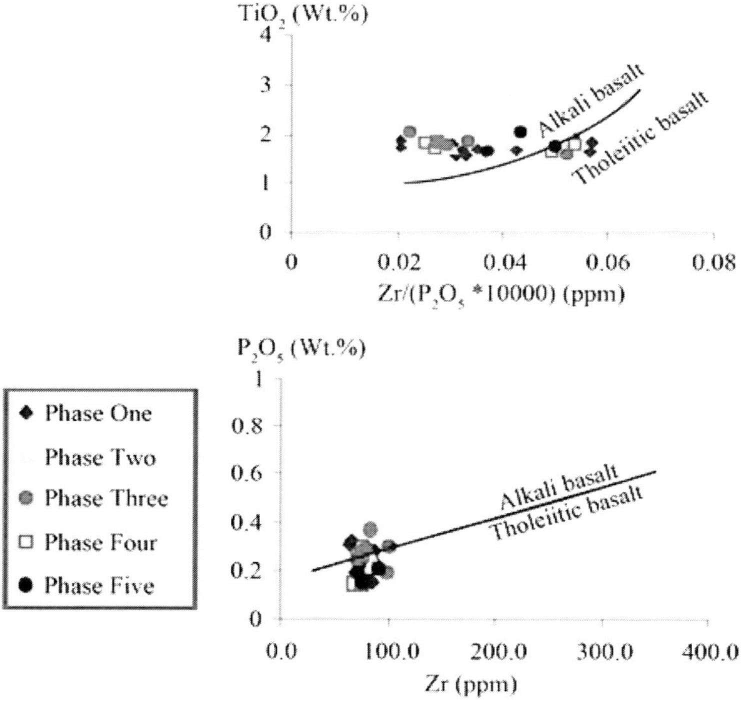

Figure 8. Minor (TiO_2, P_2O_5) and trace (Zr) element variation plots for the classification of the studied rocks as alkalic and tholeiitic basalts, after Winchester and Floyd [23] .

The P_2O_5-Zr diagram shows that the investigated samples are plotted close to the boundary of the alkali between the alkali basalt and tholeiitic basalt, whereas the TiO_2-Zr/P_2O_5 diagram shows that most of samples are situated in alkali field (Figure 8). This is expected since these diagrams use phosphorous as a principal component, which shows varying degrees of mobility, as stated by Rollinson [24] . Moreover, the Zr values of the studied samples are generally low.

The major element data contains only Fe_2O_3 as total iron. To obtain appropriate CIPW norms these values had to be recalculated into Fe_2O_3 and FeO. In the current study the ratio of Fe_2O_3/FeO = 0.2 was used (Less et al. [12] ; Peregi et al., [5]). Middlemost [20] has proposed a range of oxidation ratios (Fe_2O_3/FeO) for use with volcanic rocks drawn from geological literature; that of basalts is Fe_2O_3/FeO = 0.2, Rollinson, [24] .

There are various published programs for the calculation of the CIPW. The CIPW normative calculation in the current study was carried out by Norm4 software, written by Kurt Hollocher of the Geology Department, Union College, New York State. Several methods for the normative classification of basaltic rocks have been proposed. Yoder and Tilley [25] devised a scheme for classifying basaltic rocks by their CIPW norms using a basalt tetrahedron. Several modifications have been suggested to this classification and various schemes have been proposed by later investigators. Among these, Thompson [26] proposed a classification scheme for basalts based upon CIPW normative proportions of Ne, Ol, Di, Hy and Q. The three equilateral triangles of Thompson's diagram [26] represent basaltic and related rocks, which are respectively undersaturated, saturated and oversaturated: silica-undersaturated basalts (alkali basalts) are characterized by normative Ol + Ne; silica-saturated basalts (olivine tholeiite) are characterized by Hy + Ol; and, silica-oversaturated basalts (quartz tholeiites) are characterized by normative Q + Hy. With regard to normative minerals, the most studied samples are classified into hypersthene-normative; Wilkinson [27] considered hypersthene-normative rocks to be transitional or mild alkaline.

Busrewil and Oun [28] suggested that the transitional basalt is applied to basaltic rock that exhibits a nepheline-free norm. Three samples belonging to Phase 1, two samples from Phase 2 and two samples from Phase 3 were classified into nepheline-normative alkaline basalts (ne = 0.10 - 4.8; alkali-ol basalts), and one sample from Phase 4 was located at the boundary between ol-tholeiite and quartz tholeiite; this latter sample is near to oversaturation in SiO_2, which is a common feature of tholeiitic basalts (Figure 9).

The two main types of variation diagrams currently used are bivariate plots and triangular diagrams. A bivariate plot shows the variation between samples and identifies trends. Elements plotted along the x-axis of the diagram should be selected either to show maximum variability between samples or illustrate a particular geochemical process. Normally, in basic igneous rock MgO might be selected, Rollinson, [24] .

One of the most commonly used alternatives to the Harker diagram is MgO plot. This is most appropriate for a rock series which includes abundant mafic members, as in this case the range of SiO_2 concentrations may be small. The changes in selected major and minor element oxides are plotted against MgO for representative sam ples from the study area. The analyses exhibit an evolutionary trend representing slight increases in SiO_2, K_2O and Na_2O, along with decrease CaO, FeOt and MgO. Clearly, MgO increases with increased FeOt. However, in spite of this overall trend there is no consistent chemical pattern related to each phase.

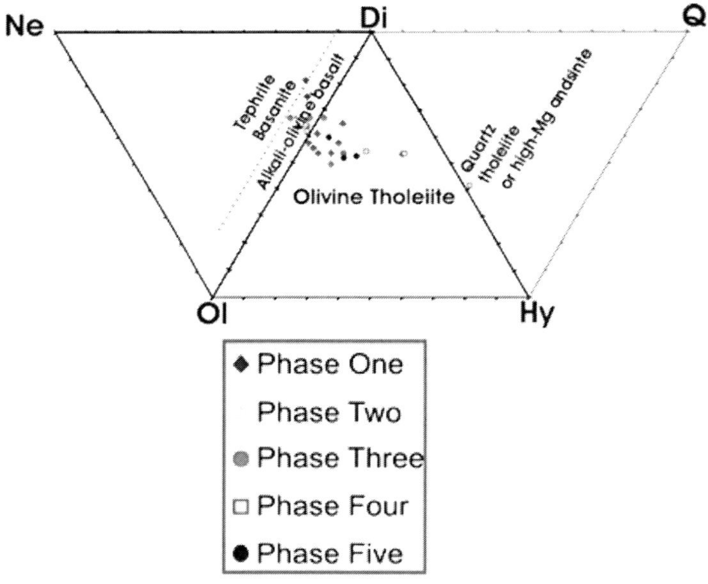

Figure 9. The classification of the studied samples according to their CIPW normative composition expressed as Ne-Ol-Di, Ol- Di-Hy, Di-Hy-Q (basalt tetrahedron, after Thompson, [26]).

Vague correlations between MgO and most other oxides in the studied samples probably suggest that these are related by different degrees of partial melting rather than by fractional crystallization. The relatively low MgO of less than 7.5 wt% and Ni (81 - 193 ppm) and Cr (238 - 361 ppm) contents indicates that the studied basaltic rocks exhibit slight to moderate fractionation of olivine and/or spinel. Consequently, the basaltic rocks are not primary magmas.

On average in the studied samples, the lava flows of Phase 5 and two samples from Phase 3 are more primitive. MgO is an important component of the solid phases in equilibrium with mafic melts and shows a great deal of variation, either as a consequence of the breakdown of magnesium phases during partial melting or as a result of their removal during fractional crystallization. Below 7% MgO-CaO correlates positively with MgO, indicating the removal of CaO-MgO from the liquid due to the crystallization of plagioclase and clinopyroxene. However, above 7% MgO and CaO-MgO correlate negatively, because this part of the trend is controlled entirely by the olivine. Inflections are most obvious where the number of fractionating minerals is small, Rollinson, [24] .

Trace element data provide a clearer view than the major and minor elements of distinct sources for the various magma types. In the current study the focus is on the trace elements, which due to enrichment and depletion are revealing about the nature of the basalt plateau. The trace element data can be used to evaluate the source composition, degree of partial melting and subsequent processes. The ratios of incompatible elements have been employed to distinguish between source reservoirs. The concentrations of the compatible trace elements (Ni, Cr, Co, Cu) are low in the studied samples, which supports our contention that they are not primary magma.

A characteristics feature of the studied volcanic rocks is their relatively constant ratios of certain incompatible trace elements (Nb/Zr and Rb/Zr) (Table 4). Relative constancy of ratios of incompatible elements provides strong evidence of a common source (homogenous). Also, the (Zr/Nb) ratios of studied basaltic rocks are less than 10, which indicates an ocean island basalt (OIB) source, Winter [22] .

Crustal contamination would certainly alter the ratios of Zr/Nb and Hf/Ta. The studied mafic samples have similar ratios, so the effect of crustal contamination on the studied basaltic rocks is probably minor, Baker [29] .

Several variants of the spider diagram have been used in the literature, with different elements and different normalization schemes. The order of the elements along the abscissa may vary slightly between different authors. Spider diagrams are flexible, and a variety of elements and normalizations have been displayed using them, Winter, [22] . In the current study, the comparison of the geochemical composition of the different rock samples, representative of different phases, with the MORB-normalized trace element patterns was used by Pearce [30] .

Figure 10 compares the incompatible elements of each sample with the most abundant igneous rocks on the planet. The large ionic lithophile (LIL) elements are on the left-hand side of the diagram and the high field strength (HFS) elements are on the right. Both are arranged in order of increasing incompatibility away from the margins, so that the most incompatible elements are just left of the centre of the diagram, Winter [22] .

The MORB-normalized trace element patterns of representative samples of different volcanic phases are given in Figure 10. Based on these diagrams, the patterns show a fairly close similarity in the shape and concentration range. The diagrams show a slightly different concentration of some incompatible trace elements within samples of the same phase, and also between samples representative of different phases, but they showed nearly identical behaviour for the less incompatible trace elements (Hf, Sm, Ti, Yb).

The studied basaltic rocks showed more peaks and troughs, reflecting the variable behaviour of the elements involved. Basalts from the study area are clearly enriched by the whole spectrum of more incompatible elements relative to MORB samples (normalized values greater than one). They show a broad central hump in which both LIL (K, Ba), and HFS (Ta) elements also reveal slight negative anomalies in Rb, Nb and Zr. The rocks display similar patterns of the peaks and troughs, which strongly suggests they share a common parent and subsequent processes, Winter [22] . Thus, based on this, the basalts could have a common genesis.

Table 4. Ratios between some trace elements of the studied mafic rocks that could be used for their petrogenesis.

	Phase One											Phase Two	
Sample No.	97	19	12	13	17	5	49	50	52	53	54	64	23	80
Nb/Zr	0.135	0.116	0.135	0.134	0.106	0.144	0.136	0.137	0.139	0.121	0.117	0.120	0.138	0.143
Rb/Zr	0.078	0.056	0.063	0.074	0.063	0.075	0.070	0.095	0.069	0.084	0.062	0.064	0.067	0.065
	Phase Two			Phase Three			Phase Four					Phase Five		
Sample No.	94	95	35	43	51	55	30	33	69	92	108	86	90	87
Nb/Zr	0.129	0.120	0.125	0.126	0.128	0.122	0.127	0.120	0.145	0.129	0.142	0.127	0.128	0.151
Rb/Zr	0.065	0.056	0.066	0.048	0.065	0.051	0.092	0.086	0.081	0.083	0.062	0.081	0.085	0.075

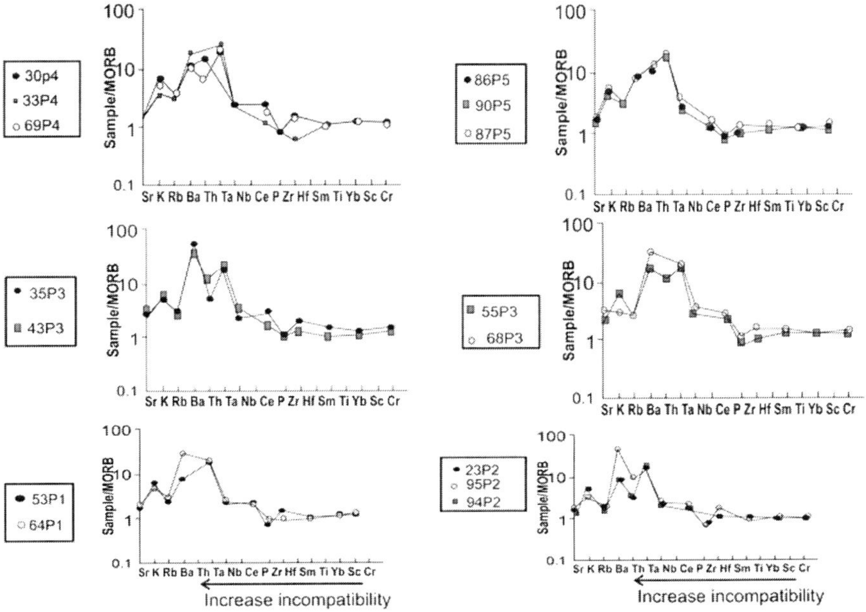

Figure 10. MORB-normalized patterns for selected samples from the study area, representative of different phases from Phases 1 to 5 (normalized data is greater than one). The Ba concentration of the studied samples shows extreme variation in some samples due to very high sensitivity to secondary processes.

The basaltic samples from different volcanic phases show a fairly similar character to each other in their geochemical behaviour. Consequently, one phase cannot be distinguished from another based on their trace element contents.

From the consistent negative slopes, it can be deduced that the slope of the studied basaltic rocks are distinct from MORB slope and appear to originate in the lower enriched mantle reservoir, although a very low degree of partial melting may also produce LREE-enriched melts from primordial or slightly depleted sources, Cox et al. [19] . The pattern is regarded as typical of melts generated from relatively undepleted mantle material.

Spikes at LIL (Rb, Ba, K) may reflect enrichment, perhaps due to contamination of magma by continental crust. Negative Nb anomalies are also characteristic of the continental crust and may be an indicator of crustal involvement in magma processes, Rollinson [24] . Enriched LIL and the slight alkaline of the studied basaltic rocks suggest derivation from an enriched mantle source, Aboazom [31] .

The high LIL/HFS pattern and negative Nb anomaly is a feature of many intracontinental volcanics, suggesting this pattern may be also a feature of the subcontinental lithospheric mantle (Fitton et al. [32] ; Hawesworth [33]).

As the degree of partial melting increases, the proportion of a more incompatible element with respect to a less incompatible element decreases as it is progressively diluted, so that (as illustrated by Figure 11) the La/Nb ratio should decrease and the La/Zr ratio should increase. The extent of partial melting should thus increase from the upper left to the lower part of the mantle partial melting curve, Winter [22] .

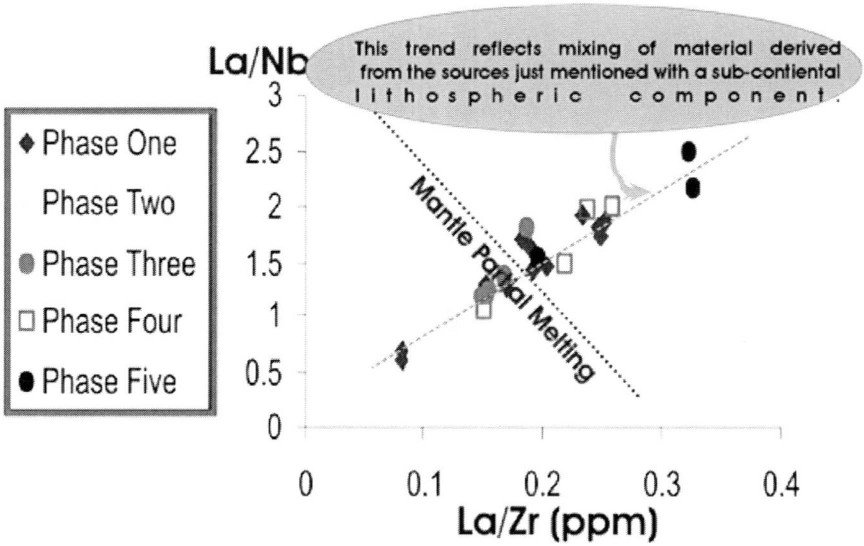

Figure 11. La/Zr vs La/Nb for the studied basalt samples, after Hopper et al. [34] .

The distinct phases suggest a single source and probably similar degrees of partial melting. The positive trend in Figure 12 reflects the mixing of material derived from the source, Hopper et al. [34] .

The elements of the first transition series (Ti, V, Cr, Mn, Fe, Co, Ni, Cu, Zn) vary in geochemical behaviour, and most of them are considered to be compatible elements. Transitional element plots have been mainly used with basalts as a means of exploring the geochemical behaviour of the first transitional elements, Rollinson [24] .

Figure 12 shows consistent trends for most studied samples representative of different phases on chondrite- normalized plots. They show progressive depletion from Ti to Ni and have a positive Ti anomaly and negative Cr anomaly. Cu and Zn are more variable and their concentrations may not reflect the original igneous values, (Longmuir et al. [35] . Anomalies in Ni and Cr concentrations may reflect the role of olivine (Ni) and clinopyroxene or spinel (Cr); Ti anomalies indicate the role of Fe-Ti oxides as opaque minerals.

Despite differences in the major element chemistry, the pronounced and very close patterns of similarity of trace elements of all the samples representative of the different phases certainly suggest there is a common source for the mantle peridotite beneath Al-Haruj province.

3.1. Comparison of the Studied Mafic Rocks with the Mid-Oceanic Ridge (OIB)

A broad spectrum of trace elements for the studied basalts rock can be compared using composite (average) OIB, as proposed by Sun [37] (Figure 13). The patterns of the studied basaltic rocks are similar in the shape to OIB, but vary in the concentration range of some incompatible trace elements. OIB enrichment in LIL (Sr, K, Rb) thus suggests a somewhat depleted mantle source relative to OIB, Winter [22] .

3.2. The Results of Correspondence Analysis

There is a large body of literature on the statistical treatment of geochemical data. Correspondence analysis is a descriptive and exploratory technique designed to analyse multi-way tables containing some measures of correspondence between rows (observations) and columns (variables). The results provide information that is similar in concept to those produced by correlation techniques. The term "inertia" in correspondence analysis can be thought of as a measurement of variation in data.

Obviously, the final goal of correspondence analysis is to find a graphical interpretation (meaning) from extracted dimensions (dimension 1 on the x-axis and dimension 2 on the y-axis). Thus, one can make the interpretation that the great similarity between the average of the elements of each phase with regard to the correlation between most HFS, LIL and transitional metal elements is more in Phase 4. However, as has been indicated, it is appropriate to make general statements about the nature of the dimensions (Figure 14).

The variance is explained by coordinates (dimensions) of more than 98% (96.03% of inertia on dimension 1 and 2.192% of inertia on dimension 2), which implies the relation between the average of the different phases and relation of elements between them and the relation between the elements and average of the five phases are highly significant, and so it can be stated with great confidence.

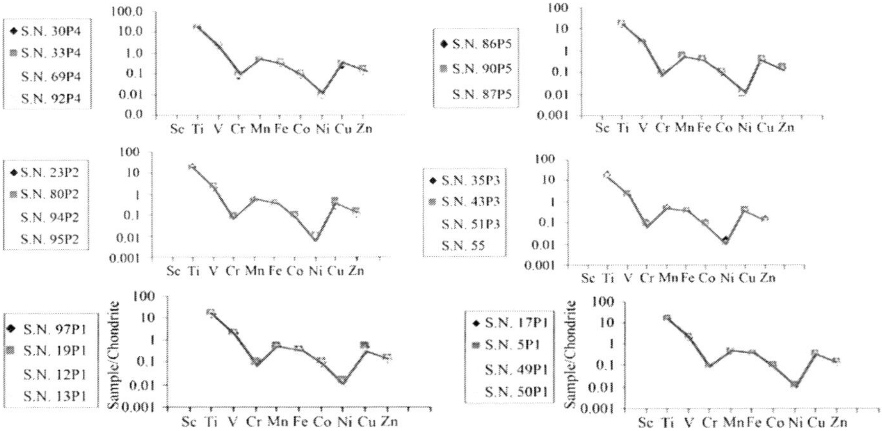

Figure 12. Transitional metal concentrations in most studied samples representative of the five phases, normalized to chondritic values, showing the same behaviour of transitional metal of different phases. The normalized values are taken from Kay and Hubbard [36] .

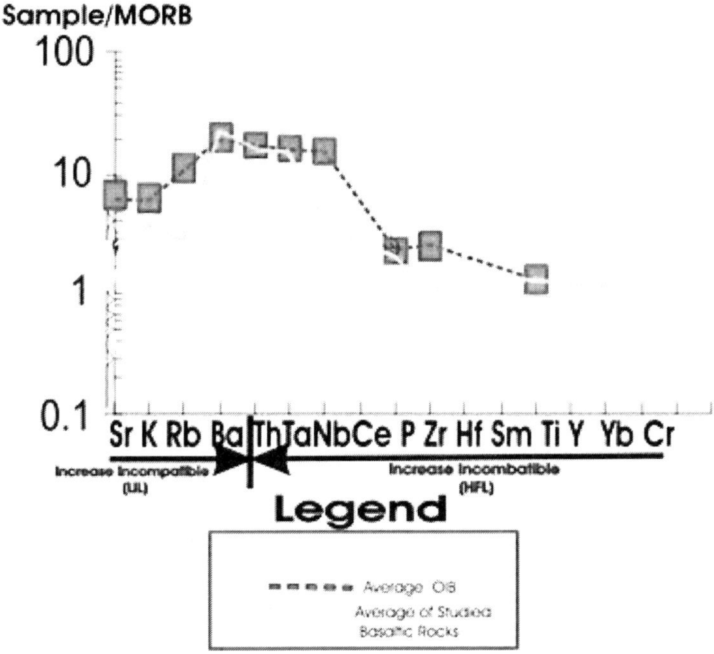

Figure 13. MORB-normalized spider diagram for average OIB, data from Rollinson, [24] and average representative analyses of the studied samples.

Judging the quality of the average of five phases expresses the value and importance of each phase in the current study, so in Table 5 it can be clearly observed that all phases are of highly significant importance. The correspondence analysis is useful for drawing attention to the likelihood that all phases were generated by sharing a common source.

3.3. Tectonic Environments

The approach of using geochemical analyses to try to discriminate magmas into different tectonic settings is described by several authors, and is probably best attributed to Pearce ([38] [39]), Pearce and Cann ([40] [41]), Pearce et al. ([42] [43]), Pearce and Norry [44] and Wood et al. [45] . They accepted that the major, minor and trace elements of basalts should reflect their geotectonic position, because the geochemistry of basalts is dependent on the crustal nature of the area in which they were formed.

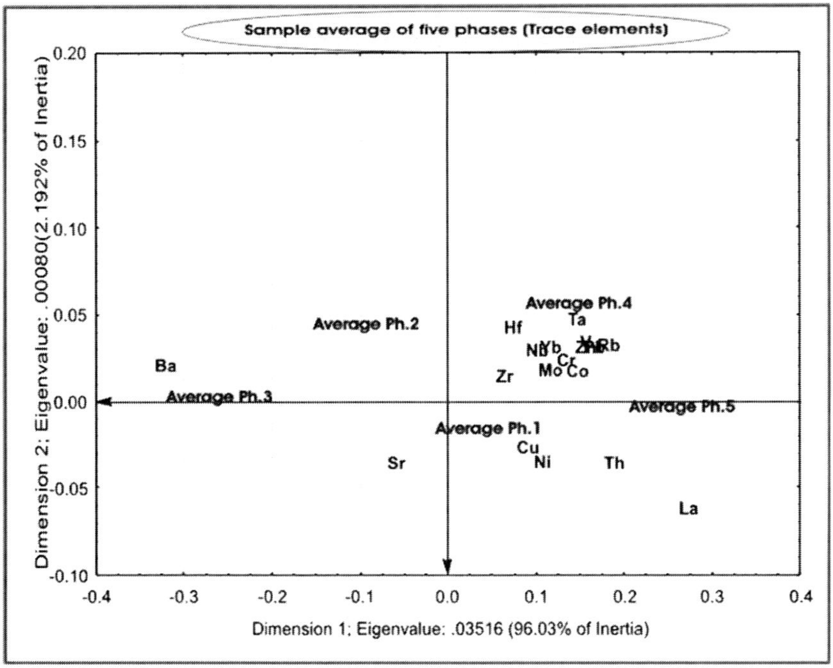

Figure 14. Correspondence analysis for trace elements of the average five phases presented in the area investigated, plotted on coordinates (on dimension 1 and dimension 2).

Table 5. The results of correspondence analysis for the average of the five phases.

Aver. Phases	Colum	Coordinate	Coordinate	Quality
Average Ph. 1	1	0.052836	−0.032650	0.818930
Average Ph. 2	2	−0.093243	0.028226	0.904941
Average Ph. 3	3	−0.262192	−0.0125540	0.995548
Average Ph. 4	4	0.135643	0.042508	0.970742
Average Ph. 5	4	0.261218	−0.020566	0.991675

Floyd and Winchester [46] found that basalts with similar chemistry may be produced in different tectonic regimes, since their compositions reflect the nature of the particular mantle and the melting processes operating rather than the regime in which they are found. Probably, this often makes the chemical designation of ancient basalt less certain.

Wood et al. [47] considered the effects of crustal contamination which were not considered by Pearce and Cann [41] and yet which are likely to have a significant influence in the appearance of continental flood basalts and may be a contributory factor in their persistence in classification, Rollinson [24] . There are a large number of discrimination diagrams applicable to basalts and basaltic andesites, which use trace elements and minor and major elements to distinguish the different environments of basalt. Of course, the investigator should not rely wholly on any single technique or plot, Winter [22] .

Pearce [38] developed a method that classified basalts in term of their tectonic setting on the basis of their major element patterns. He calculated the discriminant function based upon the eight major oxides and presented discriminant diagrams to identify mid-oceanic ridge basalt (MORB), within plate basalt, oceanic island basalt (OIB), continental flood basalt (CFB), calc-alkaline basalts (CAB), island arc tholeiites (IAT) and shoshonites (SB).

Two discriminant functions, F1 and F2, were identified as follows:

$$F1 = +0.0088SiO_2 - 0.0774TiO_2 + 0.0102Al_2O_3 + 0.0066FeO - 0.0017MgO$$
$$- 0.0143CaO - 0.0155Na_2O + 0.0007K_2O$$

$$F2 = -0.0130SiO_2 - 0.0185TiO_2 - 0.0129Al_2O_3 - 0.0134FeO - 0.0300MgO$$
$$- 0.00204CaO - 0.0481Na_2O - 0.071K_2O$$

It is clear that the majority of the samples investigated are plotted in the field of within-plate basalts (Figure 15).

Wood [47] proposed the Th-Hf-Ta ternary discrimination diagram based upon the immobile (HFS) elements. Using this classification (Figure 16(a)), the studied basaltic rocks plot closer to the fields of within-plate basalts. The elements Th, Hf, and Ta existed in very low concentrations in basalts and cannot accurately determined by XRF analysis.

The advantage of using minor elements as discriminants between basalts is that they are readily detected and more accurately determined than trace elements, Rollinson [24] .

Pearce et al. [42] found in an empirical study that it was possible to discriminate between oceanic and continental basalts with a single straight boundary line on a TiO_2-K_2O-P_2O triangular diagram (Figure 16(b)). The field oceanic basalts included MORB and OIB. The studied basaltic rocks are plotted around the line separating the two fields.

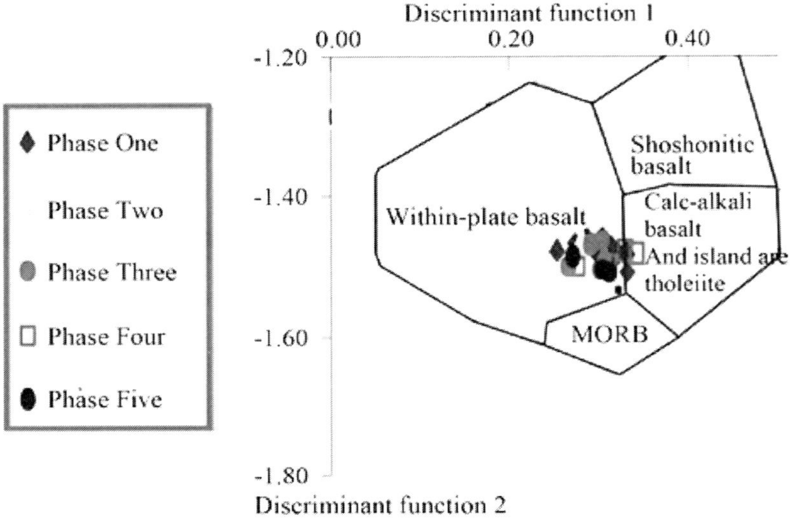

Figure 15. Major element discrimination diagram for basalts, after Pearce [38] , showing the most studied samples plot in the field of within-plate basalts.

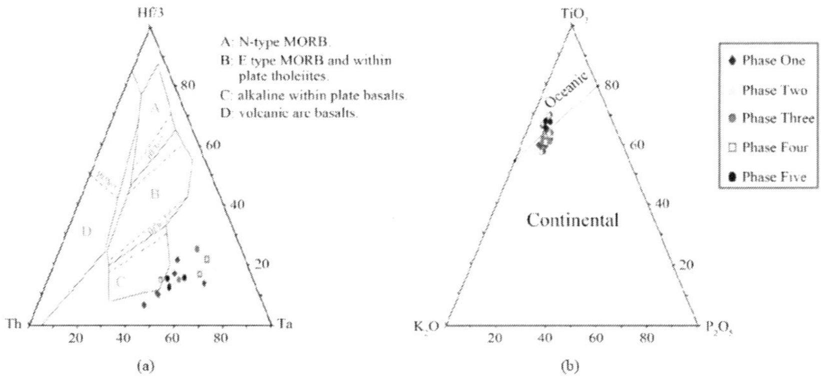

Figure 16. (a) The Th-Hf/3-Ta ternary discrimination diagram for basalts (after Wood, [47]); the broken lines indicate transitional zones between basalts types. (b) Triangular discrimination diagram between oceanic and continental fields, after Pearce et al. [42] .

4. SUMMARY AND DISCUSSIONS

The tertiary volcanic rocks of Jabal Al Haruj represent the largest volcanic province in Libya covering an area of ≈ 45,000 km^2, which originated during several phases of eruptions that were closely associated in space and time.

The study area is located in north-eastern Jabal Al Haruj al Aswad, an area which has not been studied in a thorough manner so far. Nevertheless, the work of Klitzsch [13] gave research details showing the distribution of six major phases and one minor phase. The most important study in this respect was done by Busrewil and Suwesi [2] on the Al Haruj al Aswad. The authors gave a detailed volcanological study of the area and distinguished six volcanic phases.

The division of the volcanic phases used in the present work is based on similar principles to those developed by the earlier authors, which are as follows:

Colour variation of the lava flows of different phases on Landsat TM images.

Degree of weathering of the rocks.

Accessibility of lava fields.

Morphological and lithological characteristics observable in the field.

Superposition of the lava flows and drainage pattern.

Field evidence has revealed that each of these criteria are complementary and cannot be used separately, Busrewil and Suwesi [2] . Field studies included geologic mapping and the identification of mode of occurrence of the volcanics. According to modern age dating, the volcanic activity in the area started in the Late Miocene and lasted until the Late Pleistocene. The first five of the six volcanic phases of the Al Haruj volcanic province are present in the study area. The geological setting, the petrography, the geochemistry and tectonic setting of these occurrences are examined in this study. The following is a summary of the above-mentioned aspects and the main conclusions obtained during the present work.

Field geological mapping was carried out using Landsat TM multispectral satellite imagery. This imagery formed the base of the geological map, which was completed by the addition of topographic contours, locality names and drainage pattern of wadis, as well as other field criteria. The study area is remote, rather inaccessible and extremely difficult terrain. Due to this, the field work was carried out in parallel with the Industrial Research Centre (IRC), Tripoli during their regional geological mapping of SE Libya in March 2006.

The topographic form of the old volcanic phases is highly modified, while the forms of more recent phases are usually fresh and very well preserved. In general, each basalt flow unit begins at the bottom with the most massive part, characterized by a fine to micro-vesicular texture, and it evolves upwards into an obviously vesicular texture. The basalts are dark to medium and pale gray with a porphyritic appearance. A fresh sample was taken from the massive part of each flow unit.

Characteristic features of Phase 1 are the blocky, rough, subangular to subrounded rocky structure and have smooth surfaces, massive and less vesicular. On the satellite image it is distinguished by a bluish colour.

Phase 2 covers a limited geographical extent and outcrops in small scattered localities in parts of the study area. Lava flows from this phase tend to disintegrate into small blocks and slabs, less than 50 cm in diameter. The vesicles are often elongated in the direction of flow. The presence of lamination is pronounced by the alteration of the variable concentration of vesicles. The accessibility of the area covered by Phase 2 is particularly bad. On the satellite image it is discriminated by a greyish to reddish blue colour, and an almost homogenous surface without any readily recognizable features.

The lava field of Phase 3 covers a large area with subangular blocks that form the rough type of morphology. The blocks are coated by thin dark brown crust providing a rather characteristic field appearance. The fresh broken surface of the rock is black or dark grey. The rocks are readily recognizable on the satellite image by their brown tone or hue, and the hilly morphology dotted by number of large balta (kipukas). The kipuka, or according to their Arabic name balta, are closed shallow depressions within the lava flows surrounded by lava tongues. In Phase 3 the geomorphological dimension of baltas may vary.

The exposures of Phase 4 are widespread in the central Al Haruj al Aswad and in the study area. From the central part of Al Haruj al Aswad large volumes of Phase 4 lava flow poured downward in a more or less radial pattern. The flows are distinguished by their morphology and the preservation of surface features. Lava flowing down pre-existing wadis acquired an elongated form as they filled the wadis.

Phase 4 overlies the Maradah Formation in some places, but the contact is not always clear due to talus covers, but in other places the contact is quite sharp. This phase is represented by dark grey to black vesicular lavas in the upper part with polygonal columnar joints and dense masses in the lower part. This flow is characterized by hilly morphology, pressure ridges and steep distinctly fresher lava scarps. On the satellite image, it has a black colour and features a large number of baltas.

Phase 5 is represented by a well-preserved volcanic form with the freshest of rocks, a vesicular form, and the conspicuous features of the pahohoe structure. The rocks of this phase cover a relatively small area confined to the south-west part of the study area. The excellent preservation and appearance of this phase reflects the younger age of these basaltic flows. On the satellite image the rocks are dark black with dots, representing small baltas of 5 to 20 m in size.

The drainage pattern developed on the surface of basaltic lava flows shows systematic alteration with the age of the basaltic lava flows, Busrewil and Suwesi [2] . From the Landsat TM image, the drainage frequently decreases as the flows of basaltic rock become progressively younger, as Phase 5 has a less dense drainage pattern.

The Libyan volcanic provinces are spatially related to two structural trends: early Palaeozoic structural elements of a prevalent NNW-SSE direction and by late Palaeozoic to Mesozoic structures with a general ENE- WSW trend. These elements represent the regional structures that have played an important role in the geological evolution of Libya, Peregi et al. [5] . Klitzsch [8] suggested that the basalts of Al Haruj al Aswad erupted at the crossing zone of NW trending early Palaeozoic elements (south Al Haruj uplift) and the NE trending late Paleozoic elements (Tibesti-Sirt uplift), and they probably also used some of the upper Cretaceou-Tertiary faults as conduits.

According to Busrewil and Suwesi [2] , the major structural trend of the central part of Al Haruj al Aswad volcanic province is controlled by the alignment of spatter cones and the distribution of the more recent lava flows (Phases 5 and 6) along NW-SE and NE-SW directions. Herein, the present area is located to the north- eastern part of Al Haruj al Aswad, where the data taken from the investigated area reveal that Phases 4 and 5 are controlled by the NE-SW direction, which is coincident with the pattern of the most distinct structures in the region (Figure 17).

From the direction of flow of the more recent lava (Phases 4, 5, 6) and from the alignment of cones, it is suggested that the situation of the Al Haruj lava field is related to tensional lines, one parallel to the Dur al Qussah fault system and the other running along the prolongation of the Hun graben, Busrewil and Suwesi [2] . These lines probably represent extensional stresses reactived by Proterozoic basement faults in the direction NW-SE, NNW-SSE, NE, NNE and E-W during the late Cenozoic, reflecting the evolution of the Red Sea, EL-Akhal [48] . The rising magma erupted along the fault planes cutting across the lower most resistant areas at the cross points of these tensional lines. The lava field and its associated volcanoes are thus considered to have resulted from a rather continuous release of pressure in the main central feeder. After the activity waned in Al Haruj volcanic province (after the Pleistocene), and the area was cut by a conspicuous rejuvenated NW-SE set of faults, Busrewil and Suwesi [2] .

The different rock varieties of the study area are petrographically described in 42 thin sections showing certain petrographic features signifying the different varieties. Mineralogically, most of the studied basaltic rocks contain olivine as an essential mineral together with clinopyroxenes (augite), plagioclase (labradorite to bytownite) feldspars and opaque accessory minerals (probably ilmenite). The recorded olivines are unzoned and either fresh or altered.

Phases 1, 3 and 4 contain a higher modal olivine content and olivine phenocrysts that are larger in size than those from Phases 2 and 5.

The volcanic trends (Phases 4 and 5) are well documented in the area investigated by the continuation of the NE-SW structural trend.

Therefore, the studied rocks can be considered as olivine bearing or olivine basaltic rocks, since they contain a significant amount of modal olivine.

The degree of alteration of the olivines in the studied rocks varies greatly; alteration is manifested around the edges and along the cracks of Phases 3 and 4. On the other hand, highly altered basaltic rocks are found in Phases 1 and 2. The alteration products are either partial or complete to deep reddish iddingsite. Clinopyroxene (augite) is detected in the investigated rocks: it has a maximum extinction angle of 44° to 51°, and is characterized by a pale brown colour, and is subhedral to anhedral and prismatic in shape. Pleochrism is not observed in most samples.

The rocks from Phase 5 are characterized by low plagioclase modes (labradorite to bytownite) and clinopy roxene (augite) and high contents of vesicles, amygdales and cryptocrystalline to glassy groundmass and microphenocrysts of olivine, which are fresher when compared to rocks from other phases. Plagioclase laths are the most abundant, representing 37% to 48% of the matrix in most of the studied samples. The composition of the plagioclase varied between An 58% to An 78% (labradorite to bytownite). They are both pilotaxitically and randomly arranged, and sometimes they exhibit a trachytic arrangement around olivine phenocrysts. Brown glass and cryptocrystals fill the interstices between laths of plagioclase. The rocks exhibit a porphyritic texture with phenocrysts embedded within the groundmass, whereas the groundmass shows intergranular, intersertal, ophitic and subophitic textures. Olivine phenocrysts occur sometimes as clots displaying a glomerporphyritic texture.

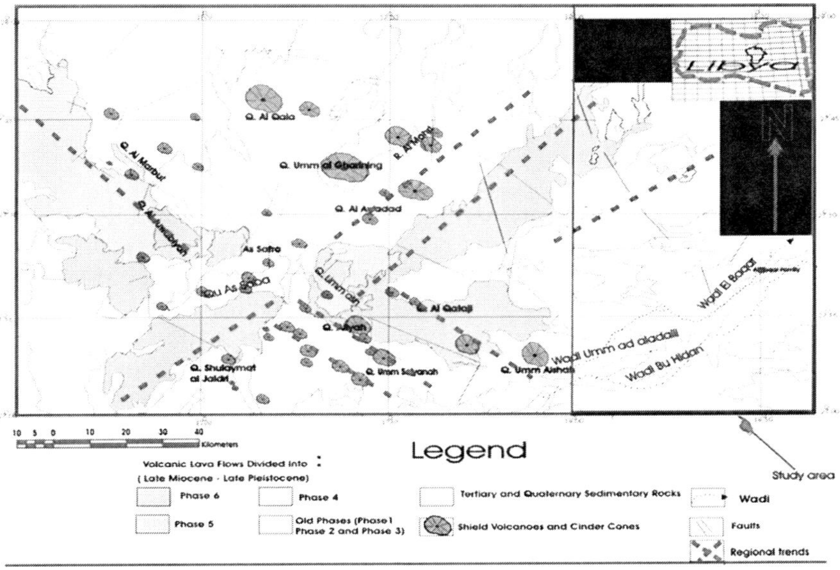

Figure 17. The structural alignment of the central part of al Haruj al Aswad volcanic province (Phases 5 and 6) and a scatter of cones following NE-SW and NW-SE trends, Busrewill and Suwesi [2] .

The order of crystallization of the minerals in the rocks from the various phases is similar. Phenocrysts of olivine result from a slower cooling beneath the surface in the magma chamber prior to eruption. Therefore, they indicate the presence of enough time to equilibrate with the melt at mantle depth. Upon eruption, the remaining liquid crystallizes to form fine laths and tabular or equant crysts comprising the groundmass and resulting in a dense network of plagioclase microlites and pyroxene granules, with opaque oxides of a small size. Glass may solidify at late stage, filling spaces between minerals and forming an intersertal texture, as in Phase 5.

Inclusions of opaque minerals should have formed at an earlier stage than the plagioclase (labradorite to bytownite) that enveloped them. In the ophitic texture, laths of plagioclase (labradorite to bytownite) enveloped by larger pyroxene (augite) is commonly interpreted as indicating that the clinopyroxene (augite) formed later, Winter, [22] . An anhedral or skeletal form of olivine indicates that their formation is in the later or final stages of rock crystallization, Winter [22] . From petrographical and mineralogical criteria alone, it is evident that the investigated basaltic rocks have more olivine phenocrysts than the pyroxene (augite). The groundmass typically consists of interlocking plagioclase (Labradorite to Bytwonite) laths with stubby prisms of clinopyroxene (augite) granules of olivine and opaque minerals.

This petrography study concluded that the five phases are typical alkaline basalts, which supports the findings of the study of Busrewil and Wadsworth [1] . Busrewil and Suwesi [2] suggested that they are a "subalkaline basaltic series which show some similarities except that they are more clearly subophitic with the presence of more opaque minerals". Consequently, the studied mafic volcanic rocks representative of the five phases are probably transitional between alkaline and subalkaline series. But, it must be emphasized that it is necessary to establish their geochemical character to define their petrogenetic affinity within the framework that will establish the magma type. The difference between the various phases is due to the circumstances of magma solidification, not a manifestation of major variances in mineralogical and chemical composition.

The findings of the present study are a positive supplement to previous works such as that by Peregi et al. [5] , which suggested that "the petrographic similarity of the mafic rocks that should be noticed, cannot be distinguished frequently in terms of volcanic phases and flow units based on their petrographic characteristics. The difference sometimes may be more significant in samples taken from the surfaces and deeper part of the same lava flow, than between the different volcanic rock phases."

Again, the above study agrees with that by Farahat et al. [14] who stated that "differences in groundmass textures may be related to differential degrees of cooling and crystallization at various levels of the outpouring of magma."

The geochemistry of the basaltic rock under consideration has been revealed through the analysis of 29 samples for their major and trace elements determined quantitatively by Philips (2404) X-ray fluorescence (XRF). Several geochemical parameters and various diagrams were adopted to recognize the chemical characteristics of these rocks, to evaluate their chemistry within the framework of established magma types and to determine the tectonic environment of their phases.

The major elements of the investigated rocks and the content of their trace elements demonstrates that all the phases are similar in chemical composition. In terms of different methods of recent chemical classification based on major, minor and trace elements, the investigated basaltic rocks have a subalkaline to alkaline affinity. If the term "transitional basalts" is accepted, the investigated rocks are best named as "transitional".

The most studied samples are hypersthene-normative. Wilkinson [49] considered hypersthene-normative rocks to be transitional or mild alkaline, and other samples were classified as nepheline-normative or alkaline basalts, and one sample was located at the boundary between olivine tholeiite and quartz tholeiite, which could represent a distinctive feature of tholeiitic basaltic rock.

In the current study, the focus has been on trace elements, which are generated as enriched or depleted and reveal more about the nature of basaltic rocks. Concentrations of compatible trace elements are low in the studied samples, which supports the contention that they are not primary magma. A characteristic feature of the studied volcanic rocks is their relatively constant ratios of certain incompatible trace elements such as Nb/Zr and Rb/Zr. The constancy of incompatible element ratios provides strong evidence that they were generated from a common source. The Zr/Nb ratios of the basaltic rocks are less than 10, which also indicates an OIB source. The studied samples have similar ratios of Zr/Nb and Hf/Ta, so the effect of crustal contamination on the studied basaltic rocks is probably minor.

Based on MORB normalized trace element patterns of representative samples of different volcanic phases, the patterns show fairly close similarities in their shape and concentration range, whereas they are characterized by very slightly different concentrations of some incompatible trace elements within samples of the same phase and also between samples representative of different phases, but those showed a nearly similar pattern in the less incompatible trace elements.

The studied samples of the different phases, which were enriched in all incompatible trace elements over MORB (normalized values greater than one) in proportion to the degree of incompatibility, also showed slightly positive anomalies in K, Ba and Ta and slightly negative anomalies in Rb, Nb and Zr. The rocks display similar patterns of peaks and troughs, strongly suggesting that they share a common parent and subsequent processes, Winter [22] .

The basaltic samples from the different volcanic phases showed remarkable similarities with each other, but with slight geochemical variations in some incompatible elements. Consequently, the phases cannot be distinguished from one another on the basis of their trace element contents.

Spikes at Pb, Ba and K may reflect an enrichment component, perhaps due to their contamination by continental crust, Winter, [22] . Enriched LIL and the slight alkaline of the studied basaltic rocks could suggest a derivation from an enriched mantle source. The high LIL/HFS pattern and negative Nb anomaly is a feature of much intracontinental volcanism, suggesting a distinctive feature of the subcontinental lithospheric mantle. The studied samples tend to be depleted in Ni and Cr, along with Mg# (Mg/(Mg + Fet))

(52 - 62), suggesting the fractionation of olivine and spinel prior to eruption.

Despite differences in the major element chemistry, the pronounced similarities in the pattern of trace element between different phases certainly suggest that there is a common source of the mantle peridotite beneath the Al Haruj volcanic province.

The results of correspondence analysis are explained by the variance on the coordinates (dimensions) of more than 98% (96.03% of inertia on dimension 1 and 2.192% of inertia on dimension 2), which implies that the relation between the average of the different phases is highly significant. The quality of the average of the five phases expresses the value and importance of each phase, so it can clearly be noticed that all phases are of very significant importance.

The basalts of the area representing all phases were formed in an intracontinental (within plate) tectonic setting and they represent alkali to tholeiitic basalts (transitional basalt), most of them having undergone slight to moderate fractionation. The studied basaltic rocks were plotted around the separating line between the two fields (oceanic and continental), Pearce et al., [42] ; Figures 5-14(b)). Probably the volcanic activity responsible for these basalts occurred in a zone of initial continental rifting.

The trace element distribution patterns of the studied rocks is similar to OIB, which probably represents basaltic melts from the enriched asthenospheric mantle source, often associated with an uprising mantle plume, Peregi et al. [5] .

Comparison of the Tertiary Volcanics from Different Parts of Libya

Generally, Libya is situated on the Mediterranean foreland of the African shield, which extends over a platform of cratonic basins. The major diastrophic disturbances include the Caledonian and Hercynian disturbances during the Cretaceous, middle Tertiary (Oligocene through to Miocene) and Holocene times, causing uplifts, subsidence, tilting, faulting and intrusions. However, the effects of major diastrophic events were generally broad, and there are very few compressional folds, Goudarzi [50] .

Apart from the different ages of the four principal volcanics in Libya, much similarity can be detected among them, particularly in the mechanism and mode of eruption. The common criterion for Tertiary volcanics is that volcanism occurs mainly along well-defined old structural elements. Those structural elements are basically developed from three structural periods characteristic of the middle part of North Africa, Klitzsch [51] .

The basement underneath the Tertiary volcanism areas is bound to the overlap of older structural elements forming the boundaries of the major current basins in Libya. The overburden sequence of those areas are of the minimum thickness, and occasionally the Precambrian crystalline rocks are exposed in limited and comparatively small areas such as in Jabal al Hasawinah, west of Jabal Nuqay, and the Tibesti area (Busrewil and Oun [28] ; El-Makhrouf [52] ; Goudarzi [50]). The volcanic fields of the four principal provinces in Libya are strongly correlated to the current basement elevated areas. Consequently, it seems to be that the attenuation of the crust within specific areas during Paleozoic and Mesozoic times is one of the major contributory factors that facilitated the outpouring in the volcanic provinces in Libya during the Tertiary Period.

The Gharyan volcanic province (GVP) is the smallest of the principal Tertiary volcanic areas in Libya, and it forms a generally continuous sheet in the central part of Jabal Nafusah. The others are Jabal as Swada, Jabal Al Haruj and Jabal Nuqy. These volcanic fields represent the Tertiary-Quaternary outpouring of volcanic activity and define a NNW-SSE linear trend with a decrease in age from NNW to SSE. The GVP lies at the intersection of two major structures, the NNW trending Tripoli-Tibesti uplift formed in Caledonian times, and the E-W trending "Nafusah uplift" formed during the Hercynian orogeny (Anketell and Ghellali [53] ; Goudarzi [50] ; Mikbel [54]).

The volcanic field of Jabal al Hasawinah is related to the intersection of the early Paleozoic NNW-SSE Tripoli-Tibesti uplift and the Hercynian ENE-WSW Al Qarqaf uplift, (Busrewil and Oun, [28] ; Jurak, [55]). Further NE of the Jabal al Hasawinah volcanic field, there is the Al Qarqaf uplift crossing the early Paleozoic NNW-SSE structural element of the Al Haruj uplift; Jabal as Swada is spatially related to this area (Woller and Fediuk, [9]).

The situation of the Al Haruj lava field, contrary to the other volcanic provinces, is related to three structural elements: the early Paleozoic Al Haruj uplift of a NNW-SSE direction, the late Paleozoic to Mesozoic NE fault system that is associated with Tibesti-Sirt uplift, and the prolongation of the Cretaceous-Tertiary Hun graben, Busrewil and Suwesi [2] ; Woller and Fediuk [9]). This may account for the greater apparent size of the Al Haruj volcanic field compared to the other volcanic provinces, Farahat et al. [14] .

Chemical analyses of major, trace and rare earth elements (REE) from Al Haruj volcanics are presented in Appendix III, El-Shaafi [6] , together with analyses of representative rocks from different parts of Libya for comparison purposes (Tables B.1-B.4). The analysed samples from representative rocks from different parts of Libya are depicted on variation diagrams. The consistent patterns of the plotted data suggest that apart from oxidation, there has been little alteration.

The subalkaline basalts from the different volcanics in Libya tend to be relatively evolved with a differentiation index (DI) extending from 29.9 in the most primitive basalts to 39.9 in the most evolved hawaiites, Busrewil and Suwesi, [2] . There are no associated less-evolved rock types exposed that might be representative of the parent magma. All the chemically analysed rocks have relatively high silica and low TiO_2 and total alkalis, and although there is some partial overlap with the alkali basalts of the main lava series, they are clearly subalkali. They could be related to other lavas generated at lower pressure and/or higher degrees of partial melting of the parent magma. Fractional crystallization at shallow levels, involving major minerals (ol, cpx, ±opaques) occurring in the rock, can account for the observed series, Busrewil and Suwesi, [2] .

For classifying the rocks on a geochemical basis, the variation of silica and total alkalis is one of the most important discrimination techniques. The IUGS total alkalis (TAS) classification of volcanic rocks of Le Bas et al. [56] is adopted and shown in Figure 18.

The Gharyan volcanics, the present study volcanics and the study made by Busrewil and Suwesi, [2] of the basaltic rocks of the central part of Al Haruj al Aswad, illustrate the suite of subalkaline basalts and alkali basalts, and most of them fall within the basalt field of the TAS diagram. The more differentiated types are simply referred to as hawaiites and mugearites (Figure 19(a)). Twenty-seven samples of the studied basaltic rocks from GVP were chemically analysed, Gafeer [57] for major and trace elements (Tables B.1-B.4, Appendix III). Fresh samples were selected as far as possible and attention was paid to the homogeneity of the samples.

The volcanics in the GVP comprise four petrologic and chronologic phases: the Old Lava Series (OLS); its subordinate, the Young Lava Series (YLS; Early-Eocene to Quaternary); phonolite-trachyite intrusions (Late Eocene-Early Oligocene); and the late volcanic centres (LVC) of the Middle-Late Miocene (see Figure 19(a);

Aboazoam et al. [58] ; Almond et al. [59] ; Busrewil and Wadsworth [1] ; Busrewil and Wadsworth [3] ; Hegazy [60] ; Piccoli [61]).

The Gharyan lava series demonstrates a subalkaline affinity, as shown by its basalt and basaltic andesite presented in Cox et al. ([19] ; Figure 19(a), Figure 19(b)), which typically contain normative hypersthene (although three of the analysed samples contain minor amounts of normative nepheline). Generally, the term “transitional” applies to basaltic rocks that exhibit a nepheline-free norm, Le Bas et al. [56] .

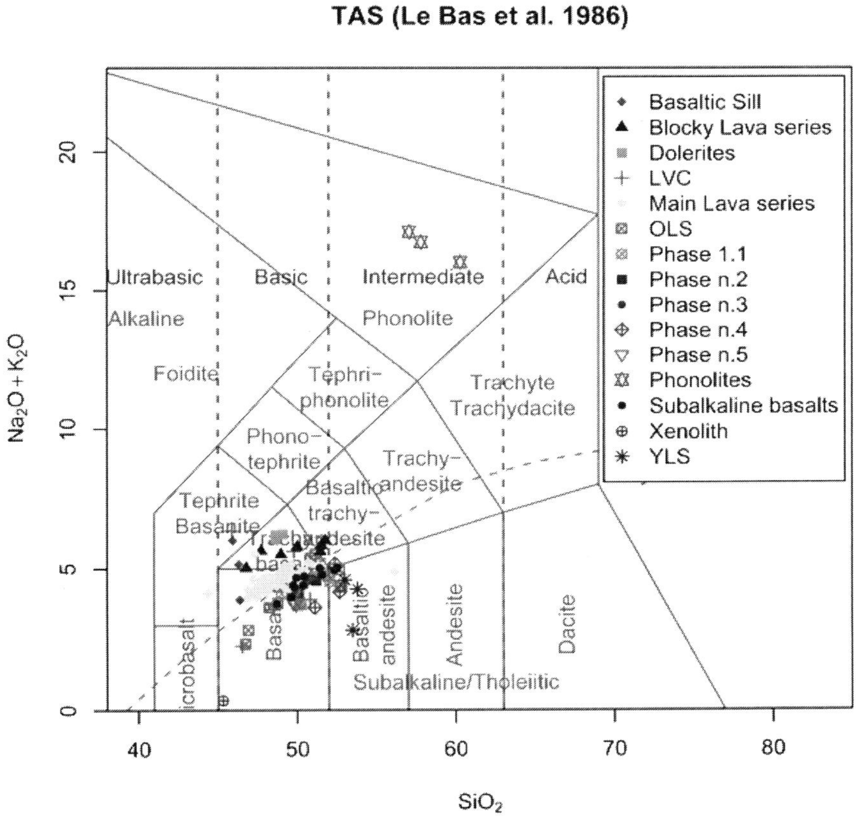

Figure 18. The chemical classification and nomenclature of volcanic rocks studied in the different volcanic provinces in Libya, using the total alkalis versus silica (TAS) diagram of Le Bas et al. [56] .

Wilkinson [49] considered hypersthene-normative rocks to be transitional or mildly alkaline, a characterization which Almond et al. [59] had used to define the suite of the lava series within the Gharyan volcanic province. Phonolites were not observed in Jabal Al Haruj al Aswad by previous investigators or in this study, but the emplacement of phonolites in Jabal Awayant, the Gharyan volcanic province and Jabal Hassawnah, Jurak [55] suggests that a genetic relationship between the basement highs, broad domal uplifts and the emplacement of phonolites exists Busrewil and Suwesi [2] .

The divisions of Al Haruj volcanics into plateau lavas, blocky lava, shield volcanoes and doleritic dyke-like masses, based on field appearance have already been established by Busrewil and Suwesi [2] . These data together with chemical observations are used to identify the various rock types present and to assess their petrogenesis. Variations of silica and total alkalis demonstrate a subalkaline affinity trending towards alkaline field (Figure 19(b); in Cox et al. [19]).

Eighty-eight samples of the studied basaltic rocks from Al Haruj al Aswad were chemically analysed by Busrewil and Suwesi [2] for major and trace elements (Tables B.1-B.4, Appendix III). Fresh samples were selected as far as possible and attention was paid to the homogeneity of the samples. Although there is similarity between the lava series of the current study data (Figure 19(c)) to counterparts in the study by Busrewil and Suwesi [2] , the above suite shows a wider compositional range as it attains a wider differentiation index (DI) than the present study.

Twenty-nine samples of the studied basaltic rocks from Al Haruj al Aswad were chemically analysed (the current study) for major and trace elements (Tables B.1-B.4, Appendix III; El-Shaafi [6]). Fresh samples were selected as far as possible and attention was paid to the homogeneity of the samples, (Figure 19(c); Cox et al. [19]).

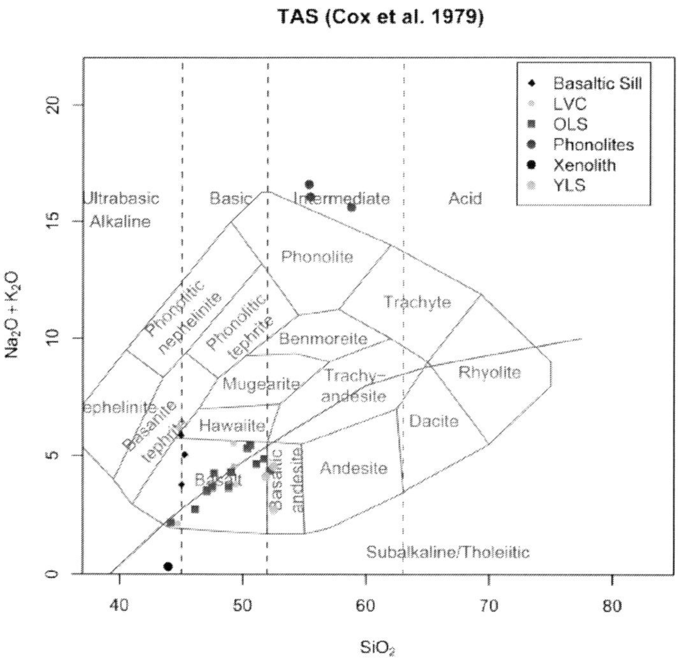

(a)

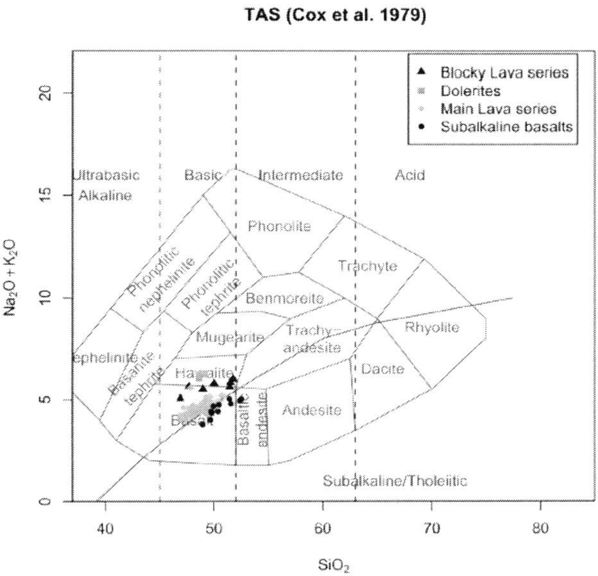

(b)

(b)

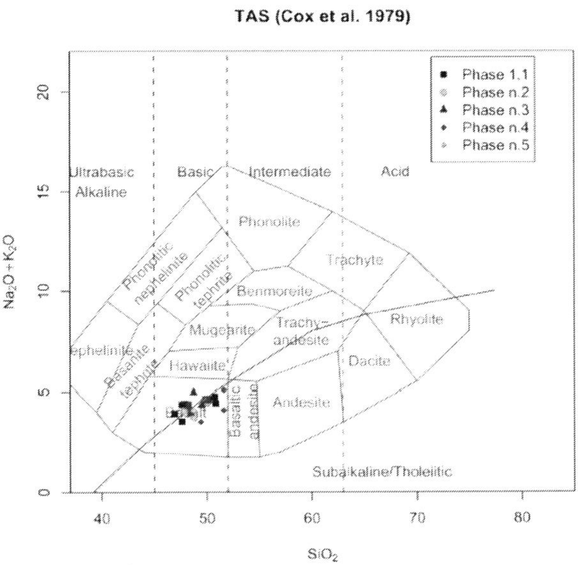

(c)

Figure 19. Alkali versus silica diagrams of Cox et al. [19] showing (a) Plot of total alkalis vs silica for Gharyan province after Gafeer [57] ; (b) Plot of total alkalis vs silica from Busrewil and Suwesi [2] ; (c) Plot of total alkalis vs silica for the area investigated (current study).

Based on geochemical data, alkali-olivine basalt and transitional basalt of the study area appear to have been emplaced continentally within plate. The use of immobile HFS elements Th-Hf-Ta is proposed by Wood [47] for basalt plots in the field of "within plate basalts" (Figure 20), whereas the most differentiated members of subalkaline transitional basalts lie in the range of "continental basalts" field on FeOt, MgO, Al_2O_3, (Figure 21; Pearce et al. [43]), which encompasses both subalkaline basalts and basaltic andesites.

The within plate tendency for the analysed basaltic rocks from various regions are pronounced in the Zr/Y-Zr diagram (Figure 22).

The chemical data for different volcanic provinces are normalized to the primitive mantle (Figure 23) in which the observed Nb anomaly is a characteristic feature of continental and oceanic alkali basalts. The contamination by crustal rocks is the main factor controlling the behaviour of K, Ba and Rb (Najafi et al., 1981, as cited in Aboazom, [31]).

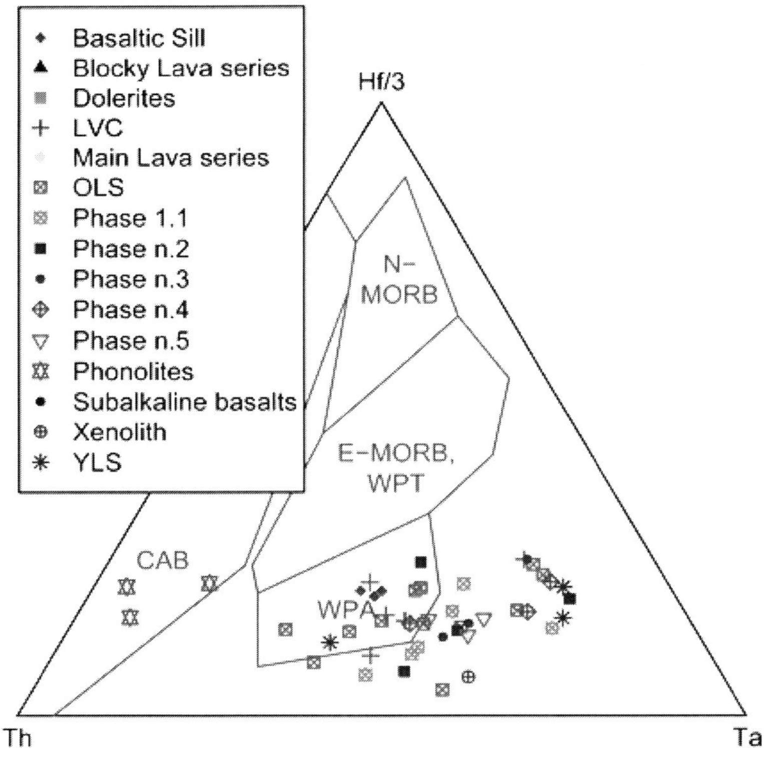

Figure 20. The Th-Hf-Ta discrimination diagram for alkali and tholeiite basalts, after Wood [47] .

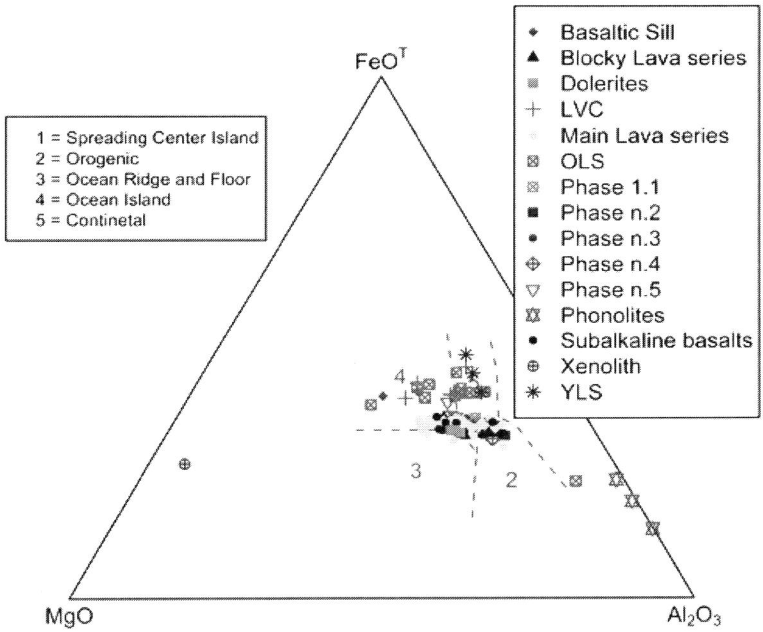

Figure 21. The MgO-FeOt-Al_2O_3 diagram, after Pearce et al. [43] , showing the discriminant boundaries between tectonic settings based on the compositional range of recent volcanics. The diagram can only used with subalkaline rocks in the silica range of 45% - 56%.

The basalts lie within the field of ocean island basalts (Figure 24; Mullen, [62]), which represents basaltic melts from an enriched asthenospheric mantle source. This characteristic has been confirmed for the major outpourings in volcanic provinces in Libya, such as Jabal as Swada, Al Haruj Al-Abiyad (Aboazoam et al., [58] ; Hegazzy, [60] ; Peregi et al., [5]).

Finally, it is interesting to ask whether the clustering in each and every plot shown in this summary illustrates that the current study represents an evolved portion or a different degree of partial melting from a common igneous source. Further research may give the answer.

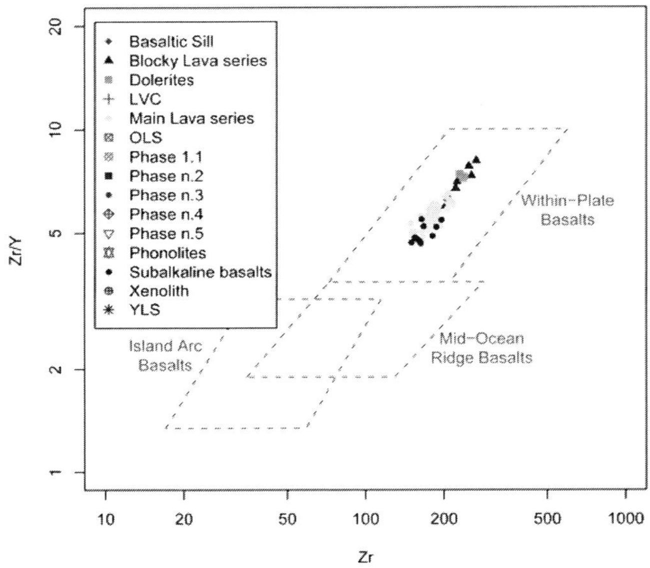

Figure 22. Discrimination diagram for basalts based upon the Zr-Zr/Y variation. The diagram has a logarithmic scale. The analysed samples of Busrewil and Suwesi [2] from the central part of Al Haruj al Aswad in the field within plate basalts (WPB; after Pearce and Norry, [44]).

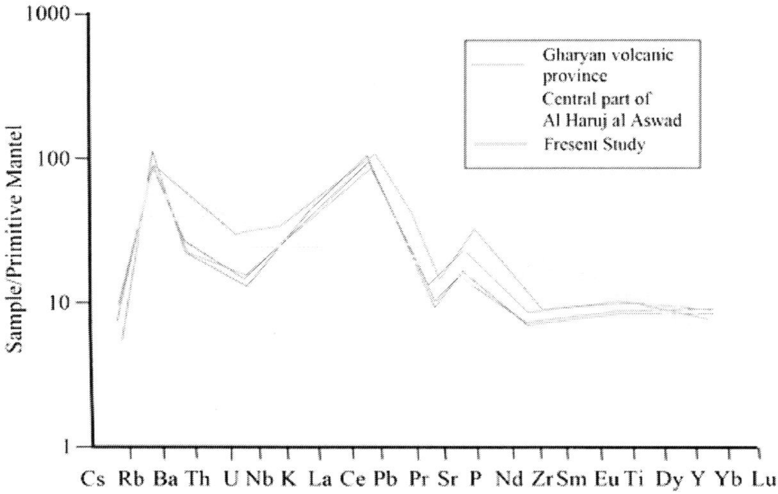

Figure 23. Trace element concentrations of some representative samples from the Gharyan volcanic province by Gaffer [57] and Al Haruj al Aswad by Busrewil and Suwesi [2] and the current study, normalized to the composition of primitive mantle.

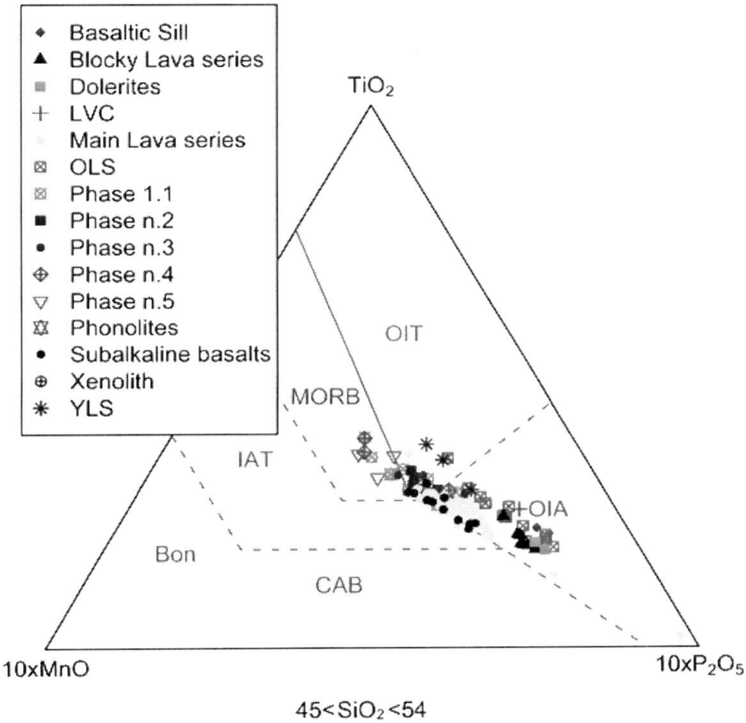

Figure 24. The MnO-TiO_2-P_2O_5 discrimination diagram for basalts and basaltic andesites of 45% - 54% SiO_2, after Mullen, [62] . The fields are: MORB; OIT?oceanic island tholeiite; OIA?oceanic island alkali basalts; CAB?is- land arc calc-alkaline basalt; IAT?island arc tholeiite; Bon?boninite.

ACKNOWLEDGEMENTS

The authors would like to express their appreciation to A. El-Arnauti and A. El-Makhrouf for their valuable remarks. We feel obliged to express gratitude to M. J. Aburkheis, the former Head of the Popular Committee for the Industrial Research Centre (IRC), and are also indebted to I. Al. Abeid, the Director of the Geological Research and Mining Department and to S. Turki, the Director of the Mapping Department for their encouragement and kind support to the present work while in the field. Thanks and appreciation must go to Rami Gebreel for his valuable help in statistics. Appreciations is expressed to the geological staff of the (IRC): A. Buagrab, A. Garbee, N. Buziad, K. Etrish and Chtronbe from Belgrade University , for their kind assistance during the field work. Thanks are also due to the crew from the IRC mapping project, SE Libya for their help and kind assistance on the camp.

N. M. Al-Hafdh, Abd El-Salam S. El-Shaafi Thanks also extend to M. El-Amawi from the Department of Earth Sciences, University of Benghazi for the coordination and communication with the Central Laboratories of the Geological Survey of Egypt to access the analyses of the samples in question.

REFERENCES

1. Busrewil, M.T. and Wadsworth, W.J. (1980) Preliminary Chemical Data on the Volcanic Rocks of Al Haruj Area, Central Libya. In: Salem, M.J. and Busrewil, M.T., Eds., The Geology of Libya, Vol. 3, Academic Press, London, 1077-1080.
2. Busrewil, M.T. and Suwesi, S.Kh. (1993) Geological Map of Libya 1:250,000, Sheet. Al Haruj Al Aswad NG 33-4. Explanatory Booklet. Industrial Research Centre, Tripoli, 95. [Citation Time(s):25]
3. Nemeth, K., Suwesi, Kh., Peregi, Z., Gulacsi, Z. and Ujszaszi, J. (2003) Plio/Pleistocene Flood Basalt Related Scoria and Spatter Cones, Rootless Lava Flows and Pit Craters, Al Haruj Al Abiyad, Libya. Geolines, 15, 107-112.
4. Busrewil, M.T. and Wadsworth, W.J. (1996) Tertiary-Quaternary Alkaline Subalkaline Magmatism in Gharayan Area: Field Aspects and Petrography. PRJ, 8, 13-23.
5. Peregi, Zs., Less, G.Y., Konrad, Gy., Fodor, L., Gulacsi, Z., Gyalog, L., Turki, S.M., Suwesi, S.Kh., Sherif, Kh. and Dalub, H. (2003) Explanatory Booklet. Geological Map of Libya 1:250,000. Sheet: Al Haruj Al Abyad NG 33-8. Industrial Research Centre, Tripoli, 248. [Citation Time(s):15]
6. El-Shaafi Abd El-Salam, S. (2008) Geochemistry and Petrology of Basic Volcanic Rocks of the NE. Part of Jabal AL-Haruj AL-Aswad, Central Libya. Unpublished MSc Thesis, University of Benghazi, Libya. [Citation Time(s):14]
7. Hallett, D. (2002) Petroleum Geology of Libya. Elsevier, Amsterdam.
8. Klitzsch, E. (2000) The Structural Development of the Murzuq and Kufra Basins: Significance for Oil and Mineral Exploration. In: Sola, M.A. and Worsley, D., Eds., Geological Exploration of the Murzuq Basin, Elsevier, Amsterdam, 143-150.
9. Woller, F. and Fediuk, F. (1980) Volcanic Rocks of Gabal as Sawda. In: Salem, M.J. and Busrewil, M.T., Eds., The Geology of Libya, Vol. 3, Academic Press, London, 1081-1093.
10. Pesce, A. (1966) Uau en Nammus: South Central Libya and Northern Chad. Excursions fürhrer der petroleum Exploration Society of Libya, Tripoli, 47-51.
11. Ade Hall, F.M., Reynolds, P.H., Dagley, P., Musset, A.G., Hubbard, T.B. and Klitzsch, E. (1974) Geophysical Studies of North African Cenozoic Volcanic Areas Al Haruj Al-Assuad, Libya. Canadian Journal of Earth Science, 11, 998- 1006.
12. Less, Gy., Turki, S.M., Suwesi, S.Kh., Peregi, L.F., Koloszar, L., Kalmar, J., Sherif, Kh., Csaszar, G., Gulasci, Z., Dalum, H. and Al Tajuri, A. (2006) Explanatory Booklet. Geological Map of Libya 1:250,000. Sheet: Waw Al Kabir NG 33-12. Industrial Research Centre, Tripoli, 295.
13. Klitzsch, E. (1968) Der Basalt Vulkanismus des Djebel Haroudj, Ostfezzan, Libya. Geologische Rundschau, 57, 585- 609.
14. Farahat, E.S., Abdel Ghani, M.S., Aboazom, A.S. and Asran, A.M.H. (2006) Mineral Chemistry of Al Haruj Low- Volcanicity Rift Basalts, Libya: Implications for

Petrogenetic and Geotectonic Evolution. Journal of African Earth Sciences, 45, 198-212.
15. Vesely, J. (1985) Geological Map of Libya 1:250,000 Sheet: Zallah NH 33-16. Explanatory Booklet. Industrial Research Centre, Tripoli, 125.
16. Woller, F. (1984) Geological Map of Libya 1:250,000, Sheet: Al Fuquaha NG 33-3. Explanatory Booklet. Industrial Research Centre, Tripoli, 123.
17. Mackenzie, W.S. and Adams, A.E. (1995) A Color Atlas of Rocks and Minerals in Thin Section. Manson, London .
18. Hatch, F.H., Wells, A.K. and Wells, M.K. (1972) Petrology of the Igneous Rocks, Fifth Impression. Thomas Murby, London .
19. Cox, K.G., Bell, J.D. and Pankhurst, R.J. (1979) The Interpretation of Igneous Rocks. Allen & Unwin, London.
20. Middlemost, E.A.K. (1980) Iron Oxidation Ratios, Norms and the Classification of Volcanic Rocks. Chemical Geology, 77, 19-26.
21. Middlemost, E.A.K. (1975) The Basalt Clan. Earth Science Reviews, 11, 337-364.
22. Winter, J.W. (2001) An Introduction to Igneous and Metamorphic Petrology. Prentice Hall.
23. Winchester, J.A. and Floyd, P.A. (1976) Geochemical Discrimination of Different Magma Series and Their Differentiation Products Using Immobile Elements. Chemical Geology, 20, 325-343.
24. Rollinson, H.R. (1993) Using Geochemical Data, Evolution, Presentation, Interpretation. Longman Scientific & Technical, Oxford/John Wiley, New York, 1-206.
25. Yoder, H.S. and Tilley, C.E. (1962) Origin of Basalt Magmas: An Experimental Study of Natural and Synthetic Rock Systems. Journal of Petrology, 3, 342-532.
26. Thompson, R.N. (1984) Dispatches from the Basalts Front. 1. Experiments. Proceedings of the Geologists' Association, 95, 249-262.
27. Wilkinson, J.F.G. (1986) Classification and Average Chemical Composition of Common Basalts and Andesites. Jour- nal of Petrology, 27, 181-194.
28. Busrewil, M.T. and Oun, K.M. (1991) Geochemistry of the Tertiary Alkaline Rocks of Jabal al Hasawnah, West Libya. In: Salem, M.J., Busrewil, M.T. and Ben Ashour, A.M., Eds., Geology of Libya, Vol. 7, Elsevier, Amsterdam, 2587- 2598.
29. Baker, B.H. (1987) Outline of the Petrology of the Kenya Rift Alkaline Province. In: Fitton, J.G. and Upton, B.J., Eds., Alkaline Igneous Rocks, Geological Society, Special Publications, London, 293-311.
30. Pearce, J.A. (1983) Lithosphere in Magma Genesis at Active Continental Margins. In: Hawesworth, C.J. and Norry, M.J., Eds., Continental Basalts and Mantle Xenoliths, Shiva.
31. Aboazom, A.S. (2005) Petrogenesis of Phase 6 of Al Haruj Al Aswad Volcanic Province, Central Libya. Sebha University Journal, 4, 96-113.
32. Fitton, J.G., James, D., Kempton, P.D., Ormerod, D.S. and Leeman, W.P. (1988) The Role of the Lithospheric Mantle in the Generation of Late Cenozoic Basic Magmas in the Western United States. J. Petrol. Special Lithosphere Issue, 331-349.
33. Hawkesworth, C.J., Hempton, P.D., Ellam, R.M. and Calsteren, P. (1990) Continental Mantle Lithosphere and Shallow Level Enrichment Processes in the Earth's Mantle. Earth and Planetary Science Letters, 96, 256-268.
34. Hopper, P.R., Bailey, D.G. and Hawkeworth, C.J. (1993) Isotopic and Geochemical Constraints on the Origin and Evolution of the Columbia River Basalt. Journal of Petrology, 34, 1203-1246.
35. Longmuir, C.H., Beneder, J.F., Bence, A.E., Hanson, G.N. and Taylor, S.R. (1977) Petrogenesis of Basalts from the Famous Area: Mid-Oceanic Ridge. Earth and Planetary Science Letters, 36, 133-156.
36. Kay, R.W. and Hubbard, N.J. (1978) Trace Elements in Ocean Ridge Basalts. Earth and Planetary Science Letters, 38, 95-116.

37. Sun, S.S. (1980) Lead Isotopic Study Volcanic Rocks from Mid-Ocean Ridges, Ocean Islands and Islands Arcs. Philosophical Transactions of the Royal Society A, 297, 409-445.
38. Pearce, J.A. (1976) Statistical Analysis of Major Element Patterns in Basalts. Journal of Petrology, 17, 15-43.
39. Pearce, J.A. (1980) Geochemical Evidence for the Genesis and Eruptive Settings of Lavas from Tethyan Ophiolites. Proceedings of the International Ophiolite Symposium, Cyprus 1979, Institute of Mining and Metallurgy, 261-272.
40. Pearce, J.A. and Cann, J.R. (1971) Ophiolite Origin Investigated by Discriminants Analysis Using Ti, Zr and Y. Earth and Planetary Science Letters, 12, 339-349.
41. Pearce, J.A. and Cann, J.R. (1973) Tectonic Setting of Basic Volcanic Rocks Determined Using Trace Element Analyses. Earth and Planetary Science Letters, 19, 290-300.
42. Pearce, T.H., Gorman, B.E. and Birkett, T.C. (1975) The TiO_2-K_2O-P_2O_5 Diagram: A Method of Discriminating between Oceanic and Non-Oceanic Basalts. Earth and Planetary Science Letters, 24, 419-426.
43. Pearce, T.H., Gorman, B.E. and Birkett, T.C. (1977) The Relation between Major Element Chemistry and Tectonic Environment of Basic and Intermediate Volcanic Rocks. Earth and Planetary Science Letters, 36, 121-132.
44. Pearce, J.A. and Norry, M.J. (1979) Petrogenetic Implications of Ti, Zr, Y and Nb Variations in Volcanic Rocks. Contributions to Mineralogy and Petrology, 69, 33-47.
45. Wood, D.A., Tarney, J., Varet, J., Saunders, A.D., Bougault, H., Joron, J.L., Treuil, M. and Cann, J.R. (1979) Geochemistry of Basalts Drilled in the North Atlantic by IPOD Leg 49: Implications of Mantle Heterogeneity. Earth and Planetary Science Letters, 42, 77-79.
46. Floyd, P.A. and Winchester, J.A. (1975) Magma Type and Tectonic Setting Discrimination Using Immobile Elements. Earth and Planetary Science Letters, 27, 211-218.
47. Wood, D.A. (1980) The Application of a Th-Hf-Ta Diagram to Problems of Tectonomagmatic Classification and to Establishing the Nature of Crustal Contamination of Basaltic Lavas of the British Tertiary Volcanic Province. Earth and Planetary Science Letters, 50, 11-30.
48. El-Akhal, H. (2004) Contribution to the Petrology, Geochemistry and Tectonic Setting of the Basalt Flows of the Umm-Qais Plateau, North Jordan.
49. Wilkison, J.F.G. (1986) Classification and Average Chemical Composition of Common Basalts and Andesites. Journal of Petrology, 27, 181-194.
50. Goudarzi, G.H. (1980) Structure: Libya. In: Salem, M.J. and Busrewil, M.T., Eds., The Geology of Libya, Vol. 3, Academic Press, London, 879-892.
51. Klitzsch, E. (1971) The Structural Development of Parts of North Africa since Cambrian Times. In: Gray, C., Ed., First Symposium on the Geology of Libya, Faculty of Science, University of Libya, Tripoli, 253-262.
52. El-Makhrouf, A.A. (1988) Tectonic Interpretation of Jabal Eghei Area and Its Regional Application to Tibesti Orogenic Belt, South Central Libya. Journal of African Earth Science, 7, 945-967.
53. Anketell, J.M. and Ghellali, S.M. (1991) A Palaeogeologic Map of the Pre-Tertiary Surface in the Region of the Jifarah Plain and Its Implication to the Structural History of Northern Libya. In: Salem, M.J., Busrewil, M.T. and Ben Ashour, A.M., Eds., Geology of Libya, Vol. 7, Elsevier, Amsterdam, 2681-2687.
54. Mikbel, S.R. (1977) Basement Configuration and Structure of West Libya. Libyan Journal of Science, 7A, 19-33.
55. Jurak, L. (1978) Geologic Map of Libya 1:250,000, Sheet: Jabal Al Hasawanah NH 33-14. Explanatory Booklet. Industrial Research Centre, Tripoli, 74-100.
56. Le Bas, M.J., LeMaitre, R.W., Strecheisen, A.L. and Zanettin, B. (1986) A Chemical Classification of Volcanic Rocks Based on the Total Alkali-Silica Diagram. Journal of Petrology, 27, 745-750.

57. Gafeer, A.S. (2007) The Petrology and Geochemistry of Gharyan Volcanic Province, NW Libya. Unpublished MSc Thesis, The University of Garyounis, Benghazi.
58. Aboazom, A.S., Asran, A.S.H., Abdel Ghani, M.S. and Farhat, E.S. (2006) Geologic and Geochemical Constraints on the Origin of Some Tertiary Alkaline Rift Volcanics, Gharyan Area, Northwestern Libya. Assiuit University Journal of Geology, 35, 25-47.
59. Almond, D.C., Busrewil, M.T. and Wadsworth, W.J. (1974) The Gharian Tertiary Volcanic Province of Tripolatania, Libya. Geological Journal, 9, 17-28.
60. Hegazzy, A.H. (1999) Tertiary Volcanics in Libya: Evidence for the Direction and Rate of the African Plate Motion. Proceedings of the Fourth International Conference on Geochemistry, Alexandria University, Egypt, 401-419.
61. Piccoli, C. (1971) Outlines of Volcanism in Northern Tripolitania. In: Gray, C., Ed., Symposium on the Geology of Libya, Fac. Sci., Univ. Padova.
62. Mullen, E.D. (1983) $MnO/TiO_2/P_2O_5$: A Minor Element Discriminant for Basaltic Rocks of Oceanic Environments and Its Implications for Petrogenesis. Earth and Planetary Science Letters, 62, 53-62.

CHAPTER 10

Experimental Study of Montmorillonite Structure and Transformation of Its Properties under Treatment with Inorganic Acid Solutions

Victoria V. Krupskaya [1,2,*], Sergey V. Zakusin [1,2], Ekaterina A. Tyupina [3,4], Olga V. Dorzhieva [1,5], Anatoliy P. Zhukhlistov [1], Petr E. Belousov [1] and Maria N. Timofeeva [6]

[1]*Belov Laboratory of Mineral Cristallochemistry, Institute of Ore Deposits, Petrography, Mineralogy and Geochemistry, Russian Academy of Science, Moscow 119017, Russia*

[2]*Department of Ecological Geology, Faculty of Geology, Lomonosov Moscow State University, Moscow 119899, Russia*

[3]*Institute of Modern Energetics and Nanotechnology Materials-IPC, Dmitry Mendeleev University of Chemical Technology of Russia, Moscow 125480, Russia*

[4]*National Research Nuclear University, Moscow 115409, Russia*

[5]*Laboratory of Physical Methods Application to the Study of Rock-forming Minerals, Geological Institute, Russian Academy of Sciences, Moscow 119017, Russia*

[6]*Research Group for Heterogeneous Catalysts for Liquid-Phase Selective Oxidations, Boreskov Institute of Catalysis SB RAS, Novosibirsk 630090, Russia*

ABSTRACT

This paper discusses the mechanism of montmorillonite structural alteration and modification of bentonites' properties (based on samples from clay deposits Taganskoye, Kazakhstan and Mukhortala, Buriatia) under thermochemical treatment (treatment with inorganic acid solutions at different temperatures, concentrations and reaction times). Treatment conditions were chosen according to those accepted in chemical industry for obtaining acid modified clays as catalysts or sorbents. Also, more intense treatment was carried out to simulate possible influence at the liquid radioactive site repositories. A series of methods was used: XRD, FTIR, ICP-AES, TEM, nitrogen adsorption, and particle size analysis. It allowed revealing certain processes: transformation of montmorillonite structure which appears in the leaching of interlayer and octahedral cations and protonation of the interlayer and –OH groups at octahedral sheets. In turn, changes in the structure of the 2:1 layer of montmorillonite and its interlayer result in significant alterations in the properties: reduction of cation exchange capacity and an increase of specific surface area. Acid treatment also leads to a redistribution of particle sizes and changes the pore system. The results of the work showed that bentonite clays retain a significant portion of their adsorption properties even after a prolonged and intense thermochemical treatment (1 M HNO_3, 60 °C, 108 h).

Keywords: engineered barriers; bentonite clays; thermochemical treatments; montmorillonite; structural modification; adsorption properties

1. INTRODUCTION

Clay minerals are widely used in various industries including radioactive waste management as a component of barrier systems for waste disposal. These systems are used for radioactive waste repositories of different levels of activity and are intended to provide safe storage for several hundreds or thousands years due to their high adsorption capacity for radionuclides and low water permeability [1,2,3].

Bentonite buffer between containers and the tunnel walls achieves several strategic outcomes at once: it provides access restriction of groundwater to the radioactive waste (RW) creating the conditions under which mass transfer between the waste and underground waters is only possible by diffusion; it suppresses migration of the radionuclides in colloidal form into the groundwater; it ensures effective sorption of radionuclides after possible depressurization of radioactive waste containers; it seals open cracks and large pores in the rocks due to the high swelling capacity, etc. [2].

One of the main components of the engineered barriers is bentonite, which contains 70–95% montmorillonite, dioctahedral specie of the smectite mineral group. Montmorillonite is a 2:1 type hydrous aluminosilicate with the octahedral sheet "sandwiched" between two tetrahedral sheets. Cation substitution in tetrahedral and mostly in octahedral sites provides a negative layer charge of about 0.2–0.5 eV. Layer charge is compensated by the introduction of exchangeable interlayer cations (Na^+, Ca^{2+}, Mg^{2+}, etc.) usually in hydrated form [4,5,6,7,8] which in turn provides adsorption sites on the inner and outer surface of the crystal. These particularities of montmorillonite structure determine specific properties of bentonite clays, especially high adsorption capacity towards heavy metals such as cesium, plutonium etc., which are commonly found in radioactive wastes.

Construction concept of waste disposal in the Russian Federation also includes the use of bentonite clay as a component of an engineered multi-barrier system [9,10]. According to current concepts, a bentonite barrier must retain its properties for over thousands of years. Thus, when analyzing the prospects of using bentonite clays, one needs to consider not only their sorption properties in their natural state, but a possible loss of sorption capacity and other parameters needed to preserve the stability of a bentonite barrier in aggressive environments. The most aggressive environment for bentonite clays is acid solutions. A number of processes take place under acid treatment and lead to a significant transformation of the structure and properties of montmorillonite bentonite clays [11,12,13,14]. A significant increase in the specific surface of the acid-modified clays promotes wide use of such material for removal of heavy metals, radionuclides and for oil refining [15,16,17,18,19]. The transformation of the structure and properties of bentonite under the influence of acids, mostly under sulfuric and hydrochloric acids, have been studied by different authors [20,21,22,23,24,25,26]. Using ultrasound and microwaves in addition to an acid treatment intensifies smectite mineral transformation [27]. In industry, acid-modified bentonites are most commonly used as catalysts [28,29,30,31,32].

In Russia, studies have been conducted to identify the most thermochemically resistant bentonite in order to determine the most appropriate bentonite clay for engineered barriers for waste disposal. The increase of the temperature in the vicinity of radioactive waste occurs due to radioactive decay. Thermochemical effects were modeled based on inorganic acid solutions (nitric and hydrochloric) at elevated temperatures. Due to its toxicity, nitric acid is not used in the chemical industry for the production of modified materials. Currently in Russia, there are plants that still dump liquid radioactive waste into deep layers of geological structures and in artificial surface repositories [33,34,35,36,37]. Nitric acid is used to prepare liquid radioactive waste for disposal. These concepts are obsolete and are being withdrawn from use. Bentonite is also used to bury surface basins, such as Lake Karachai [38].

However, in the long run, bentonite clays may be affected by aggressive fluids that are most likely to decrease insulating properties of the barrier. The aim of this study was to evaluate the mechanism of montmorillonite transformation under the acid solution treatment as well as its influence on bentonite properties.

2. MATERIALS AND METHODS

Fine clay fractions (<1 μm) separated from bentonites from Mukhortala (Buriatia, sample Mt-M) and Taganskoye (Kazakhstan, sample Mt-T1 and Mt-T2 (numbers "1" and "2" correspond to different treatments as will be shown below)) deposits were used for this study. Montmorillonite content in Mt-T and Mt-M sample were 70% and 97%, respectively. Impurities in Mt-M sample were presented by 20–30% opal C/A, which could not be removed with the sedimentation technique, and quartz (about 2–3%) in Mt-T sample.

All experiments on acid treatment were carried out in closed systems under static conditions. For this purpose, bentonite samples were placed in sealed vessels with acid solution in a weight ratio of 1:100. Vessels were in turn placed in an oven and aged for the desired time. The experiments with HCl solutions (0.125, 0.25, 0.5, 1, 3 M) were carried out at room temperature for 7 days (Mt-T;1 and M;t-M; samples) and with 1 M HNO_3 solutions (Mt-T2 sample) at 60 °C for 12, 36, 50, and 108 h. The bentonite samples from the Taganskoye deposit have quite a different composition and were named Mt-T1 for the experiment with HCl solution and Mt-T2 for the experiment with HNO_3 solution. Hydrochloric acid treatment was conducted under the conditions accepted in chemical industry for modified bentonite production. Experiments with nitric acid treatment were conducted at high temperatures in order to simulate the conditions that may occur during liquid radioactive waste disposal.

Initial and modified samples were analyzed with a series of methods. X-ray diffraction patterns were obtained with X-ray diffractometer Ultima-IV (Rigaku, Tokyo, Japan) acquired with the funding of Moscow State University Development Program (Cu-Kα radiation, semiconductor 1D detector D/Tex-Ultra, scan range 3–65° (2θ), scan speed 5°/min and step—0.02° (2θ)). Partially oriented specimens were prepared by pressing powder into a sample holder.

The chemical analyses of the solids were carried out by means of inductively coupled plasma atomic emission spectrometry (ICP-AES) using ICPE-9000 equipment (Shimadzu, Kyoto, Japan).

Excessive pressure and surface leveling leads to partial orientation of the montmorillonite particles in the specimen plane. We did not succeed in preparing samples with a good degree of particle orientation from drops of aqueous suspensions because some samples had been significantly modified after the treatment so they could not form thin films on a glass surface. The results were analyzed according to Drits, Kossovskaya [4], Moore and Reynolds [5]. Mineral composition was estimated using the Rietveld method [39] with PROFEX GUI for BGMN [40].

Fourier transform infrared spectroscopy analysis was carried out using FTIR spectrometer Vertex 80v equipped by DTGS detector and KBr beam-splitter (Bruker, Ettlingen, Germany). The adsorption spectra recordings were performed in vacuum in the 4000–400 cm^{-1} wavelength range with 256 scans for each sample and the resolution of 4 cm^{-1}. Samples were prepared as pressed KBr-pellets: 0.5 mg of sample was dispersed in 200 mg of KBr; this mixture was placed in a 13 mm pellet die and pressed in vacuum for 1 h. Spectra manipulations were performed using the OPUS 7.1 software (Bruker, Ettlingen, Germany). Baseline correction was made automatically by Concave Rubberband method with 64 baseline points and 10 iterations.

Natural montmorillonites (Mt-M, Mt-T1, Mt-T2) as well as some samples after treatment were selected based on the results of X-ray diffraction studies (Mt-M;—3.0 M HCl; Mt-T1—0.25 M and 0.5 M; HCl; 12 and 108 h) for further investigation by high-resolution transmission electron microscopy (HRTEM) using JEM-2100 with an X-Max attachment for X-ray energy dispersive analysis.

Evaluation of a specific surface area was carried out using the Analyzer Quadrasorb SI/Kr (Quantachrome Instruments, Boynton Beach, FL, USA). Adsorption was performed at the liquid nitrogen temperature (77.35 K). Nitrogen with a 99.999% purity served as an adsorbate. Helium grade 6.0 (99.9999%) was used for the volume calibration of the measuring cells. Calculation was carried out by the BET multiple-point isotherm in the range of P/P0 from 0.05 to 0.30. Samples were pre-dried in vacuum at 100 °C.

Cation exchange capacity was determined by triethylenetetramine copper complex $[Cu(Trien)]^{2+}$ adsorption method [41].

Particle size distribution was evaluated with the laser diffraction technique by Fritsch ANALYSETTE 22 NanoTec (Fritsch, Idar-Oberstein, Germany) equipped with 70 W and 36 kHz ultrasonic emitter in the size range 0.01–1000 µm Data processing was carried out by algorithm based on Fredholm integral equations.

3. RESULTS AND DISCUSSION

3.1. Transformation of the Montmorillonite Structure under Acid Treatment

Considerable crystal-chemical transformations—in particular, changes of montmorillonite micromorphology and adsorption characteristics—were detected under treatment with hydrochloric and nitric acid solutions. A variety of processes that modified structure and properties of montmorillonite particles were observed during the treatment with inorganic acid solutions: dissolution of carbonates and feldspars, destruction of the most defective phyllosilicate phases (e.g., nanosized smectites), removal of cations from the interlayer spacing, substitution of interlayer cations with oxonium ion, leaching of the octahedral cations, and finally, the complete destruction of the structure.

Studied montmorillonite samples are characterized by the heterogeneous composition of the interlayer: Ca^{2+} and $Mg^{2+}$for the Mt-M sample (d_{001} = 15.3 Å), and Ca^{2+}, $Mg^{2+}$and Na^{+} (d_{001} = 13.7 Å and d_{001} = 13.9 Å, respectively) for Mt-T1 and Mt-T2 samples (Figure 1, Table 1). Presence of Na^{+} in the montmorillonites from the Taganskoye deposit was also confirmed in previous studies [26,42,43].

Treatment of Mt-M sample with hydrochloric acid solution with various concentrations leads to partial protonation of the interlayer which is observed by a decrease in d_{00l} values and changes in basal reflections series from d_{001} = 15.3 Å, d_{003} = 5.0 Å, d_{005} = 3.0 Å to d_{001} = 13.7 Å. d_{002} = 7.2 Å, d_{003} = 4.8 Å, d_{004} = 3.6 Å (Figure 1). These changes can be related with leaching of the interlayer cations (Ca^{2+} and Mg^{2+}) and partial protonation of interlayer space i.e., substitution of interlayer cations with oxonium ion [44,45] because protonation of Ca–Mg montmorillonite proceeds much faster than that of Na-forms [43].

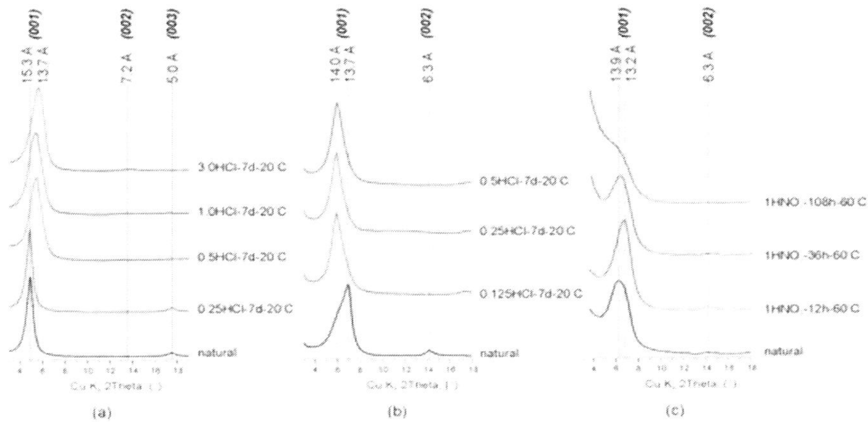

Figure 1. Fragments of the X-ray diffraction patterns of natural and treated montmorillonites: (**a**) Mt-M; (**b**,**c**) Mt-T. Abbreviations: 3.0 HCl, 1 HNO_3, type and concentration (given in mol/L) of acid; 7 d (days), 12 h (hours), time of treatment, 20, 90 °C, temperature of treatment.

Table 1. Changes of the structural parameters, textural and adsorption characteristics of montmorillonites due to acid activation (Specific Surface Area: S_{BET}, Total Pore Value: V_Σ, Cation Exchange Capacity: CEC).

Sample	Treatment			Interlayer Space d_{001} (Å)	Particle Thickness h(001) (CSR) (nm)	Number of Layers (N)	S_{BET} (m^2/g)	V_Σ (cm^3/g)	CEC (meq/100 g)
	Acid (mol/L)	T (°C)	Time Days/Hours						
					HCl				
Mt-M	-	-	-	15.3	10.7	7	77	0.211	46
	0.25	20	7 d	15.3	12.2	8	79	0.219	43
	0.5	20	7 d	14.0	7.0	5	92	0.243	-
	1.0	20	7 d	14.0	5.6	4	95	0.256	49
	3.0	20	7 d	13.7	5.5	4	101	0.328	64
Mt-T1	-	-	-	13.7	9.6	7	42	0.074	75
	0.125	20	7 d	14.0	7.0	5	45	0.084	65
	0.25	20	7 d	14.0	7.0	5	50	0.095	64
	0.5	20	7 d	14.0	5.6	4	51	0.083	70
					HNO_3				
Mt-T2	-	-	-	13.9	8.2	6	67	0.085	86
	1	60	12 h	14.0	7.0	5	110	0.114	58
	1	60	36 h	14.0	7.0	5	191	0.192	56
	1	60	108 h	13.8	5.5	4	301	0.353	40

Early, we demonstrated that the chemical composition and structural characteristics of Mt depended on the HNO_3 and HCl concentration [26,32]. Changes of chemical composition and interlayer space were negligible after modification of montmorillonite under the concentrations up to 0.5 M. These changes were noticeable only after treatment with 3.0 M acid. Here we investigated effect of exposure time of 1M HNO_3 at 60 °C on chemical composition of Mt-T2. The main results are shown in the Table 2. Experimental data point that leaching of interlayer cation is observed after the treatment with 1 M HNO_3 for 12 h.

Table 2. Chemical composition of MM-T2 natural and modified by 1 M HNO_3 at 60 °C.

Time (h)	Chemical Composition (wt %)						
	Si	Al	Fe	Mg	Ca	Na	Si/Al
-	26.4	7.8	3.9	1.9	0.6	2.4	3.4
12	27.3	7.6	3.7	1.6	trace	trace	3.6
36	28.0	6.9	3.2	1.4	0.2	trace	4.1
108	31.8	5.0	2.0	0.9	trace	trace	6.3

Profile alignment of basal reflections (001) and displacement of maximum 13.7 Å to 14.0 Å was observed after treatment of Mt-T1 sample with hydrochloric acid solution. Irregular profile shape of d_{001} reflection indicates the presence of two possible montmorillonite phases with different interlayer cations in the sample. This was also found by other researchers of the Taganskoye montmorillonite deposit [43]. Since any other impurities except quartz were not found in this sample, the described change in the XRD patterns can be attributed to the partial destruction (dissolution) of the most defective and probably nanosized particles of Na-montmorillonite.

Treatment of Mt-T sample with 1M HNO_3 solution for 12 h leads to a displacement of the d_{001} reflection from 13.9 to 13.2 Å with an increase of its intensity. This fact is explained by the dissolution of the defective part of the montmorillonite and enhancement of the interlayer ordering by partial substitution of Ca^{2+} and Mg^{2+} cations. Further treatment of the sample with nitric acid solution reduces the layer stacking ordering degree and leads to disintegration of montmorillonite 2:1 layer structure. In XRD patterns, the described process is manifested by a decrease in the peaks intensity and broadening of the basal reflections until the complete disappearance except (001) reflection. Displacement of the (001) reflection back to 13.9 Å after the 36 and 108 h of treatment can be explained by the Mg and Al octahedral cations leaching and their migration into the interlayer space of montmorillonite.

The changes in the sizes of the coherent scattering domains (CSD) along the c-axis were calculated in accordance with the Scherrer equation for the (001) reflection of natural and modified montmorillonites (Table 1). In general, the size of CSD along the c-axis corresponds with the crystallite thickness [46]. The average crystallite thicknesses of the samples equals 10.7 nm which corresponds to 7 layers (7 N), 9.6 nm (7 N), 8.2 nm (6 N) for natural Mt-M, MtM-T1 and Mt-T2 samples respectively, and to 5.5–5.6 nm (4 N) for the acid treated montmorillonites. A decrease of CSD sizes is related to a consequent disintegration of montmorillonite particles and increase of the stacking faults quantity.

The increase of the adsorbed water bands on IR spectra in the range of 4000–2500 cm^{-1} probably indicates a decrease of crystallite size (Figure 2a) after the treatment of Mt-T2 sample with HNO_3 acid for 108 h. Previous studies of Mt-M and Mt-T1 samples treated with HCl acid [26] showed the reduction of band intensity at 841–845 cm^{-1} (Al–Mg–OH), 882 cm^{-1} (Al–Fe–OH) and 914–926 cm^{-1} (Al–Al–OH), which indicates the leaching of Al^{3+}, Mg^{2+} and Fe^{3+} cations from the octahedral sites [44,45,46,47,48,49]. Reduction of the band intensities at 925 and 876 cm^{-1} is shown at Figure 2c. It corresponds to Al–Al–OH and Al–Fe–OH vibrations in the structure of octahedral sheet of montmorillonite (Mt-T2 sample) during the treatment with the nitric acid. It may indicate a leaching of Fe^{3+} and Al^{3+} from the octahedral sites.

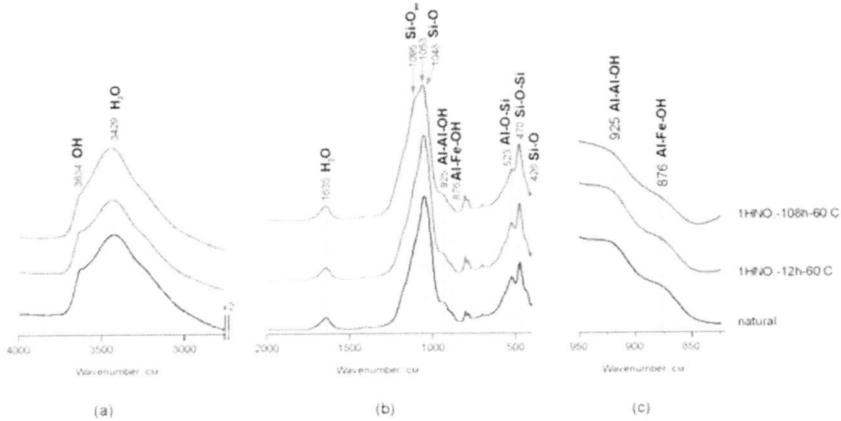

Figure 2. Areas of interest on MM-T IR spectra before and after treatment with 1 M HNO_3 60 °C: (**a**,**b**) the characteristic part of spectrum; (**c**) a magnified part of the spectra.

Leaching of the octahedral cations leads to modifications of the interaction between the octahedral and tetrahedral sheets in the 2:1 layer and, as a result, to the partial destruction of octahedral sheets. Alterations in FTIR data revealed as decreasing intensities of the bands related to octahedral Al, Fe and Mg and exchangeable Ca, Na and Mg correlate with changes in chemical composition shown above. Therefore, the structural units comprised of the tetrahedral sheet fragments are released from the 2:1 structure, which corresponds to "Si–O_{free}" vibrations—1095 cm^{-1} [11,15,21,50]. Changes in the Si–O–Si band profile within the range of 1100–1050 cm^{-1} (Figure 2b) indicate transformation of interaction within the tetrahedral sheet. In particular, displacement of the maximum in the natural sample from 1050 to 1095 cm^{-1} after treating it with 1 M HNO_3 for 108 h indicates the increase of amorphous silica content due to partial destruction of the tetrahedral sheet which is confirmed by chemical analysis data as an increase in Si content in powder samples.

Transmission electron microscopy (TEM) allows collection of data at the level of crystal lattice and helps to estimate structural transformation of montmorillonite, properties of surface and interactions between the layers during its treatment with inorganic acid solutions, and changes in montmorillonite particles' chemical composition [51]. In natural samples from the Tagansoye deposit (Figure 3a) among the laminar montmorillonite particles with a size of 1–2 μm, there is a significant amount of small and thin nano-sized particles that cover the specimen and produce grey background in micrographs. Particle identification was carried out based on microdiffraction patterns and micro-area chemical analysis. The nanosized phase is characterized by a higher content of Na as an interlayer cation and Fe is located in the octahedral sheet due to the isomorphic substitution of Al. Under hydrochloric acid with lower concentrations, as well as under the treatment with nitric acid for shorter periods, almost full dissolution of nano-sized particles as the less acid resistant phase takes place and the grey background on micrographs is not observed. Besides, these laminar particles have a lot of folded edges which are not usually observed in acid-modified montmorillonites. Bentonites from Mukhortala deposit (Mt-M) contain a lot of opal distributed in pores between thin and relatively thick laminar montmorillonite particles and also on their relatively clean surfaces (Figure 3c).

Studied samples of the natural montmorillonites are characterized by peculiar particle morphology that is typical for montmorillonites of different genesis and by a wide spread of welted edges, which can be useful for obtaining pictures of the lattices (Figure 4). Images obtained from edges of natural montmorillonite (Mt-T2 sample) show the lattice stripes that correspond to the basal planes (Figure 4a,b). The width of the areas with lattice stripes is about 23–32 nm. The value of the interlayer space for different particles varies within the range from 10.5 to 11.8–12.3 Å.

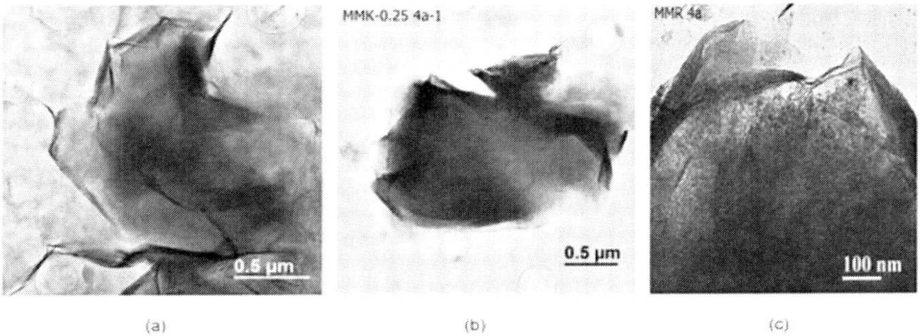

Figure 3. Transmission electron microscopic (TEM) images: (**a**) natural particles from the sample MM-T; (**b**) treated with 0.25 M HCl; (**c**) natural sample MM-M.

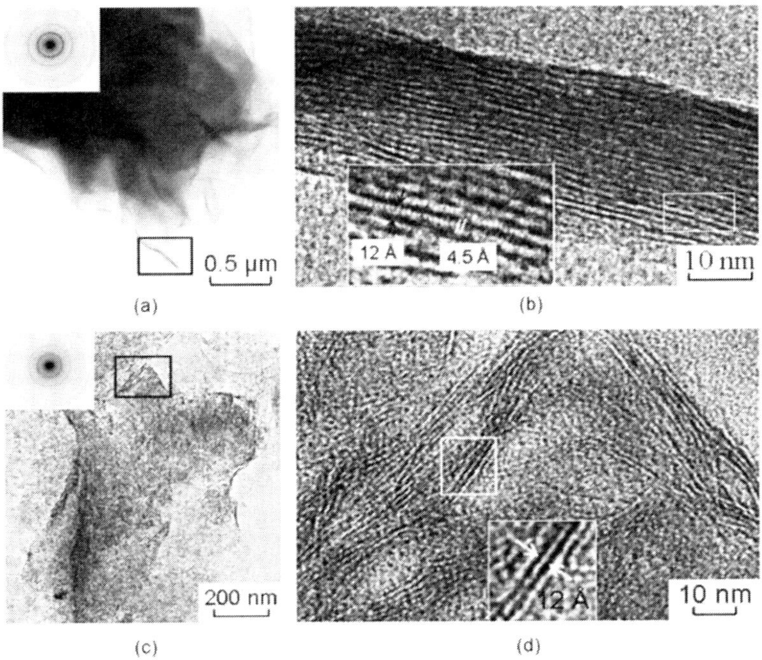

Figure 4. Transmission electron microscopic (TEM) images of natural montmorillonite particles (**a**,**b**), and treated with 1 M HNO_3 for 108 h (**c**,**d**). Microdiffraction figures are located at enlarged areas. (**b**,**d**) high resolution images from the folded edge of the montmorillonite particles (area locations are shown as black rectangles in Figure (**a**,**c**), respectively).

The lattice stripes of different particles can be pulled simultaneously, maintaining a noticeable distance between the adjacent bands or showing some curving of the surface, which indicates only a small change of interlayer space along the layers. Moreover, there are areas of the image with two-dimensional lattice stripes (Figure 4b). Basal lattice stripes are crossed by lattice stripes with an interlayer space of 4.5 Å which corresponds to planes (020) and indicates local manifestations of coherence in stacking of adjacent layers.

Microaggregate surface of Mt-T2 sample became uneven as the result of the treatment with HNO_3 for 108 h. This effect can be seen only at some fragments of the images with folded edges with typical parallel basal lattice stripes from planes (00l). It correlates to the small packs of 2–3 layers (Figure 4b) less than 25 nm long. The interlayer spacing 12–14 Å indicates the presence of initial montmorillonite. The observed pattern of distribution of the basal lattice stripes reflects the nano-level changes in the surface morphology of montmorillonite. It implies an existence of thin packages of layers where the planes (00l) are arranged perpendicularly to the surface of the particles. This effect in particular leads to an increase of its surface and the occurrence of active sites for the selective sorption of heavy metals and radionuclides.

3.2. Transformation of Montmorillonite Adsorption Properties under Acid Treatment

Specific surface area (BET) and cation exchange capacity (CEC) can be qualified as characteristics of adsorption properties of bentonite clays. CEC values of montmorillonites depend most significantly on the amount of isomorphic substitutions in octahedral sites. In turn, specific surface area is primarily controlled by degree of fineness, amount of impurities, particle charge, their ability to coagulate, etc. Mt-M sample contains a significant amount of opal which provides a relatively higher specific surface area than that of Taganskoye deposit samples (Mt-T1 and Mt-T2). Commonly, specific surface area (BET) values are in direct relationship with CEC value [52]. Increasing of BET value leads to an increase of CEC value and vice versa. In most cases, the above is true for natural soils with different mineral compositions [46]. However, the ratio of these indexes for montmorillonites of different composition may be in a more complex relationship. This dependence can be clearly seen in the Table 1 as an increase of specific surface area and a decrease in cation exchange capacity due to the acid treatment.

Coherent scattering domain sizes are commonly used to analyze the size of crystallites [46,53,54,55]. There is a cumulative index of physical-chemical activity of the most clay minerals which have correlations with specific surface area and adsorption characteristics. Thus, reduction of crystallite size should lead to an increase of specific surface area, which is shown in the study (Table 1).

At the same time, an increase of the specific surface area appears mostly in the samples which have the greatest structural changes. Thus, in the montmorillonite sample from Taganskoye deposit, treated with HNO_3 for a long period of time, the increase of the specific surface area was 301 m^2/g compared to 67 m^2/g in the natural sample. Also, considerable changes were observed in the sample with montmorillonite from the Mukhortala deposit (from 77 to 101 m^2/g, respectively).

The average pore size in all samples is approximately 5 nm and it remains constant during the experiments, while the total pore volume has been changed slightly for Mt-M and Mt-T1 samples and considerably so for Mt-T2 sample. This fact cannot be related only to the size decrease. In this way, the observed increase of specific surface area, as well as the total pore volume is related to the formation of porous structure due to the modification of montmorillonites under the acid treatment.

As a result of the treatment, a regular change is observed in the chemical composition of montmorillonite (Table 2). First of all, the reduction of Ca, Mg, Fe, and Al content and the increase of Si content is found in all samples after dissolving the nano-sized phases of montmorillonite from Taganskoye deposit. The most significant changes were found in the Mt-T2 sample treated with HNO_3 acid solution for 108 h. The increase of Si content in montmorillonite samples subjected to the intensive acid treatments connected with subsidence of amorphous silica formed during the destruction of tetrahedral sheets of 2:1 layers on the particle surface and its accumulation in the micropores.

Protonation of Al bond pairs in the octahedron transforms it from 4-coordination to 6-coordination [23] and leads to the modification of octahedral and tetrahedral sheets and to an appearance of micropores simultaneously decreasing the number of octahedral cations. Changes in the layer stacking and particles' micromorphology results in mesopore formation [14,45]. Structural transformations are shown simplified in Figure 5. Thus, there is an increase of specific surface area and pore space both capable of large cation sorption, e.g., Cs.

Critical values for describing the concepts of "micropores", "mesopores", and "macropores" vary in classifications developed for different science approaches. In engineering geology, for example, mesopore sizes are in the range of 10 to 1000 μm and micropores are from 0.1 to 10 μm [56]. V. Osipov and V. Sokolov in their morphometric investigations of soil microstructures with different composition [52] detailed the previously proposed classification of pore sizes. They divided micropores into thin

(0.1–10 μm), small (1–10 μm), and large (10–100 μm) ones, suggesting the lower limit for macropores as 100 μm.

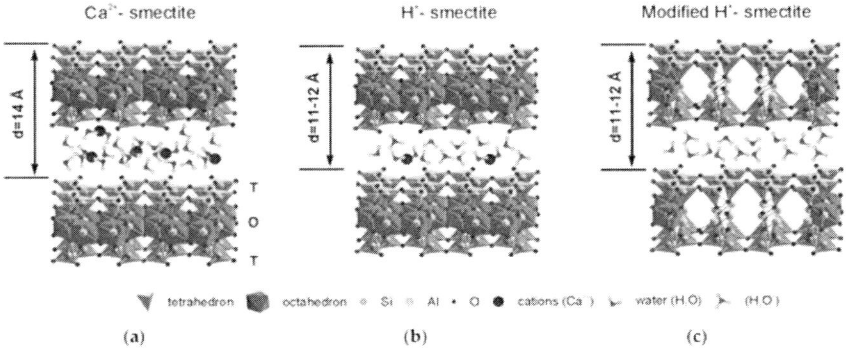

Figure 5. Schematic picture representing the structural changes in natural montmorillonite under the treatment with inorganic acid solutions: (**a**) natural Ca-montmorillonite; (**b**) partial protonation of interlayer space; (**c**) full protonation of interlayer space (H-smectite) and protonation of the OH-groups of the octahedral sheet and Al coordination change.

There are other classifications to consider. For example, the International Union for Pure and Applied Chemistry (the IUPAC) recommended to distinguish pores size into macropores (50 nm), mesopores (2 to 50 nm), and micropores (up to 2 nm) [57,58,59]. The micropores are conventionally divided into thin ultramicropores (less than 0.7 nm) and supermicropores that have an intermediate size between ultramicropore and mesopores.

In fact, all the pores in montmorillonite referred above of the size not larger than 6–9 nm for mesopores and 1–3 nm for micropores are predominantly involved in the adsorption [60]. The average pore size measured in natural and modified montmorillonites is about 5 nm which corresponds to interparticle pores according to Osipov and Sokolov [52]. At the same time, acid treatment, as shown above, leads to the appearance of pores in the structure of the layer (micropores) by partial leaching of octahedral cations, protonation of OH-groups, and changes of Al^{3+} coordination. The above process does not lead to an increase of an average pore size, however, it results in an increase of the total pore volume (Table 1).

Mesopores of a size about 5 nm and a small amount of micro-pores with a diameter of about 3 nm are predominant in the natural montmorillonite. After treatment with hydrochloric acid with concentration up to a 3.0 M at room temperature for 7 days, significant changes in the structure of the pore space were not observed, which also can be seen below in the N_2adsorption-desorption isotherms. Treatment with 1 M HNO_3 for 12 h does not lead to significant changes of the pore space either. The most significant changes were observed as a result of treatment for 108 h (Figure 6a): the amount of 5–5.5 nm mesopores significantly increases and micropores with an average diameter of 3.5 nm appear which is not observed in natural montmorillonite samples. The particle size distribution changes even at low exposure times in the experiment with 1 M HNO_3 (Figure 6b) from monomodal shape with a maximum at 50 μm to a multimodal shape with peaks at ~80, 100, 145, and 200 μm. Thus, with increased micro- and mesoporosity, particle aggregation occurs that leads to the formation of relatively large aggregates of different sizes.

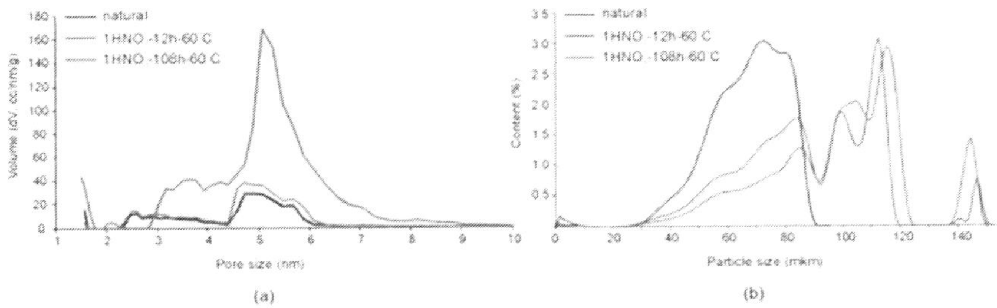

Figure 6. Changes of the porosity (**a**) and the particle size distribution (**b**) of Mt-T2 a sample due to exposure to 1 M HNO_3 at 60 °C.

Adsorption-desorption isotherms of N_2 are shown in Figure 6. As shown earlier, exposure to 0.25 and 0.5 M HCl for 7 days at room temperature results only in minor changes in the chemical composition (Table 2). The same tendency is observed in changes of N_2 adsorption-desorption isotherms for samples Mt-M and Mt-T1 (Figure 6). Mt-T1 sample was treated only at 0.5 M HCl concentrations and is not shown in Figure 7 since the changes in the adsorption-desorption rates are within the error margin of the method.

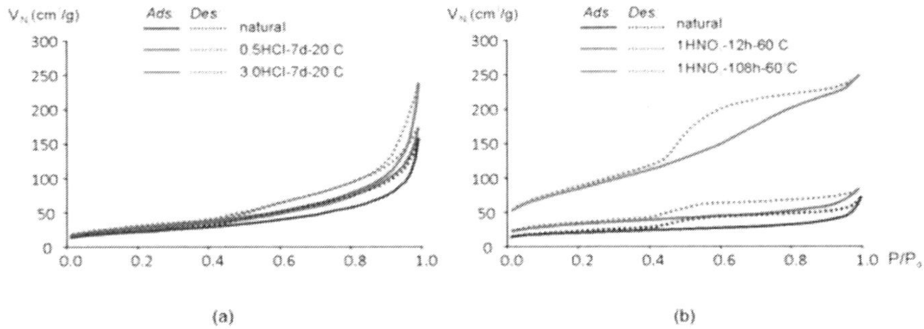

Figure 7. Isotherms of N_2 adsorption (Ads)/desorption (Des) of montmorillonites: (**a**) Mt-M; (**b**) Mt-T2.

Hysteresis loop and the shape of the isotherm is close to the type IVa [61,62], which means that the natural and acid-modified montmorillonites studied can be characterized as mesoporous adsorbents where monolayer-multilayer adsorption is followed by capillary condensation. There is a significant amount of the micropores and mesopores so N_2adsorption/desorption isotherms cannot be described by one classical type only [61].

Hysteresis loops may have a different shape and, in the study of natural montmorillonite, can be attributed to the H3-H4 type [61]. Loops of this type correspond to the interaction between nonrigid lamellar particles. Smooth slope loops correspond to filling the micro- and mesopores. Increase in steepness after the acid treatment may be explained by an increase of the pore volume and the appearance of macro-porosity. The specific hysteresis loop of the sample Mt-T2 isotherm after prolonged exposure to 1 M HNO_3 (Figure 7b) can be attributed to the type H2(b) [61]. This type of a desorption branch of the isotherm can be the result of a complex system of channel and pore formations and a partial blocking of them.

As a result of exposure to acid solutions, the hysteresis loop increases. In particular, it is noticeable after a prolonged influence of nitric acid at a high temperature (Figure 6b) when the hysteresis loop changes in a certain way.

The measured cation exchange capacity value (CEC) decreases from 86 meq/100 g in natural bentonite clay to 40 meq/100 g in bentonite samples with the longest time of the HNO_3 solutions treatment for 108 h (Table 1). Such a decrease of CEC under the treatment with inorganic acid solutions (e.g., HCl and H_2SO_4) was described by other researchers [12]. Cation exchange capacity of bentonite clays is connected not only to montmorillonite content but primarily, to such structural particularities as layer charge and its distribution among octahedral sheets. As a result, treatment with inorganic acid solutions leads to layer charge reduction due to leaching of the octahedral cations and the structural degradation; all of which leads to a decrease in the cation exchange capacity. In addition, amorphous silica appeared as a result of the destruction of tetrahedral sheets. The silica can settle on the surface of montmorillonite particles and reduce its exchange capacity.

A lot of attention in the work on acid treatment of natural bentonite clay was dedicated to the issues of activation—i.e., improving the properties of modified clays in comparison with natural clays [63,64]. An increase of specific surface area and pore volume is observed in the course of the conducted experiments (Table 1). Leaching of the cations from interlayer space and octahedral positions results in a modification of layer charge and particles in general, this in turn affects the interaction of the individual particles between each other. As a result, the destruction of large aggregates, the restructuring of smaller ones, the appearance of an uneven surface, and the appearance of micropores on the site of octahedral grid take place. All the processes described above result in an increase of the total pore size and surface area. However, opportunities for cation exchange and its capacity are reduced due to layer charge modification and protonation of interlayer space.

4. CONCLUSIONS

The study concludes that an interlayer modification occurs even at early stages of treatment with HCl and HNO_3solutions. This modification involves partial substitution of interlayer cations, especially Ca and Mg, to oxonium and partial protonation of the interlayer space. As the least stable components, natural nano-sized smectites and the most defective phase are completely dissolved in natural montmorillonite samples after acidic treatment.

Further treatment of montmorillonite structure with inorganic acid solutions (with increasing concentration and exposure period) leads to further modification of its structure accompanied by intense leaching of cations from the octahedral positions and partial penetration of leached octahedral cations into interlayer space. These transformations lead to partial protonation of OH-groups, partial destruction of the octahedral sheets, and modification of the interaction between tetrahedral and octahedral sheets which changes layer charge and the nature of interaction between adjacent layers and partial amorphization of tetrahedral sheets. A decrease in the layer charge leads to a decrease in CEC.

As a result of structural transformation, specific surface area increases. Also, volume and pore diameters increase as well as micropores of a 3.5 nm diameter occur which is not typical for natural montmorillonites. As a consequence, particle size distribution changes. Due to the rise in micro- and mesoporosity, adsorption capacity of the acid-treated bentonite clays still remains high.

Modification of structural and adsorption characteristics with the acid treatment can be used to simulate behavior of the engineered barrier properties for repositories of radioactive and industrial wastes, especially in the case of dealing with liquid radioactive wastes.

ACKNOWLEDGMENTS

This work was financially supported by the Russian Science Foundation (Project #16-17-10270). We thank Boris Pokid'ko for his help with analytical approach, and we also thank the "Bentonite" company for providing us with natural Bentonite clay from the Taganskoe deposit.

AUTHOR CONTRIBUTIONS

Victoria V. Krupskaya prepared material, conceived and designed the experiments, and wrote the paper; Sergey V. Zakusin performed the experiments by HNO_3 treatment and measured samples by X-ray diffraction and CEC; Ekaterina A. Tyupina analyzed specific surface area and porosity; Olga V. Dorzhieva collected data by infrared spectroscopy; Anatoliy P. Zhukhlistov studied samples by electron microscopy; Maria N. Timofeeva performed the experiments by HCl treatment acid treatment; and Petr E. Belousov prepared <1 μm fractions and measured CEC.

CONFLICTS OF INTEREST

The authors declare no conflict of interest.

REFERENCES

1. Pusch, R.; Knutsson, S.; Al-Taie, L.; Mohammed, M.H. Optimal ways of disposal of highly radioactive waste. Nat. Sci.2012, 4, 906–918.
2. Sellin, P.; Leupin, O.X. The use of clay as an engineered barrier in radioactive-waste management—A review. Clays Clay Miner. 2013, 61, 477–498.
3. Laverov, N.P.; Velichkin, V.I.; Omelianenko, B.I.; Yudincev, S.V.; Petrov, V.A.; Bichkova, A.V. Part 5: Changes of the environment and climate. In Isolation of Spent Nuclear Materials: Geological and Geochemical Bases; IGEM RAS, IFZ RAS: Moscow, Russia, 2008; p. 280. (In Russian)
4. Drits, V.A.; Choubar, C. X-ray Diffraction by Disordered Lamelar Structure. X-ray Diffraction by Disordered Lamelar Structure; Springer-Verlag: Berlin/Heidelberg, Germany, 1990; p. 371.
5. Moore, D.M.; Reynolds, R.C., Jr. X-ray Diffraction and the Identification and Analysis of Clay Minerals, 2nd ed.; Oxford University Press: Oxford, UK, 1997; p. 378.
6. Guggenheim, S.; Adams, J.M.; Bain, D.C.; Bergaya, F.; Brigatti, M.F.; Drits, V.A.; Formoso, M.L.L.; Gala, N.E.; Kogure, T.; Stanjek, H. Summary of recommendations of nomenclature committees. Relevant to clay mineralogy: Report of the Association Internationale Pour L'etude des Argiles (AIPEA) Nomenclature Committee for 2006. Clays Clay Miner.2006, 54, 761–772.
7. Brindley, G.W.; Brown, G. (Eds.) Crystal Structures of Clay Minerals and Their X-ray Identification; Mineralogical Society: London, UK, 1980.
8. Wilson, M.J. Rock-Forming Minerals. Sheet Silikates: Clays Minerals; The Geological Society: London, UK, 2013; p. 724.
9. Laverov, N.P.; Yudintsev, S.V.; Kochkin, B.T.; Malkovsky, V.I. The Russian strategy of using crystalline rock as a repository for nuclear waste. Elements 2016, 12, 253–256.
10. Gupalo, T.A.; Kudinov, K.G.; Jardine, L.J.; Williams, J. Development of a Comprehensive Plan for Scientific Research, Exploration, and Design: Creation of an Underground Radioactive Waste Isolation Facility at the Nizhnekansky Rock Massif. In Proceedings of the Waste Management 2005 Symposium, Tucson, AZ, USA, 27 February–3 March 2005.
11. Komadel, P. Chemically modified smectites. Clay Miner. 2003, 38, 127–138.
12. Tomić, Z.P.; Antić Mladenović, S.B.; Babić, B.M.; Poharc Logar, V.A.; Đorđević, A.R.; Cupać, S.B. Modification of smectite structure by sulfuric acid and characteristics of the modified smectite. J. Agric. Sci. 2011, 56, 25–35.
13. Okada, K.; Arimitsu, N.; Kameshima, Y.; Nakajima, A.; MacKenzie, K.J.D. Solid acidity of 2:1 type clay minerals activated by selective leaching. Appl. Clay Sci. 2006, 31, 185–193.
14. Kumar, P.; Jasra, R.V.; Bhat, T.S.G. Evolution of Porosity and Surface Acidity in Montmorillonite Clay on Acid Activation. Ind. Eng. Chem. Res. 1995, 34, 1440–1448.
15. Komadel, P.; Madejova, J. Chapter 7.1: Acid activation of clay minerals. In Handbook of Clay Science; Developments in Clay Science; Bergaya, F., Theng, B.K.G., Lagaly, G., Eds.; Elsevier: Amsterdam, The Netherlands, 2006; Volume 1, p. 263287.
16. Carrado, K.A.; Komadel, P. Acid activation of bentonites and polymer-clay nanocompo-sites. Elements 2009, 5, 111–116.
17. Dubikova, M.; Cambier, P.; Sucha, V.; Caplovicova, M. Experimental soil acidification. Appl. Geochem. 2002, 17, 245–257.
18. Pagano, T.; Sergio, M.; Glisenti, L.; Diano, W.; Grompone, M.A. Use of pillared montmorillonites to eliminate chlorophyll from rice bran oil. Ing. Quim. 2001, 19, 11–19.
19. Tyupina, E.A.; Magomedbekov, E.P.; Tuchkova, A.I.; Timerkaev, V.B. The sorption refinement of liquid organic radioactive waste for Cs-137. Adv. Mater. Spec. Issue 2010, 8, 329–333. (In Russian).

20. Vicente, M.A.; Suarez Barrios, M.; Lopez Gonzalez, J.D.; Banares Munoz, M.A. Characterization, surface area, and porosity analyses of the solids obtained by acid leaching of a saponite. Langmuir 1996, 12, 566–572.
21. Komadel, P. Structure and chemical characteristics of modified clays. In Natural Microporous Materials in Environmental Technology; Misealides, P., Macasek, F., Pinnavaia, T.J., Colella, C., Eds.; Kluwer: Alphen aan den Rijn, The Netherlands, 1999; pp. 3–18.
22. Tkac, I.; Komadel, P.; Muller, D. Acid-treated Montmorillonites—A Study by 29Si and 27Al MAS NMR. Clay Miner.1994, 29, 11–19.
23. He, H.; Guo, J.; Xie, X.; Lin, H.; Li, L. A microstructural study of acid-activated montmorillonite from Choushan, China. Clay Miner. 2002, 37, 337–344.
24. Timofeeva, M.N.; Panchenko, V.N.; Gil, A.; Zakusin, S.V.; Krupskaya, V.V.; Volcho, K.P.; Vicente, M.A. Effect of structure and acidity of acid modified clay materials on synthesis of octahydro-2H-chromen-4-ol from vanillin and isopulegol. Catal. Commun. 2015, 69, 234–238.
25. Novikova, L.; Belchinskaya, L.; Krupskaya, V.; Roessner, F.; Zhabin, A. Effect of acid and alkaline treatment on physical-chemical properties of surface of natural glauconite. Sorpt. Chromatogr. Process. 2015, 15, 730–740.
26. Timofeeva, M.N.; Volcho, K.P.; Mikhalchenko, O.S.; Panchenko, V.N.; Krupskaya, V.V.; Tsybulya, S.V.; Gil, A.; Vicente, M.A.; Salakhutdinov, N.F. Synthesis of octahydro-2H-chromen-4-ol from vanillin and isopulegol over acid modified montmorillonite clays: Effect of acidity on the Prins cyclization. J. Mol. Catal. A Chem. 2015, 398, 26–34.
27. Franco, F.; Pozo, M.; Cecilia, J.A.; Benítez-Guerrero, M.; Lorente, M. Effectiveness of microwave assisted acid treatment on dioctahedral and trioctahedral smectites. The influence of octahedral composition. Appl. Clay Sci. 2016, 120, 70–80.
28. Adams, J.M. Synthetic organic chemistry using pillared, cation-exchanged and acid-treated montmorillonite catalysts—A review. Appl. Clay Sci. 1987, 2, 309–342.
29. Brown, D.R. Review: Clays as catalyst and reagent support. Geol. Carpath. Ser. Clays 1994, 45, 45–56.
30. Bovey, J.; Jones, W. Characterization of Al-pillared acid-activated clay catalysts. J. Mater. Chem. 1995, 5, 2027–2035.
31. Timofeeva, M.N.; Panchenko, V.N.; Volcho, K.P.; Zakusin, S.V.; Krupskaya, V.V.; Gil, A.; Mikhalchenko, O.S.; Vicente, M.A. Effect of acid modification of kaolin and metakaolin on Brønsted acidity and catalytic properties in the synthesis of octahydro-2H-chromen-4-ol from vanillin and isopulegol. J. Mol. Catal. A Chem. 2016, 414, 160–166.
32. Timofeeva, M.N.; Panchenko, V.N.; Krupskaya, V.V.; Gil, A.; Vicente, M.A. Effect of nitric acid modification of montmorillonite clay on synthesis of sotketal from glycerol and acetone. Catal. Commun. 2017, 90, 65–69.
33. Tokarev, I.V.; Rumynin, V.G.; Zubkov, A.A.; Pozdnyakov, S.P.; Polyakov, V.A.; Kuznetsov, V.Y.U. Assessment of the long-term safety of radioactive waste disposal: 1. Paleoreconstruction of groundwater formation conditions. Water Resour. 2009, 36, 206–213.
34. Zubkov, A.A.; Balakhonov, B.G.; Sukhorukov, V.A.; Noskov, M.D.; Kessler, A.G.; Zhiganov, A.N.; Zakharova, E.V.; Darskaya, E.N.; Egorov, G.F.; Istomin, A.D. Radionuclide distribution in a sandstone injection zone in the course of acidic liquid radioactive waste disposal. Dev. Water Sci. 2005, 52, 491–500.
35. Rybalchenko, A.; Pimenov, M.; Kostin, P. Injection Disposal of Hazardous and Industrial Wastes, Scientific and Engineering Aspects. In Deep Injection Disposal of Liquid Radioactive Waste in Russia; Academic Press: New York, NY, USA, 1998; p. 780.
36. Rybalchenko, A.I.; Pimenov, M.K.; Kurochkin, V.M.; Kamnev, E.N.; Korotkevich, V.M.; Zubkov, A.A.; Khafizov, R.R. Deep Injection Disposal of Liquid Radioactive Waste in Russia. Dev. Water Sci. 2005, 52, 13–19.

37. Utkin, S.S.; Linge, I.I. Decommissioning strategy for liquid low-level radioactive waste surface storage water reservoir. J. Environ. Radioact. 2016.
38. Malkovsky, V.I.; Dikov, Y.P.; Asadulin, E.E.; Krupskaya, V.V. Influence of host rocks on composition of colloid particles in groundwater at the Karachai Lake site. Clay Miner. 2012, 47, 391–400.
39. Post, J.E.; Bish, D.L. Rietveld refinement of crystal structures using powder X-ray diffraction data. Rev. Mineral. Geochem. 1989, 20, 277–308.
40. Doebelin, N.; Kleeberg, R. Profex: A graphical user interface for the Rietveld refinement program BGMN. J. Appl. Crystallogr. 2015, 48, 1573–1580.
41. Czímerová, A.; Bujdák, J.; Dohrmann, R. Traditional and novel methods for estimating the layer charge of smectites. Appl. Clay Sci. 2006, 34, 2–13.
42. Zakusin, S.V.; Krupskaya, V.; Dorzhieva, O.V.; Zhuhlistov, A.P.; Tyupina, E.A. Modification of the adsorption properties. Sorpt. Chromatogr. Processes 2015, 15, 280–289.
43. Fineevich, V.P.; Allert, N.A.; Karpova, T.R.; Dupliakin, V.K. Composite nanomaterials based on acid-activated montmorillonite. Russ. Chem. J. 2007, 4, 69–74. (In Russian).
44. Tyagi, B.; Chudasama, C.D.; Jasra, R.V. Determination of structural modification in acid activated montmorillonite clay by FT-IR spectroscopy. Spectrochim. Acta Part A 2006, 64, 273–278.
45. Rhodes, C.N.; Brown, D.R. Catalytic activity of acid-treated montmorillonite in polar and nonpolar reaction media. Catal. Lett. 1994, 24, 285–291.
46. Shlikov, V.G. X-ray Analysis of Mineral Composition of Fine-Grained Soil; GEOS: Moscow, Russia, 2006; p. 175. (In Russian)
47. Madejova, J.; Komadel, P. Baseline studies of The Clay Minerals Society Source Clays: Infrared methods. Clays Clay Miner. 2001, 49, 410–432.
48. Russell, J.D.; Fraser, A.R. Clay Mineralogy: Spectroscopic and Chemical Determinative Methods; Wilson, M.J., Ed.; Chapman & Hall: London, UK, 1996; pp. 11–67.
49. Pentrák, M.; Czímerová, A.; Madejová, J.; Komadel, P. Changes in layer charge of clay minerals upon acid treatment as obtained from their interactions with methylene blue. Appl. Clay Sci. 2012, 55, 100–107.
50. Komadel, P. Acid activated clays: Materials in continuous demand. Appl. Clay Sci. 2016, 131, 84–99.
51. Dong, H.; Peacor, D.R. TEM observations of coherent stacking relations in smectite, i/s and illite of shales: Evidence for Macewan crystallites and dominance of 2M1 polytypism. Clays Clay Miner. 1996, 44, 257–275.
52. Osipov, V.I.; Sokolov, V.N. Clays and Its Properties. Composition, Structure and Properties Formation; GEOS: Moscow, Russia, 2013; p. 576. (In Russian)
53. Simc, V.; Uhlík, P. Crystallite size distribution of clay minerals from selected Serbian clay deposits. Geoloski Anali Balkanskoga Poluostrva 2006, 67, 109–116.
54. Kotarba, M.; Srodon, J. Diagenetic evolution of crystallite thickness distribution of illitic material in Carpathian shales, studied by the Bertaut-Warren-Averbach XRD method (MudMaster computer program). Clay Miner. 2000, 35, 383–391.
55. Mystkowski, K.; Srodon, J. Mean thickness and thickness distribution of smectite crystallites. Clay Miner. 2000, 35, 545–557.
56. Trofimov, V.T.; Korolev, V.A.; Voznesensky, E.A.; Golodovskaya, G.A.; Vasil'chuk, Y.K.; Ziangirov, R.S. Soil Science, 6th ed.; MSU: Moscow, Russia, 2005; p. 1024. (In Russian)
57. Rouquerolt, J.; Avnir, D.; Fairbridge, C.W.; Everett, D.H.; Haynes, J.H.; Pernicone, N.; Ramsay, J.D.F.; Sing, K.S.W.; Unger, K.K. Recommendations for the characterization of porous solids. Pure Appl. Chem. 1994, 66, 1739–1758.
58. Karnauhov, A.P. Adsorption. Texture of Dispersive and Porous Materials; Nauka: Novosibirsk, Russia, 1999; p. 470. (In Russian)
59. IUPAC. Manual of Symbols and Terminology. Pure Appl. Chem. 1972, 31, 577.
60. Churakov, S. Mobility of Na and Cs on Montmorillonite Surface under Partially Saturated Conditions. Environ. Sci. Technol. 2013, 47, 9816–9823.

61. Thommes, M.; Kaneko, K.; Neimark, A.V.; Olivier, J.P.; Rodriguez-Reinoso, F.; Rouquerol, J.; Sing, K.S.W. Physisorption of gases, with special reference to the evaluation of surface area and pore size distribution (IUPAC Technical Report). Pure Appl. Chem. 2015, 87, 1051–1069.
62. Novikova, L.; Ayrault, P.; Fontaine, C.; Chatel, G.; Jérôme, F.; Belchinskaya, L. Effect of low frequency ultrasound on the surface properties of natural aluminosilicates. Ultrason. Sonochem. 2016, 31, 598–609.
63. Kheok, S.C.; Lim, E.E. Mechanism of palm oil bleaching by montmorillonites clay activated at various acid concentrations. J. Am. Oil Chem. Soc. 1982, 59, 129–131.
64. Morgan, D.A.; Shaw, D.B.; Sidebottom, T.C.; Soon, T.C.; Taylor, R.S. The function of bleaching earth in the processing of palm, palm kernel and coconut oils. J. Am. Oil Chem. Soc. 1985, 62, 292–299.

CHAPTER 11

The Gboko Limestone, Yandev, Benue State, Nigeria: Geology, Geochemistry and Industrial Potentials

Anuba Basil Ofulume[1], Sabinus Ikechukwu Ibeneme[1,2*], Donatus Maduka Orazulike[1], Ibrahim Vela Haruna[3], Sani Aishatu[4], Diugo Okechukwu Ikoro[1], Stephen Ikpendu Nwankwo[5], Nnaemeka Oluchukwu Ezetoha[1], Joseph Azi Bulus[6]

[1]*Department of Geology, Federal University of Technology, Owerri, Nigeria*

[2]*GEM Executive Education, Grenoble Ecole De Management, Rue Pierre Sêmard Grenoble, France*

[3]*Department of Geology, Federal University of Technology, Yola, Nigeria*

[4]*Department of Geology, University of Maiduguri, Maiduguri, Nigeria*

[5]*Department of Architecture, Federal University of Technology, Owerri, Nigeria*

[6]*Department of Geology, University of Jos, Jos, Nigeria*

ABSTRACT

The Cretaceous shallow marine Gboko limestone, Yandev, Nigeria is a component of the sedimentary fill of the 800 km NE-SW trending Benue Trough, Nigeria. The limestone is made up of thin bedded to massive limestone beds interspersed with laminated grey shale having foraminifera as the dominant fossil. The limestone has both mud supported and grain supported texture, and micrites constitute about 75% of the limestone. Bulk chemical composition analysis of the limestone reveals average $CaCO_3$ of 92.41% and a range of 77.50% - 99.00%. Mineralogical impurities include quartz, dolomite, pyrrhotite, fluorapatite etc. Trace elements concentration analysis was carried out using Energy Dispersive X-ray (EDXRF) spectrophotometry and showed the following trace elements: Mn (841.3 ppm), Sr (444.6 ppm), Fe (470 ppm), Zn (114.6 ppm) and Pb (116.4 ppm). Calcining the limestone in a laboratory muffle furnace at 1050°C for 90 minutes produced a compact, soft burnt porous and reactive lime that does not crumble into fines. The lime so produced neither meets the requirements of the Steel Making Shop (SMS) of the Ajaokuta Steel Plant nor could it be used in the growing sugar refining industry in Nigeria. It can however be used in the food and the food by-products industry, environmental, agricultural and petroleum industries etc. The raw stone remains a major source of raw materials for cement manufacture for the ever expanding building industry.

Keywords: Benue Trough, Calcination, Fluxing, Geochemistry, Gboko, Limestone

1. INTRODUCTION

Nigeria in the comity of nations is rightly tagged developing country, a third world economy. The nation is definitely desirous of shedding this "developing" toga and putting on the enamored appellation of developed economy. In fact she wants to be among the top 20 developed economies by the year 2020. Attainment of this goal can only be via industrialization. Is it realizable? Yes, we can. There are abundant human and mineral resources. Many nations that do not have raw materials but have depended solely on importation have attained industrialized status. In this regard Japan and South Korea are noteworthy. But a situation where the raw materials are abundant as is the situation in Nigeria, only what is needed is the political will and focus on the part of government so as to harness the resources, exploit and utilize the raw materials to produce goods and services to uplift the standard of living and boost the Gross Domestic Product (GDP). This calls for very sound economic and political leadership to put right the priorities of the nation and acquire the needed technological expertise to utilise the abundant resources to produce goods both for export and local consumption. Nigeria is a vast market for diverse goods and services and given adequate time and focus will be satisfied and room for export created.

Limestone and its burnt derivative lime are very important raw materials for the industry. Lime on its own has more than two hundred applications and is reputed to be the largest bulk industrial chemical. Limestone is of course a sine- quanon in the making of cement—a very vital ingredient in the construction industry especially housing, bridges and factories. The task ahead is to examine the role this monomineralic rock specifically the Gboko limestone can play in enhancing the developmental economy of Nigeria.

2. GEOLOGY OF GBOKO LIMESTONE, YANDEV, BENUE STATE, NIGERIA

The Benue Basin is an elongate NE-SW oriented sedimentary rift that extends inland for approximately 800 km from the Nigerian coastline. The Gboko limestone (Figure 1) actually exposed at Yandev quarry is a component of the sedimentary fill of the Benue trough. The limestone is Cretacious in age and is regarded as being part of Asu River Group of [1] . The Bima Sandstone forms the basal fill of the basin and consists of the braided river, lacustrine and deltaic clastics. [2] has given a detailed description of the stratigraphic sequence of deposition at the Yandev quarry. The rock unit consists of sandstone, limestone, and shales that were laid down during the initial phase of the first tectonic cycle of [3] . The section is made up of thin bedded to massive limestone bed interspersed with laminated grey shale. Foraminifera dominate the fossils. The limestone can be divided into lower and upper units. While the lower unit is composed of limestone with minor shale parting the upper unit is made of shale/limestone alternations. [2] reported that the limestones have both mud supported and grain supported textures in a ratio 3:1 and that micrites constitute 75% of the limestone. While the lower unit is composed of limestone with mudstone-wackestone texture and oncolite and mollusk packstone the upper unit is an alternation of limestone with mudstone-wackestone-packstone texure and shales. The limestone is inferred to have been deposited in a shallow coastal hypersaline lagoon. The lagoon is interpreted to have been formed from marine incursion into a localized lacustrine body during the first transgression into the Benue Trough.

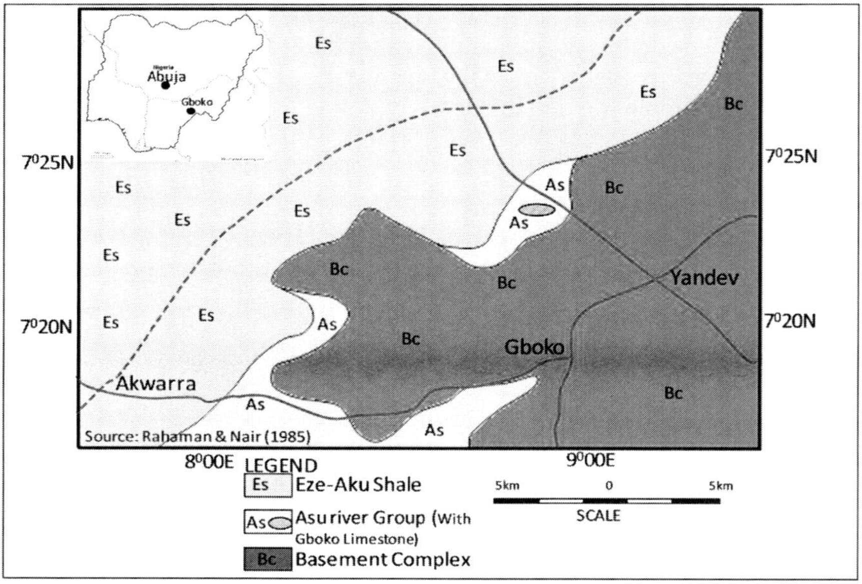

Figure 1. Geological map of Gboko area showing Gboko limestone outcrop at about 7 km Northwest of Yandev.

3. MATERIALS AND METHODS

3.1. Geochemistry

The major elements abundance in the representative samples of the limestone were determined using Atomic Absorption Spetrophotometry (AAS). After digestion of the samples the concentration of Si, Ti, Al, Fe, Mn, Mg, Ca, Na, K, P, and S in the limestone samples were measured with a single element light source. Calibration for each element was achieved using four AAS standard solutions of known concentrations. Elemental values were recalculated to oxides using appropriate conversion factors. Summary of the chemical results is presented in Table 1. Determination of trace elements concentrations in the samples was carried out using Energy Dispersive X-ray Fluorescence (EDXRF). Pellets 19 mm diameter were prepared from ground samples of less than 125 microns. Then measurements were performed using an annular 25mCi^{109}Cd as the excitation source that assists Ag-K X-rays in which case all elements with lower characteristic excitation energies were accessible for detection in the samples. Quantitative analyses of the samples were carried out using the

Emission-Transmission (E-T) method. Table 2 shows some trace elements concentrations in the Gboko limestone.

Table 1. Major oxide composition of the Gboko Limestone Samples.

Oxides	GB1	GB2	GB3	GB4	GB7	GB9	GB10	GB13	GB15	GB20
SiO_2	12.26	4.14	5.41	3.28	6.98	4.82	3.52	10.88	6.02	5.21
TiO_2	0.48	0.64	0.59	0.75	0.78	0.60	0.72	0.64	0.67	0.71
Al_2O_3	0.84	0.32	0.43	0.66	0.64	0.77	0.60	0.74	0.72	0.68
FeO	0.69	0.20	0.23	0.30	0.13	0.91	0.73	0.51	0.77	0.73
MnO	0.84	1.09	0.08	0.09	1.13	0.09	0.10	0.12	0.08	0.11
MgO	0.88	1.01	1.26	0.91	0.96	1.29	1.38	1.30	1.23	1.45
CaO	43.40	53.90	51.94	52.64	48.99	52.92	55.44	50.58	53.36	54.32
K_2O	0.65	0.45	0.49	0.42	0.81	0.54	0.60	0.50	0.42	0.79
P_2O_5	0.006	0.01	0.008	0.006	0.001	0.007	0.01	0.008	0.01	0.008
SO_3	0.16	0.09	0.04	0.12	0.03	0.38	0.32	0.35	0.22	0.31

Average $CaCO_3$% = 92.41; Range $CaCO_3$ = 77.50% - 99%; Silica is the dominant impurity followed by Magnesia.

The low Sr content of ancient limestones like the Gboko limestone cannot be due to the variations in the mSr/mCa ratio of sea waters from which the limestones were precipitated for the mSr/mCa ratio of sea waters has remained remarkably constant throughout the Phanerozoic era ([4] [5]). Although ancient limestones are considerably depleted in Sr they exhibited a wide range of Sr content from a low of 50 ppm to a high of 1000 ppm [5] . The Sr in Gboko limestone has a mean value of 444.6 ppm and a range of 198 - 694 ppm. The mean value of Sr is quite close to the [6] low Sr value (less than 400 ppm) for shallow water limestone. The Mn content averages 841.3 ppm and has a range of 545 - 1760 ppm. Considering therefore the 510 ppm mean value of Mn for shallow water carbonates and 2100 ppm (mean) for deep marine limestones of [7] the Gboko limestone with the aforementioned mean values of Sr and Mn is considered a shallow marine limestone based on trace element geochemical consideration.

Table 2. Some Trace Element Concentrations (ppm) in Gboko Limestone.

Sample No	Mn (ppm)	Sr (ppm)	Zn (ppm)	Pb (ppm)	Fe (ppm) × 10^3
GB1	1760	193	88	97	6.54
GB2	851	256	102	113	1.61
GB3	667	213	102	108	1.79
GB4	740	369	107	270	2.04
GB5	545	205	95	91	5.55
GB6	733	389	132	96	8.86
GB7	880	266	122	136	1.52
GB8	658	499	86	88	5.97
GB9	743	694	95	122	7.11
GB10	843	692	153	117	5.73
GB11	953	348	157	115	2.81
GB12	545	450	70	77	9.17
GB14	943	497	124	108	4.85
GB15	638	689	94	107	6.06
GB16	973	628	124	121	3.19
GB17	829	569	127	120	3.52
GB18	977	637	120	111	5.20
GB19	823	346	143	114	2.17
GB20	927	509	138	102	5.7
Mean	841.3	444.6	114.6	116.4	4.70

3.2. Mineralogy

X-ray diffraction analysis was done on pulverised (1 micron) bulk samples of the limestone using Philips P.W 1800 X-ray diffractometer. Cu-K radiation was employed. The diffraction peaks were measured and plotted by the aid of a computer. Minerals were recognized by the use of their d-values picked from X ray powder data cards. The corresponding 2theta values for d-value were traced in the diffractograms and peaks marching them were labelled (Figure 2). Mineralogical impurities associated with the limestone are dolomite, pyrrhotite, albite, fluorite and fluorapatite. Dolomite is a mineralogical impurity and source of MgO whose level is critical in the use of limestone for cement making. If a critical level is exceeded it could swell and cause disruption of structure.

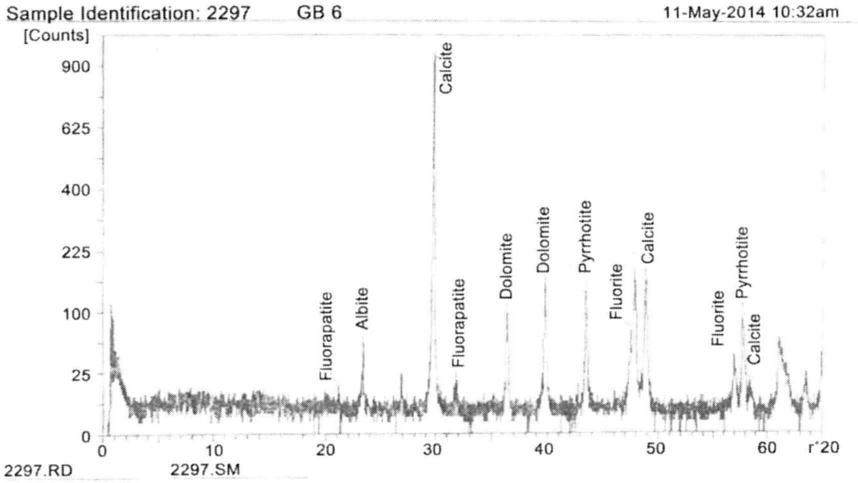

Figure 2. XRD Trace of Gboko Limestone.

Burning the Gboko limestone to produce lime is by heating the limestone either in kilns or laboratory muffle furnace. Lime production is very important because very many applications of the limestone is in the form of lime prepared from it. Lime manufacturers choose quality raw stone because quality of the lime is dictated by quality of the limestone and by its manufacturing processes. A compact highly reactive lime is desired by the industries. A general requirement for commercial lime production is a rock with minimum of 95% of $CaCO_3$ [7] but BIS (IS: 10345-1582) prescribed CaO 52% minimum (i.e. 92.86% $CaCO_3$) for use in lime manufacturing for fluxing in steelmaking. Gboko limestone chemically has average 92.41% $CaCO_3$ and ranges from 77.50% - 99.00% $CaCO_3$. This can be accepted as meeting the BIS requirement and therefore shall be calcined and the technological quality of the resultant lime tested for suitability of production of commercial lime for varied applications.

3.3. Crystal Structure

The physical characteristics of the various limestones determine the quality of the lime product. The texture of the limestone is a factor in the successful calcination of the stone into lumps of lime. Coarse grained limestones are prone to fracturing and crumbling into fines when heated. They can be disintegrated to dust by the heat of the calcination process, a situation dreaded by lime manufacturers.

On the other hand, the fine grained limestones can easily be calcined to form lumps of lime because the crystals resist the temperature stress .The Gboko limestone is fine-grained (Figure 3) and is therefore favourably disposed to calcination into lumps of lime.

3.4. Calcination

$$CaCO_3 \xrightarrow[\Delta]{} CaO(\text{lime}) + CO_2$$

As hinted earlier, calcination is an important factor in quality lime production. The objective always is to produce a compact, softburned, highly reactive lime. To achieve this, the limestone has to be calcined in laboratory muffle furnace through a range of temperatures at varied retention times and the resultant lime then tested technologically for strength and reactivity to determine the best burning parameters. Samples of the Gboko limestone trimmed to size (2 cm diameter) were put in crucibles and then shock calcined. This implies they were introduced in the furnace only when furnance has attained and stabilized at desired calcination temperatures of 950°C, 1000°C, 1050°C, and 1100°C.

Figure 3. Representative Sample of Gboko Limestone.

Generally larger stone sizes are more difficult to calcine uniformly and require more time. This is because dissociation always penetrates gradually from the surface to the interior of the stone [8] . To expel the CO_2 from such large stone high temperatures are necessary to generate sufficient CO_2 pressure in the interior of the crystal lattice for the escape of the gas. Frequently these high temperatures overburn the surface of the stone causing excessive shrinkage, which narrows and closes the pores. [9] prescribed that the ideal stone to calcine for optimum quality, uniformity and thermal efficiency would be of small size (1.25 cm) and be of uniform size and shape.

3.5. Temperature and Retention Time in the Furnance

Since the temperature and time can easily be controlled, they were varied so as to determine the best calcination parameters for the Gboko limestone by testing the quality of the lime product using technological tests such as weight loss, decrepitation, loss on ignition of lime, mechanical strength, and reactivity and by inference porosity and surface area of the lime. The calcination duration was 60, 90, and 120 minutes.

3.6. Weight Loss and Loss on Ignition of Lime

After calcining at the desired temperature and time, the lime products were allowed to cool to room temperature in a dessicator before weighing accurately to determine the weight loss during calcinations. The loss in weight represents the weight of CO_2 driven off during the burning and this is expressed as percentage of the original weight. The loss on ignition of lime should not be confused with weight loss or loss in ignition of limestone. Loss on ignition is done to determine if calcination was complete. After calcining at the appropriate temperature the lime was allowed to cool in a dessicator and thereafter about two grammes weighed again accurately to four places of decimal, placed in a crucible and then calcined further at 1100°C. It is cooled to room temperature in a dessicator and then weighed again. The difference in weight was expressed as a percentage of the original weight to give the loss in ignition of lime at the particular calcined temperature (Table 3).

3.7. Decrepitation and Mechanical Strength

The various calcines 950°C/60minutes, 1000°C/90minutes etc were cooled to room temperature in a desicator then poured onto 1.18 sieves and shaken gently. The quantity of minus 1.18 sieve material generated was measured and expressed as percentage of the original weight are then expressed as percent decrepitation (Table 3).

Screen analysis was carried out on the calcined stone. After shock calcining the samples at the appropriate temperature the lime products were subsequently weighed and then poured on to a set of sieves of aperture sizes 13.20 mm, 9.50 mm, 4.75 mm, 2.36 mm, 1.18 mm and 0.60 mm and then given an extended shaking with hands for about 10 minutes. Screening the stones for 10 minutes gives an indication of the tendency of the stone to decrepitate in the kiln [9] . The result is presented in Table 4. From the test one can see that Gboko limestone does not show any tendency to decrepitate in the kiln judging from the minus 0.6 mm material.

Table 3. Results of technological tests on the above lime.

Temp (°C) Calcine	wt loss (%)	L.O.I	Decrepitation
950	42.82		
1000	43.41		
1050	43.80	0.47	1.0
1100	43.87		

Levels of LOI in Basic Oxygen Furnace (BOF) is generally less than 3% [11].

Table 4. Mechanical strength screening test result.

Aperture	Gboko lst	Jakura marble
13.20 mm	93.46	16.82
9.50 mm	9.68	10.00
4.75 mm	5.88	8.11
2.36 mm	4.60	7.32
1.18 mm	4.12	6.31
0.60 mm	0.96	5.60

The less than 0.6 mm sieve material for Gboko limestone is less than 1.0 grammes. This is not significant compared to 5.6 gm of Jakura marble which generates fines. The result is an indication that the lime will not generate fines in the kiln and during handling.

3.8. Reactivity of the Gboko Lime

The maximum temperature reached through the exothermic reaction of quicklime with water is a good indicator of the quality of the lime in terms of the available CaO [10] . The ASTM specified the use of 100 gm of lime in 400 ml of distilled water at 25°C. The sample was prepared to pass 2.36 sieve as quickly as possible and then left in a desicator to cool to room temperature before slaking test using a thermos flask, thermometer, and a stop clock, and a rod as stirrer, a pestle and a mortar for crushing the quicklime. Stirring was done continuously and reading of temperature taken at 30 seconds interval. Readings were recorded until less than 0.5°C temperature change was noted in each of three consecutive readings were taken. The initial temperature was subtracted from the final temperature to obtain the total temperature rise. Also the initial temperature was subtracted from the temperature at 2 minutes to obtain temperature rise in 2 minutes (Table 5). Suitable curves showing temperature rise as the ordinate and time as the abscissa were then plotted (Figure 4). The results were reported as temperature rise in Celsius after 2 minutes.

Table 5. Results of the Reactivity with water (slaking test).

Calcine	Temp (°C) after 2 mins
950°C/90mins	13
1000°C/90mm	21
1050°C/90mins	57
1100°C/90mins	16
950°C/40mins	22

Optimum reactivity is defined as >40°C rise after 2 minutes; Therefore 1050°C/90minutes calcine yielded a highly reactive lime that is a softburned, porous lime with high surface area and that does not decrepitate.

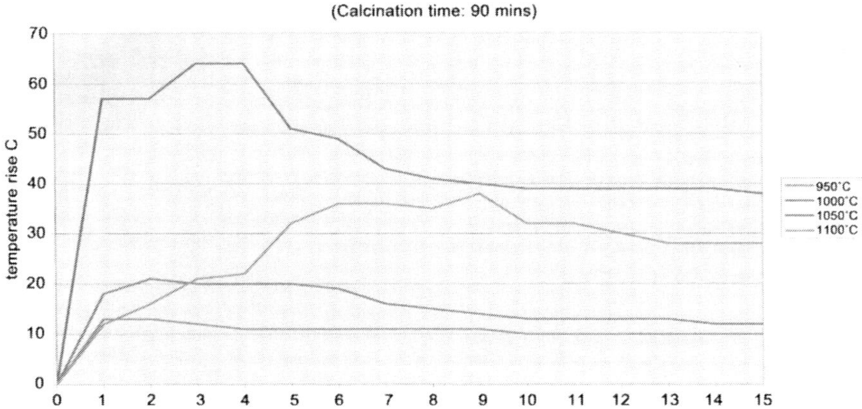

Figure 4. Reactivity Graph of Gboko Lime Time (Min). Activity Defined by temperature rise (°C) after 2 minutes.

4. RESULTS AND DISCUSSION

All parameters did not meet the specification except CaO (Table 6). Gboko limestone cannot be used in the growing Nigerian Sugar industry.

Table 7 presents the British Standard Institute guidelines for flux limestone for use in steel plants. The Gboko limestone can be utilized in grades 1 and II for lime manufacturing and for use in steel making. Both quicklime and hydrated lime are widely used in flotation or recovering of many non ferrous ores in particular copper ore flotation in which lime acts as a depressant (settling aid) and maintains proper alkalinity in the flotation circuit.

However, Gboko limestone does not meet the requirements for the SMS (steel making shop) limestone for the BOS (Basic Oxygen Steelmaking) shop of the Ajaokuta Steel Plant where the purity of the limestone must be greater than 95.5% and silica less than 1.0% (Table 8).

From Table 8, the sulphur content of Gboko limestone (0.20%) far exceeds the less than 0.05% required for steel making at Ajaokuta. Consequently this limestone will not be used in the Ajaokuta steel industry, Nigeria.

Table 6. (BSI(IS:3204-1978) Specification of limestone for use in sugar manufacture.

Parameter	BST	Gboko lst
LOI	44.0	0.47
SiO_2 (max)	2.0	6.25
CaO {min}	50.0	51.7
MgO (max)	1.0	1.17
CO_2 (min)	41.0	-
$Al_2O_3 + Fe_2O_3$(max)	1.5	3.57

Table 7. BS1 Guideline for limestone for use as flux in steel industry.

Characteristics	Grade l	Grade II
CaO	52% min	50%min
MgO	2% max	3%max
Acid insoluble	2% max	4% max out of which SO_2 should not exceed 1.8%
Preferred size	60 mm, 25 mm not exceed 5%	6 mm not exceed 10%

Table 8. Average Chemical Composition of Gboko Limestone compared with Ajaokuta SMS grade lst.

Parameters	Gboko lst	Ajaokuta SMS grade lst
$CaCO_3$	92.41	95.5
$MgCO_3$	1.17	1.2
SiO_2	6.25	1.0
Al_2O_3	0.64	1.5
Fe_2O_3	0.52	1.5
S	0.20	0.05
P	0.007	0.06
LOI	41.75	42.44

5. POTENTIAL APPLICATIONS

5.1. Environmental Use of Gboko Lime

Lime is useful in the sewage biosolids and sludges. Both quicklime and hydrated lime have been in use for biological organic wastes for more than 100 years. Lime neutralizes acid wastes generated in industry thereby impeding corrosion and protecting the natural environment. Lime removes silica, manganese, fluorides, iron and other impurities from water. Most importantly lime is used in the treatment of drinking water, softening, pH adjustment, coagulation. The Gboko lime can be of use in the treatment of hazardous wastes currently generated or abandoned materials for example, the hazardous waste that was imported and dumped in Koko, Delta state, Southern Nigeria.

5.2. Gboko Lime in Food Industry

In the dairy industry, lime water is often added to the cream separated from whole milk so as to reduce acidity prior to pasteuralization when butter is produced. Lime is also beneficial in preparation of the common types of baking powder, and the preservation of fruits and vegetables. The need for the setting up of dairy industry in the northern states especially Benue state that prides itself as the foods basket of the nation is highlighted with Gboko lime as adequate and available compliment for this food industry (Figure 5).

5.3. Gboko Lime for Agricultural Purposes

Gboko lime or the ground limestone can be useful in pH adjustment of agricultural soils, but the focus seems to be the use of lime in conjuction with nitrogen fertilizers because the lime allows reduced usage of fertilizers resulting in reduction of nitrogen leaching from the soils. Gboko lime can also serve useful purpose in compositing, poultry litter and pesticides and certain fertilizers (Figure 6).

5.4. Gboko Lime in Petroleum Industries

Various uses of lime in the petroleum industry are captured graphically in Figure 7. The Gboko lime will surely boost operation in the Nigerian petroleum industry.

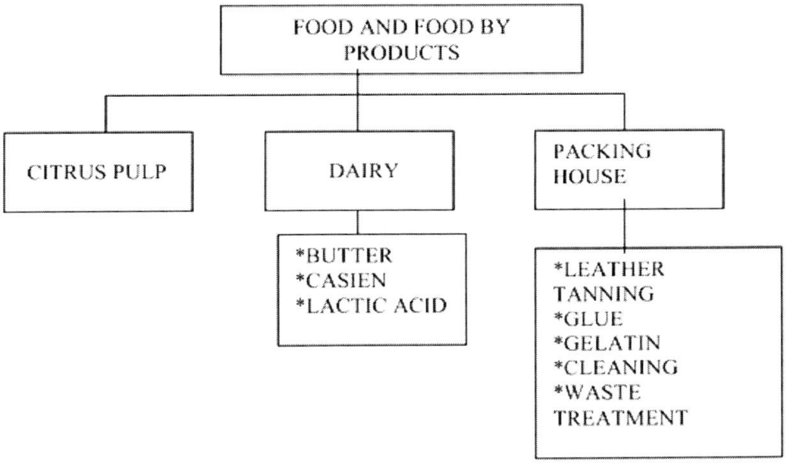

Figure 5. Graphical presentation of uses of lime in food industry.

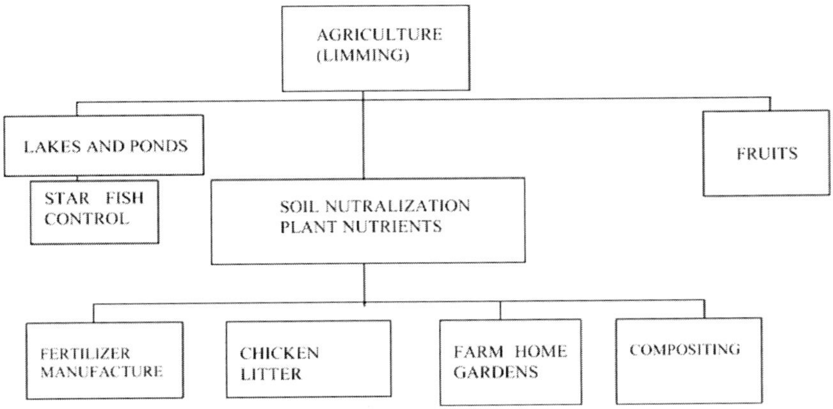

Figure 6. Graphical presentation of uses of lime in agriculture.

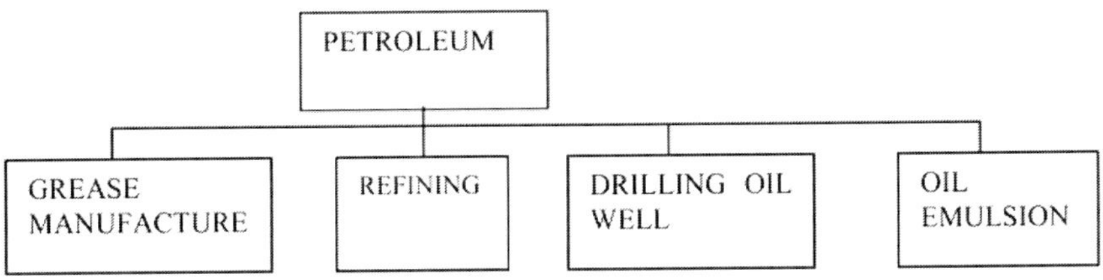

Figure 7. Graphical presentation of uses of lime in petroleum industry.

Table 9. Average Chemical Composition of Gboko Limestone.

Major Oxide	Average % Composition
SiO_2	6.25
TiO_2	0.65
Al_2O_3	0.48
Fe_2O_3	0.52
MgO	1.16
CaO	51.75
MnO	0.37
K_2O	0.57
P_2O_5	0.007
SO_2	0.20

Range $CaCO_3$ % = 77.50 - 99.00; average $CaCO_3$% = 92.41; LOI = 43.41; Total = 105.367.

5.5. Gboko Limestone for Cement Making

The Gboko limestone has for many years been the source of limestone for quality cement manufacture at the Benue cement factory at Yandev. The output capacity of the factory has received a tremendous boost by the acquisition of the factory by the Dangote Group. The Group presently is the greatest cement maker in the whole of Africa and owns cement factories in many African countries. Cement making by and large remains the major use of the Gboko limestone, and the diversification of its application is called for with the introduction or incoporation of lime manufacturing facilities in the factory at Yandev. Table 9 shows the average chemical composition of Gboko Limestone.

6. CONCLUSIONS

The Cretaceous Gboko limestone is made up of thin bedded to massive limestone beds interspersed with laminated grey shale. It has foraminifera as the dominant fossil. The limestone is shallow marine based on the trace elements analysis. The bulk chemical composition reveal $CaCO_3$range of 75.50% - 99.00% averaging 92.41%. Technological tests show that the limestone can be calcined to produce a compact soft burned, reactive lime that does not crumble into fines. The lime chemically fails to meet the requirements for fluxing application in the Steel Making Shop (SMS) of the Ajaokuta Steel Plant and for use in sugar refining industry in Nigeria.

However, it can be utilized in the food and the food by-products industry, environmental, agricultural and petroleum industries. Cement manufacturing by Dangote cement remains the main user of the raw stone.

REFERENCES

1. Reyment, R.A. (1965) Aspects of Geology of Nigeria. Ibadan University Press, Ibadan, 145.
2. Nair, K.M. and Ramanatham, R.M. (1989) Sedimentology, Stratigraphy and Paleogeographic Significance of Lower Cretaceous Gboko Limestone, Nigeria. Journal Mining and Geology, 21, 203-210.
3. Murat, K.C. (1972) Stratigraphy and Paleogeography of the Cretaceous and Lower Tertiary in Southern Nigeria. In: Dessauvagie, T.F.J. and Whiteman, A.J., Eds., African Geology, University of Ibadan, 251-266.
4. Lowestam, H.A. (1961) Mineralogy O18/O16 Ratios and Srontium and Magnessium Content of Recent and Fossil Brachiopods and Their Bearing on History of the Oceans. Journal of Geodesy, 69, 241-260.
5. Morrow, D.W. and Mayers, I.R. (1978) Simulation of Limestone Diagenesis—A Model Based on Strontium Depletion. Canadian Journal of Earth Sciences, 369-377.
6. Bausch, W.M. (1968) Outline of Distribution of Strontium in Marine Limestone. In: Muller, G. and Friedman, G.M., Eds., Recent Developments in Carbonate Sedimentology Central Europe, Berlin-Heidelberg, Springer, New York, 106-115.
7. Lutke, F. (1976) Sedmentologische and Geochemiche un arsucheungen zur zennese der flinzfazies in zur (givet and oberden) dilsch. Geol. Ges, 127, 499-508.
8. Ofulume, A.B. (2008) The Geology, Geochemistry and Economic Potentials of Some Nigerian Limestones and Marble Resources. Unpublished Ph.D. Thesis ATBU Bauchi, 142.
9. Boynton, R.S. (1980) Chemistry and Technology of Lime and Limestone. Wiley Inter Science, 482.
10. Ruskulis, O. (1997) Methods for Testing Lime in the Field. http://www.itdg.org./docs/technicalinformationservice/testinglime
11. Anderson, L.C. and Vernon, J. (1971) Quality and Production of Lime for Basic Oxygen and Steel Making. The Quarry Managers Journal Institute of Quarrying Translation, 169-175.

CHAPTER 12

Geological Controls on Mineralogy and Geochemistry of an Early Permian Coal from the Songshao Mine, Yunnan Province, Southwestern China

Ruixue Wang [1,2]

[1]*College of Geoscience and Surveying Engineering, China University of Mining and Technology (Beijing), Beijing 100083, China; Tel.: +86-10-8232-0638*

[2]*State Key Laboratory of Coal Resources and Safe Mining, China University of Mining and Technology (Beijing), Beijing 100083, China*

ABSTRACT

This paper discusses the content, distribution, modes of occurrence, and enrichment mechanism of mineral matter and trace elements of an Early Permian coal from Songshao (Yunnan Province, China) by means of coal-petrological, mineralogical, and geochemical techniques. The results show that the Songshao coal is characterized by high total and organic sulfur contents (3.61% and 3.87%, respectively). Lithium (170.39 μg/g) and Zr (184.55 μg/g) are significantly enriched in the Songshao coal, and, to a lesser extent, elements such as Hg, La, Ce, Nd, Th, Sr, Nb, Sn, Hf, V, and Cr are also enriched. In addition to Hg and Se that are enriched in the roof and floor strata of the coal seam, Li, La, Ce, Pr, Nd, Sm, Gd, Y, Cd, and Sb are slightly enriched in these host rocks. Compared to the upper continental crust, rare earth elements and yttrium in the host rocks and coal samples are characterized by a light-REE enrichment type and have negative Eu, positive Ce and Gd anomalies. Major minerals in the samples of coal, roof, and floor are boehmite, clay minerals (kaolinite, illite, and mixed layer illite-smectite), pyrite, and anatase. Geochemical and mineralogical anomalies of the Songshao coal are attributed to hydrothermal fluids, seawater, and sediment-source rocks.

Keywords: coal; mineral; elements; Early Permian; genetic types

1. INTRODUCTION

Yunnan is one of the most coal-rich provinces in Southern China. Coal resources are mainly concentrated in the east and south of Yunnan. The Songshao Mine is located in the eastern part of Yunnan Province (Figure 1). Previous studies have focused on geologic structure, coal-bearing sequences, and coal quality of the Songshao Mine [1]. This paper aims to discuss geological controls on mineralogy and geochemistry of Early Permian coal from the Songshao Mine. Studies on mineral matter in coal are important because the process of coal formation, including peat accumulation, the interaction of the organic matter with basinal fluids, sediment diagenesis, and sometimes synsedimentary volcanic inputs, may result in enrichment of mineral matter in coal. Therefore, investigations on the mineral matter in coal could help to better understand the process of coal formation [2,3,4]. From an economic perspective, mineral matter in coal can serve as carriers for some valuable elements (e.g., Ga, Al, rare earth elements and yttrium) that have been recovered, or have such potential, from coal combustion wastes [5,6,7,8]. Geochemical anomalies in coals from Eastern Yunnan have previously been reported; for instance, Zhou et al. [9] reported trace-element geochemistry of altered volcanic ash layers (tonsteins) in Late Permian coal-bearing sequences in Eastern Yunnan and Western Guizhou Provinces; Dai et al. [10] described modes of occurrence and origin of quartz and chamosite in Xuanwei coals. Dai et al. [11] found a new type of Nb(Ta)–Zr(Hf)–REE–Ga polymetallic deposit of volcanic origin in the Late Permian coal-bearing strata of Eastern Yunnan; Dai et al. [12] reported on the mineralogical and geochemical compositions of Late Permian C2 and C3 coals (both medium-volatile bituminous) from the Xinde Mine in Xuanwei, Eastern Yunnan. Geological factors controlling these geochemical and mineralogical anomalies have previously been analyzed [3,13,14,15,16]. For example, Dai et al. [13] documented mineralogical and geochemical anomalies of Late Permian coals from the Fusui coalfield, which were caused by influences from terrigenous materials and hydrothermal fluids.

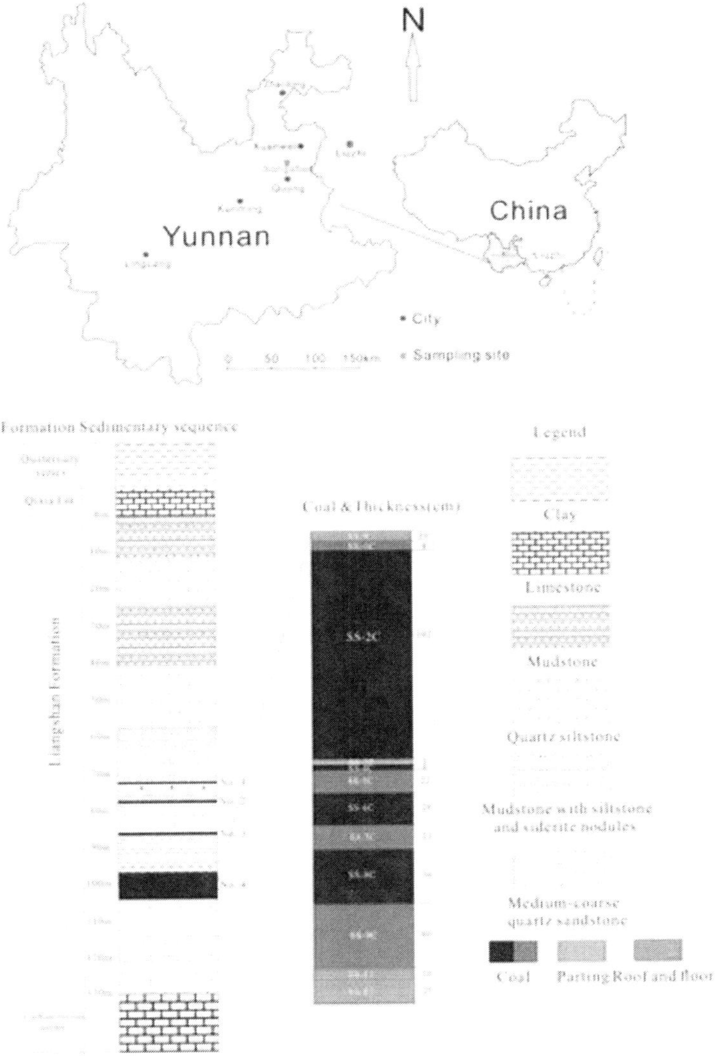

Figure 1. Location and sedimentary sequence of the Songshao Mine, Eastern Yunnan Province, China.

In the present study, new data on the Early Permian Coal from Songshao are reported with the aim to: (1) investigate geochemical and mineralogical compositions and modes of occurrence of elements and minerals; and (2) discuss the geological factors influencing mineralogical and geochemical anomalies.

2. GEOLOGICAL SETTING

The sedimentary sequences in the Songshao Mine include Upper Carboniferous strata, Lower Permian Liangshan Formation, Lower Permian Qixia Formation, and the Quaternary system (Figure 1).

The Liangshan Formation comprises the coal-bearing seam, which underlies limestone of the Qixia Formation, and conformably overlies Carboniferous limestone. The lower part of the Liangshan Formation consists of coarse quartzose sandstones; the middle part is mainly composed of mudstones, interlayered with coal seams; and the upper part of the formation consists mainly of mudstones interbedded with quartzose sandstones. The thickness of the Liangshan Formation is about 131 m, and the coal seams are numbered 1 to 4. Only the No. 4 coal seam is minable and contains several gritty mudstone partings. The thickness of the No. 4 coal varies from 2 to 6 m. Samples, taken from No. 4 coal bed were numbered as SS-R, SS-1C, SS-2C, SS-3P, SS-4C, SS-5C, SS-6C, SS-7C, SS-8C, SS-9C, SS-F1 and SS-F2 (SS-R is roof; SS-3P is parting, SS-F1 and SS-F2 are floors and others are coals).

The roof of No. 4 coal is made up of grey black carbonaceous mudstone, and the floor consists of quartzose sandstones. The Upper Carboniferous Formation is made up of dolomite limestone. The Qixia Formation consists of medium-thick and thick limestone. The Quaternary system is composed of colluvial deposits.

3. SAMPLES AND ANALYTICAL PROCEDURES

The samples selected for this study were collected from the faces of the mined coal seams in the Songshao Mine (No. 4 coal; Figure 1). All collected samples were immediately stored in plastic bags to minimize contamination and oxidation.

All the analyses were conducted at the State Key Laboratory of Coal Resources and Safe Mining (China University of Mining and Technology, Beijing, China). Proximate analysis was performed according to Chinese

Standard GB/T 212-2008 [17]. The total sulfur and forms of sulfur were conducted following Chinese Standard GB/T 214-2007 [18] and GB/T 215-2003 [19], respectively. Mean random vitrinite reflectance (R_r) and maceral composition followed ISO 7404-5: 2009 [20]. Percentages of major-element oxides were determined by X-ray fluorescence (XRF) spectrometry (Thermofisher ARL Advant'XP+, ThermoFisher Scientific, Waltham, MA, USA). Trace-element concentrations were determined by inductively coupled plasma mass spectrometry (ICP-MS, Thermofisher X series II, Thermo Fisher Scientific), with exception of Hg and F. Mercury was determined using a Milestone DMA-80 Hg analyzer (Milestone, Sorisole, Italy), with a detection limit of 0.005-ng Hg, 1.5% relative standard deviation (RSD), and 0–1000 ng linearity for the calibration [21]. Fluorine was determined by pyrohydrolysis in conjunction with an ion-selective electrode, following the method described in Chinese National Standard GB/T 4633-1997 (1997) [22]. The mineralogy was determined by X-ray powder diffraction (XRD, Rigaku, Tokyo, Japan) plus Siroquant™ (Sietronics Pty Ltd, Canberra, Australia), optical microscopic observation, and a Field Emission-Scanning Electron Microscope (FE-SEM, FEI Quanta™ 650 FEG, Hillsboro, OR, USA), in conjunction with an energy-dispersive X-Ray spectrometer(EDS, Genesis Apex 4, EDAX Inc., Mahwah, NJ, USA), and these analytical procedures were described by Dai et al. [13].

4. RESULTS

4.1. Ultimate and Proximate Analyses and Coal Rank

The vitrinite reflectance (1.24%) and volatile matter (30.91 wt %, daf) of the coal bench samples (Table 1) indicate a middle- to high-volatile bituminous rank, according to the ASTM classification (ASTM D388-12, 2012) [23]. The coal is a medium-ash and high-sulfur coal according to Chinese Standards GB/T 15224.1-2004 [24] (coals with ash yield 20.01%–30.00% are high-ash coal) and GB/T 15224.2-2004 [25] (coals with total sulfur content >3% are high-sulfur coal). Organic sulfur accounts for most of the otal sulfur (Table 1). However, sample SS-1C has a higher content of pyrite sulfur than other coal benches.

Table 1. Proximate and ultimate analyses (wt %), and forms of sulfur (wt %) in Songshao coal.

Sample	R_r	M_{ad}	A_d	VM_{daf}	$S_{t,d}$	$S_{s,d}$	$S_{p,d}$	$S_{o,daf}$	N_{daf}	C_{daf}	H_{daf}
SS-R	nd	nd	79.88	nd	5.35	nd	nd	nd	nd	nd	nd
SS-1C	1.35	1.40	33.35	29.84	13.03	0.31	8.06	6.99	0.70	75.50	4.04
SS-2C	1.21	0.64	19.22	30.68	3.85	0.05	0.71	3.83	1.10	85.25	4.84
SS-3P	1.17	0.47	47.19	77.10	1.68	0.20	0.04	2.72	0.62	64.71	3.16
SS-4C	1.26	0.63	9.67	27.66	3.78	bdl	0.11	4.09	1.16	87.57	4.73
SS-5C	1.29	0.59	26.35	28.84	2.86	bdl	0.05	3.84	1.21	84.75	4.98
SS-6C	1.23	0.52	24.20	29.81	2.83	bdl	0.07	3.68	1.11	84.97	4.93
SS-7C	1.41	0.44	17.84	28.31	3.26	0.01	0.10	3.83	0.98	86.05	4.82
SS-8C	1.22	0.58	32.77	31.62	2.81	bdl	0.47	3.55	1.11	83.36	5.18
SS-9C	1.24	0.52	29.85	29.87	3.16	0.07	0.25	4.04	1.01	83.63	4.97
SS-F1	nd	nd	50.33	nd	1.53	nd	nd	nd	nd	nd	nd
SS-F2	nd	nd	80.78	nd	0.45	nd	nd	nd	nd	nd	nd
WA	1.24	0.60	23.80	30.91	3.61	0.04	0.62	3.87	1.07	84.32	4.88

M, moisture; A, ash yield; VM, volatile matter; C, carbon; H, hydrogen; N, nitrogen; S_t, total sulfur; S_s, sulfate sulfur; S_p, pyrite sulfur; S_o, organic sulfur; ad, air-dry basis; d, dry basis; daf, dry and ash-free basis; WA, weighted average for coals (weighted by thickness of sample interval); bdl, below detection limit; nd, not detected.

4.2. Geochemical Composition

4.2.1. Major Element Oxides

The major-element oxides are mainly represented by SiO_2 and Al_2O_3 (Table 2). When considered on a whole-coal basis, Songshao coal contains higher proportions of Al_2O_3, SiO_2, and, to a lesser extent, MgO, TiO_2, and K_2O, than the average values for Chinese coals reported by Dai et al. [26]; other major element oxides, however, are either lower than, or close to, corresponding average values for Chinese coals. The SiO_2/Al_2O_3 ratios for the coals are much lower than those of common Chinese coals (1.42) [26] and the theoretical SiO_2/Al_2O_3 ratio of kaolinite (1.18).

The content of Fe_2O_3 is significantly higher in the roof and sample SS-1C than in other samples. The content of CaO in the parting SS-3P is particularly high. Based on the variation in concentrations through the seam section, two groups of major element oxides can be classified:

Group 1 includes SiO_2, TiO_2, Al_2O_3, and K_2O, all of which show a saw-like distribution, similar to the ash yield variation through the seam section. These oxides are enriched in the host rocks (roof and floor), but are at a low level in the coal benches and parting (Figure 2).

Table 2. Percentages of major-element oxides in the Songshao coal, parting, roof and floor rocks.

Sample	Th (cm)	LOI	SiO_2	TiO_2	Al_2O_3	Fe_2O_3	MnO	MgO	CaO	Na_2O	K_2O	P_2O_5	SiO_2/Al_2O_3
SS-R	10	20.12	40.27	0.86	28.27	6.41	0.01	1.03	0.35	0.05	2.15	0.06	1.42
SS-1C	8	66.65	9.08	0.2	8.73	13.83	0.01	0.13	0.14	0.04	0.11	0.2	1.04
SS-2C	192	80.78	8.35	0.46	8.07	0.9	0	0.14	0.37	0.03	0.1	0.12	1.03
SS-3P	5	52.81	0.76	0.02	1.6	0.59	0.08	0.26	36.97	0.02	0	0.01	0.48
SS-4C	4	90.33	3.07	0.13	4.8	0.2	0	0.18	0.41	0.19	0.03	0.03	0.64
SS-5C	22	73.65	11.55	0.44	13.08	0.19	0	0.36	0.1	0.04	0.19	0.03	0.88
SS-6C	29	75.8	10.36	0.42	12.28	0.22	0	0.3	0.07	0.02	0.19	0.02	0.84
SS-7C	23	82.16	7.72	0.29	8.86	0.13	0	0.27	0.1	0.02	0.13	0.03	0.87
SS-8C	50	67.23	14.57	0.57	15.75	0.61	0	0.4	0.11	0.03	0.33	0.06	0.93
SS-9C	60	70.15	13.72	0.54	13.74	0.66	0	0.31	0.08	0.02	0.42	0.03	1
SS-F1	10	49.67	25.24	1.12	22.83	0.19	0	0.21	0.07	0.01	0.39	0.03	1.11
SS-F2	25	19.22	39.79	1.65	37.45	0.46	0	0.31	0.05	0.01	0.64	0.06	1.06
SS Coal	-	76.5	10.24	0.46	10.56	0.95	0	0.23	0.23	0.03	0.19	0.08	0.97
Chinese Coal [a]	n.d.	n.d.	8.47	0.33	5.98	4.85	0.02	0.22	1.23	0.16	0.19	0.09	1.42
CC	n.d.	n.d.	1.21	1.4	1.77	0.2	0.15	1.06	0.19	0.18	1.01	0.88	0.68

Th, thickness; LOI, loss on ignition; CC, weighted average for coal samples; [a] From Reference [26]; n.d., no data.

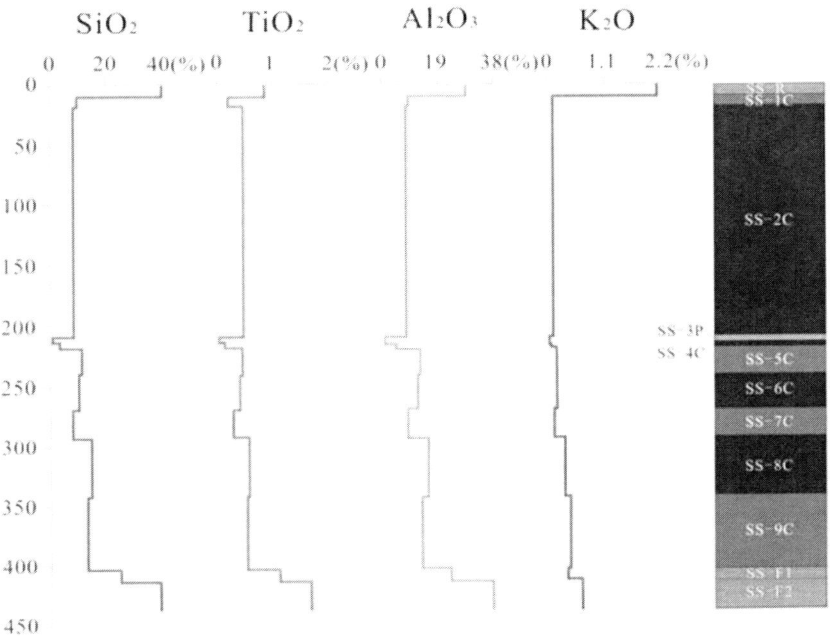

Figure 2. Variations of SiO_2, TiO_2, Al_2O_3 and K_2O through the seam section.

Group 2 consists of Na_2O, MgO, P_2O_5, MnO, and Fe_2O_3. The concentrations of these oxides do not show distinct variation through the seam section.

4.2.2. Trace Elements

The trace-element contents of Songshao coal are listed in Table 3. Compared to the averages for world hard coals [27], and based on the enrichment classification of trace elements in coal by Dai et al. [28], a number of trace elements are enriched in Songshao coal. Lithium has a concentration coefficient (CC: ratio of element concentration in Songshao coals vs. world hard coals) >10. Zirconium displays CC values between 5 and 10. Elements, including Hg, Sr, Nb, Sn, Hf, V, Cr, Se, Th, and some light rare earth elements (La, Ce, and Nd), are slightly enriched in the coal (CC = 2–5). However, Rb, Ba, Bi, Mo, and Sb, are depleted in the coal (CC < 0.5). The concentrations of the remaining elements (0.5 < CC < 2) are close to the corresponding averages for world hard coals [27].

Table 3. Trace-element concentrations (μg/g) in the Songshao coal, parting, and roof and floor rocks.

Sample	SS-R	SS-1C	SS-2C	SS-3P	SS-4C	SS-5C	SS-6C	SS-7C	SS-8C	SS-9C	SS-F1	SS-F2	World	SC	C1	UCC	PC	C2
Li	160.70	84.29	140.24	17.10	72.32	170.02	184.16	124.26	231.48	257.89	844.97	944.01	14.00	170.39	12.17	20.00	116.01	5.80
Be	3.45	3.29	0.98	0.37	0.78	1.88	1.87	1.30	1.89	1.92	3.24	3.19	2.00	1.41	0.71	3.00	1.62	0.54
F	670.30	138.20	131.09	49.00	154.90	176.81	175.67	158.68	204.91	174.03	165.51	293.06	82.00	153.84	1.88	-	348.55	-
Sc	25.53	25.02	3.81	1.61	7.10	5.07	6.41	4.13	4.06	6.87	10.47	30.63	3.70	5.03	1.36	11.00	4.57	0.42
V	231.22	383.34	57.25	9.74	34.93	76.15	72.88	55.08	88.81	99.12	231.22	120.60	28.00	75.55	2.70	60.00	59.02	0.98
Cr	149.02	87.56	35.44	6.44	19.58	41.68	36.16	34.45	43.83	54.30	149.02	111.25	17.00	40.26	2.37	35.00	32.18	0.92
Co	20.24	19.44	5.76	6.38	3.70	3.28	2.23	4.35	2.78	2.92	20.24	2.46	6.00	4.73	0.79	10.00	5.17	0.52
Ni	55.07	34.39	14.59	11.27	6.26	20.50	14.34	15.10	23.40	27.39	55.07	38.61	17.00	18.28	1.08	20.00	17.84	0.89
Cu	60.63	20.57	23.98	3.82	23.53	34.69	32.71	34.61	47.51	32.42	60.63	17.88	16.00	29.80	1.86	25.00	25.00	1.00
Zn	150.53	9.04	2.42	11.40	0.60	2.88	2.93	3.77	9.96	7.09	150.53	8.69	28.00	4.46	0.16	71.00	12.65	0.18
Ga	50.30	15.91	7.43	1.10	3.98	10.80	9.00	10.02	11.37	14.08	36.39	24.71	6.00	9.46	1.58	17.00	8.26	0.49
Ge	1.46	0.84	1.64	0.15	0.24	0.65	0.88	1.32	1.11	0.98	0.86	1.28	2.40	1.29	0.54	1.60	1.68	1.05
As	9.34	12.31	2.87	1.06	2.11	1.53	1.40	1.51	2.79	4.02	2.18	2.25	9.00	2.96	0.33	2.00	3.81	1.91
Se	7.22	9.17	4.16	0.94	4.60	5.53	6.18	4.08	7.01	6.22	3.90	2.61	1.60	5.18	3.24	0.08	3.92	47.23
Rb	96.10	3.71	1.97	0.09	0.55	2.77	2.71	1.97	4.43	4.02	2.94	4.34	18.00	2.69	0.15	112.00	8.02	0.07
Sr	104.84	588.93	493.21	180.81	139.62	22.06	22.96	16.11	31.17	17.25	19.01	10.42	100.00	267.14	2.67	350.00	209.03	0.60
Zr	209.44	227.44	219.24	7.06	40.47	151.48	136.14	93.08	178.18	168.14	422.17	589.72	36.00	184.55	5.13	190.00	132.90	0.70
Nb	18.43	4.15	8.89	0.94	2.56	8.21	8.15	5.82	8.51	7.56	21.85	32.61	4.00	8.10	2.03	25.00	6.68	0.27
Mo	0.69	2.08	0.39	0.14	0.22	1.77	0.50	0.57	0.80	0.73	0.88	1.23	2.10	0.62	0.29	1.50	1.13	0.76
Ag	0.97	1.29	1.18	bdl	0.17	0.83	0.76	0.44	0.96	0.97	0.97	2.54	bdl	1.00	bdl	0.05	0.65	13.07
Cd	0.53	0.56	0.29	0.05	0.19	0.30	0.28	0.23	0.63	0.50	0.53	0.67	0.20	0.36	1.81	0.10	0.31	3.05
In	0.19	0.10	0.04	0.13	0.02	0.05	0.04	0.05	0.06	0.09	0.25	0.13	0.04	0.05	1.30	-	0.05	-
Sn	4.97	5.12	2.54	38.34	0.68	2.02	1.51	2.50	2.16	6.26	8.68	5.73	1.40	3.45	2.46	5.50	2.73	0.50
Sb	0.97	0.62	0.26	0.12	0.23	0.37	0.30	0.18	0.46	0.51	0.47	0.66	1.00	0.33	0.33	0.20	0.57	2.83
Cs	15.02	0.69	0.36	0.14	0.12	1.51	1.14	0.77	2.08	2.72	3.13	3.44	1.10	1.09	0.99	3.70	1.09	0.29
Ba	131.56	77.11	24.02	6.63	5.59	10.54	7.81	6.70	15.72	16.17	10.38	5.04	150.00	19.47	0.13	550.00	64.86	0.12
La	100.65	29.82	5.99	25.27	34.61	36.14	23.61	45.01	36.70	60.67	102.44	57.28	11.00	34.27	3.12	30.00	83.12	2.77
Ce	185.72	58.81	12.06	41.35	77.86	80.91	54.56	95.52	88.70	120.63	240.78	110.15	23.00	72.30	3.14	64.00	185.05	2.89
Pr	18.18	5.23	1.41	4.04	6.98	7.77	5.11	8.88	8.64	14.30	17.93	12.74	3.40	6.70	1.97	7.10	15.97	2.25
Nd	68.89	18.92	5.82	14.89	25.38	29.32	19.84	32.67	33.62	54.74	68.81	47.63	12.00	24.91	2.08	26.00	60.98	2.35
Sm	11.82	3.02	1.10	2.61	3.95	4.93	3.44	5.13	5.92	9.47	11.40	7.43	2.20	4.10	1.86	4.50	10.09	2.24
Eu	2.44	0.53	0.21	0.46	0.71	0.88	0.59	0.87	1.07	1.59	1.73	1.25	0.43	0.73	1.70	0.90	1.59	1.77
Gd	11.63	3.42	1.30	2.90	4.55	5.37	3.51	5.37	6.23	9.76	10.35	7.14	2.70	4.44	1.65	3.80	9.50	2.50
Tb	1.10	0.38	0.16	0.36	0.48	0.59	0.38	0.55	0.72	1.20	1.26	0.80	0.31	0.49	1.57	0.60	1.14	1.90
Dy	5.30	2.14	0.87	1.97	2.61	3.10	1.98	2.68	3.91	6.74	6.77	4.69	2.10	2.61	1.24	3.50	6.30	1.80
Y	25.96	10.69	5.05	10.31	13.23	15.34	9.04	12.34	18.96	37.75	56.89	25.90	8.40	12.79	1.52	22.00	45.75	2.08
Ho	1.02	0.41	0.16	0.37	0.47	0.56	0.35	0.47	0.71	1.23	1.19	0.91	0.57	0.48	0.85	0.80	1.14	1.42
Er	3.20	1.26	0.48	1.12	1.40	1.67	1.00	1.34	2.08	3.62	3.49	3.02	1.00	1.45	1.45	2.30	3.42	1.49
Tm	0.44	0.18	0.06	0.15	0.18	0.22	0.13	0.17	0.27	0.45	0.44	0.43	0.30	0.20	0.65	0.30	0.44	1.46
Yb	3.26	1.28	0.40	1.08	1.27	1.56	0.88	1.16	1.88	3.06	2.96	3.19	1.00	1.38	1.38	2.20	3.03	1.38
Lu	0.46	0.17	0.05	0.15	0.16	0.20	0.12	0.15	0.25	0.40	0.39	0.45	0.20	0.18	0.91	0.30	0.40	1.34
Hf	5.34	3.83	5.56	0.21	1.12	4.25	3.83	2.48	4.76	4.75	10.00	13.78	1.20	4.81	4.01	5.80	3.55	0.61
Ta	1.28	0.36	0.70	0.29	0.17	0.59	0.59	0.38	0.42	0.27	1.41	2.15	0.30	0.55	1.83	2.20	0.46	0.21
W	5.17	0.24	1.20	0.14	0.16	0.56	0.50	0.62	0.58	0.66	2.65	11.90	0.99	0.87	0.88	2.00	0.91	0.46
Hg	0.26	1.97	0.17	0.02	0.12	0.14	0.13	0.10	0.21	0.24	0.17	0.24	0.10	0.21	2.11	0.02	0.17	8.63
Tl	0.79	3.45	bdl	bdl	bdl	bdl	bdl	bdl	0.02	0.03	0.79	0.02	0.58	0.07	0.12	0.80	0.25	0.31
Pb	34.89	11.87	10.11	0.97	6.79	15.73	17.38	9.81	18.75	21.37	34.89	7.12	9.00	13.65	1.52	20.00	12.03	0.60
Bi	1.80	0.04	0.21	0.60	bdl	0.23	0.19	bdl	0.13	1.06	1.03	1.84	1.10	0.32	0.29	-	0.59	-
Th	25.72	16.97	10.34	0.99	5.22	7.90	9.60	7.37	10.36	11.75	20.28	38.40	3.20	10.16	3.17	10.70	7.74	0.72
U	5.62	3.91	2.25	0.48	1.18	3.70	3.24	2.92	3.96	4.09	8.27	8.69	1.90	2.94	1.55	2.00	2.58	1.29

World, world hard coals, from Ketris and Yudovich [27]; SC, weighted average for coal samples; C1, concentration coefficient of coal samples; UCC, upper continental crust; PC, weighted average for host rock; C2, concentration coefficient of host rock; -, without the data item.

Particularly, a number of trace elements are enriched in sample SS-1C. Trace elements with a concentration coefficient >10 include Hg (CC = 19.73) and V (CC = 13.69); trace elements with a CC of 5–10 include Sc, La, Ce, Pr, Sm, Eu, Th, Li, Sr, Se, Zr, Cr, and Tl.

Trace elements in the roof and floor rocks with a concentration coefficient (CC = ratio of element concentration in Songshao coal vs. UCC) higher than 10 include Hg and Se. Lithium shows a CC of 5–10. Lanthanum, Ce, Pr, Nd, Sm, Gd, Y, Cd, and Sb are slightly enriched in the roof and floor samples (CC = 2–5).

The anomalous geochemical composition of Songshao coal is represented by ratios of some trace elements. On average, Songshao coal has higher Nb/Ta, Zr/Hf, and especially Li/Rb values, as well as lower values of Ba/Sr and Rb/Cs than the averages for the world hard coals. From top (roof) to bottom (floor), the ratios of Ba/Sr and Nb/Ta show similar distribution patterns, which gradually increase; the ratio of Rb/Cs, however, shows an adverse trend.

4.2.3. Rare Earth Elements

Rare earth elements (REE) in coal are generally associated with minerals, especially clay minerals and phosphate, which are generally associated with ash yield [29]; in some cases, heavy rare earth elements are associated with the organic matter in coal [30]. A threefold classification of REE was used for this study: Light (LREE: La, Ce, Pr, Nd, and Sm), medium (MREE: Eu, Gd, Tb, Dy, and Y), and heavy (HREE: Ho, Er, Tm, Yb, and Lu) REE [31]. Accordingly, in comparison with the upper continental crust [32], three enrichment types are identified [31]: L-type (light-REE; $La_N/Lu_N > 1$), M-type (medium-REE; $La_N/Sm_N < 1$ and $Gd_N/Lu_N > 1$), and H-type (heavy REE; $La_N/Lu_N < 1$).

With the exception of sample SS-9C, the REE patterns in the coal benches are characterized by L-REE enrichment, Eu negative anomalies, and Gd-maximum (Gd reaches to the peak of the patterns) (Figure 3). In particular, sample SS-1C has the highest REE concentration relative to other coal samples and does not show an Eu anomaly.

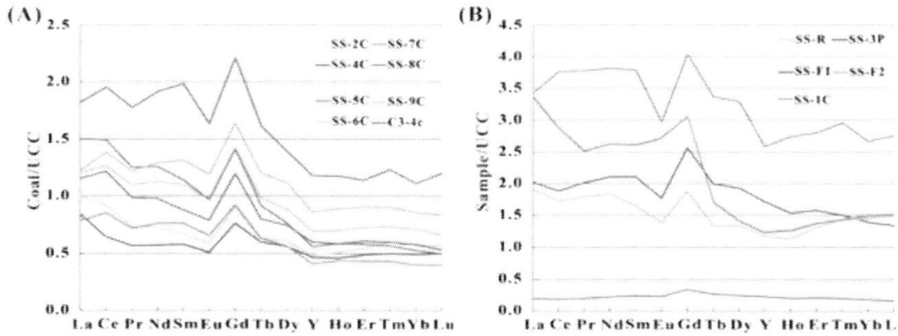

Figure 3. Distribution patterns of rare earth elements (REE) of samples in the Songshao coal mine. (**A**) Plots for the Songshao coal in comparison with C3-4c from the Xinde Mine. (**B**) Plots of SS-1C, roof and floor, and parting from the Songshao Coal Mine. REE are normalized by upper continental crust (Taylor and McLennan [32]).

The lower REE abundance of the parting is attributed to its lithological composition, which is represented by 97% calcite (Table 4). According to previous studies, the content of REE in limestone is lower in comparison with other lithology, such as clay and oil shale [33,34]. The REE distribution pattern for the parting is characterized by an M-REE enrichment type and almost no Eu anomalies.

Table 4. Low temperature ash (LTA) yields of coal samples and mineral compositions (%) of LTAs, parting, roof and floor determined by XRD and Siroquant. "-"means below the detection limit of Siroquant analysis.

Sample	LTA	Kaolinite	Illite	I/S	Marcasite	Pyrite	Calcite	Anatase	Boehmite	Diaspore	Brucite
SS-R	-	54.5	39.2	-	-	5.1	-	1.2	-	-	-
SS-1C	47.2	41.3	-	-	12.7	44.6	-	0.9	-	-	-
SS-2C	22.9	55.2	17.9	-	-	-	6.1	1.3	3.4	14.9	1.1
SS-3P	73.8	-	-	-	-	-	97.7	-	2.1	0.1	0
SS-4C	11.0	22.6	17.3	14.7	-	-	4.7	1.7	34.3	4.3	0.4
SS-5C	31.9	50.8	9.1	14.6	-	-	-	0.6	24.9	-	-
SS-6C	23.7	60.5	6.7	1.7	-	-	-	0.9	30.3	-	-
SS-7C	20.7	51.8	6.1	-	-	-	-	0.1	34.2	7.6	0.3
SS-8C	39.0	53.7	14.8	-	-	-	-	0.4	25.3	5.1	0.7
SS-9C	35.1	63.6	14.7	-	-	-	-	-	20.9	-	0.4
SS-F1	-	83.6	6.3	-	-	-	-	0.7	9.4	-	-
SS-F2	-	78.5	7.7	-	-	-	-	1.0	12.9	-	-

In addition, the floor and roof samples have similar REE distribution patterns, with weak negative Eu anomalies, and L- and M-REE enrichment types.

REE in all the samples are characterized by a Gd-maximum. The distribution patterns of REE in Songshao coal are similar to that in sample C3-4c from the Xinde Mine in Eastern Yunnan, China [12]. Sample C3-4c is characterized by M-type REE spectra with a Gd-maximum (Figure 3); such patterns in coal are typical of a great deal of acid water, including high pCO_2-waters in coal basins [35,36].

4.3. Mineralogical Composition

4.3.1. Mineral Phases

The proportions of each crystalline phase, identified from the X-ray diffractograms of the coal LTA (low temperature ash), parting, roof, and floor samples, are given in Table 4. The phases identified in the coal LTAs include kaolinite, illite, I/S (mixed layer of illite and smectite), marcasite, pyrite, calcite, anatase, boehmite, diaspore, and a trace amount of brucite. Additionally, stannite, fluorapatite, apatite, zircon, halotrichite, and some REE-bearing minerals, including xenotime, florencite, and silicorhabdophane, are also identified in sample SS-7C by SEM-EDS analysis.

Calcite is the dominant mineral in parting. Other minerals including boehmite and diaspore are also identified from parting LTA.

Minerals in the roof sample (SS-R) include kaolinite, illite, pyrite, and small proportions of anatase, quartz, and gypsum. The floors samples (SS-F1 and SS-F2) are composed of kaolinite, illite, anatase, and boehmite.

4.3.2. Comparison between Mineralogical and Chemical Compositions

The chemical composition of the (high-temperature) coal ash calculated from the XRD and Siroquant analyses of each LTA or roof and floor samples are listed in Table 4. The two data sets, derived respectively from the XRD and the XRF data, have been compared and are presented as *X–Y* plots (Figure 4), with a diagonal line on each plot indicating where the points would fall if the estimates from the two different techniques were equal. The points for SiO_2, and Al_2O_3 in Figure 4 plot close to the equality line, suggesting that the XRD results are generally compatible with the chemical analysis data.

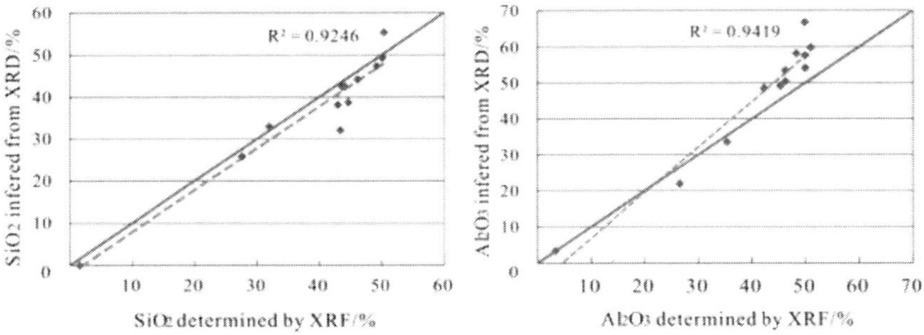

Figure 4. Comparison of observed normalized Si and Al oxide percentages from chemical analysis (*X*-axis) to oxide percentages for sample ash inferred from XRD analysis data (*Y*-axis). The diagonal line in each plot indicates equality.

4.3.3. Modes of Mineral Occurrence

Clay Minerals

Clay minerals, including kaolinite, I/S (mixed-layer illite-smectite), illite, and smectite, are identified using XRD and SEM-EDS. They normally occur in the matrix of organic matter of the coal (Figure 5A). Kaolinite is the dominant mineral both in the coals and the roof and floor strata. In some cases, kaolinite distributes along the bedding (Figure 5B), and sometimes it occurs in detrital forms. Other modes of occurrence of kaolinite include fracture- and cell-fillings (Figure 5C) of authigenic origin [37]. I/S, illite, and smectite distribute along the bedding planes (Figure 5B); occur as cell-fillings (Figure 5D,E); or distribute around boehmite (Figure 5F).

Boehmite, Disapore, Brucite

Oxyhydroxide minerals, such as boehmite, disapore, and brucite, were identified in Songshao coal. High boehmite content (mean 4.33%) was identified in Songshao coal, occurring as fracture-fillings (Figure 5B), small lumps (Figure 5F), cell-fillings (Figure 5D,E), and bead-like block embedded in the clay minerals (Figure 5F). The lumps show different shapes and variable sizes, from a few to one hundred micrometers. The surface of the boehmite in the coal is much smoother than that of the clay minerals (Figure 5F) and the relief of boehmite is higher under the polarizing microscope (Figure 5G). Minerals associated with the boehmite in the coal include goyazite, rutile, zircon, and Pb-bearing minerals (galena, clausthalite, and selenio-galena) [38]. Additionally, zircon is also identified in Songshao coal.

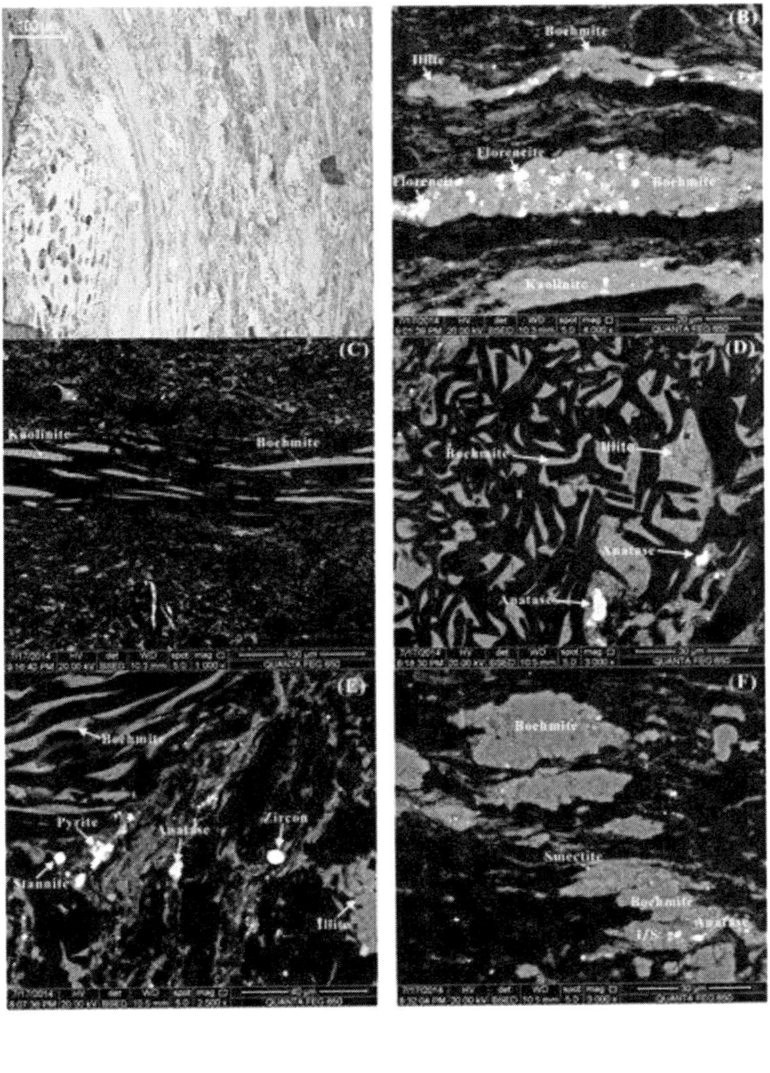
(A)
(B)
Boehmite
Illite
Florencite
Florencite
Boehmite
Kaolinite
(C)
Kaolinite
Boehmite
(D)
Boehmite
Illite
Anatase
Anatase
(E)
Boehmite
Pyrite
Anatase
Zircon
Stannite
Illite
(F)
Boehmite
Smectite
Boehmite
I/S
Anatase

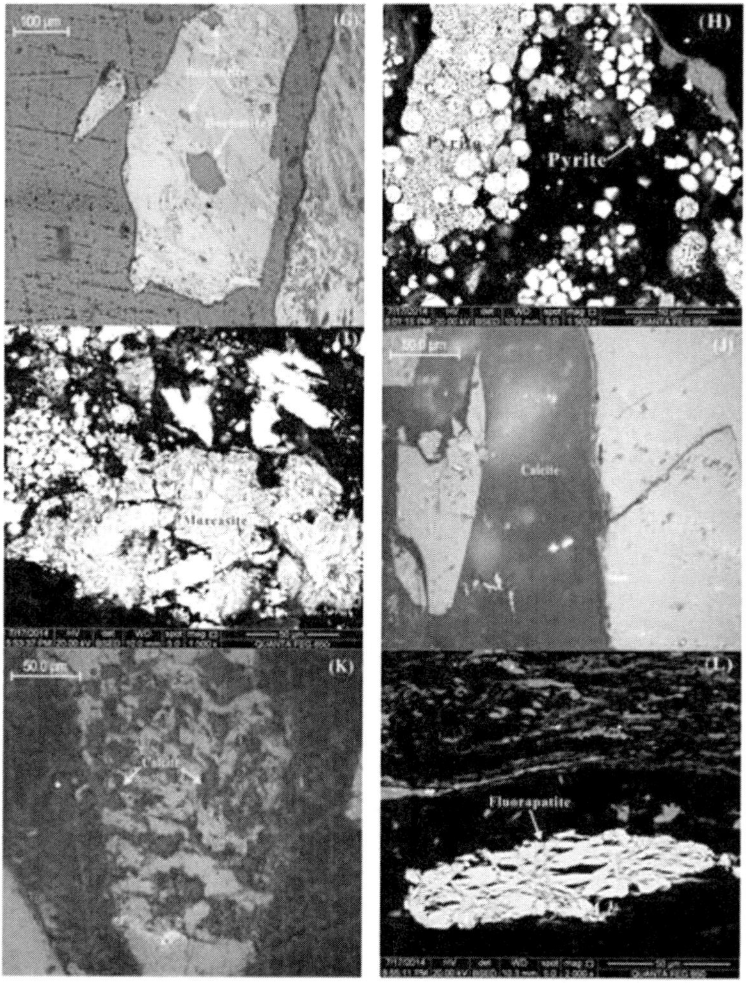

Figure 5. Minerals in the Songshao coal. (**A**) clay minerals in SS-9C; (**B**) kaolinite, illite, boehmite and florencite in SS-7C; (**C**) cell-filling kaolinite and boehmite in SS-7C; (**D**) cell-filling boehmite, illite, and anatase in SS-7C; (**E**) cell-filling boehmite, ditrital zircon, anatase, pyrite and stannite in SS-7C; (**F**) boehmite occurs small lumps and bead-like block embedded in the clay minerals and ditrital anatase in SS-7C; (**G**) boehmite in SS-9C; (**H**) framboidial pyrite in SS-1C; (**I**) plate, block and radial form marcasite in SS-1C; (**J**) vein-filling calcite in SS-3P; (**K**) cell-filling calcite in SS-3P; (**L**) fracture-filling florencite in SS-7C. B-F, H, I, L, SEM and back-scattered electron images; (**A**), (**G**), (**J**), (**K**), reflected light.

Pyrite and Marcasite

Pyrite is only detected in samples SS-R and SS-1C using the XRD technique. A trace amount of pyrite is also detected in sample SS-7C using SEM-EDS. Pyrite in sample SS-1C occurs as framboidal massive (Figure 5H) and euhedral crystal forms (Figure 5H), or distributes along the bedding (Figure 5E). Marcasite is only detected in sample SS-1C and coexists with pyrite, occurring as plate, block and radial forms (Figure 5I). This kind of marcasite may be of a syngenetic origin.

Calcite

Calcite is detected in samples SS-2C, SS-3P, and SS-4C using XRD and it accounts for 72.1% in the parting (SS-3P). Calcite occurs as vein- (Figure 5J), fracture- or cells-fillings (Figure 5K), indicating an epigenetic origin.

Antase and Zircon

Antase occurs as discrete particles in clay minerals (Figure 5E,F) or as cell-fillings (Figure 5D). Zircon occurs as detrital particles of terrigenous origin (Figure 5E).

The mode of occurrence of zircon indicates a detrital material of terrigenous origin. Zircon may be a pyroclastic mineral in some tonsteins [4,39] or it may occur as detrital material derived from the sediment source region.

Apatite, Halotrichite, and Stannite

Apatite in coal is classified as fluorapatite and zwiesellite. Fluorapatite is a common mineral in coal. It looks like bamboo leaves filling in the fracture (Figure 5L), indicating an epigenetic chemical deposition. A spot of halotrichite was also observed in sample SS-7C. The copiapite-group of minerals is one of the most common Fe-sulfates reported from many coals, such as coals from the Jaintia Hills coalfield in Meghalaya, India [40]. A trace of stannite was detected using SEM-EDS (Figure 5E).

REE-Bearing Minerals

REE-bearing minerals are at a low concentrations, below the detection limit of the XRD and Siroquant analysis, but were observed under SEM-EDS in sample SS-7C. These minerals are Xenotime, florencite, rhabdophane, and silicorhabdophane.

Phosphate minerals, such as apatite, monazite, and, in some cases, xenotime, zircon, and some clay minerals, are usually the carriers of thorium in coal [41]. In this study, thorium was not detected, but Dy was detected in xenotime.

Florencite occurs as fracture-fillings, coexisting with kaolinite (Figure 5B). However, the small size of these particles makes it difficult to obtain a representative EDS spectrum.

5. FACTORS CONTROLLING ENRICHMENT OF TRACE ELEMENTS AND MINERALS IN SONGSHAO COAL

It has been reported that seven factors control the enrichment of elements and minerals in Chinese coals: The sediment-source rocks, low-temperature hydrothermal fluids, marine environments, volcanic ash, magmatic fluids, submarine exhalation, and groundwater [3]. The first three factors are likely to be responsible for the geochemical and mineralogical anomalies of Songshao coal.

5.1. Input from Sediment-Source Region

As described above, Songshao coal is enriched in Al, Si, Ti, Li, Zr, Hf, Th, Sr, and V. Aluminum and Si are largely contained in clay minerals. Two modes of kaolinite occurrence in Songshao coal were identified: Detrital kaolinite of terrigenous origin and cell-filling kaolinite of authigenic origin. Anatase, which occurs as discrete particles in the clay mineral matrix, is the host of Ti and was probably derived from sediment-source sources. Lithium generally occurs in minerals, or can be absorbed by bauxite minerals, which can then be named as a Li deposit [42] if Li is highly concentrated. Moreover, Li may occur in silicate minerals, because it tends to be absorbed by clay minerals, which are formed during a weathering process [43]. It was reported that the Langdaisa Li-bearing claystone of the Permian Liangshan Formation contains 0.12%–0.74% Li [44,45]. Langdaisa is located in the Liuzhi District of Guizhou and Liuzhi is situated to the northeast of the Songshao Mine, indicating the same sediment source for the Langdaisa Li-bearing claystone and Songshao coal. Thorium is not easily altered by the processes of weathering and transportation, and, thus, it can serve as an effective indicator for deducing the source region. Thorium is generally absorbed by clay and more likely occurs in bauxite [43]. In addition, thorium can occur in apatite, xenotime, or zircon in coal [41]. In this study, minerals including apatite, xenotime, and zircon have been identified in sample SS-7C by SEM-EDX; however, thorium was not found to occur in these minerals. The correlation coefficients of Zr-Th and Zr-Hf are 0.814 and 0.974, respectively (Figure 6). Zircon is the carrier of Zr and also contains Th and Hf. The mode of zircon occurrence indicates a detrital material of terrigenous origin. Elements Hf, Zr, and Li show a similar variation tendency through the seam section, suggesting a same source (Figure 6).

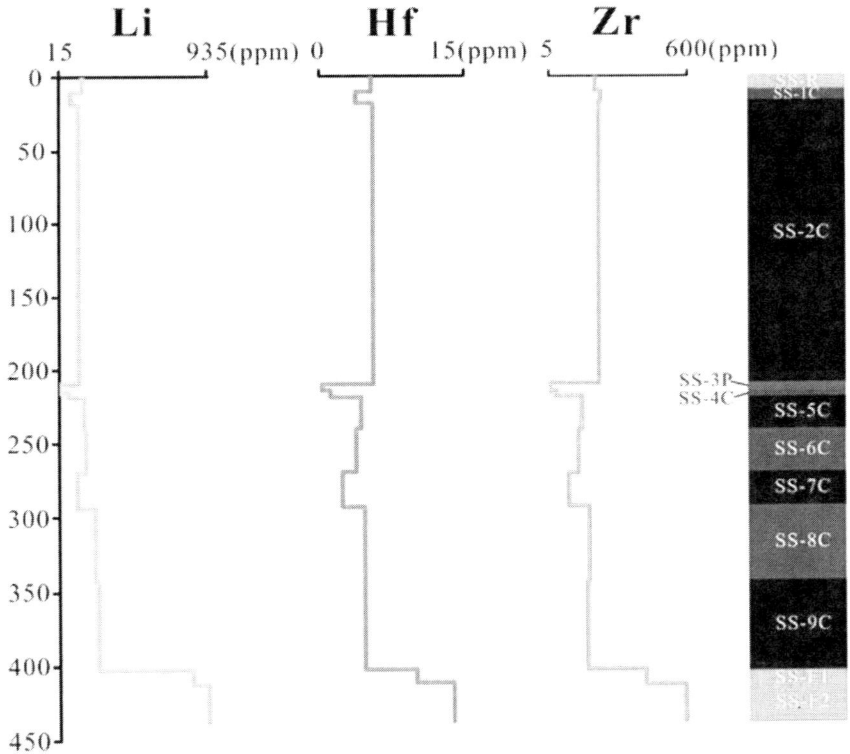

Figure 6. Variations of Li, Hf, and Zr through the seam section.

Seredin and Dai [31] suggested that the L-type distribution of REE in coal are attributed to terrigenous or tuffaceous origins during peat accumulation stage. In the first case, the REE may be transported as colloidal and ionic forms from uplift containing magmatic rocks enriched in light REE, such as granite and carbonatite [31]. The light REE enrichment type in Songshao coal indicates a terrigenous origin. A similar instance occurred in the No. 6 Coal seam in the Jungar deposit in Inner Mongolia, China [46]. REE are present in Songshao coal as REE-rich minerals, such as xenotime and florencite.

5.2. Hydrothermal Fluid Influences

Hydrothermal fluid is another critical factor responsible for the trace-element and mineral anomalies in the Songshao coal, parting, and roof and floor rocks. The hydrothermal fluid includes syngenetic hydrothermal solutions injected into the peat mire during peat accumulation. Mineralogical evidence of hydrothermal influence on Songshao coal includes:

(1) Pyrite and calcite occur as micro-veinlets (Figure 5H,J). Multi-generation pyrite occurs in sample SS-1C. Framboidal pyrite is of an early diagenetic origin [47]. It occurs as lenses and spheres and the aggregates are composed of microcrystalline paticles. The framboids are commonly coated by late-formed sulfides. Meanwhile, sample SS-1C is enriched in As, Se, and Hg. According to previous research [48,49,50,51], some trace elements, for instance, arsenic and mercury, are concentrated in the late stage of hydrothermal fluids. In some cases, arsenic was syngenetically derived from hydrothermal solutions [6]. Particularly, the enrichment of arsenic was associated with late-stage pyrite that coats early framboids and with microscale faulting. This phenomenon also occurred in the West Virginia coals from the Appalachian region, USA [52]. Therefore, hydrothermal fluid may be responsible for the enrichment of trace elements in sample SS-1C.

(2) Clay minerals, boehmite and calcite occur as cell-fillings (Figure 5B,C,K). Under favorable conditions, solutions that contained soluble Al and Si entered the coal basin, filled and then were deposited in the cells during peat accumulation.

(3) High boehmite content (mean 4.33%) in Songshao coal is not commonly observed; the content of boehmite in the coal present in this study is close to that (mean 6.1%) in the No. 6 Coal from the Jungar Coalfield [46] and from some mines of the Daqingshan Coalfield [53]. In addition, boehmite may be the result of low-temperature hydrothermal fluids, generally associated with zeolites [54]. However, zeolites have not been observed in the present study. The occurrence mode of boehmite in Songshao coal is similar to that in the No. 6 Coal in the Jungar Coalfield [46] and the Daqingshan Coalfield [53]. The boehmite in Jungar coals is the major carrier of Ga and Th, which are significantly enriched in the coal (44.8 μg/g Ga; 17.8 μg/g Th; on a whole coal basis). However, trace elements were not enriched with boehmite in Songshao coal. The difference in element abundance between boehmites in the coals from the different coalfields was probably due to different origins of boehmite. An explanation for high boehmite content is that in the humid–warm climatic conditions, rock in sediment source suffered from weathering

and erosion, which can cause Al to be concentrated in weathering crust. Colloidal solution containing Al removed from the crust flowed into the peat mire, and boehmite was formed by compaction and dehydration of Al colloidal solution during peat accumulation and early diagenesis.

(4) The presence and mode of occurrence of xenotime, florencite, rhabdophane and silicorhabdophane in the coal present in this study indicate that these minerals were deposited from hydrothermal fluids at syngenetic or early diagenetic stages. Rhabdophane, florencite, and silicorhabdophane are the major carriers of light rare earth elements in the samples of the present study. These minerals in coal are generally derived from hydrothermal fluids, and they are also the major REE carriers in some other REE-rich coals as well [31,54,55,56,57,58,59,60].

(5) S positively correlates with Hg and Fe, with correlation coefficients $r_{S\text{-}Hg} = 0.85$ and $r_{S\text{-}Fe} = 0.88$ (Figure 7), respectively. Fluorine is slightly enriched in Songshao coal with CC = 1.88. Syngenetic hydrothermal solutions may be rich in Hg and F, and this could lead to enrichment of Hg in upper and lower portions of the coal seam and high F concentrations throughout the whole seam section. Hydrothermal fluids leading to enrichment of Hg and F has also been reported in some coals from other coalfields [61,62,63,64]. Fluorine-bearing fluorapatite was found to occur as fracture-fillings in sample SS-7C. Epigenetic hydrothermal fluids led to the enrichment of F, Fe, Se and S in the roof rock. Owing to the injection of epigenetic hydrothermal fluids into the coal seam, the concentrations of epithermal-associated elements (e.g., F, Se, S, and Hg) are highly elevated near the contact between the coal bench and roof strata.

5.3. Marine Environments

Because Songshao is located in the area of arenaceous and argillaceous facies in litoral zone (Figure 8), the peat swamp of the coal had been subjected to a marine environment, as indicated by the following evidence:

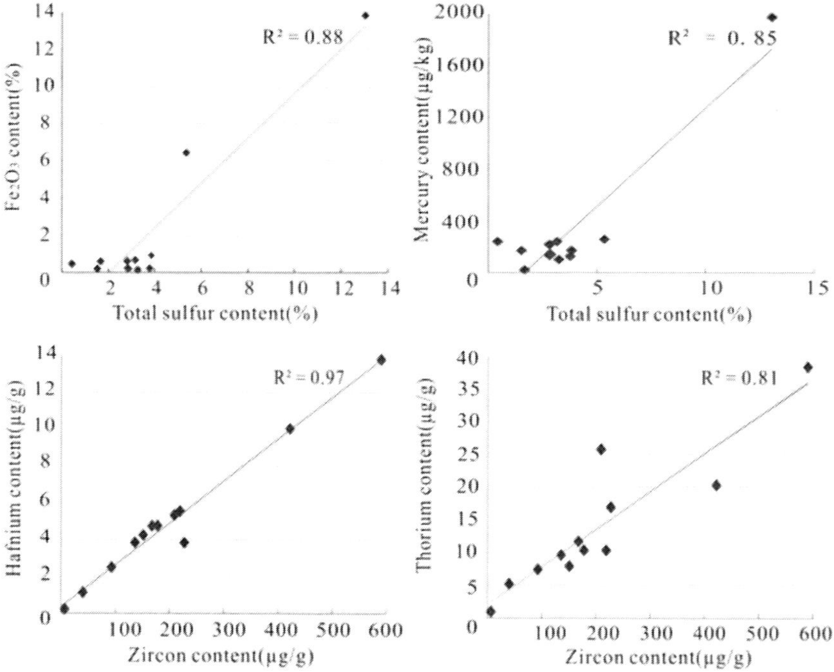

Figure 7. Correlation coefficients between two elements.

(1) A number of researches showed that coals formed in marine-influenced environment are generally enriched in some elements, including S, B, V, Br, Rb, Sr, Mo, and U [65,66,67,68,69], because these elements are enriched in seawater in comparison with fresh water. Songshao coal, as described above, is enriched in V and Sr.

(2) Songshao coal is a high-sulfur coal with average total sulfur and organic sulfur contents 3.61% and 3.87%, respectively. Previous work showed that sulfate in seawater that flooded peat swamps is the major sources of sulfur for medium- and high-sulfur coals [70]. Some studies also indicated that submarine exhalation can also lead to sulfur enrichment in coal [71]. The elevated content of sulfur in Songshao coal is attributed to seawater that injected into the peat swamp during peat accumulation. Sulfates in the seawater were reduced by anaerobic sulfate-reducing bacteria. Various forms of sulfur, which decrease in interstitial water, react with iron ions and organic matters. Then, sulfur-bearing minerals (pyrite mostly) and organic sulfide formed [72].

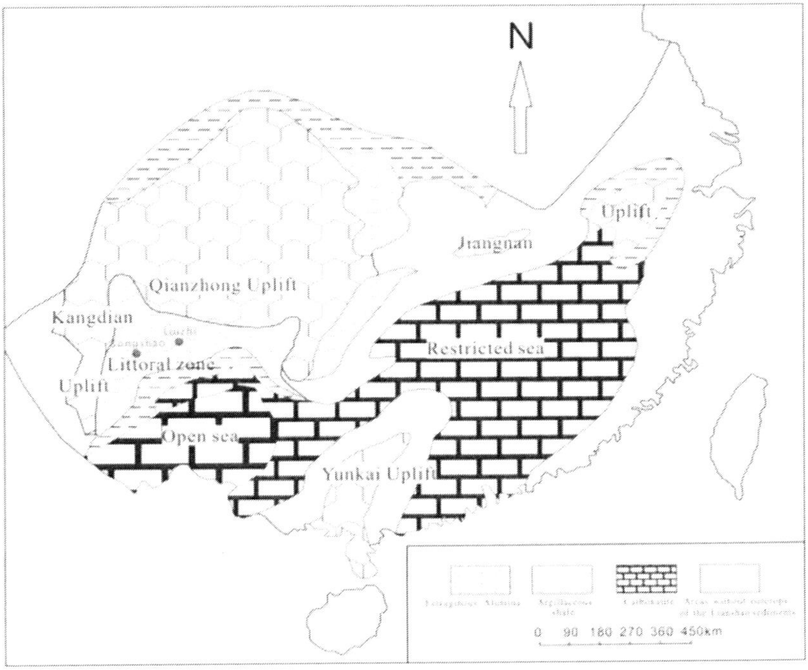

Figure 8. Sketch map of lithological facies and paleogeography in the study area during the Liangshan Stage. Modified from [73,74].

6. CONCLUSIONS

Songshao coal is medium- to high-volatile bituminous in rank, with high total and organic sulfur contents. Major-element oxides are mainly represented by SiO_2 and Al_2O_3. Trace elements, including Li, Zr, Hg, Sr, Nb, Sn, Hf, V, Cr, Se, Th, and some light REE, are enriched. The REE enrichment patterns in the coal seam are L-types (light REE enrich) and are characterized by negative Eu anomalies and a Gd-maximum. The mineral phases identified in the coal LTAs include kaolinite, illite, I/S (mixed layer illite-smectite), pyrite, calcite, anatase, boehmite, diaspore, and some REE-bearing minerals. Clay minerals occur in the matrix of organic matter. Boehmite in the coal occurs as fracture-fillings, small lumps, as cell-fillings, or as bead-like blocks embedded in the clay minerals. Pyrite occurs as framboidal forms; calcite occurs as fracture-fillings or cell-fillings. Other minerals occur as debris in the samples.

High total and organic sulfur contents and the enrichment Sr and V are attributed to a marine influence. Hydrothermal fluids are responsible for the high boehmite content (mean 4.33%) and its modes of occurrence; the presence and modes of REE-bearing minerals, pyrite and calcite; the enrichment of Hg and F in the coal, roof and floor strata rocks. Songshao coal and the roof and floor strata rocks are also enriched in SiO_2, Al_2O_3, Li, Zr, and light rare earth elements, which were derived from the sediment source.

ACKNOWLEDGMENTS

This research was supported by the National Key Basic Research Development Program (973 Program, No. 2014CB238902). Special thanks are given to Shifeng Dai and Xiaolin Song for providing samples and constructive suggestion for this paper. Shifeng Dai also helped with SEM-EDX experiments. Peipei Wang, Qin Zhu, Shaohui, Jia and Jihua Sun helped the determination of major and trace elements.

CONFLICTS OF INTEREST

The author declares no conflict of interest.

REFERENCES

1. Yunnan Coal Geology Prospecting Team No. 143. Report on Coal Geological Exploration of Songshao Mine; Yunnan Coal Geology Prospecting Team No. 143: Qujing, China, 1996.
2. Dai, S.; Wang, X.; Zhou, Y.; Hower, J.C.; Li, D.; Chen, W.; Zhu, X.; Zou, J. Chemical and mineralogical compositions of silicic, mafic, and alkali tonsteins in the late Permian coals from the Songzao Coalfield, Chongqing, Southwest China. Chem. Geol. 2011, 282, 29–44.
3. Ren, D.; Zhao, F.; Dai, S.; Zhang, J.; Luo, K. Geochemistry of Trace Elements in Coal; Science Press: Beijing, China, 2006; pp. 40–59. (In Chinese)
4. Ward, C.R. Analysis and significance of mineral matter in coal seams. Int. J. Coal Geol. 2002, 50, 135–168.
5. Wang, X.; Dai, S.; Ren, D.; Yang, J. Mineralogy and geochemistry of Al-hydroxide/oxyhydroxide mineral-bearing coals of Late Paleozoic age from the Weibei coalfield, southeastern Ordos Basin, North China. Appl. Geochem. 2011, 26, 1086–1096.
6. Dai, S.; Wang, X.; Seredin, V.V.; Hower, J.C.; Ward, C.R.; O'Keefe, J.M.K.; Huang, W.; Li, T.; Li, X.; Liu, H.; et al. Petrology, mineralogy, and geochemistry of the Ge-rich coal from the Wulantuga Ge ore deposit, Inner Mongolia, China: New data and genetic implications. Int. J. Coal Geol. 2012, 90, 72–99.

7. Dai, S.; Jiang, Y.; Ward, C.R.; Gu, L.; Seredin, V.V.; Liu, H.; Zhou, D.; Wang, X.; Sun, Y.; Zou, J.; et al. Mineralogical and geochemical compositions of the coal in the Guanbanwusu Mine, Inner Mongolia, China: further evidence for the existence of an Al (Ga and REE) ore deposit in the Jungar Coalfield. Int. J. Coal Geol. 2012, 98, 10–40.
8. Dai, S.; Seredin, V.V.; Ward, C.R.; Jiang, J.; Hower, J.C.; Song, X.; Jiang, Y.; Wang, X.; Gornostaeva, T.; Li, X.; et al. Composition and modes of occurrence of minerals and elements in coal combustion products derived from high-Ge coals. Int. J. Coal Geol. 2014, 121, 79–97.
9. Zhou, Y.; Bohor, B.F.; Ren, Y. Trace element geochemistry of altered volcanic ash layers (tonsteins) in Late Permian coal-bearing formations of eastern Yunnan and western Guizhou Provinces, China. Int. J. Coal Geol. 2000, 44, 305–324.
10. Dai, S.; Tian, L.; Chou, C.-L.; Zhou, Y.; Zhang, M.; Zhao, L.; Wang, J.; Yang, Z.; Cao, H.; Ren, D. Mineralogical and compositional characteristics of Late Permian coals from an area of high lung cancer rate in Xuan Wei, Yunnan, China: Occurrence and origin of quartz and chamosite. Int. J. Coal Geol. 2008, 76, 318–327.
11. Dai, S.; Zhou, Y.; Zhang, M.; Wang, X.; Wang, J.; Song, X.; Jiang, Y.; Luo, Y.; Song, Z.; Yang, Z.; et al. A new type of Nb (Ta)–Zr (Hf)–REE–Ga polymetallic deposit in the late Permian coal-bearing strata, eastern Yunnan, southwestern China: possible economic significance and genetic implications. Int. J. Coal Geol. 2010, 83, 55–63.
12. Dai, S.; Li, T.; Seredin, V.V.; Ward, C.R.; Hower, J.C.; Zhou, Y.; Zhang, M.; Song, X.; Song, W.; Zhao, C. Origin of minerals and elements in the Late Permian coals, tonsteins, and host rocks of the Xinde Mine, Xuanwei, eastern Yunnan, China. Int. J. Coal Geol. 2014, 121, 53–78.
13. Dai, S.; Zhang, W.; Ward, C.R.; Seredin, V.V.; Hower, J.C.; Li, X.; Song, W.; Wang, X.; Kang, H.; Zheng, L.; et al. Mineralogical and geochemical anomalies of late Permian coals from the Fusui Coalfield, Guangxi Province, southern China: Influences of terrigenous materials and hydrothermal fluids. Int. J. Coal Geol. 2013, 105, 60–84.
14. Dai, S.; Zhang, W.; Seredin, V.V.; Ward, C.R.; Hower, J.C.; Song, W.; Wang, X.; Li, X.; Zhao, L.; Kang, H.; et al. Factors controlling geochemical and mineralogical compositions of coals preserved within marine carbonate successions: A case study from the Heshan Coalfield, southern China. Int. J. Coal Geol. 2013, 109, 77–100.
15. Dai, S.; Luo, Y.; Seredin, V.V.; Ward, C.R.; Hower, J.C.; Zhao, L.; Liu, S.; Zhao, C.; Tian, H.; Zou, J. Revisiting the late Permian coal from the Huayingshan, Sichuan, southwestern China: Enrichment and occurrence modes of minerals and trace elements. Int. J. Coal Geol. 2014, 122, 110–128.
16. Li, B.; Zhuang, X.; Li, J.; Querol, X.; Font, O.; Moreno, N. Geological controls on mineralogy and geochemistry of the Late Permian coals in the Liulong Mine of the Liuzhi Coalfield, Guizhou Province, Southwest China. Int. J. Coal Geol.2016, 154, 1–15.
17. Standardization Administration of China; General Administration of Quality Supervision, Inspection and Quarantine of the China. Chinese National Standard GB/T 212-2008; Proximate Analysis of Coal; Standand Press of China: Beijing, China, 2008; pp. 3–5. (In Chinese)
18. Standardization Administration of China; General Administration of Quality Supervision, Inspection and Quarantine of the China. Chinese National Standard GB/T 214-2007; The Determination Methods of Total Sulfur in Coal; Standand Press of China: Beijing, China, 2007; pp. 1–5. (In Chinese)
19. Standardization Administration of China; General Administration of Quality Supervision, Inspection and Quarantine of the China. Chinese National Standard GB/T 215-2003; The Determination Methods of Form Sulfur in Coal; Standand Press of China: Beijing, China, 2003; pp. 1–5. (In Chinese)
20. International Organization for Standardization. Methods for the Petrographic Analysis of Bituminous Coal and Anthracite-Part 5: Method of Determining Microscopically the Reflectance of Vitrinite (ISO 7404-5); International Organization for Standardization (ISO): Geneva, Switzerland, 2009.

21. Dai, S.; Hower, J.C.; Ward, C.R.; Guo, W.; Song, H.; O'Keefe, J.M.K.; Xie, P.; Hood, M.M.; Yan, X. Elements and phosphorus minerals in the middle Jurassic inertinite-rich coals of the Muli Coalfield on the Tibetan Plateau. Int. J. Coal Geol. 2015, 144, 23–47.
22. Standardization Administration of China; General Administration of Quality Supervision, Inspection and Quarantine of the China. Chinese National Standard GB/T 4633-1997; Determination of Fluorine in Coal; Standard Press of China: Beijing, China, 1997; pp. 1–5. (In Chinese)
23. ASTM International. Standard Classification of Coals by Rank; ASTM D388-12; ASTM International: West Conshohocken, PA, USA, 2012.
24. Standardization Administration of China; General Administration of Quality Supervision, Inspection and Quarantine of the China. Chinese National Standard GB/T 15224.1-2004; Classification for Quality of Coal-Part 1: Ash; Standard Press of China: Beijing, China, 2004; pp. 1–2. (In Chinese)
25. Standardization Administration of China; General Administration of Quality Supervision, Inspection and Quarantine of the China. Chinese National Standard GB/T 15224.2-2004; Classification for Quality of Coal-Part 2: Sulfur Content; Standard Press of China: Beijing, China, 2004; pp. 1–2. (In Chinese)
26. Dai, S.; Ren, D.; Chou, C.-L.; Finkelman, R.B.; Seredin, V.V.; Zhou, Y. Geochemistry of trace elements in Chinese coals: A review of abundances, genetic types, impacts on human health, and industrial utilization. Int. J. Coal Geol. 2012, 94, 3–21.
27. Ketris, M.P.; Yudovich, Y.E. Estimations of Clarkes for Carbonaceous biolithes: World averages for trace element contents in black shales and coals. Int. J. Coal Geol. 2009, 78, 135–148.
28. Dai, S.; Seredin, V.V.; Ward, C.R.; Hower, J.C.; Xing, Y.; Zhang, W.; Song, W.; Wang, P. Enrichment of U–Se–Mo–Re–V in coals preserved within marine carbonate successions: Geochemical and mineralogical data from the Late Permian Guiding Coalfield, Guizhou, China. Miner. Deposita 2015, 50, 159–186.
29. Dai, S.; Li, D.; Chou, C.-L.; Zhao, L.; Zhang, Y.; Ren, D.; Ma, Y.; Sun, Y. Mineralogy and geochemistry of boehmite-rich coals: new insights from the Haerwusu Surface Mine, Jungar Coalfield, Inner Mongolia, China. Int. J. Coal Geol.2008, 74, 185–202.
30. Dai, S.; Chekryzhov, I.Y.; Seredin, V.V.; Nechaev, V.P.; Graham, I.T.; Hower, J.C.; Ward, C.R.; Ren, D.; Wang, W. Metalliferous coal deposits in East Asia (Primorye of Russia and South China): A review of geodynamic controls and styles of mineralization. Gondwana Res. 2016, 29, 60–82.
31. Seredin, V.V.; Dai, S. Coal deposits as potential alternative sources for lanthanides and yttrium. Int. J. Coal Geol. 2012, 94, 67–93.
32. Taylor, S.R.; McLennan, S.M. The Continental Crust: Its Composition and Evolution; Blackwell Scientific Publications: Palo Alto, CA, USA, 1985; p. 312.
33. Zamanian, H.; Ahmadnejad, F.; Zarasvandi, A. Mineralogical and geochemical investigations of the Mombi bauxite deposit, Zagros Mountains, Iran. Chem. Erde Geochem. 2016, 76, 13–37.
34. Fu, X.; Wang, J.; Zeng, Y.; Tan, F.; Feng, X. REE geochemistry of marine oil shale from the Changshe Mountain area, northern Tibet, China. Int. J. Coal Geol. 2010, 81, 191–199.
35. Shand, P.; Johannesson, K.H.; Chudaev, O.; Chudaeva, V.; Edmunds, W.M. Rare Earth Element Contents of High pCO2 Groundwaters of Primorye, Russia: Mineral Stability and Complexation Controls. In Rare Earth Elements in Groundwater Flow Systems; Johannesson, K.H., Ed.; Springer: Dordrecht, The Netherlands, 2005; Volume 7, pp. 161–186.
36. Dai, S.; Graham, I.T.; Ward, C.R. A review of anomalous rare earth elements and yttrium in coal. Int. J. Coal Geol.2016, 159, 82–95.
37. Ward, C.R. Minerals in bituminous coals of the Sydney Basin (Australia) and the Illinois Basin (USA). Int. J. Coal Geol.1989, 13, 455–479.
38. Hrinko, V. Technological, chemical, and mineralogical characteristics of bauxites and country rocks near Drienovec. Miner. Slovaca 1986, 18, 551–555.

39. Spears, D.A. The origin of tonsteins, an overview, and links with seatearths, fireclays and fragmental clay rocks. Int. J. Coal Geol. 2012, 94, 22–31.
40. Sahoo, P.K.; Tripathy, S.; Panigrahi, M.K.; Equeenuddin, S.M. Geochemical characterization of coal and waste rocks from a high sulfur bearing coalfield, India: Implication for acid and metal generation. J. Geochem. Explor. 2014, 145, 135–147.
41. Finkelman, R.B. Environmental Aspects of Trace Elements in Coal. In Modes of Occurrence of Environmentally-Sensitive Trace Elements in Coal; Swaine, D.J., Goodarzi, F., Eds.; Springer: Dordrecht, The Netherlands, 1995; Volume 2, pp. 24–50.
42. Chen, P.; Chai, D. Sedimentary Geochemistry of Carboniferous Bauxite Deposits in Shanxi Massif; Shanxi Science and Technology Press: Shanxi, China, 1997; pp. 1–194. (In Chinese)
43. Liu, Y.; Cao, L.; Li, Z.; Wang, H.; Chu, T.; Zhang, J. Element Geochemistry; Geological publishing press of Beijing: Beijing, China, 1984; pp. 125–136. (In Chinese)
44. Qian, D.D. History of China's Deposit Discovery, Guizhou Volume; Geological Publishing House: Beijing, China, 1996; pp. 177–182. (In Chinese)
45. Wang, D.; Li, P.; Qu, W.; Yin, L.; Zhao, Z.; Lei, Z.; Wen, S. Discovery and preliminary study of the high tungsten and lithium contents in the Dazhuyuan bauxite deposit, Guizhou, China. Sci. China Earth Sci. 2013, 56, 145–152.
46. Dai, S.; Ren, D.; Chou, C.-L.; Li, S.; Jiang, Y. Mineralogy and geochemistry of the No. 6 coal (Pennsylvanian) in the Junger coalfield, Ordos basin, China. Int. J. Coal Geol. 2006, 66, 253–270.
47. Wilkin, R.T.; Barnes, H.L. Formation processes of framboidal pyrite. Geochim. Cosmochim. Acta 1997, 61, 323–339.
48. Dai, S.; Hou, X.; Ren, D.; Tang, Y. Surface analysis of pyrite in the No. 9 coal seam, Wuda Coalfield, Inner Mongolia, China, using high-resolution time-of-flight secondary ion mass-spectrometry. Int. J. Coal Geol. 2003, 55, 139–150.
49. Diehl, S.F.; Goldhaber, M.B.; Hatch, J.R. Modes of occurrence of mercury and other trace elements in coals from the warrior field, Black Warrior Basin, Northwestern Alabama. Int. J. Coal Geol. 2004, 59, 193–208.
50. Zhang, J.; Ren, D.; Zheng, C.; Zeng, R.; Chou, C.-L.; Liu, J. Trace element abundances in major minerals of Late Permian coals from southwestern Guizhou province, China. Int. J. Coal Geol. 2002, 53, 55–64.
51. Dai, S.; Yang, J.; Ward, C.R.; Hower, J.C.; Liu, H.; Garrison, T.M.; French, D.; O'Keefe, J.M.K. Geochemical and mineralogical evidence for a coal-hosted uranium deposit in the Yili Basin, Xinjiang, northwestern China. Ore Geol. Rev. 2015, 70, 1–30.
52. Diehl, S.F.; Goldhaber, M.B.; Koenig, A.E.; Lowers, H.A.; Ruppert, L.F. Distribution of arsenic, selenium, and other trace elements in high pyrite Appalachian coals: Evidence for multiple episodes of pyrite formation. Int. J. Coal Geol.2012, 94, 238–249.
53. Dai, S.; Zou, J.; Jiang, Y.; Ward, C.R.; Wang, X.; Li, T.; Xue, W.; Liu, S.; Tian, H.; Sun, X.; et al. Mineralogical and geochemical compositions of the Pennsylvanian coal in the Adaohai Mine, Daqingshan Coalfield, Inner Mongolia, China: Modes of occurrence and origin of diaspore, gorceixite, and ammonian illite. Int. J. Coal Geol. 2012, 94, 250–270.
54. Dai, S.; Ren, D.; Li, S.; Chou, C.-L. A discovery of extremely-enriched boehmite from coal in the Junger coalfield, the Northeastern Ordos Basin. Acta Geol. Sin. 2006, 80, 294–300. (In Chinese)
55. Dai, S.; Yan, X.; Ward, C.R.; Hower, J.C.; Zhao, L.; Wang, X.; Zhao, L.; Ren, D.; Finkelman, R.B. Valuable elements in Chinese coals: A review. Int. Geol. Rev. 2016.
56. Hower, J.C.; Eble, C.F.; Dai, S.; Belkin, H.E. Distribution of rare earth elements in eastern Kentucky coals: Indicators of multiple modes of enrichment? Int. J. Coal Geol. 2016, 160–161, 73–81.
57. Zhao, L.; Dai, S.; Graham, I.; Wang, P. Clay mineralogy of coal-hosted Nb-Zr-REE-Ga mineralized beds from Late Permian strata, eastern Yunnan, SW China: Implications for palaeotemperature and origin of the micro-quartz. Minerals 2016, 6, 45.

58. Johnston, M.N.; Hower, J.C.; Dai, S.; Wang, P.; Xie, P.; Liu, J. Petrology and Geochemistry of the Harlan, Kellioka, and Darby Coals from the Louellen 7.5-Minute Quadrangle, Harlan County, Kentucky. Minerals 2015, 5, 894–918.
59. Dai, S.; Liu, J.; Ward, C.R.; Hower, J.C.; French, D.; Jia, S.; Hood, M.M.; Garrison, T.M. Mineralogical and geochemical compositions of Late Permian coals and host rocks from the Guxu Coalfield, Sichuan Province, China, with emphasis on enrichment of rare metals. Int. J. Coal Geol. 2015.
60. Liu, J.; Yang, Z.; Yan, X.; Ji, D.; Yang, Y.; Hu, L. Modes of occurrence of highly-elevated trace elements in superhigh-organic-sulfur coals. Fuel 2015, 156, 190–197.
61. Dai, S.; Wang, P.; Ward, C.R.; Tang, Y.; Song, X.; Jiang, J.; Hower, J.C.; Li, T.; Seredin, V.V.; Wagner, N.J.; et al. Elemental and mineralogical anomalies in the coal-hosted Ge ore deposit of Lincang, Yunnan, southwestern China: Key role of N2-CO2-mixed hydrothermal solutions. Int. J. Coal Geol. 2015, 152, 19–46.
62. Dai, S.; Liu, J.; Ward, C.R.; Hower, J.C.; Xie, P.; Jiang, Y.; Hood, M.M.; O'Keefe, J.M.K.; Song, H. Petrological, geochemical, and mineralogical compositions of the low-Ge coals from the Shengli Coalfield, China: A comparative study with Ge-rich coals and a formation model for coal-hosted Ge ore deposit. Ore Geol. Rev. 2015, 71, 318–349.
63. Dai, S.; Li, T.; Jiang, Y.; Ward, C.R.; Hower, J.C.; Sun, J.; Liu, J.; Song, H.; Wei, P.; Li, Q.; et al. Mineralogical and geochemical compositions of the Pennsylvanian coal in the Hailiushu Mine, Daqingshan Coalfield, Inner Mongolia, China: Implications of sediment-source region and acid hydrothermal solutions. Int. J. Coal Geol. 2015, 137, 92–110.
64. Eskenazy, G.; Dai, S.; Li, X. Fluorine in Bulgarian coals. Int. J. Coal Geol. 2013, 105, 16–23.
65. Hower, J.C.; Eble, C.F.; O'Keefe, J.M.K.; Dai, S.; Wang, P.; Xie, P.; Liu, J.; Ward, C.R.; French, D. Petrology, Palynology, and Geochemistry of Gray Hawk Coal (Early Pennsylvanian, Langsettian) in Eastern Kentucky, USA. Minerals 2015, 5, 592–622.
66. Dai, S.; Ren, D.; Tang, Y.; Shao, L.; Li, S. Distribution, isotopic variation and origin of sulfur in coals in the Wuda coalfield, Inner Mongolia, China. Int. J. Coal Geol. 2002, 51, 237–250.
67. O'Keefe, J.M.K.; Bechtel, A.; Christanis, K.; Dai, S.; DiMichele, W.A.; Eble, C.F.; Esterle, J.S.; Mastalerz, M.; Raymond, A.L.; Valentim, B.V.; et al. On the fundamental difference between coal rank and coal type. Int. J. Coal Geol. 2013, 118, 58–87.
68. Goodarzi, F.; Swaine, D.J. Paleoenvironmental and Environmental Implications of the Boron Content of Coals; Geological Survey of Canada: Calgary, AB, Canada, 1994; Volume 471, p. 82.
69. Shao, L.; Jones, T.; Gayer, R.; Dai, S.; Li, S.; Jiang, Y.; Zhang, P. Petrology and geochemistry of the high-sulphur coals from the Upper Permian carbonate coal measures in the Heshan Coalfield, southern China. Int. J. Coal Geol. 2003, 55, 1–26.
70. Dai, S.; Xie, P.; Jia, S.; Ward, C.R.; Hower, J.C.; Yan, X.; French, D. Enrichment of U-Re-V-Cr-Se and rare earth elements in the Late Permian coals of the Moxinpo Coalfield, Chongqing, China: Genetic implications from geochemical and mineralogical data. Ore Geol. Rev. 2016.
71. Dai, S.; Ren, D.; Zhou, Y.; Chou, C.-L.; Wang, X.; Zhao, L.; Zhu, X. Mineralogy and geochemistry of a superhigh-organic-sulfur coal, Yanshan Coalfield, Yunnan, China: Evidence for a volcanic ash component and influence by submarine exhalation. Chem. Geol. 2008, 255, 182–194.
72. Chou, C.-L. Sulfur in coals: A review of geochemistry and origins. Int. J. Coal Geol. 2012, 100, 1–13.
73. Jin, Y.; Fang, R. Early Permian brachiopods from the Kuangshan Formation in Luliang County, Yunnan, with notes on paleogeography of South China during the Liangshan Stage. Acta Palaeontol. Sin. 1985, 24, 216–228.
74. Zhao, L.; Dai, S.; Graham, I.T.; Li, X.; Liu, H.; Song, X.; Hower, J.C.; Zhou, Y. Cryptic sediment-hosted critical element mineralization from eastern Yunnan Province, southwestern China: Mineralogy, geochemistry, relationship to Emeishan alkaline magmatism and possible origin. Ore Geol. Rev. 2016.

CHAPTER 13

The Role of Soil Mineralogy, Geochemistry and Grain Size in the Development of Mediterranean Badlands: A Review

Vito Summa[1] and Maria Luigia Giannossi[1]

[1] *Institute of Methodologies for Environmental Analysis of the National*

1. INTRODUCTION

It has long been recognized that the local Mediterranean climate, tectonics and human impact interact to determine the gross morphology and surface conditions of this landscape. However, attention has recently been given to the explanatory role of lithology, in particular sediment size and clay mineralogy, in explaining the badland formation [1-9].

For instance, on *biancane* sites, Battaglia et al. [10] found clay fractions to be significantly high. These sites have been reported to possess also high percentage of clay minerals in particular in the smectitic group.

Additionally, for these clay minerals, high exchangeable sodium on the exchange complex promotes dispersion (deflocculation) of the clays. The exchangeable sodium percentage (ESP), sodium adsorption ratio (SAR), sodium percentage (PS) and total dissolved salts (TDS) are commonly used to measure the dispersive state.

This chapter aims to contribute to the international framework of research on water erosion processes, and to identify critical emerging erosional risk factors. It focuses particularly on experimental research on material properties that could be the promoter of soil erosion processes.

Results show that many components of soil erosional response, such as soil dispersivity, badlands development or surface and subsurface processes like crusting or pipes, are strongly affected by spatially variable and temporally dynamic soil properties.

2. SOIL DEGRADATION

When land is degraded, its productivity is reduced and many other eco system services are deleteriously affected. Land degradation may be primarily caused by natural processes, related to the characteristics of the given land resources and ecosystems. However, human activities often accelerate these degradation processes, leading to a rapid decline in the quality and quantity of the land resources and the ecosystem services flowing from these. Drylands are fragile and particularly susceptible to land degradation.

The United Nations Convention to Combat Desertification (UNCCD) defines land degradation in the context of drylands as: "a reduction or loss, in arid and semi-arid and dry sub-humid areas, of the biological or economic productivity and complexity of rainfed cropland, irrigated cropland, or range, pasture, forest and woodlands resulting from land uses or from a process or combination of processes, including processes arising from human activities and habitation patterns" [11].

Land degradation is caused by a variety of complex interrelated degradation processes. These can be grouped into three major land degradation types, each of which can be subdivided according to a specific sub-set of degradation processes, namely:

1. Soil degradation;
2. Vegetation degradation;
3. Water resources degradation.

Vegetation degradation involves a combination of processes that may be natural, notably climate change which may lead to a loss of certain species and habitats, reduced biomass due to reduced moisture availability, or encroachment by invasive species. However, vegetation degradation is generally induced by human activity, through the over use or mismanagement of forests, grazing and croplands, uncontrolled burning or introduction of pests and diseases.

Degradation of water resources in terms of quantity, quality and flow regime will lead to reduced productivity of the aquatic system in terms of fish and other useful aquatic species and products. It also affects the availability of clean drinking water for consumption by humans, livestock and wildlife.

Soil degradation is defined as the decline in soil quality caused through its misuse by human activity [12]. Degradation or decline of soil quality may occur due to physical or chemical processes triggered off by natural phenomena, or induced by humans through misuse of land resources. Processes such as soil erosion, nutrient run-off, water logging, desertification or compaction, may give examples of physical degradation processes, while acidification, organic matter loss, salinization, nutrient depletion by leaching, or toxicants accumulation, are all processes that can be classified as being agents and indicators of chemical degradation of soil.

2.1. Soil Degradation Types and Processes

Soil degradation occurs when there is a decline in the productive capacity of the soil as a result of adverse changes in its biological, chemical, physical and hydrological properties and/or attributed to the removal of soil through erosion by water or wind or by mass movement. Sheet, rill and gully erosion by water, also the scouring and re-deposition of soil by wind and landslides are some of the most visible symptoms of soil degradation, but other less visible forms of degradation of soil properties are even more widespread and sometimes more serious, notably depletion of nutrients and soil organic matter decline.

3. SOIL EROSION

Soil erosion is a major form of land degradation. It comprises various processes that are described separately below. However, any one of these processes may occur in the same locality, either in combination or at different times of year.

Soil erosion by water is often quite widespread and can occur in all parts of drylands where rainfall is sufficiently intense for surface runoff to occur. This category includes processes such as splash, sheet, rill and gully erosion. Splash erosion is commonly the first stage of water erosion and occurs when rain drops fall onto the bare soil surface. Their impact can break up surface soil aggregates and splash particles into the air. As water runs over the soil surface it has the power to pick up particles released by splash erosion and the capacity to detach particles from the soil surface. This may result in sheet erosion, where soil particles are removed from the whole soil surface on a fairly uniform basis.

Where runoff becomes concentrated into channels, rill and gully erosion may result. Rills are small rivulets of such a size that they can be ploughed over with farm machinery. Gullies are much deeper (often being several metres deep and wide) and form a physical impediment to the movement across the slope of farm machinery, even people and livestock [13].

Soil erosion by wind is also widespread throughout drylands that are exposed to strong winds. It includes both the removal and re-deposition of soil particles by wind action and the abrasive effects of moving particles as they are transported. In areas with extensive loose, sandy material, wind erosion can lead to the formation of mobile sand dunes that cause considerable economic losses through engulfing adjacent farm land, pastures, settlements, roads and other infrastructure [14].

Gravitational erosion tends to be more localised in regions with steep, rocky slopes and in mountain ranges. On sloping land when soil is saturated, its weight increases and the downward forces of gravity will induce a relatively large down-slope movement of soil and / or rocks (e.g. landslides, slumps, earth flows and debris avalanches). This mass movement of material may be very rapid and involve large volumes of soil, but is usually limited to isolated and localised events. Landslides may be natural events, however, their frequency and severity is likely to greatly increase following deterioration or loss of the natural vegetative cover by logging, overgrazing and / or clearing for cultivation [5, 15-17].

Freeze/thaw erosion is restricted to high altitude areas and areas with cold climates. It occurs when water in the topsoil initially freezes and expands, then melts, damaging topsoil structure and enabling loosened surface soil particles to be carried away in melt water runoff. It is primarily a natural process rather than one which is accelerated by particular human activities [15, 17-18].

This chapter covers only the assessment of soil erosion by water.

3.1. Soil Erosion Landscape: Badlands

The term badlands is currently used for areas of unconsolidated sediments or poorly consolidated bedrock, with little or no vegetation. They are useless for agriculture because of their intensely dissected landscape.

They appear to offer in a miniature spatial scale and a shortened temporal scale many of the processes and landforms exhibited by more normal fluvial landscapes, including a variety of slope forms, bedrock or alluvium-floored rills and washes, and flat alluvial expanses similar to large-scale pediments.

Although badlands evoke an arid image, they can develop in nearly any climate in soft sediments where vegetation is absent or disturbed. General reviews of badlands and badland processes are provided by Campbell [19] and Bryan and Yair [20], including discussions of the climatic, geologic, and geographic setting of badlands, sediment yields, host rock and regolith variations among badlands and field measurements of processes.

Badland landscapes are typically asymmetrical (Figure 1). The sunny aspects show impoverished or null vegetation cover because of the strong control on water availability effected by radiation, whereas the shady aspects may bear a vegetation cover close to 100% [21-23]. Steep slopes and gullies do not allow the formation of a developed soil because erosion processes are either frequent and/or intense.

FIGURE 1. Typical badland form at Pisticci (Basilicata Southern Italy) with southern eroded side and northern vegetated side

Calanchi and *biancane* are considered peculiar forms of badlands evolution [10].

The term *calanchi* describes the dendritic network of slope forms created on a single hillslope scarp (Figure 2). An individual *calanco* is defined by knife-edged ridges, separating small hydrographic drainage networks with horseshoe-shaped headwalls [24-25]. *Biancane* are small, conical or dome-shaped forms up to 20 m high (Figure 2), which may occur singly or in groups [9, 26-28].

Calanchi is the result of a rill erosion. Rill erosion is the removal of soil by water from very small but well defined, visible channels or streamlets where there is a concentration of overland flow [29]. In general, rill erosion is more serious than sheet erosion, and it is most accentuated when intense storms occur in watersheds or sites with high runoff-producing characteristics, loose, and grading operations.

Rill erosion is often described as the intermediate stage between sheet and gully erosion, and occurs by a concentration of runoff or low points through the soil.

Gully erosion could be considered as an advanced stage of rill erosion, where surface channel gullies (intermittent stream channels larger than rills) have been eroded to the point where they cannot be smoothed over by normal tillage operations.

Underground (groundwater) erosion is the removal of soil caused by groundwater seepage or movement towards a free face. It is also known as piping and occurs as a result of bank drainage or, in general, when seepage forces exceed intergranular stresses or cohesive forces [29]. Pipes can form in the downstream side of earth dams, gully heads, streambanks, and slopes where water exits from the ground. Once a cavity (pipe) forms, it is able to enlarge quickly since the flow follows the path of low flow resistance.

FIGURE 2. Typical morphological features of the landscape forms in Mediterranean area (Aliano, Basilicata – Southern Italy)

3.2. Compositional Controls of Badlands Occurrence

The main factor controlling badland formation is the particular character of the rocks or other materials which form the base for the interaction of weathering and erosion processes [19]. However, the existence of other risk factors such as climatic condition, human activities, geomorphological exposition, structural features, encourages the intensification of erosion and development of morphological features of the landscape forms [27, 30-36].

Soils in badlands deserve special attention, because soils are the inter-phase between the lithosphere and the atmosphere, and so constitute one of the key elements either favoring or restricting the initiation of badland formation.

When soils are resilient against erosion processes, gullies do not form; however, when soils, either because of their particular ground cover, i.e. sparse vegetation, and/or intrinsic properties, cannot withstand erosive forces, the topsoil is eroded and deep gullies develop, which may give rise to badlands if the underlying material is also erosion-sensitive.

Consequently, the characteristics of the materials underlying soils are crucial for the development of true badlands.

Lithology is a major factor for badland production, and is probably of greater importance than tectonics, climate, topography or land use [4, 6, 19, 37-38].

The general characteristics of a soil, regolith or geological formation that favours badland relief are the unconsolidated or very poorly cemented material of clay and silt, sometimes with soluble minerals such as gypsum or halite [39]. Specific characteristics, like structure, mineralogy, physical and chemical properties, may play either a primary or secondary role in material disintegration and badland development.

In the Mediterranean area most parent materials are essentially silt-dominant, with clay as the second particle size, while sand is generally very poorly represented.

Texture depends on four factors: particle-size distribution, grain shape, degree of crystallinity and relationship among grains [40]. Of these, particle-size distribution plays the key role in susceptibility for material disintegration and erosion: the larger the range of particle sizes, the higher the degree of packing, and hence the greater resistance to breakdown processes.

Conversely, the narrower the particle-size distribution, the higher the susceptibility for material disintegration, piping and, consequently, for badland development [40-41].

Besides textural properties, porosity is the second most important physical property [42-43]. If suitable macropores are available within the material for enlargement, dispersion can encourage the rapid enlargement of subsurface pipes [44-45], a process sometimes referred to as piping or tunneling.

Infiltration is defined as the process by which water enters the soil. Its rate depends on soil type, soil structure and soil water content [46]. Infiltration is important for reducing run-off and consequent erosion. Increased soil compaction and loss of surface structure (reduced aggregation) are the main factors in reducing water infiltration rates in soils. Such rates are normally dependent upon the occurrence of large pores occupying the upper surfaces of the soil; therefore they depend on soil texture in the first place [47].

Soil bulk density is defined as the mass of soil per unit volume in its natural field state, including air space and mineral matter, plus organic substance. High values of bulk density may restrict the movement of surface waters through the soil, leading to a loss of nutrients by leaching. It may also increase erosion rates. Bulk density measurements are very important for assessing soil quality, since root growth and penetration of soil, together with the ease of soil aeration, are largely controlled by this factor [48].

Geotechnical properties provide another important control for erosion: Atterberg limits (for consistency), swelling, and slaking behavior are considered in many badland studies [10, 49-50].

Certain minerals play an essential role in the breakdown of some rocks at near surface conditions.

Some minerals are important because they may become soluble, like all soluble salts (halite), but also moderately soluble like sulphates (gypsum) or carbonates (calcite and dolomite), especially when they can be dissolved because of the small size of their constitutive particles and/or solvent characteristics [51].

Some other minerals; like clay minerals (smectite in particular), can absorb water in amounts several times their dry weight, with consequent volume increases. Wetting-drying alternations in materials with expandable clays cause the formation of nets of deep cracks and may also lead to the formation of a shallow layer of loose expanded regolith fragments, usually called popcorn [5].

It was found that the percentage presence of the swelling clays in the overall material mass is very important. Where clay percentages are high, the material mass is rendered impermeable on swelling, encouraging surface wash erosion and reducing infiltration. Where clay percentages are low, the deflocculation of the clay fraction merely destructures a material already lacking in other sources of cohesion, encouraging subsurface erosion. Given the presence of a suitable hydraulic gradient through a site, this distinction will separate materials that are dispersive but which do not develop large pipes from those that do.

Clay dispersion is a physico-chemical process relevant to erosion processes, particularly to the development of pipes. Materials (soils, regoliths or rocks) with a potential to disperse are those which contain a high exchangeable sodium percentage (ESP), saturating part of the exchangeable cations of their clays. This percentage is considered to be critical when higher than 13.

To predict the tendency of materials to pipe, Faulkner et al. [52-55] explored the effectiveness of the relationship between electrical conductivity (EC) and SAR (sodium adsorption ratio), originally used by Rengasamy et al. [54]. Whilst this improves diagnosis over the use of ESP or SAR values alone, it seems that this analysis is also insufficient in itself to distinguish between badland surfaces in terms of their morphology.

The relationship between pH and SAR can be used to indicate the extent of material buffering as dispersivity changes.

3.3. Erosion Prediction

In the analysis of the processes connected to the land degradation, many models were developed in the past with the objective to give a qualitative and quantitative solution to the problem of the soil loss and erosion. This estimation of those phenomena is particularly difficult due to the number of variables to consider and their typologies including both natural, i.e. soil nature, vegetation and rainfall, and anthropic ones, as the many options for management practices and land use. Models about this evaluation are often based on empirical or process-based analysis and the synthetic equations used to describe the phenomenon are necessarily complex because they have to include the interactions of all the parameters. Besides, the complexity of the erosion processes, and the need for huge data banks to compile the many algorithms which are included in the models, are also technical problems to consider in the analysis plan. Anyway many erosion prediction models are available: event-based or long-term models, empirical or physically based models, on a basin or plot scale, which have been improved in the last few decades. One of the best examples in the estimation of long-term average annual soil loss from arable lands, is the Universal Soil Loss Equation (USLE) model.

The Universal Soil Loss Equation (USLE), developed by Wischmeier, Smith, and others in the 1960, predicts the long term average annual rate of erosion on a field slope based on rainfall pattern, soil type, topography, crop system and management practices.

In the applications related to the analysis of erosion processes, the USLE equation was replaced by the Revised Universal Soil Loss Equation (RUSLE), which has a similar structure (that is a black-box factor empirical model), but with more sophisticated inputs and it is designed for operation on personal computers

Because there is a wide discrepancy between predicted and observed erosion rates, models are better as research tools than as public policy and regulatory instruments or for prescriptive design measures for constructed landforms. But some models may provide useful guidance for the design engineer if adequately calibrated and verified for local conditions and if the design accounts for the uncertainty.

4. INFLUENCE OF SOIL FEATURES ON DEVELOPING *CALANCHI* EROSIONAL LANDFORMS IN SOUTHERN ITALY (BASILICATA)

The soil erosion risk is widespread in the Mediterranean. Some areas of Italy are an excellent example of soil erosion risk. Soil erosion vulnerability in the Basilicata region (Southern Apennines - Italy) is mostly represented by water erosion forms. *Calanchi* and *biancane* are two typical erosion landforms present in the Basilicata region and this area is a key reference for the international studies of water erosion processes [3, 8, 30, 56-58].

4.1. Geological and Climatic Settings

The studied sites are part of a well-known area of present desertification [1, 7, 20, 27, 30, 56-65], and is located in the far south of the Apennines (Figure 3).

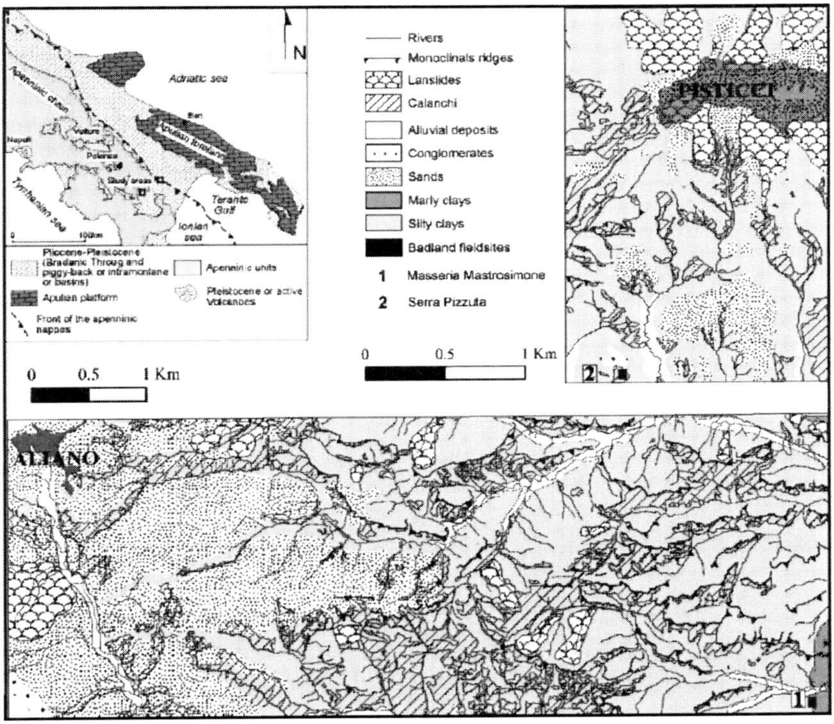

FIGURE 3. Geologic map showing the location of the study areas [adapted from Piccarreta et al., 2006]

The Pisticci site is a hilly area between the river Basento and the seasonal stream Salandrella, both deeply carved in the late Pliocene-Calabrian Sub-Apennine clays of the Bradanic Foredeep [61].

The Sub-Apennine clays are clays and silty clays, sometimes sandy clays, typically blue in colour. The lithological uniformity of the pelitic facies is attenuated by thin sand-silt and tuff layers (centimetric scale).

The Aliano site is situated in the heart of the Basilicata Appenines and is part of the north-eastern area of the Plio-Pleistocene Sant' Arcangelo basin.

The formations were affected by the uplift of the eastern margin of the Apennine chain during the upper Pliocene and Post-Calabrian ages [66]. The whole area shows tectonic aspects due to movements which have prevailing vertical components. This is clearly caused by the features of the sedimentary layers, from sub-horizontal to gently dipping. Generally speaking the stratification is still not well-distinguished. The stratigraphic sequence and the basin structure have been studied in depth [67-74].

In the study areas the Plio-Pleistocene clays are the most diffused lithology to be found. These are 500 to 900 m thick [27, 75] and consist of marly and silty clays with a middle–high plasticity [27, 71].

Pieri et al. [76] redefined the Plio-Pleistocene successions by subdividing them into four depositional sequences, such as the Late Pliocene – Middle Pleistocene on the basis of the stratigraphic, sedimentologic and structural features of the deposits outcropping in the northern basin. Each cycle, several hundred metres thick, represents one or more depositional systems (alluvial, marine-deltaic and lacustrine).

Patacca and Scandone [73] suggest a new structural architecture for the Southern Appenines, especially for the Plio-Pleistocene Foredeep/thrust-sheet-top deposits. According to these authors, the Plio-Pleistocene thrust-sheet-top and Foredeep deposits were subdivided into two depositional sequences both of which are governed by tectonic processes active in the mountain chain (P1-2 - lower-upper Pliocene - and Q1-2 – lower-middle Pleistocene - thrust-related depositional sequences).

Details about the thrust-related depositional sequence in the Southern Appenines and the relative systems tracts, including the characteristic stratigraphic signatures, are supplied in Patacca and Scandone [72].

The area studied is characterized by extremely widespread erosion mainly affected by the lithological features of their soils. The overall geomorphological development of this area resulted from periodic intensive erosion, which began in the Late Pleistocene and continued during the Holocene due to tectonic movements, climatic changes and related sea-level fluctuations [77].

These two sites have been chosen because they are represent by two different badlands areas.

In the Pisticci area there are "typical" badlands, a representation of the usual morphology of the semi-arid Mediterranean area – characterized both by an eroded slope facing south and a non-eroded (covered) slope facing north [56, 58, 61, 65]. Erosion was studied on sediments of Sub-Apennine clays, unvegetated slopes (SE-facing scarp slopes up to 35°-40°) with high rates of erosion, labelled "Pisticci, eroded", and opposite vegetated slopes (NW-facing scarp slopes up to 20°) labelled "Pisticci, non-eroded" (Figure 4).

FIGURE 4. Photos from Pisticci studied area: "typical" badlands on sediments of Sub-Apennine clays, characterized both by an eroded slope facing south (scarp slopes up to 35°-40°) and a non-eroded (covered) slope facing north (scarp slopes up to 20°).

Also in the Aliano area there are slopes with features of erosion common to clayey-silty rocks exposed to the south-east as well as adjacent slopes having the same exposure (south-east), but showing a different erosional action. A partially-vegetated covering can be found (Figure 5).

FIGURE 5. Photos from Aliano studied area: slopes with clayey-silty rocks exposed to the south-east as well as adjacent slopes having the same exposure (south-east), but showing a different erosional action.

The annual average rainfall for the Pisticci area 1923–2000, is about 645 mm [78-79]. The most abundant precipitation is in Autumn and Winter; Summer is the driest season [80].

In the Aliano area, from 1955 to 2000, the annual mean precipitation was 738 mm (st.dev. 174mm), mainly concentrated from the months of October to January. In the same area the minimum and maximum values recorded over a period of 46 years are, respectively, 367mm and 1090 mm [34].

In both areas, the climate is typically Mediterranean, characterized by warm and dry Summers with temperatures averaging 26-27°C with a maximum as high as 39°C, and cold and rainy Winters with temperatures averaging 8-10°C in January.

4.2. Materials and Methods

For each slope, eroded and non eroded (in all case studies), samples were collected in order to represent the several litho-pedological levels. Since vegetated soils resist breakdown and crusting [58], within the eroded slope, the crust was only differentiated with respect to the substrate and was defined as existing at 0-2 cm depth. Below the crust, samples were labelled "substrate". For each eroded and non-eroded slope, three different profiles were sampled: top, middle, and bottom (Figure 6).

Detailed grain size analyses were carried out by laser diffraction, a Malvern MasterSizerE laser particle-sizer with a 100-mm lens, which identifies grain-size intervals from 0.5 to 100 μm. For mineralogical analysis the clay fraction (<2 μm) was separated by means of fractioned sedimentation in accordance with Stokes' law.

Mineralogical analyses were carried out by X-ray diffraction (XRD) on a Rigaku D/Max-2200/ Pc powder diffractometer (theta-theta configuration, Cu Kα radiation). Quantitative mineralogical data were obtained according to Barahona [81], and the results were checked by means of a comparison with chemical data.

Chemical bulk-rock elements were measured by X-ray fluorescence (XRF) on a Philips PW 1480/10 spectrometer with Cr radiation. Recommendations made by Franzini et al. [82] were applied in order to correct matrix effects by using international geological standards.

pH measurements were made according to the procedure indicated in Italian law no.79 [83]. Dried samples were mixed with distilled water (ratio 1:2.5) and the mixture was then stirred and pH measured.

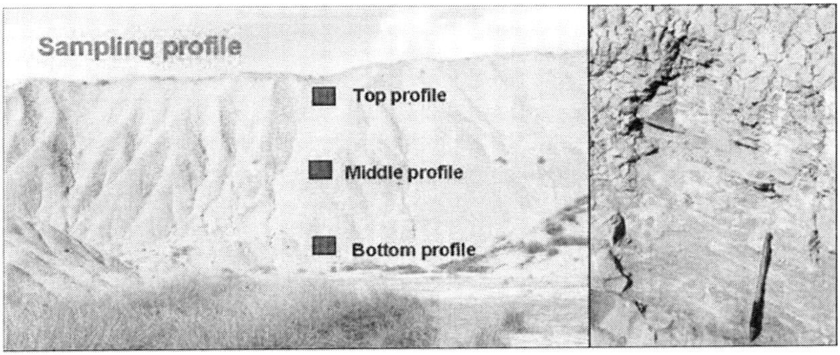

FIGURE 6. . Sampling profile

Selected soluble salt concentrations (Na^+, K^+, Mg^{2+}, and Ca^{2+}) were measured by ion chromatography [52], and sodium adsorption ratio (SAR) and Exchangeable Sodium Percentage (ESP) calculated according to the formula:

SAR=NaMg+Ca2√SAR=NaMg+Ca2

ESP=NaCEC100ESP=NaCEC100

Pearson's correlation coefficients and Student's T-test were calculate to quantify the relationship between variables.

4.3. Results and Discussion

The mineralogical, geochemical and grain-size composition features of these slopes has been determined to find common risk factors for the different areas.

Only a few grain-size parameters, mineralogical and geochemical features discriminate the eroded and non eroded substrates [8]. The water erosion phenomena is present where the fine fraction is abundant (more evident in Aliano than in Pisticci). This can be explained by a reduction of permeability in eroded soils while the non eroded ones are more stable with respect to the weathering phenomena, as they are more permeable.

Crusts represent the more weathered and modified part of eroded sides, but their grain size and chemical features resemble non eroded materials better than their own substrate. Such a similarity can be depicted as an auto-stabilization process of superficial portion of eroded slopes [e.g. 53, 84]. Chemical data enable discrimination between eroded and non-eroded slopes in all case studies.

pH, SAR (sodium adsorption ratio), TDS (total dissolved salts) and PS (percentage of sodium) are distinctive parameters for both eroded and non-eroded slopes. On average, eroded substrates are higher in pH, SAR and PS than non-eroded ones. The ESP (exchangeable sodium percentage) of the eroded slope has a higher value than the non-eroded one [8].

The results of this study show that, even if geological and geomorphological differences exist between the two areas, common erosion risk factors can be characterized.

4.3.1. Geomorphological and structural observations

In both study areas, the topography has a gentle dip and morphology is expressed as a typical monoclinal landscape. However, the causes for the monoclinal topography differ in the two regions (Figure 7).

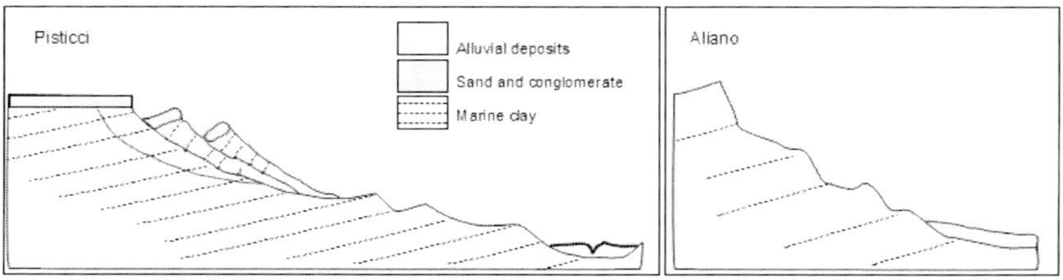

FIGURE 7. Schematic diagrams of the studied sites [33, modified]

The Aliano site has been interpreted as a simple monoclinal system, whereas at the Pisticci site, landslides are particularly widespread on the South-East facing hillslopes [33-34, 85].

Although the monoclinal morphology has differing origins in the two areas, in both settings, the existing primary and secondary network of fractures and joints appears to influence the genesis and development of surface drainage [33-34, 61].

In Aliano, down-valley structural features have been focussed around the weaker parts of the structural sequence of marine clays, although in the case of Pisticci, since these are failure planes not lithological features, these lineaments are more discontinuous in their down-valley pattern as might have been imagined.

In both settings, the rapidity of the geomorphic processes on the relatively steeper scarp slopes generally prohibits vegetation from securing a stable function.

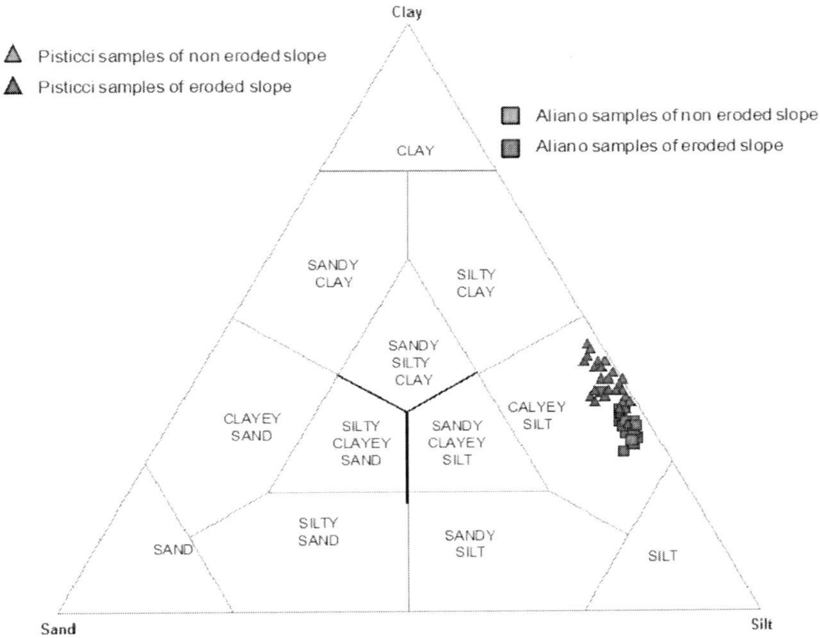

FIGURE 8. Granulometrical classification diagram

4.3.2. Granulometrical and mineralogical properties

After grain size analyses, all the substrate samples studied were found to be of the clayey-silt type, which is a typical substrate facilitating the formation of *calanchi*, as suggested by Battaglia et al. [10]. The grain size diagram [86] does not distinguish the eroded from the non-eroded slopes (Figure 8) neither does the soil erodibility nor the soil-quality diagram, from the CORINE Land cover [87]. However, more detailed grain-size distribution is shown in Figures 4 and 5, and gives further information, as it deals with the lower coarse fraction (>63 μm) of eroded substrates in all profiles ($r=0.783$, $p<0.000$). Comparing the profiles of the two slopes, further grain size discrimination is achieved due to the fact that the non-eroded profile of Aliano has larger course fraction (>16 μm, Figure 9) instead the non eroded profile of Pisticci are enriched in 4–63 μm fraction (Figure 10).

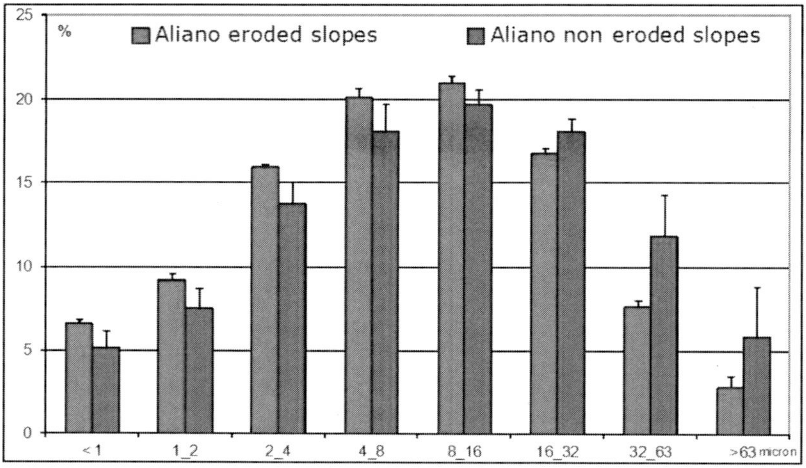

FIGURE 9. Average granulometrical composition of eroded and non eroded samples of Aliano

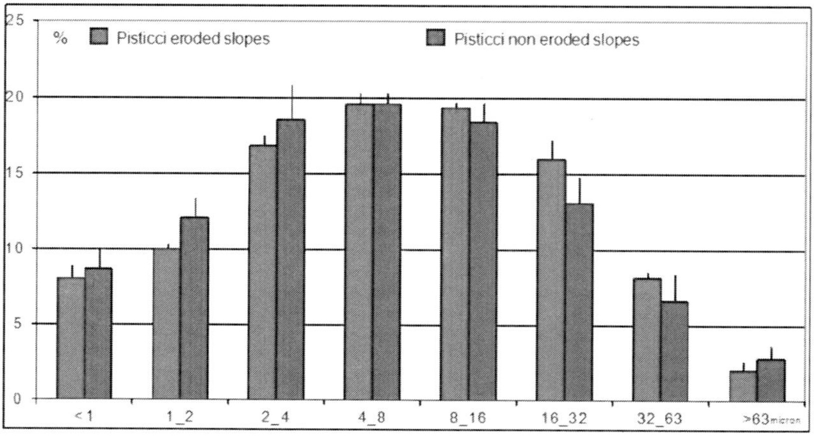

FIGURE 10. Average granulometrical composition of eroded and non eroded samples of Pisticci

In both cases, the granulometric characteristics of the crust of the eroded slope are comparable with those of the substrate not eroded, as demonstrated by a linear correlation coefficient R close to 1 ($p<0.000$). This means that after erosion the most delicate part of the slope (the crust) becomes less dispersive as a sort of auto-stabilization process.

The micromorphological information on some samples of the eroded side show three distinct domains (Figure 11).

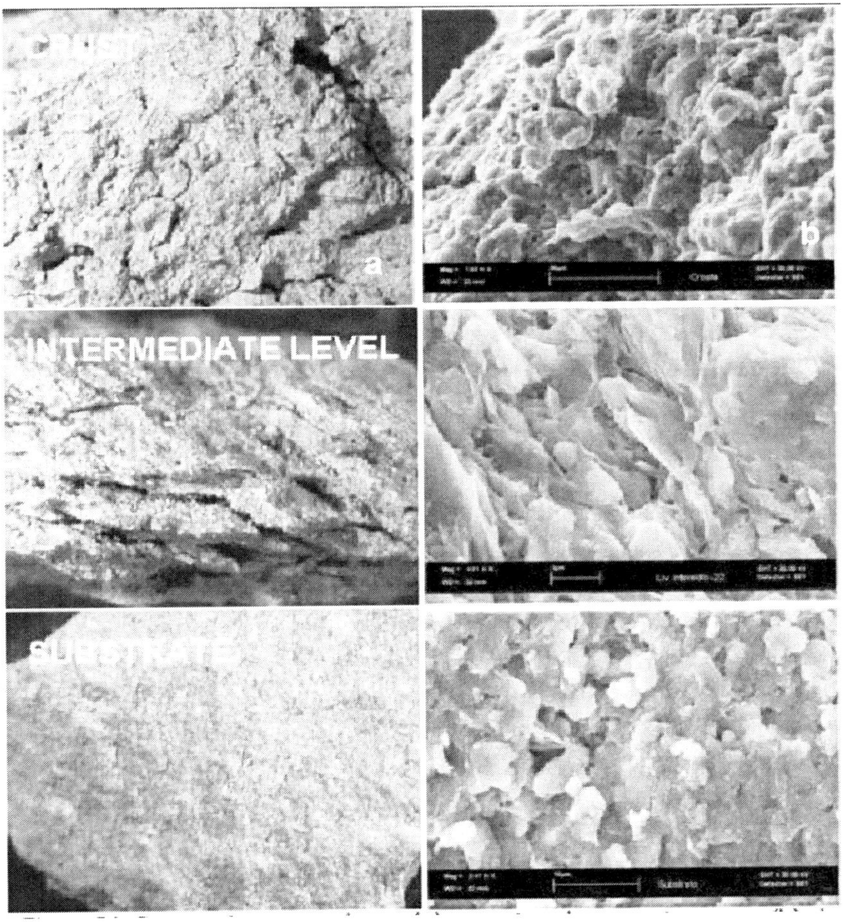

FIGURE 11. Sample of the eroded side at optical (a) and scanning electron microscopy (b)

Below 20 cm the fine-grained dense sub-zone displays a massive structure and is relatively impermeable. Immediately above this dense sub-zone, a zone of isorientate structure with low porosity is present (2–20 cm).

The top 2 cm of the profile appears dispersive with high porosity. Infiltration process dominates, due to high permeability. The high porosity of the top 2 cm of the soil suggests that the most important hydraulic activity is restricted under this depth, at the intermediate level.

Both the eroded and non-eroded slopes in Aliano and Pisticci show a fairly comparable mineralogical composition of the samples on account of their mineralogical phases and quantity. So, it was not possible to define a systematic trend distinguishing bulk rock mineralogy with erosive features.

The mineralogical assemblage of the samples consists of quartz, calcite and feldspars (Table 2), with some difference in quantity for the two sites. Dolomite is always present but in lower concentrations; traces of gypsum and hematite occasionally occur at low levels. Among the clay minerals, illite is the most abundant (on average 50% of the clay fraction), while chlorite, kaolinite and mixed-layer illite–smectite generally having lower concentrations (Tab. 2). The amount of kaolinite is higher in the eroded slopes ($r=0.829$) than in the non-eroded ones ($r=-0.703$). The quantity of illite in the eroded slope is lower than that in the non-eroded one.

4.3.3. Chemical properties

Some parameters are found to be higher in the eroded substrates than in the non-eroded ones (Figure 12) ($p<0.000$). Also the crusts of eroded slopes differ from the substrates and the values of these three parameters increase with depth.

The total dissolved salts (TDS) can turn out to be quite distinctive in a comparison between the eroded and non-eroded sides. The sodium adsorption ratio (SAR) is a slightly different expression of the importance of Na in the soluble cation composition with respect to PS. pH values are alkaline (from 8.0 to 9.1).

The sodium effect of clay stability is often expressed as ESP (exchangeable sodium percentage, [3, 8, 55]), and is closely related to the Na available for cation-exchange from the clays. ESP–depth diagrams between the eroded and non-eroded slopes turned out to be extremely differentiating (Figure 13).

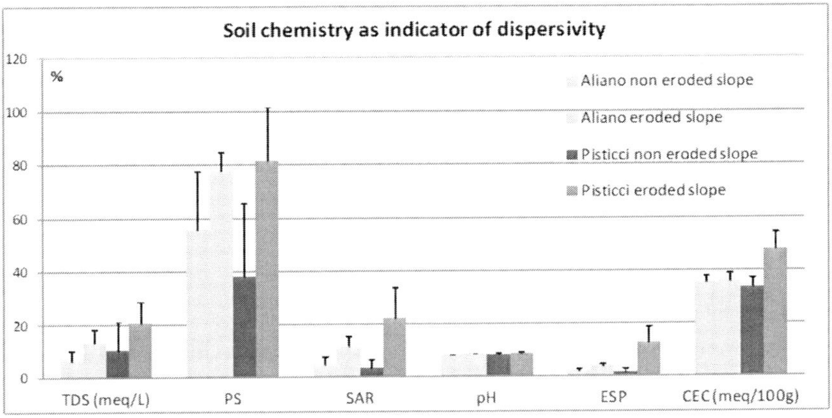

FIGURE 12. Comparison among some indicators of dispersivity (TDS= Total dissolved salts; PS= Sodium percentage; SAR= Sodium Adsorption ratio; ESP= Exchangeable sodium percentage; CEC= Cation exchange capability)

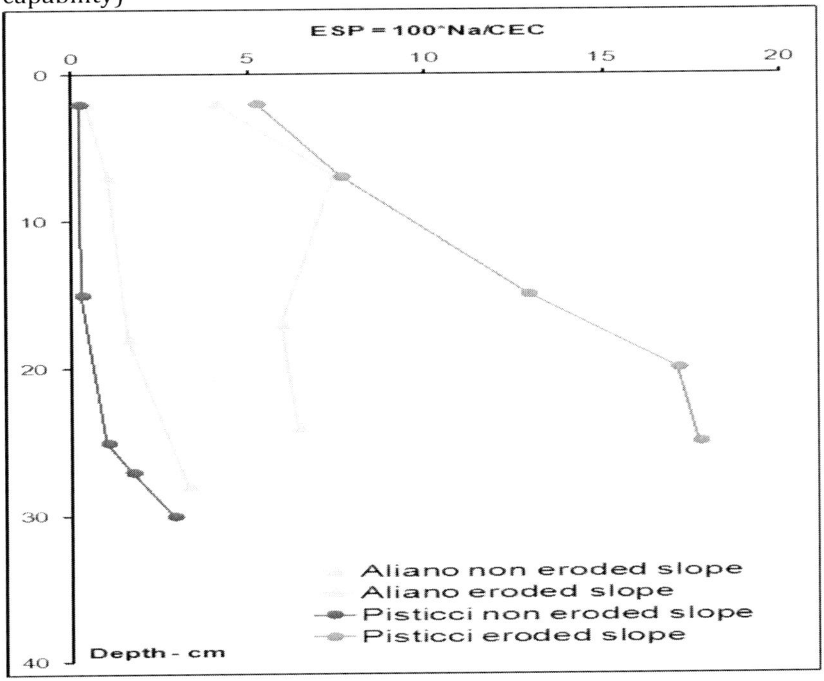

FIGURE 13. Exchangeable sodium percentage (ESP) versus soil depth

The various relations between these chemical parameters, which define soil susceptibility to dispersion, can also predict the performance of the surface layers and subsoil [10, 53-55, 88-89]. According to this approach the substrate of the eroded and non eroded slopes can be tentatively distinguished by using the general subdivision reported in the PS–TDS–SAR diagram of Figure 14 [8, 10, 89-90].

As shown in the diagram (Figure 14), all the samples of the non-eroded slopes have a high degree of variability as they are to be found in all the three classification zones.

The eroded slopes are mainly included in a dispersive zone, except for some samples included in the overlap zone. The dispersive properties of the latter samples are not so clear. In some cases, this shows a tendency of some portions of eroded slopes, which generally correspond to the topmost part of the slope, towards geochemical stabilization.

Other diagrams such as ESP-pH (Figure 15) can also be effective for distinguishing the eroded from non-eroded slopes [3, 8, 90]. The reason why ESP is a better discriminator may be due to the fact that the composition of the exchange complex is an intrinsic soil property.

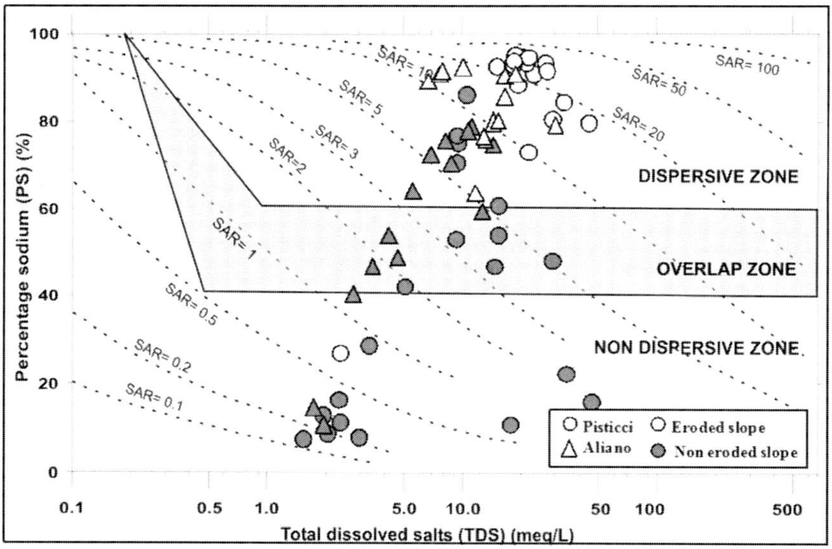

FIGURE 14. Relationships between sediment dispersivity and pore water composition (expressed through the PS, TDS and SAR parameters), as established by Sherard et al.[87]

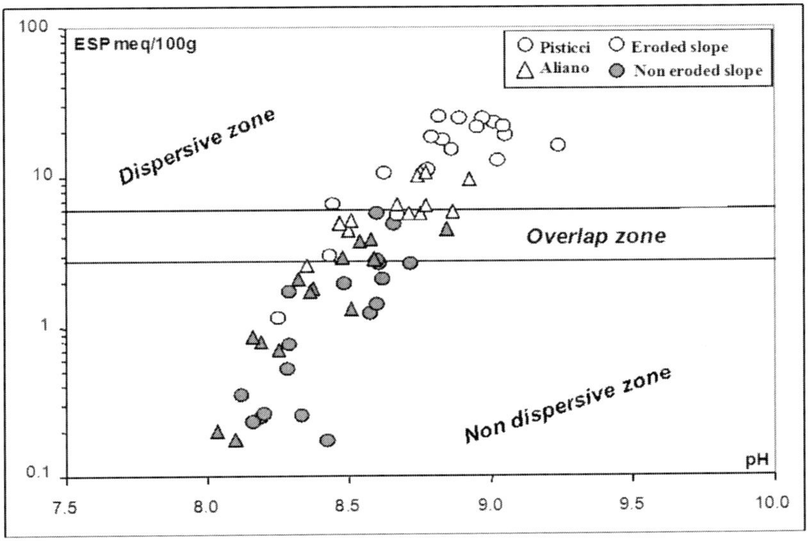

FIGURE 15. Correlation between ESP and pH for the prediction of dispersive behaviour of soil

An interesting feature that arises from many reported diagrams (Figures 14–15) is the anomalous plotting of crust samples in the eroded slope compared with other eroded profiles (white symbol in the non dispersive zone). Comparing the crusts with eroded substrates, crusts are clearly characterized by lower dispersivity parameters. The uppermost substrate samples of the non-eroded slope also follow the same trend as shown by crusts, suggesting that severely weathered portions of slope tend to reach a stable condition due to strong decreases in SAR, PS and ESP.

This effect, as mentioned by Faulkner et al. [53], is severe for the upper profile of the eroded slope, so that crust samples often plot in the non-dispersive field. This sort of 'auto-stabilization' process has been mentioned by several authors [e.g. 53, 84, 91] dealing with Na leaching.

As can be observed the eroded slope of Pisticci and that of Aliano have similar compositional characters with chemical characteristics associated with highly dispersive soils, regardless of exposure, or other geomorphological or climate factors.

4.4. Conclusion

This research has demonstrated that certain physico-chemical properties of the local sodic Plio-Pleistocene clays influence the different erosional processes in the two study slopes in Basilicata in fundamental ways.

Clay materials in the middle and base of the slopes retain a dispersive character. Only a few grain-size parameters discriminate eroded from non-eroded substrates. The water erosion phenomena is present where the fine fraction is abundant. This can be explained with a reduction of permeability in eroded soils while the non-eroded ones are more stable.

Erosional risk factors can be found in granulometrical features, in particular an higher fine fraction (< 16 μm for Aliano slopes and more abundant 8-16 μm for Pisticci slopes) promote erosional phenomena.

From a chemical perspective, a higher value of pH, SAR, PS parameters and above all ESP are an indication of intensive erosional processes.

The substrate of eroded and non eroded slopes can be discriminated by classification diagrams using some chemical parameters (SAR, PS and TDS) as dispersivity descriptors.

A better separation of substrates is obtained using other diagrams such as the ESP-pH.

If for the Pisticci site the exposure has always been considered one of the main erosion factors, by comparison with the slopes of Aliano we can understand how the intrinsic characteristics of the soil are crucial for the development of the erosion process.

The granulometric, mineralogical and chemical characters of the non-eroded slopes of Aliano (facing South) are comparable with those of the slopes of Pisticci (facing North).

The two study sites also have another common feature that it is possible to extend to all badlands domains: the auto-stabilization process.

This process has been identified thanks to the physico-chemical properties of the two monitored badland sites and the pH / SAR relationship that shows the tendency of the crust to auto-stabilise, confirming they are really effective signature sites.

Also the granulometrical similar composition of crusts and non-eroded substrates can be interpreted as an auto-stabilization process of superficial portion of eroded slopes.

5. CONCLUDING REMARKS

Badlands are a typical landform of greatly dissected fine-grained materials. *Calanchi* and *biancane* are considered peculiar forms of badlands evolution. Their formation is closely related to the physico-chemical properties of the soil, climatic and geomorphologic conditions, and human activities.

Studies reviewed have shown that badlands have been investigated for their peculiar features and processes in the frame of landscape evolution, spatial and temporal distributions, denudation rates, effects of man's activity, and erosion risk assessment or mitigation, but several aspects of their genesis and evolution are still unclear.

Various factors and geomorphic processes seem to interplay in *calanchi* genesis and evolution, in relation to local microclimatic conditions, geological features and land use changes.

Some parameters, however, are not yet sufficiently considered: grain size, physical and mineralogical characteristics of clay sediments, soil chemistry, and chemical features of surface and underground water.

The results of the presented study show that the erosional mechanism involves morphological and geographic exposure and climatic elements as well as grain size, mineralogy, chemistry and exchangeable processes of soils. They are important characteristics of eroded soil to give a further contribution to the issue of *calanchi* genesis, in an attempt to integrate pre-existing studies with new risk factors.

It's possible to define erosional risk factors as granulometrical, mineralogical from a chemical perspective. These indicators of soil erodibility risk can be applied in different erosional development and can be used to update the current model for erosion prediction in term of soil-erodibility factor K in the RUSLE equation.

ACKNOWLEDGEMENT

Erosional and salinization study were supported by fund from Programma Operativo FESR Basilicata 2007/2013, "Pro-Land Project".

The authors gratefully acknowledge comments by Paula Preston on the English revision of this chapter.

REFERENCES

1. Alexander D. Difference between 'calanchi' and 'biancane' badlands in Italy. In Badland Geomorphology and Piping, Bryan R, Yair A (eds). Geobooks: Norwich; 1982; p. 71–88.
2. Bryan RB, Yair A, Hodges WK. Factors controlling the initiation of runoff and piping in Dinosaur Provincial Park badlands, Alberta, Canada. Z. Geomorph. Suppl. Bd. 1978; 29: 151-68.
3. De Santis F, Giannossi ML, Medici L, Summa V, Tateo F. The influence of physic-chemical material properties on different badland development (Aliano, Southern Italy). Catena. 2012; 81: 172–81
4. Gerits J, Imeson AC, VersUraten JM, Bryan RB. Rill development and badland regolith properties. Catena Suppl. 1987; 8: 141-60.
5. Hodges WK, Bryan RB. The influence of material behaviour on runoff initiation in the Dinosaur Badlands, Canada. In: Bryan RB, Yah A. (Eds.), Badland Geomorphology and Piping. Geoabstracts, Norwich. 1982; p. 13-46.
6. Imeson AC, Verstraten JM. Rills on badland slopes: a physico-chemically controlled phenomenon. Catena Suppl. 1988; 12: 139-50.
7. Imeson AC, Kwaad FJ, Verstraten JM. The relationship of soil physical and chemical properties to the development of badlands in Morocco. In Badland Geomorphology and Piping, Bryan R, Yair A (eds). Geobooks: Norwich. 1982; p. 47–71.
8. Summa V, Tateo F, Medici L, Giannossi ML. The role of mineralogy, geochemistry and grain size in badland development in Pisticci (Basilicata, Southern Italy). Earth Surf. Process. Landf. 2007; 32: 980–97.
9. Torri D, Bryan RB. Micropiping processes and biancane evolution in southeast Tuscany, Italy. Geomorphology. 1997; 20: 219–35.
10. Battaglia S, Leoni L, Sartori F. Mineralogical and grain size composition of clays developing calanchi and biancane erosional landforms. Geomorphology. 2002; 49: 153–70.
11. UNCCD. United Nations Convention to combat desertification in countries experiencing serious drought and/or desertification, particularly in Africa. A/AC.241/27, Paris; 1994.
12. Barrow CJ. Land Degradation: Development and Breakdown of Terrestrial Environments.Cambridge: Cambridge University Press; 1991.
13. Cooke RU, Doornkamp JC. Geomorphology in environmental management. An introduction. Clarendon press, Oxford; 1974.
14. EEA-European Environment Agency. SOER 2005 – The European environment – State and outlook 2005. EEA, Copenhagen; 2005.
15. Clotet N, Gallart F, Balasch J. Medium term erosion rates in a small scarcely vegetated catchment in the Pyrenees. Calena Supplement. 1988; 13:37-47.
16. Gerits JJP, Imeson AC, Verstraeten JM. Chemical thresholds and erosion in saline and sodic materials. In F. Lopez-Bermudez and J. Thornes (Eds) Estudios Sobre la Geomorfologìa del Sur de Espana. Murcia; 1986: p. 71-4.
17. Oostwoud Wijdenes DJ, Ergenzinger P. Erosion and sediment transport on steep marly hillslopes, Draix, Haute-Provence, France. An experimental field study. Catena. 1998; 33: 179-200.
18. Regués D, Pardini G, Gallart F. Regolith behaviour and physical weathering of clayey mudrock as dependent on seasonal weather conditions in a badland area at Vallcebre, Eastern Pyrenees. Catena. 1995; 25:199-212.
19. Campbell LA. Badlands and badland gullies. In D.S.G. Thomas (Ed.) Arid Zone Geomorphology. Belhaven/Halsted Press, London; 1989: p. 159-83.
20. Bryan R, Yair A (eds). Badland Geomorphology and Piping. Geobooks, Norwich; 1982.

21. Kirkby MJ, Atkinson K, Lockwood J. Aspect, vegetation cover and erosion on semi-arid hillslopes. In J. Thornes (Ed.) Vegetation and Erosion. Wiley, Chichester; 1990: p. 25-39.
22. Solé A, Calvo A, Cerdà A, Làzaro R, Pini R, Barbero J. Influence of micro-relief patterns and plant cover on runoff related processes in badlands from Tabernas (SE Spain). Catena. 1997; 31: 23-38.
23. Cantòn Y. Efectos hidrologicos y geomorfologicos de la cubierta y proprietades del suelo en paisaje de carcavas. Unpublished PhD thesis, Universidad de Almeria, Spain; 1999.
24. Alexander DE. Calanchi – accelerated erosion in Italy. Geography. 1980; 65: 95–100.
25. Farabegoli E, Agostini C. Identification of calanco, a badland landform in the northern Apennines, Italy. Earth Surface Processes and Landforms. 2000; 25: 307–18.
26. Calzolari C, Ungaro F. Characterization and quantitative spatial analysis of geomorphic features of a badland (biancane) area, central Italy. Catena. 1998; 31: 237–56.
27. Clarke ML, Rendell HM. The impact of the farming practice of remodelling hillslope topography on badland morphology and soil erosion processes. Catena. 2000; 40: 229–50.
28. Torri D, Colica A, Rockwell D. Preliminary study of the erosion mechanisms in biancane badland (Tuscany, Italy). Catena. 1994; 23: 281–94.
29. Gray DH, Sotir RB. Biotechnical Soil Bioengineering Slope Stabilization: A Practical Guide for Erosion Control. John Wiley & Sons, New York, NY; 1996.
30. Farifteh J, Soeters R. Factors underlying piping in the Basilicata region, southern Italy. Geomorphology. 1999; 26: 239–51.
31. Farifteh J, Soeters R. Origin of biancane and calanchi in East Aliano, southern Italy. Geomorphology. 2006; 77: 142–52.
32. Regues D, Guardia R, Gallart F. Geomorphic agents versus vegetation spreading as causes of badland occurrence in a Mediterranean subhumid mountainous area. In: Badlands in changing environments. Catena. Special Issue. 2000; 40 (2): 173–87.
33. Piccarreta M, Faulkner H, Bentivenga M, Capolongo D. The influence of physico-chemical material properties on erosion processes in the badlands of Basilicata, Southern Italy. Geomorphology. 2006; 81: 235–51.
34. Piccarreta M, Capolongo D, Boenzi F, Bentivenga M. Implications of decadal changes in precipitation and land use policy to soil erosion in Basilicata, Italy. Catena. 2006; 65: 138–51.
35. Della Seta M, Del Monte M, Fredi P, Palmieri EL. Direct and indirect evaluation of denudation rates in Central Italy. Catena. 2007; 71: 21-30.
36. Kasanin-Grubin M, Bryan R. Lithologic properties and weathering response on badland hillslopes. Catena. 2007; 70: 68–78.
37. Calvo A, Harvey AM, Paya-Serrano J. Processes interactions and badland development in SE Spain. In M. Sala, J.L. Rubio and J.M. Garcia-Ruiz (Eds) Soil Erosion Studies in Spain. Geoforma Ed., Logrono;1991: p. 75-90,
38. Calvo A, Harvey AM, Paya-Serrano J, Alexander RW. Response of badland surfaces in SE Spain to simulated rainfall. Cuaternario y geomorfologia. 1991; 5: 3-14.
39. Scheidegger AE, Schumm SA, Fairbridge RW. Badlands. In R. Fairbridge (Ed.) Encyclopedia of Geomorphology. Dowden, Hutchinson and Ross, Inc., USA. 1968: p. 43-8.
40. Terzaghi K, Peck RE. Soil Mechanics in Engineering Practice, 2nd Edn. John Wiley & Sons, NewYork; 1967.
41. Taylor RK, Smith TJ. The engineering geology of day minerals: swelling, shrinking and mudrock breakdown. Clay Minerals. 1986; 21: 235-60.
42. Imeson A. Investigating volumetric changes in clayey soils related to subsurface water movement and piping. Zeitschrift fur Geomorphologie Supplement Band. 1986; 60: 115-30.

43. Solé A, Josa R, Pardini G, Aringhieri R, Plana F, Gallart F. How mudrock and soil physical properties influence badland formation at Vallcebre Pre-Pyrenees (NE Spain). Catena. 1992; 19: 287-300.
44. Benito G, Gutierrez M, Rancho C. The influence of physico- chemical properties on erosion processes in badland areas, Ebro basin, N–E Spain. Z. Geomorphol. N. F. 1993; 37: 199– 214.
45. Gutierrez M, Sancho C, Benito G, Sirvent J, Desir G. Quantitative study of piping processes in badland areas of the Ebro basin, NE Spain. Geomorphology. 1997; 20: 237–53.
46. Lowery B, Hickey W. J, Arshad M.A, Lal R. Soil water parameters and soil quality. In: Doran J.W, Jones A.J, editors. Methods for assessing soil quality. Madson, WI; 1996: p. 143-55.
47. Hillel D. Introduction to Soil Physics. Academic Press, Inc. San Diego; 1982.
48. Arshad MA, Lowery B, Grossman B. Physical tests for monitoring soil quality. In: Doran, J.W., Jones, A.J. (Eds.), Methods for Assessing Soil Quality. SSSA Special Publication No. 49. SSSA, Madison, WI; 1996: p. 123–41.
49. Rienks SM, Botha GA, Hughes JC. Some physical and chemical properties of sediments exposed in a gully donga in northern KwaZulu-Natal, South Africa and their relationship to the erodibility of the colluvial layers. Catena. 2000; 39: 11–31.
50. Igwe CA, Ejiofor N. Structural stability of exposed gully wall in Central Eastern Nigeria as affected by soil properties, International Agrophysics. 2005; 19: 215-22.
51. Cantòn Y, Domingo F, Solé-Benet A, Puigdefábregas J. Hyrological and erosional response of a badlands system in semiarid SE Spain. Journal of Hydrology. 2001; 252: 65–84.
52. Faulkner H, Spivey D, Alexander R. The role of some site geochemical processes in the development and stabilisation of three badland sites in Almerìa, Southern Spain. Geomorphology. 2000; 35: 87–99.
53. Faulkner H, Alexander R, Wilson BR. Changes to the dispersive characteristics of soils along an evolutionary slope sequence in the Vera badlands, southeast Spain: implications for site stabilisation. Catena. 2003; 50: 243–54.
54. Faulkner H, Alexander R, Teeuw R, Zukowskyj P. Variations in soil dispersivity across a gully head displaying shallow sub-surface pipes, and the role of shallow pipes in rill initiation. Earth Surfaces Processes and Landforms. 2004; 29: 1143–60.
55. Rengasamy P, Greene RSB, Ford GW, Mehanni AH. Identification of dispersive behaviour and the management of red–brown earths. Australian Journal of Soil Research. 1984; 22: 413–31.
56. Del Prete M, Bentivenga M, Amato M, Basso F, Tacconi P. Badland erosion processes and their interactions with vegatation: a case study from Pisticci, Basilicata, southern Italy. Geografia Fisica e Dinamica Quaternaria. 1997; 20: 147–55.
57. Rendell HM. Soil erosion and land degradation in southern Italy. In Desertification in Europe, Fantechi R, Margaris NS (eds). Commission of European Communities, Brussels; 1986: p. 184–93.
58. Robinson DA, Phillips CP. Crust development in relation to vegetation and agricultural practice on erosion-susceptible, dispersive clay soils from central and southern Italy. Soil Till. Res. 2001; 60: 1–9.
59. Liberti M, Simoniello T, Carone MT, Coppola R, D'Emilio M, Macchiato M. Mapping badland areas using LANDSAT TM/ETM satellite imagery and morphological data. Geomorphology. 2009; 106: 333-43.
60. APAT. Stato dell'Ambiente: Sezione Geosfera. 2003; 7: 352–61. Available from: http://www.apat.gov.it/site/it-IT/APAT//Pubblicazioni/Annuario_dei_dati_ambientali//Documento/stato_ambiente_2003_10.html#Sommario
61. Del Prete M, Bentivenga M, Coppola L, Rendell H. Aspetti evolutivi dei reticoli calanchivi a sud di Pisticci. Convegno Nazionale Giovani Ricercatori di Geologia Applicata. Geol. Romana. 1994; 30: 295–306.

62. Rendell HM. Clay hillslope erosion rates in the Basento Valley, S. Italy. Geografiska Annaler. 1982; 64A: 141–47.
63. Vittorini S. La degradazione in un campo sperimentale nelle argille plioceniche della Val d'Era (Toscana) e i suoi riflessi morfogenetici. Rivista Geografica Italiana. 1971; 78: 142–69.
64. Vittorini S. Ruscellamento, deflusso ipodermico ed erosione nelle argille plastiche. Rivista Geografica Italiana. 1979; 86(3): 338–46.
65. Guerricchio A, Melidoro G. Fenomeni franosi e neotettonici nelle argille grigio azzurre calabriane di Pisticci (Lucania) con saggio di cartografia. Geol. Appl. E Idrogeol. 1979; 14: 105–38.
66. Boezi F, Palmentola G, Valduga A. Note Illustrative della Carta Geologica d'Italia Foglio 200. Servizio Geologico Italiano; 1971.
67. Vezzani L. Il bacino plio-pleistocenico di S. Arcangelo (Lucania). Atti dell'Accademia Gioenia di Scienza Naturali in Catania S6. Supplemento di scienze geologiche. 1967; 18: 207–27.
68. Lentini F. Le Unità Sifilidi della Val d'Agri (Appennino Lucano). Geol. Romana. 1979; 18: 215–24.
69. Caldara M, Loiacono F, Morlotti E, Pieri P, Sabato L. I depositi Pliopleistocenici della parte nord del Bacino di S. Arcangelo (Appennino Lucano): Caratteri geologici e Paleoambientali. Mem. Soc. Geol. Ital. 1988; 41: 391–410.
70. Carbone S, Catalano S, Lazzari S, Lentini F, Monaco C. Presentazione della carta geologica del bacino del fiume Agri (Basilicata). Memorie della Società Geologica Italiana. 1991; 41: 109–20.
71. Pieri P, Sabato L, Loiacono F, Marino M. Il Bacino di piggyback di Sant'Arcangelo: evoluzione tettonico-sedimentaria. Boll. Soc. Geol. It. 1994; 113: 465–81.
72. Patacca E, Scandone P. Late thrust propagation and sedimentary response in the thrust belt-foredeep system of the Southern Appenines (Pliocene–Pleistocene). In: Vai, G.B., Martini, I.P. (Eds.), Anatomy of an Orogen: The Appenines and Adjacent Mediterranean Basins. Kluwer Academic Publ; 2001: p. 401–40.
73. Patacca E, Scandone P. Constraints on the interpretation of the CROP-04 seismic line derived from Plio-Pleistocene foredeep and thrust-sheet-top deposits (Southern Appenines, Italy). Boll.Soc.Geol.It. (Ital.J.Geosci.), Spec. Issue. 2007;. 7: 241–56.
74. Zavala C, Mutti E. Stratigraphy of the Plio-Pleistocene Sant'Arcangelo Basin Basilicata, Italy. Atti Riunione Gruppo Sedimentologia del C.N.R. Catania; 1996: p. 279–82.
75. Lentini F, Vezzani L. Note illustrative del Foglio 506 S. Arcangelo. V. di 46, I.R.P.I., Cosenza; 1974.
76. Pieri P, Sabato l, loiacono F. Carta Geologica del Bacino di Sant'Arcangelo (tra il Torrente Sauro e il Fiume Agri). Ed. Paternoster, Matera; 1993.
77. Verstappen HTh. Geomorphology of the Agri valley, Southern Italy. ITC Journal. 1983; 4: 291–301.
78. Piccarreta M, Capolongo D, Boenzi F. Trend analysis of precipitation and drought in Basilicata from 1923 to 2000 within a Southern Italy context. International Journal of Climatology. 2004; 24: 907–22.
79. Piccarreta M. Aspetti Evolutivi della Morfogenesi Calanchiva nelle Argille Pliopleistoceniche della Basilicata, PhD Thesis. Università degli Studi di Bari; 2005.
80. Piccarreta M, Capolongo D, Bentivenga M, Pennetta L. Influenza delle precipitazioni e dei cicli umido–secco sulla morfogenesi calanchiva in un'area semi-arida della Basilicata (Italia Meridionale). Geografia Fisica e Dinamica del Quaternario, Supplement. 2005; VII: 281–89.
81. Barahona Fernandez E. Arcillas de ladrilleria de la privincia de Granada: Evaluacion de algunos ensayos de materias primas. Ph.D. thesis, Univ. Granada, Spain; 1974.
82. Franzini M, Leoni L, Saitta M. Revisione di una metodologia analitica per fluorescenza-X, basata sulla correlazione completa degli effetti di matrice. Rendiconti della Società Italiana di Mineralogia e Petrologia. 1975; 31: 365–78.

83. Ministero delle Risorse Agricole, Alimentari e Forestali (MiRAAF). etodi Ufficiali di Analisi Chimica del Suolo. Roma; 1992.
84. Alexander RW, Harvey AM, Calvo A, James PA, Cerda A. Natural stabilisation mechanism on badlands slopes: Tabernas, Almerìa.In Environmental Change in Drylands (Biogeographical and Geomorphological Perspectives), Millington AC, Pye K (eds). Wiley: Chichester; 1994: p. 85–111.
85. Guerricchio A, Melidoro G. New views on the original of the badlands in the Plio-Pleistocenic clays of Italy. Poc. IV Congr. IAEG; 1982: p. 2.
86. Pettijohn FJ. Sedimentary Rocks, 3rd edn. Harper and Row, New York; 1975.
87. APAT. La realizzazione in Italia del progetto Corine Land Cover 2000. APAT, Rapporti. 2005; 36: p. 86.
88. Gerits J, Imeson AC, Verstraten JM, Bryan RB. Rill development and badland regolith properties. In: Bryan, R.B. (Ed.), Rill Erosion: Processes and Significance: Catena, Suppl. 1997; 8: 141–60.
89. Sherard JL, Dunningan LP, Decker RS. Identification and nature of dispersive soils. J Geotech. Eng. Div. 1976; 102: 287–301.
90. Sotelo RR. Identificaciòn de arcillas erodibles dispersivas utilizzando ensayos agronòmicos de suelos. Cience & Tecnica. Comunicaciones Cientificas y Tecnològicas. 1999; 1: 200–4.
91. Harvey AM. The role of piping in the development of badlands and gully systems in south-east Spain. In: Bryan, R., Yair, A. (Eds.), Badland Geomorphology and Piping. Geobook, Norwich; 1982: p. 317–35.

CHAPTER 14

Platinum-Group Minerals and Other Accessory Phases in Chromite Deposits of the Alapaevsk Ophiolite, Central Urals, Russia

Federica Zaccarini [1,*], Evgeny Pushkarev [2], Giorgio Garuti [1] and Igor Kazakov [3]

[1]Department of Applied Geosciences and Geophysics, University of Leoben, Peter Tunner Str.5, A 8700 Leoben, Austria

[2]Institute of Geology and Geochemistry, Ural Division of Russian Academy of Sciences, Str. Pochtovy per. 7, 620151 Ekaterinburg, Russia

[3]Joint Stock Company Ural Geological Survey Expedition, str. Vaynera 55, 620014 Ekaterinburg, Russia

ABSTRACT

An electron microprobe study has been carried out on platinum-group minerals, accessory phases, and chromite in several chromite deposits of the Alapaevsk ophiolite (Central Urals, Russia) namely the Bakanov Kluch, Kurmanovskoe, Lesnoe, 3-d Podyony Rudnik, Bol'shaya Kruglyshka, and Krest deposits. These deposits occur in partially to totally serpentinized peridotites. The microprobe data shows that the chromite composition varies from Cr-rich to Al-rich. Tiny platinum-group minerals (PGM), 1–10 μm in size, have been found in the chromitites. The most abundant PGM is laurite, accompanied by minor cuproiridsite and alloys in the system Os–Ir–Ru. A small grain (about 20 μm) was found in the interstitial serpentine of the Bakanov Kluch chromitite, and its calculated stoichiometry corresponds to $(Ni,Fe)_5P$. Olivine, occurring in the silicate matrix or included in fresh chromite, has a mantle-compatible composition in terms of major and minor elements. Several inclusions of amphibole, Na-rich phlogopite, and clinopyroxene have been identified. The bimodal Cr–Al composition of chromite probably corresponds to a vertical distribution in the ophiolite sequence, implying formation of Cr-rich chromitites in the deep mantle, and Al-rich chromitites close to the Moho-transition zone, in a supra-subduction setting. The presence of abundant hydrous silicate inclusions, such as amphibole and phlogopite, suggests that the Alapaevsk chromitites crystallized as a result of the interaction between a melt enriched in fluids and peridotites. Laurite and cuproiridsite are considered to be magmatic in origin, i.e., entrapped as solid phases during the crystallization of chromite at high temperatures. The sulfur fugacity was relatively high to allow the precipitation of Ir-bearing sulfides, but below the Os–OsS_2 buffer. The alloys in the system Os–Ir–Ru are classified as secondary PGM, i.e., formed at low temperature during the serpentinization process. The $(Ni,Fe)_5P$ phase is the first occurrence of a Ni-phosphide in terrestrial samples. Its composition indicates that it may be a new mineral. However, the small size has, so far, prevented a crystallographic study to support this conclusion.

Keywords: chromitite; platinum group minerals; silicate inclusions; Ni-phosphide; ophiolite; Alapaevsk; Urals; Russia

1. INTRODUCTION

The Uralian orogenic belt, which extends 2500 km from the Polar Urals to the Aral Lake, is considered the boundary between Europe and Asia. The belt formed because of the collision between the East European continental platform to the west and the Asian plate to the east. The process of collision started in the Paleozoic and ended in the early Jurassic. This tectonic event brought to the surface many mafic–ultramafic complexes (Figure 1) occurring in both the east and west slopes of the Ural Mountains [1].

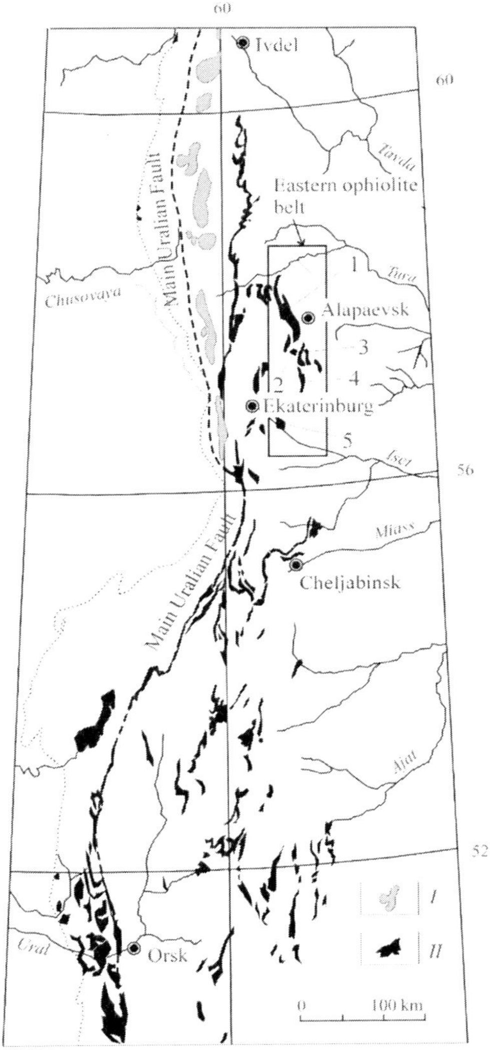

Figure 1. The distribution of mafic–ultramafic complexes in the Urals (excluding Circumpolar and Polar sectors of the Urals) (simplified after [2]). I—dunite-clinopyroxenite-gabbro massifs of the Ural Platinum Belt; II—ophiolite complexes. Eastern ophiolite belt: 1—Alapaevsk massif; 2—Ostaninsky massif; 3—Rezhevskoy massif; 4—Bazhenovsky massif; 5—Kluchevskoy massif. Rectangle shows the northern end of the Eastern ophiolite belt where the Alapaevsk massif is located.

The mafic–ultramafic complexes can be grouped into three main types: (1) harzburgite–dunite–volcanic associations considered to be sections through the upper mantle–crust transition of Palaeozoic ophiolites; (2) concentrically zoned dunite–clinopyroxenite–gabbro intrusions derived from deep magmatic activity in island arc (Ural–Alaskan type) complexes and (3) lherzolite–harzburgite blocks representing weakly depleted mantle derived rocks that formed the floor during overthrusting of huge allochthonous or huge ophiolite massifs that were true oceanic crust [2,3,4,5,6].

Since the beginning of the last century, the Urals have become known for their importance in the exploration and recovery of chromium. Chromite deposits containing more than 100 Mt of ore reserve have been discovered in the Kempirsai massif of Kazakhstan and in the complexes of Voykar Sininsky and Ray-Iz located in the Polar Urals, Russia ([6] and references therein). The giant chromite ore bodies of the Urals occur exclusively in the peridotite of subduction-related ophiolites [7]. However, chromitites varying in size from potential ore deposits down to sub-economic concentrations of chromite also occur, associated with various types of Uralian ophiolites ([6] and references therein). Ophiolitic chromitites in general, as well as those occurring in the Urals, show a wide compositional range from Al-rich to Cr-rich chromite. This compositional variation is a function of many factors such as the composition of the parental magmas, the nature of their mantle sources, the vertical distribution from the deeper part of the upper mantle to the mantle–crust transition zone, and different geodynamic settings. It is known that the ophiolitic chromitites may contain sub-economic concentrations of platinum-group elements (PGE), with particular enrichment of IPGE (Os + Ir + Ru) over the PPGE (Rh + Pt + Pd). The mineralogical expression of the PGE content is platinum-group minerals (PGM) that form tiny phases, generally less than 10 μm in size and unevenly disseminated in the chromitite. In this contribution we present the results of a new, detailed study of chromite composition and the first data on accessory minerals such as silicates, PGM, and a rare NiFeP compound found in certain chromite deposits that occur in the Alapaevsk ophiolite, located in the Central Urals. We use this data to model the tectonic setting and genesis of the chromitites and their host ophiolite and to better understand the mechanisms of the precipitation of PGM and other rare accessory phases.

2. GEOLOGICAL BACKGROUND OF THE ALAPAEVSK OPHIOLITE

The northern segment of the Eastern Uralian ophiolite belt extends along the 61th meridian for about 200 km and comprises the following ophiolite massifs, listed from north to south: Alapaevsk, Ostaninsky, Rezhevskoy, Bazhenovsky, Kluchevskoy (Figure 1), as well as other smaller bodies. The Alapaevsk [8] and Kluchevskoy [9] massifs contain economic and sub-economic chromite deposits. The Bazhenovsky massif hosts one of the largest asbestos deposits in the world, containing about 6 billion tons of chrysotile asbestos, and from the Rezhevskoy massif a few economic nickel deposits in the weathering crust have been mined [10,11]. The Alapaevsk ophiolite massif is one of the largest in the Eastern belt, and is located about 150 km northeast from Ekaterinburg and approximately 15 km west of the city of Alapaevsk (Figure 1). The massif consists of a plate-like body 55 km long and 6–13 km wide, covering an area of about 500 km^2. The Alapaevsk complex is tectonically in contact with an island arc sequence composed of metavolcanics and sediments of Devonian-Carboniferous age (Figure 2). The sequence overlaps the Alapaevsk ophiolite as a tectonic nappe in the central and southern parts. The eastern and western tectonic contacts with the country rocks dip to the center of the massif. Talc-carbonate metasomatic rocks have been recognized along the contacts [10]. The Alapaevsk ophiolite consists mainly of peridotites and gabbro. Geophysical data have shown that the thickness of peridotites varies from 0.5 up to 2 km. Harzburgite and small amounts of lherzolite, dunite and chromitite (Figure 2) have been described in the Alapaevsk complex.

The degree of serpentinization of the Alapaevsk ophiolite varies widely, but in general it is very high and peridotites are often completely transformed to serpentinite. Wehrlite and pyroxenite are rare and they are mostly found in the gabbro intrusion, cutting the peridotite in the northern apex of the massif (Figure 2). Sometimes they form small veins in peridotite. On the basis of the concordant U-Pb age of zircons found in the chromitites, the age of the Alapaevsk ophiolite massif is 588 ± 16 Ma [12]. The gabbro has a similar Sm–Nd age of 579 ± 42 Ma [12]. An Upper Paleozoic collisional granite massif (Figure 2) intrudes the Proterozoic metamorphic rocks westward of the Alapaevsk massif. Close to the contact with the ultramafic rocks, the granite intrusion forms metasomatic phlogopite rocks that contain beryl and emerald [10,13]. Smaller Upper Paleozoic granite intrusions cut the peridotite along its western contact (Figure 2). Several amphibole-asbestos deposits occur in the central and eastern part of the Alapaevsk massif. They have been intensively mined in the past [10,11].

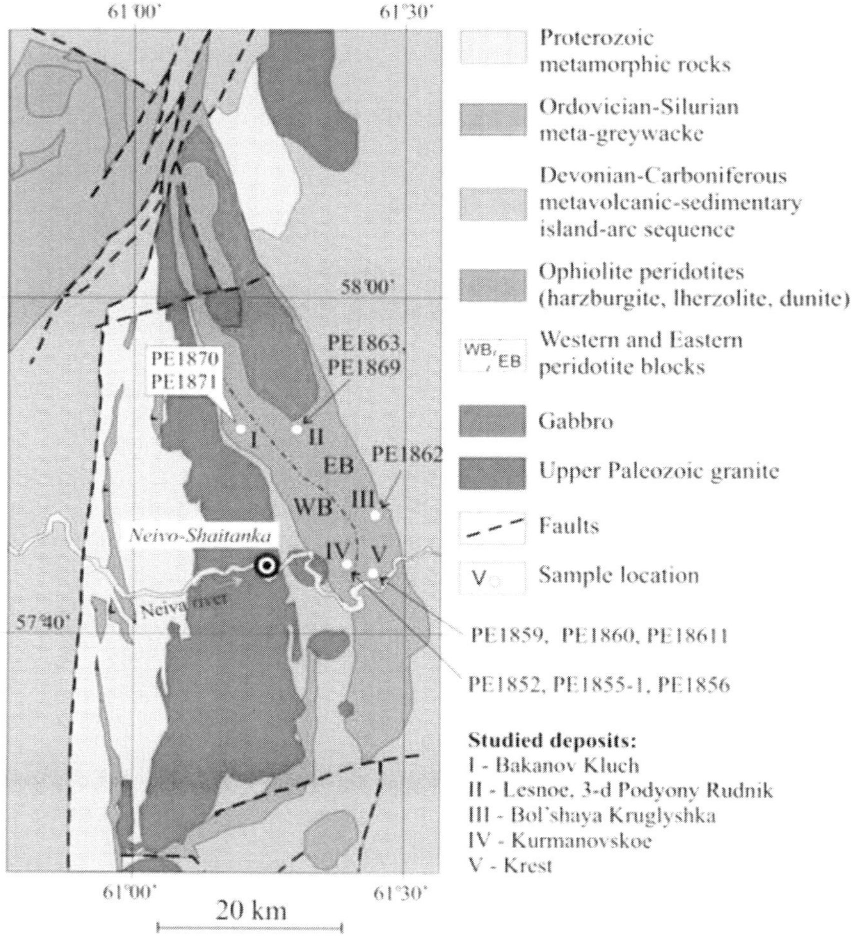

Figure 2. Simplified geological map of the Alapaevsk ophiolite massif showing the location of the chromite deposits, redrawn after [10].

3. PREVIOUS WORK AND DESCRIPTION OF THE CHROMITE DEPOSITS

From the end of the 19th century until 1929 about 80 small Al-rich chromite deposits were discovered and mined in the Alapaevsk massif. This activity yielded more than 70 thousand tons of ore. In the second half of the 20th century several new chromite deposits, comprising Cr-rich ore, were discovered and exploited [8]. Some of these deposits contain more than 1 million tons of ore. More recently, the composition of chromite from the Alapaevsk chromite deposits was described and the presence of two types of chromite ore, Cr-rich and Al-rich, demonstrated [14,15,16,17]. A recent study [17] has demonstrated that the peridotite massif can be divided into Western and Eastern blocks (Figure 2). The Western block mostly consists of highly depleted dunite–harzburgite and contains Cr-rich chromite deposits. The Eastern block is composed of more homogeneous harzburgite that hosts Al-rich chromite deposits. In this contribution the following chromite deposits have been studied: Bakanov Kluch (samples PE1870 and PE1871) and Kurmanovskoe (samples PE1852, PE1855-1 and PE1856) from the Eastern block, and Lesnoe (sample PE1863), 3-d Podyony Rudnik (sample PE1869), Bol'shaya Kruglyshka (sample PE1862), and Krest (samples PE1859, PE1860, and PE1861) collected in the Western block (Figure 2). The samples were collected in the open pits (Figure 3) and they represent schlieren, banded, massive, and disseminated chromitite (Figure 4). Only the deposits of Bakanov Kluch and Lesnoe are currently mined and the deposit of Krest is under exploration. The other deposits are exhausted. The Bakanov Kluch chromite deposit comprises six ore bodies approaching 300 m in length and extending 300 m in depth, with a thickness varying from 1 to 4 m. Chromitite shows a disseminated texture and the estimated resource is about 15 thousand tons. The mineralized area in the Kurmanovskoe chromite deposit is >1 km long and 50–250 m wide. It consists of four large ore bodies and several smaller ones. The biggest ore body is about 800 m long and 250 m deep, with a thickness of 1–10 m. Disseminated chromitites are very common, accompanied by rare massive chromitites. Resources vary between 20 and 250 thousand tons. In the Lesnoe chromite deposit, four ore bodies have been identified. They are 25–73 m long, 70 m deep, and 0.3–4 m thick. Disseminated and massive chromitite occur in the deposit and the resource is estimated to be about 75 thousand tons. The 3-d Podyony Rudnik chromite deposit consists of nine blocks that contain several ore bodies of disseminated and massive chromitites. The ore zone is about 600 m long and 40 m wide. Some ore bodies approach 100 m in length and 120 m in depth, with a thickness of 3–12 m. The Bol'shaya Kruglyshka chromite deposit consists of one ore body, about 130 m long, and 50 m deep, with a thickness of 0.3–2.5 m. The body consists of massive and disseminated chromitite, that is, totally dismembered by faulting. Resources are about 7 thousand tons. The Krest deposit forms a small body of massive and disseminated chromite with a cruciform shape. It has a length of several meters and thickness of 0.5 m, dipping sub-vertically. The source of this deposit is not known.

Figure 3. Overview of the chromite deposits of the Alapaevsk massif, Central Urals. Kurmanovskoe chromite deposit: (**A**) general view of open pit and (**B**) the location of the studied sample; Krest chromite deposit: (**C**) the exploration trench and (**D**) vertical ore body of massif chromite in dunite; (**E**) open pit of the Bol'shaya Kruglyshka chromite deposit; (**F**) open pit of the 3-d Podyony Rudnik chromite deposit; (**G**) open pit of the Lesnoe chromite deposit. Arrows indicate sample locations.

Figure 4. Different textural types of the chromite ores from the Alapaevsk massif, Central Urals. Chromite ore of the Kurmanovskoe deposit (number IV on map) (**A–C**). (**A**) Banded (sample PE1852) and massive (samples PE1855-1 and PE1856); (**B**) ores disseminated ore; (**C**) small schlieren of massive chromitite; (**D**) schlieren type of chromitite from the Bakanov Kluch chromite deposit (sample PE1870) (Ion map).

4. ANALYTICAL TECHNIQUES

Chromite, silicates, PGM, and other accessory minerals were analyzed by electron microprobe using a Superprobe Jeol JXA 8200 (JEOL, Tokyo, Japan) at the Eugen F. Stumpfl Laboratory at the University of Leoben, Austria, using both ED and WD systems. Back scattered electron (BSE) images were obtained using the same instrument. During the analyses of chromite and silicates, the electron microprobe was operated in the WDS mode, with an accelerating voltage of 15 kV of and beam current of 10 nA. The analysis of Na, Mg, K, Al, Si, Ca, Ti, V, Cr, Zn, Mn, Fe, and Ni were obtained using the Kα lines, and were calibrated on natural chromite, rhodonite, ilmenite, albite, pentlandite, wollastonite, kaersutite, sphalerite, and metallic vanadium. The following diffracting crystals were used: TAP for Na, Mg, and Al; PETJ for K, Si, and Ca; and LIFH for Ti, V, Cr, Zn, Mn, Fe, and Ni. The peak and backgrounds counting times were

20 and 10 s, respectively, for the major elements. They were increased to 40 and 20 for trace elements, such Ca, Ni, and Mn in olivine. The detection limits were automatically calculated by the microprobe software and they are listed in the following as ppm: Na (150), Mg (100), K (80), Al (150), Si (250), Ca (50), Ti (250), V (150), Cr (200), Zn (200), Mn (100), and Ni (150). The amount of Fe^{3+} in chromite was calculated assuming the ideal spinel stoichiometry. About 500 analyses of chromite were performed on 11 samples and a selection of them is listed in Table A1 (see Appendix A). Representative analyses of silicates are presented in Table A2 and Table A3 (see Appendix A). PGM and other accessory minerals were located by scanning polished sections under a reflected light microscope at 250–800 magnification. The accessory minerals with a size less than 5 μm were only qualitatively analyzed by EDS. The larger accessory minerals were quantitatively analyzed in the WDS mode, at 20 kV accelerating voltage and 10 nA beam current, and beam diameter of about 1 micron. The peak and background counting times were 15 and 5 s, respectively. The Kα lines were used for S, As, Fe, Ni, and P; Lα for Ir, Ru, Rh, Pd, and Pt, and Mα for Os. The reference materials were pure metals for the six PGE (Ru, Rh, Pd, Os, Ir, and Pt), synthetic NiS, natural pyrite, and niccolite for Ni, Fe, S, and As; and apatite for P. The following diffracting crystals were selected: PETJ for S and P; PETH for Ru, Os, and Rh; LIFH for Fe, Ni, Ir, and Pt; and TAP for As. Automatic corrections were performed for interferences involving Ru–Rh and Rh–Pd. Analyses of accessory minerals are listed in Table A4 and Table A5 (see Appendix A). Platinum was never detected and it was removed from Table A4 (see Appendix A).

5. THE ALAPAEVSK CHROMITITES

5.1. Texture and Mineral Chemistry of the Chromite

Under the microscope, most of the samples consist of massive chromitite, containing a modal proportion less than 10% of interstitial silicate (Figure 5A–D).

In the samples PE1852 from Kurmanovskoe deposit, PE1859 from Krest, PE1855-1 Kurmanovskoe, and PE1870 from Bakanov Kluch the interstitial silicates are more abundant, reaching about 30% by volume (Figure 5E,F). The massive chromitite shows, in some cases, a brecciated texture (Figure 5A,B) and it is cut by cracks filled with interstitial silicates (Figure 5C,D) and pull-apart fractures (Figure 5E,F). Chromite is generally fresh, showing the typical ferrian–chromite alteration along grain boundaries and cracks (Figure 5C–E). Silicates in the matrix are mostly altered to chlorite, talc, and serpentine. Interstitial olivine has been found only in the sample PE1855-1 collected in the Kurmanovskoe deposit. Selected analyses obtained on unaltered cores of chromite grains are listed in Table A1 and plotted in the binary diagrams of Figure 6A–C.

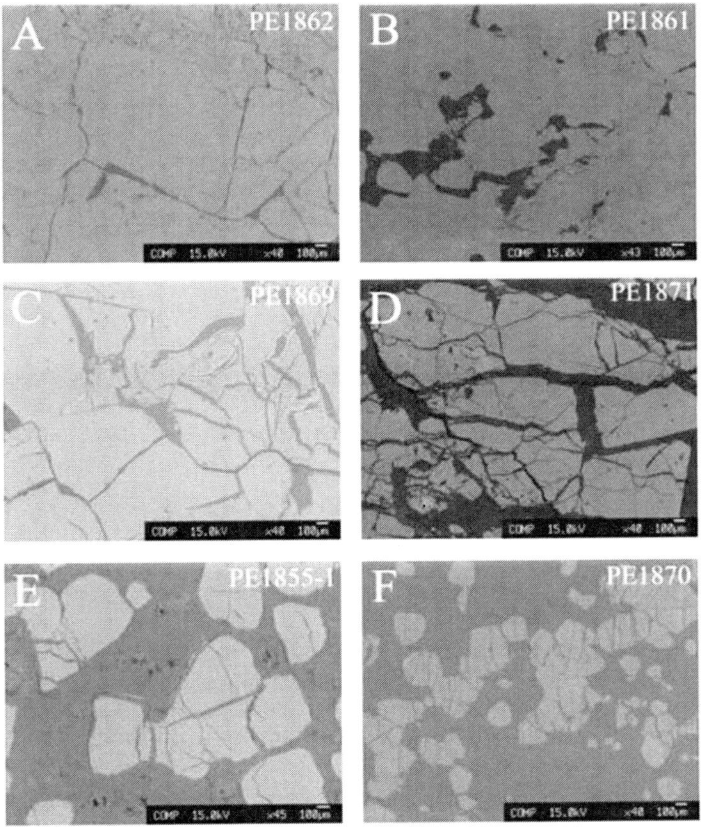

Figure 5. Back-scattered electron images of polished sections from the Alapaevsk chromitites (light grey = chromite, dark grey = altered silicates). (**A**) Massive and partially brecciated chromite from Bol'shaya Kruglyshka; (**B**) massive chromite from Krest; (**C**) massive chromite from 3-d Podyony Rudnik showing ferrian chromite alteration (lighter gray) developed along cracks and fissures; (**D**) massive chromite with ferrian chromite alteration (lighter gray) from Bakanov Kluch; (**E**) massive chromite with about 40% of altered silicates from Kurmanovskoe; (**F**) massive chromite with about 40% of altered silicates showing pull-apart texture from Bakanov Kluch.

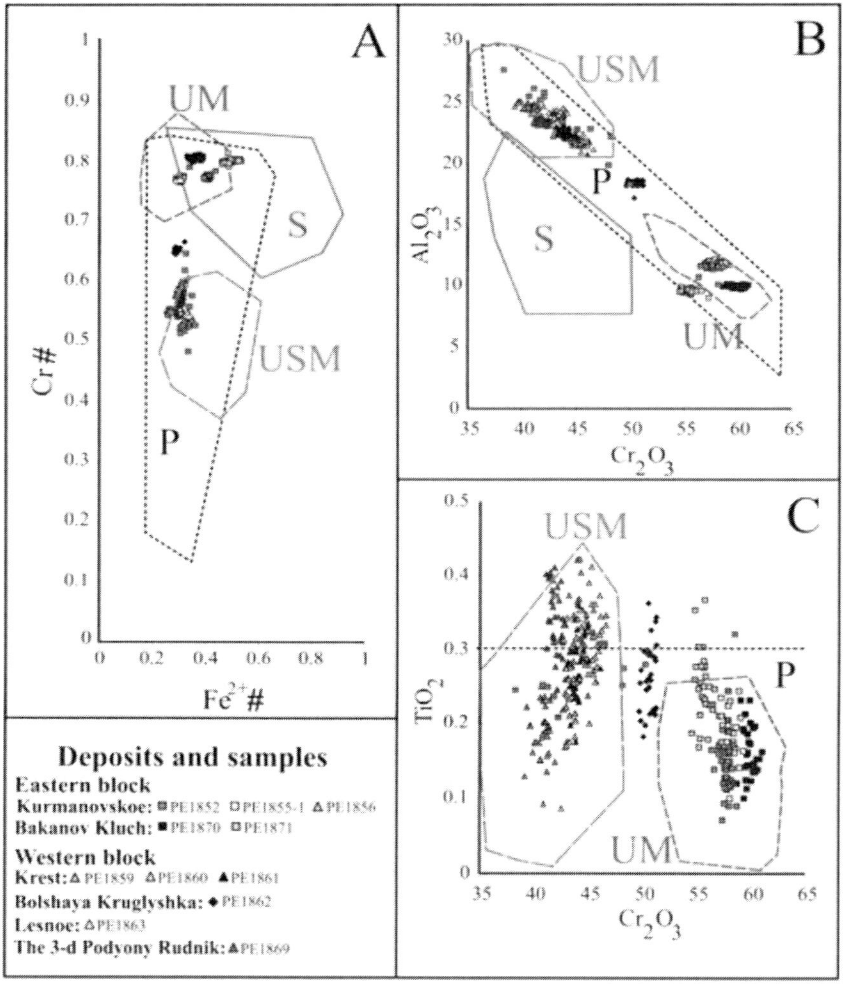

Figure 6. Composition of unaltered chromite from the Alapaevsk chromitite. (**A**) Variation of the chromium number, Cr# = Cr/(Cr + Al), and bivalent iron number, $Fe^{2+}\# = Fe^{2+}/(Fe^{2+} + Mg)$; (**B**) negative correlation of Al_2O_3 versus Cr_2O_3; (**C**) variation of Cr_2O_3 and TiO_2. P = field of podiform chromitites; S = field of stratiform chromitites; UM = field of the Uralian chromitites hosted in the ophiolitic mantle; USM = field of the Uralian chromitites hosted in the supra-Moho cumulates [6,18,19,20]. Red square = literature data from [21].

The Alapaevsk chromitites plot in the podiform field on the Cr# versus Fe^{2+}# binary diagram (Figure 6A) and the compositions form an almost continuous trend from Cr-rich (Cr# up to 0.83), to intermediate and Al-rich (minimum Cr# = 0.48). The variation range in Fe^{2+}# is 0.24–0.38 in the chromitites from the Western block and 0.39–0.56 in those from the Eastern block (Figure 6A). A good negative correlation is observed between Cr_2O_3 and Al_2O_3 (Figure 6B), consistent with the Al–Cr substitution predominant in podiform chromitites compared to the stratiform type [18,19,20]. According to the same diagram, the chromitites from Bakanov Kluch and Kurmanovskoe, located in the Western block, are Cr-rich, having a Cr_2O_3 content higher than 50 wt % and Al_2O_3 below 15 wt %. Al-rich chromitites with Cr_2O_3 content lower than 50 wt % and Al_2O_3 higher than 15 wt % have been found in the 3-d Podyony Rudnik, Krest, Lesnoe deposits (Eastern block) and Kurmanovskoe deposit (Western block). The chromitite from Bol'shaya Kruglyshka of the Western block displays an intermediate composition with Cr_2O_3 content about 50 wt % and Al_2O_3 higher than 15 wt %. These data are in agreement with those previously reported by [21] from Alapaevsk chromitites.

The Kurmanovskoe deposit (Eastern block) is the only one in which both Cr-rich and Al-rich chromitites have been found (Figure 6B). The diagrams of Figure 6 show that most of the Cr-rich chromites analyzed in the Alapaevsk chromitites plot in the field of those hosted in the mantle sequence of selected Uralian ophiolites [6 and reference therein]. The composition of the Al-rich chromites displays similarity with the chromitites found in the cumulate sequence [6 and references therein]. TiO_2 is generally lower than 0.3 wt % in most of the samples, as is typical for podiform chromitites (Figure 6C). However, several samples contain up to 0.42 wt % of TiO_2 (Figure 6C).

5.2. Accessory Minerals and Silicate Inclusions

Several PGM have been found in the samples collected in the Kurmanovskoe and Krest deposits. On the basis of their chemical composition, laurite, cuproiridsite, and alloys of Os–Ru–Ir and Ir–Os–Ru have been identified. They occur as small grains (less than 10 μm in size) in the following textural positions: (1) included in fresh chromite crystals (Figure 7A–C); (2) in the contact between chromite and the silicate matrix (Figure 7D); or (3) associated with altered minerals, such as ferrian–chromite (Figure 8A), awaruite, Fe-oxides, and chlorite (Figure 8C,D).

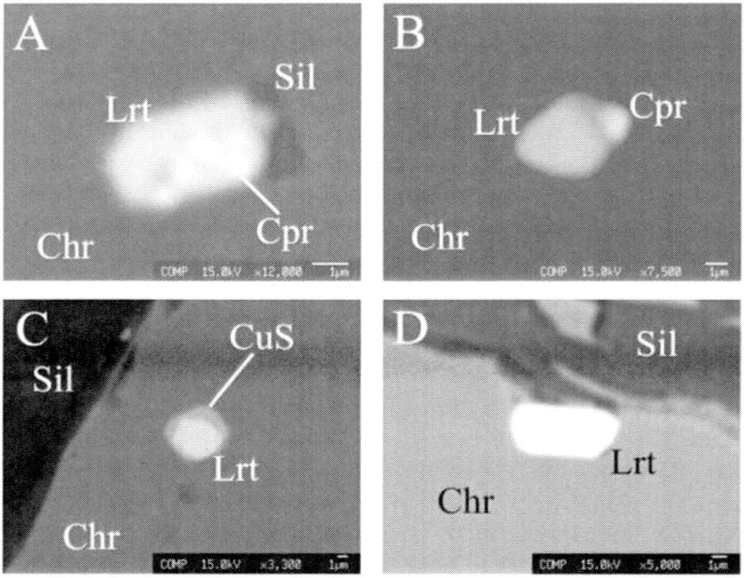

Figure 7. Back-scattered electron (BSE) images of magmatic PGM associated with the chromitites of Kurmanovskoe, from Alapaevsk ophiolite. (**A,B**) Crystals composed of laurite and cuproiridsite included in fresh chromite; (**C**) laurite associated with a Cu-sulfide, included in fresh chromite; (**D**) single phase laurite in crack. Abbreviations: Chr = chromite; Lrt = laurite; Cpr = cuproiridsite, Sil = silicates; CuS = Cu-sulfide.

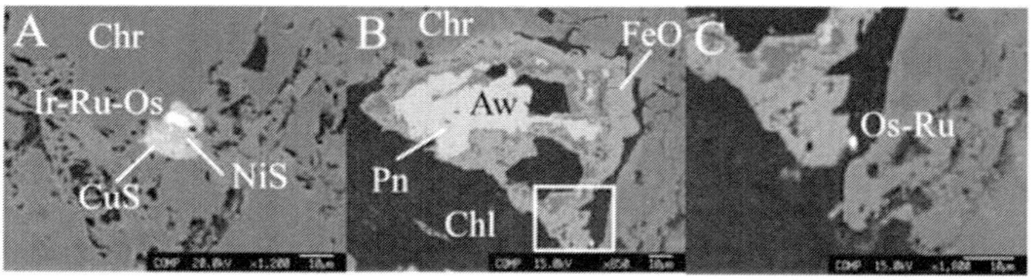

Figure 8. Back-scattered electron (BSE) images of secondary PGM associated with the chromitites from Alapaevsky. (**A**) Irregular alloy of Ir, Ru, Os, Ni-sulfide and Cu-sulfide in contact with ferrian chromite, from the chromitite of Kurmanovskoe; (**B**) complex grain composed pentlandite, awaruite, Fe-oxide or hydroxide in contact with ferrian-chromite and chlorite (Krest); (**C**) enlargement of (**B**) showing an alloy composed of Os and Ru. Abbreviations: Ir–Ru–Os = iridium alloy; Aw = awaruite; Pn = pentlandite; Fe–O = Fe-oxide or hydroxide; Chl = chlorite; Os–Ru = osmium alloy; other abbreviations = see Figure 7.

Laurite is the most abundant PGM and it forms polygonal grains, most of them included in fresh chromite (Figure 7A–C). Only one laurite grain was found in the contact between chromite and a crack filled with chlorite (Figure 7D). Laurite occurs as single phase (Figure 7D) or associate with cuproiridsite (Figure 7A,B) and with a Cu-sulfide (Figure 7C). The composition of laurite has been plotted in the Ru–Os–Ir triangle (Figure 9) and compared with those reported from relevant Uralian podiform chromitites. The laurite from Alapaevsky displays a limited compositional range of Ru–Os substitution with a minimal amount of Ir (Figure 9) compared with laurite in the chromitites from the Kluchevskoy [9], Kempirsai [22], Ray–Iz [23] and Voikar–Sininsky [24] ophiolites. Cuproiridsite was too small to be quantitatively analyzed. It is exclusively associated with laurite as an inclusion in fresh chromite crystals (Figure 7A,B). Laurite and cuproiridsite have been found in the Cr-rich chromitites of Kurmanovskoe. One irregular grain composed of Ni-sulfide, Cu-sulfide and an Ir–Ru–Os alloys was found in the ferrian-chromite of the Kurmanovskoe chromitite (Figure 8A). Due to its small size, it was qualitatively analyzed by EDS. An Os–Ru alloy was qualitatively analyzed in the chromitite of Krest. It occurs as a small bleb, about 2 µm in size, in a complex grains composed of pentlandite, awaruite, Fe-oxides, and chlorite (Figure 8D).

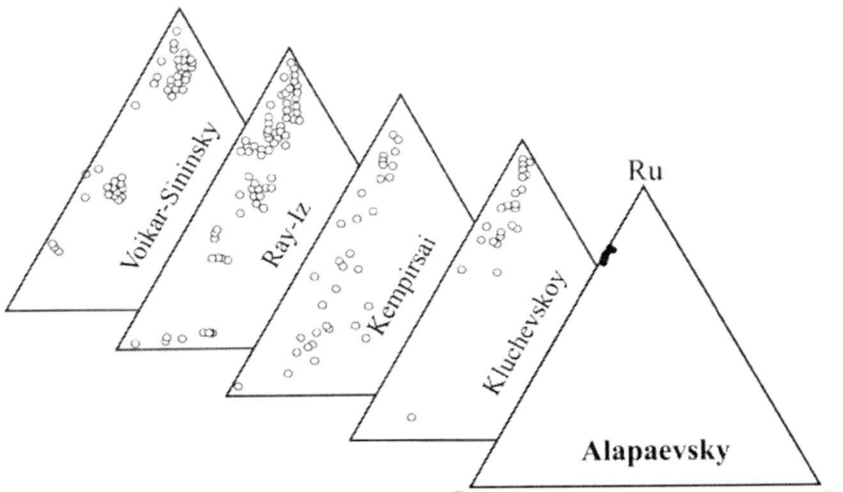

Figure 9. Plots (at.%) of the compositions of laurite from chromitites of Alapaevsk (present study) compared with those analyzed in selected ophiolitic chromitites of the Urals (data from [9,20,21,22]).

A small grain (about 20 μm) yellow in color under a reflected-light microscope, was found in the interstitial serpentine of the Bakanov Kluch chromitite (Figure 10). According to the quantitative electron microprobe analyses (Table A5), it contains Ni as the most abundant element (88.05–89.89 wt %), P (9.41–10.19 wt %) and minor amounts of Fe (1.13–1.3 wt %). The ideal calculated stoichiometry, based on the average of 10 analyses, is $(Ni,Fe)_5P$ (Figure 11, Table A5) (see Appendix A). Its composition has been plotted in the binary diagram presented in Figure 11 and compared with the only five Ni-phosphides accepted by the International Mineralogical Association (IMA) as mineral species, namely nickel phosphide [25,26]), melliniite [27]), halamishite, negevite, and transjordanite [28]. The diagram shows that the analyzed $(Ni,Fe)_5P$ in the Alapaevsk chromitite contains the highest value of Ni and it shows the highest Ni/P ratio.

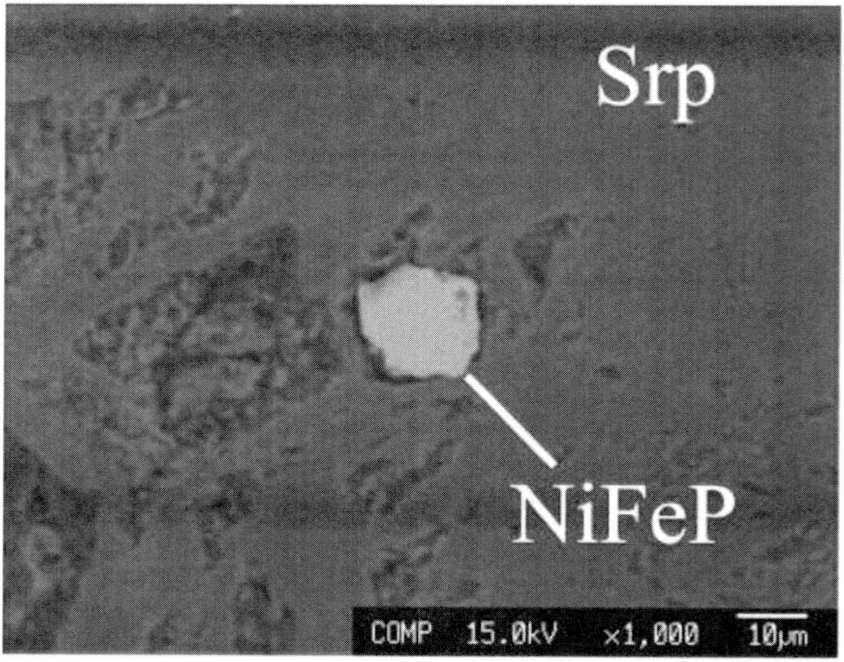

Figure 10. Back-scattered electron (BSE) image of a Ni–Fe phosphide found in the interstitial serpentine of the Bakanov Kluch chromitite.

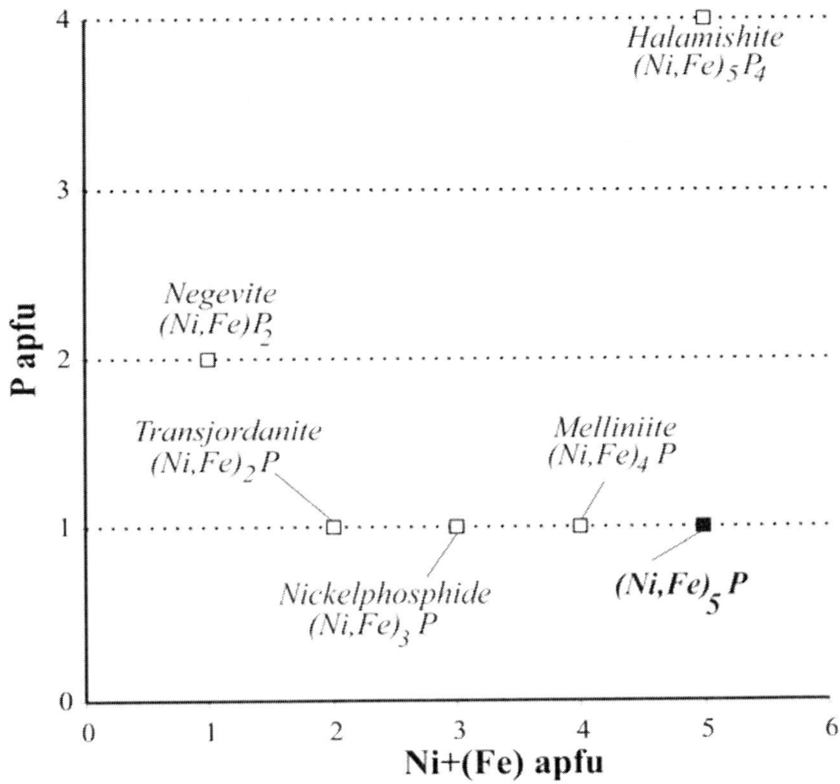

Figure 11. Binary diagram of P versus Ni + (Fe), apfu (atoms per formula unit), showing the composition of the Ni-phosphide found in the Bakanov Kluch chromitite (black square), compared with those of the minerals with Ni > Fe described in the system Ni–Fe–P (open squares). Literature data from [25,26,27,28].

Although most of the silicate in the Alapaevsk chromitite has been totally altered to serpentine, talc, and chlorite, a few grains of fresh olivine have been found in the Cr-rich chromitites from Kurmanovskoe (Western block). Olivine occurs as irregular grains located among the chromite crystals, or forming polygonal crystals (up to about 100 μm) included in the fresh chromite (Figure 12A,B).

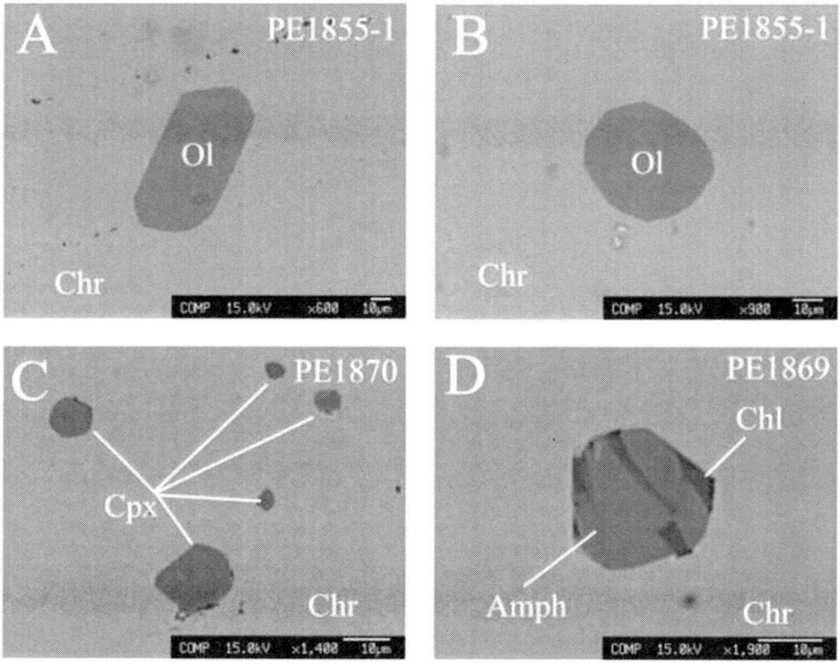

Figure 12. Back-scattered electron (BSE) images of silicates included in the Alapaevsk chromitites. Polygonal olivine in the chromitite from Kurmanovskoe (**A**,**B**); grains of clinopyroxene in the chromitite from Bakanov Kluch (**C**); inclusion composed of amphibole and chlorite found in the chromitite form 3-d Podyony Rudnik (**D**). Abbreviations: Ol = olivine; Cpx = clinopyroxene; Amph = amphibole; other abbreviations = see Figure 7 and Figure 8.

Amphibole, clinopyroxene, Na-rich phlogopite, and chlorite have been also found as small inclusions in fresh chromite crystals (Figure 12C,D). Olivine is characterized by a narrow compositional range with 95.75% to 96.16% forsterite in the interstitial olivine, and 96.83% to 97.06% forsterite in the olivine included in chromite (Table A2). The Ni, Mn and Ca contents are similar in interstitial olivine and olivine included in chromite (Figure 13). The NiO wt % content varies between 0.3 and 0.45 (Table A2). The highest value of MnO (0.15 wt %) was encountered in one interstitial olivine (Figure 13B), whereas all the other olivine contains 0.01 to 0.09 wt % MnO (Table A2). In most of the olivine, Ca is below 900 ppm and its minimum value of 400 ppm was found in the interstitial olivine (Table A2). The Cr_2O_3 content is higher in the included olivine (0.47–0.75 wt %) than in the interstitial olivine (0.06 –0.19 wt %) (Table A2).

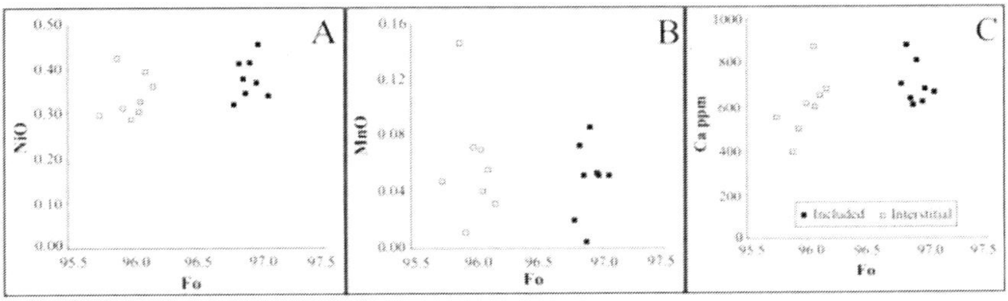

Figure 13. Variation of NiO (**A**), MnO (**B**) and CaO (**C**) as a function of forsterite mol % in the olivine in the Cr-rich chromitites from Kurmanovskoe.

Amphibole included in fresh chromite has been analyzed in the Al-rich chromitites from Krest and 3-d Podyony Rudnik and in Cr-rich chromitite from Bakanov Kluch (Figure 14A). With the exception of one amphibole grain found in the Bakanov Kluch deposit that plots in the field of tschermakite, all the analyzed amphiboles can be classified as magnesiohornblende (Figure 14A). The analyzed amphiboles contain Al_2O_3 from 9.19 wt % to 10.01 wt % and CaO from 11.46 wt % to 12.34 wt %. They show appreciable amounts of Cr_2O_3 (up to 3.29 wt %) and Na_2O (up to 3.73 wt %). The content of TiO_2 is variable, ranging from 0.41 wt % to 2.04 wt %. The amounts of MnO, NiO, and K_2O are always below 1 wt % (Table A3).

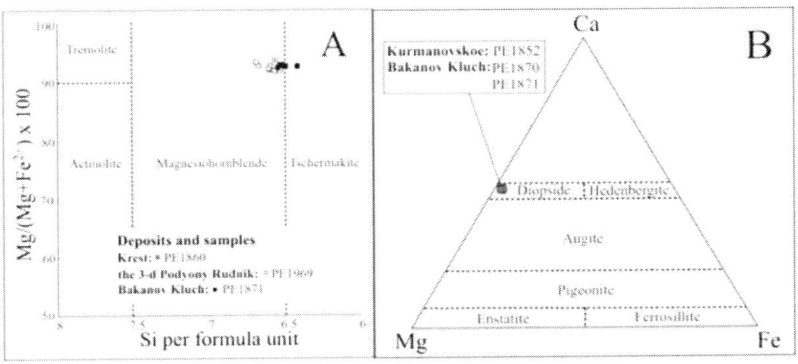

Figure 14. Compositional variations of silicates included in chromite from the Alapaevskychromitites. Binary diagram for the classification of the amphibole (**A**); ternary diagram for the classification of the pyroxenes (**B**). Simplified after [29].

Inclusions of clinopyroxene occur in the chromite of the chromitites from Kurmanovskoe and Bakanov Kluch (Figure 14B). These have a similar composition within the diopside field (Figure 14B). They contain Cr_2O_3 (up to 1.91 wt %) and very low amounts of MnO, NiO, TiO_2, Na_2O, and K_2O (Table A3). Small inclusions of phlogopite have been identified by qualitative analysis.

6. DISCUSSION AND CONCLUSIONS

6.1. Chromite Composition and Silicate Inclusion: Their Application to Define the Tectonic Setting of the Host Chromitite and Its Genesis

Chromite is widely accepted to be an important petrogenetic indicator and it can be used to define the geodynamic setting in which chromitites and their host rocks formed [29,30,31,32,33,34,35,36,37,38,39]. According to the morphology of the ore body and the chromite composition, chromitites belong to two main groups: (1) stratiform chromitites that form layers of massive chromite associated with mafic–ultramafic cumulates in the lower part of continental layered intrusions and supra-Moho cumulate sequences of ophiolites and (2) podiform chromitites that consist of lenses, pods, and schlieren, mainly occurring in the mantle section of ophiolites, orogenic lherzolites, and in the dunitic portion of Ural–Alaskan complexes [6,29,30,31,32,33,34,35,36]. In the ophiolites of the Urals, both the mantle-hosted podiform chromitites as well as the supra-Moho stratiform chromitites have been described ([6] and references therein). According to [6] (and references therein), chromitites hosted in the Uralian ophiolites are characterized by distinctive chromite compositions. In particular, most of the podiform chromitites hosted by ophiolitic mantle rocks and a few examples of supra-Moho layered chromitites are Cr-rich and Ti-poor (TiO_2 less than 0.3 wt %) [6]. In contrast, the majority of the supra-Moho chromitites are Al-rich and characterized by a variable content of TiO_2 (0.15 wt % to 0.8 wt %). Confirming the previous work by [8,14,15,16,17,21], the composition of the Alapevsky chromite indicates the coexistence of the Al-rich and Cr-rich chromitites, a common feature of many ophiolitic belts worldwide [22,36,37,38,39,40,41,42,43,44,45,46,47,48,49,50].

The bi-modal distribution of Al-rich and Cr-rich chromitite in the Urals ophiolites has been interpreted using two models. The first implies a change in the geodynamic setting, with the Al-rich chromitite generated in the mid-ocean ridge, whereas the Cr-rich chromitite forms in a supra-subduction zone (SSZ) setting. The second model suggested that the bimodal distribution was due to a vertical zoning in the ophiolite sequence, with the Cr-rich chromitite located in the deep mantle section and the Al-rich chromitite occurring higher in the succession, close to or above the Moho transition [6,22].

To better understand the origin of Al-rich chromitites and Cr-rich chromitites in the Alapaevsk ophiolite, the TiO_2content of chromite has been plotted versus Al_2O_3 (Figure 15). In Figure 15A, one group of analyses falls in the Al_2O_3 poor field of spinels from the SSZ peridotite and a second group of samples clusters in the overlapping field between SSZ and MORB chromites. A few samples are slightly enriched in TiO_2. The diagram of Figure 15B shows that the low-Ti, Cr-rich chromitites from Bakanov Kluch and Kurmanovskoe fall in the field of Island Arc Basalts (IAB). The Al-rich chromitites display compositional similarity with those reported from chromitites in supra-Moho cumulate sequences of the Urals (data from [6]). According to the TiO_2–Al_2O_3 relationships, we can deduce that the Alapaevsk chromitites formed in an SSZ, in agreement with [15,17], implying formation of Cr-rich chromitites in the deep mantle, and the Al-rich close to the Moho transition. The high-Cr chromitites are assumed to have formed by reaction of boninitic melts with residual harzburgite. The nature of the melt parental to Al-rich chromitites is uncertain. They may have formed from a single batch of boninitic melt evolving during its ascent to the shallow upper mantle. Alternatively, the Al-rich chromitites might have resulted from mixing between boninite and MORB-type melts. The composition of olivine can be used, with caution, to distinguish olivine formed deep in the mantle from olivine crystallized in the crust by mafic and ultramafic magmas [51,52]. Olivine is a common accessory mineral in most chromitites, including those of Alapaevsk, although it rarely survives the serpentinization. As a consequence, few grains of fresh olivine generally occur in the silicate matrix of the chromitites. Some occur as inclusions in unaltered chromites, reflecting their magmatic origin. In the Alapaevsk chromitite both types of olivine, i.e., interstitial and included, have been recognized in the Kurmanovskoe deposits. The composition of olivine in chromitite is strongly influenced by the element partitioning with chromitite. Several analyses reported from the Urals have shown that olivine displays a wide range of forsterite molar % with Ni, Mn, and Ca contents according to the type of host chromitite [6]. In particular, most of the olivine associated with ophiolitic mantle chromitites displays high (94–98 mol %) forsterite and NiO between 0.35 and 1.15 wt %. MnO wt % and Ca (ppm) are very low. All of the analyzed olivine grains, interstitial and included, from Kurmanovskoe chromite deposit are characterized by values of forsterite, NiO, MnO, and Ca (Figure 13) very

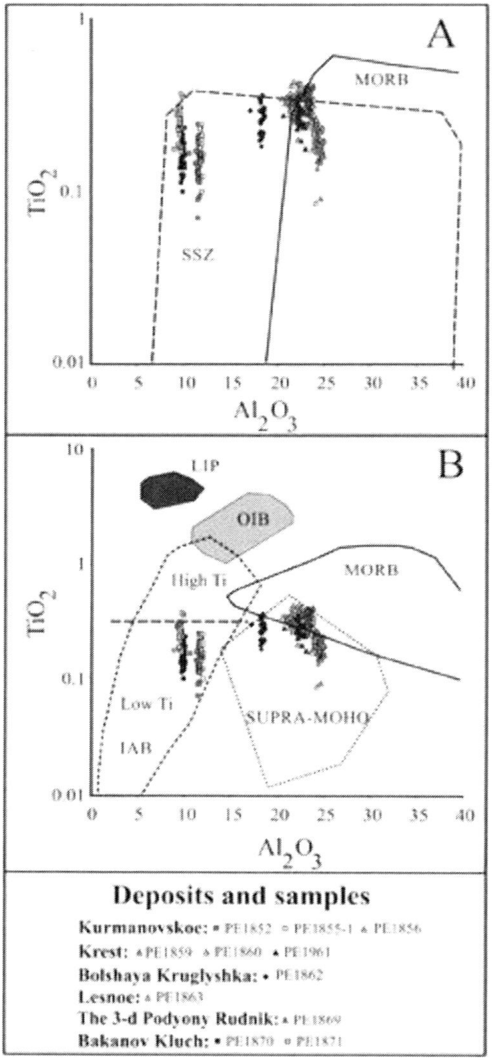

Figure 15. TiO_2–Al_2O_3 relationships in chromitites of Alapaevsk. (**A**) Composition of the chromitites, compared to the field of spinels from the supra-subduction zone (SSZ) and mid-ocean ridge (MOR) mantle peridotites (compositional fields from [34]); (**B**) the fields for IAB (island arc basalt), OIB (oceanic island basalt), MORB (mid-ocean ridge basalt), LIP (large igneous province) basalt, and supra-Moho (SUPRA-MOHO) chromitite from the Urals are from [6,36]. The horizontal dashed line at TiO_2 = 0.30 wt % separates the high-Ti IAB (high-K calc-alkaline suite) from the low-Ti IAB (boninitic, tholeiitic).

similar to those reported from the ophiolitic mantle-hosted chromitites from the Urals [6]. This observation suggests that the olivine from Kurmanovskoe and its host chromitites were generated in the mantle section of the Alapaevsk ophiolite as previously reported by [15,17].

One of the most accepted mechanisms of chromite crystallization to form large volumes of chromitites in ophiolitic mantle rocks invokes chromitite formation by melt–rock interaction. According to this model, chromitites represent the filling of magmatic conduits [53] and formed by reaction between ascending mantle melts and country-rock residual peridotite [54,55,56,57,58]. The same mechanism can be applied to the formation of the Alapaevsk chromitites. Several authors, using both mineralogical observations and experimental work, have demonstrated that melts that generate chromitites are generally enriched in water and fluids [54,55,56,57,58,59,60,61,62]. The suite of solid inclusions reported from the Alapaevesk chromitites, such as amphibole and phlogopite enriched in NaO, indicates the presence of a melt with a relatively high water activity and carrier of some incompatible elements. Therefore, we can conclude that the Alapaevesk chromitites formed because of the interaction between melts enriched in fluids, water, and incompatible elements and mantle peridotite, a common feature during the formation of ophiolitic chromitites [54,55,56,57,58,59,60,61,62].

6.2. Origin and Evolution of the PGM from High to Low Temperature

The PGM found in the Alapaevsk chromitites consist of IPGE-rich phases, the common mineral assemblage of the chromitites hosted in the mantle sequence of ophiolite complexes. On the basis of their mode of occurrence, (included in fresh chromite crystal or associated with altered minerals), the PGM described in the Alapaevsk chromitites can be classified as: (1) primary PGM that formed in the high-temperature magmatic stages during the precipitation of the host chromite and (2) secondary PGM that originated during low-temperature processes. The primary PGM in the Alapaevsk chromitites consists of laurite and cuproiridsite (Figure 7A–C), while alloys in the Ir–Os–Ru system (Figure 8A–C) are considered to be secondary. The magmatic PGM have been found only in the Cr-rich chromitites from Kurmanovskoe deposits. The secondary PGM occur in both the Cr-rich and Al-rich chromitites of Kurmanovskoe and Krest deposits. The origin of the primary PGM inclusions in the Alapaevsk chromitites can be explained by a crystallization process controlled by relative stability of IPGE alloys and sulfides as function of sulfur fugacity and temperature (Figure 16). The sulfur fugacity is expected to increase with decreasing temperature in magmatic systems. As a consequence, at high temperatures and low sulfur fugacity alloys of IPGE are the first PGM to precipitate, followed by laurite and finally by erlichmanite and Ir-bearing sulfides (Figure 16). The primary laurite of Alapaevsk shows a narrow variation of Os and Ru (Figure 9), well below the Os–OsS_2 buffer, as typical of laurite precipitated at high temperature and relatively low sulfur fugacity. Nevertheless, the absence of magmatic Os–Ir–(Ru) alloys, the presence of cuproiridsite (Figure 7A,B), and the laurite–Cu sulfide association (Figure 7C) suggest that the sulfur fugacity was relatively high during the precipitation of the PGM in the magmatic stage of the chromitites from Alapaevsk, although lower than those reported in the Uralian chromitites from Kluchevskoy [9], Kempirsai [22], Ray–Iz [23] and Voikar–Sininsky [24], in which abundant erlichmanite has been analyzed, thus confirming a precipitation of the PGM under higher sulfur fugacity (Figure 9 and Figure 16). The secondary PGM assemblage consists of very tiny grains (less than 4 μm) of Ir–Ru–Os and Os–Ru alloys (Figure 8). The Ir–Ru–Os alloy occurs in an irregular grain that consists of a Ni-sulfide and a Cu-sulfide (Figure 8A). The Os–Ru alloy occurs in a complex of grains composed of pentlandite, awaruite, Fe-oxides, and chlorite (Figure 8C,D). Very likely, the secondary IPGE alloys formed at low temperature, during the serpentinization process that affected the Alapaevsk ophiolite.

6.3. Natural Minerals in the Ni–Fe–P System

Natural phases in the Ni–Fe–P system are rare, consisting of only five minerals with Fe > Ni, namely allabogdanite $(Fe,Ni)_2P$, barringerite $(Fe,Ni)_2P$, murashkoite FeP, schreibersite $(Fe,Ni)_3P$, and zuktamrurite FeP_2, [28,63,64,65]; and five minerals with Ni > Fe, such as halamishite Ni_5P_4, melliniite $(Ni,Fe)_4P$, negevite NiP_2, nickelphosphide $(Ni,Fe)_3P$, and transjordanite Ni_2P [25,26,27,28]. All these minerals have been found in meteorites, with the exception of schreibersite, which was also reported in reduced differentiated lenses in basalts and in hydrothermal replacement in petrified wood (http://mindat.org). Pratesi et al. [27], based on the investigation of the crystal structure of selected minerals in the Ni–Fe–P system, argued that some of these compounds, characterized by high symmetry and atomic coordination, may crystallize in a high-pressure environment, such as the core of the Earth. The $(Ni,Fe)_5P$ from the Alapaevsk chromitite represents the first terrestrial occurrence of a Ni phosphide. Its chemical composition reveals that it contains the most Ni among the minerals known in the Ni–Fe–P system (Figure 11) and it represents a new mineral species.

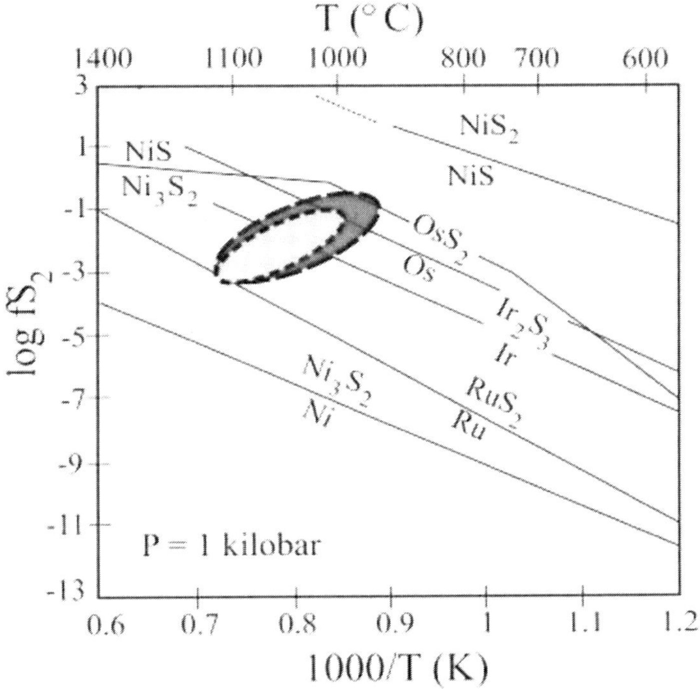

Figure 16. Metal-sulfide equilibrium curves for Ru, Os, Ir, and Ni as function of sulfur fugacity, expressed as log fS_2 and temperature. Yellow and blue fields show the conditions prevailing in mantle-hosted chromitites from ophiolites of the Urals. Yellow field = present work, blue field = [9,22,23,24].

ACKNOWLEDGMENTS

The University Centrum for Applied Geosciences (UCAG) is thanked for the access to the Eugen Friedrich Stumpfl electron microprobe laboratory. Field work and preliminary investigations were supported by Russian Foundation of Basic Research (RFBR) grant No. 16-05-00508-a. We are very grateful to John Bowles and two anonymous referees for constructive comments that improved the manuscript. Many thanks are due to the editorial staff of *Minerals* and to the guest editor Maria Economou-Eliopoulos.

AUTHOR CONTRIBUTIONS

Federica Zaccarini wrote the paper and analyzed the minerals by electron microprobe. Evgeny Pushkarev collected the samples and provided the geological and field information as well as the access to the Russian bibliography. Giorgio Garuti contributed to the elaboration and interpretation of the data and to the final revision of the manuscript. Igor Kazakov collected the samples and provided the information about the studied deposits.

CONFLICTS OF INTEREST

The authors have no conflict of interest to declare.

REFERENCES

1. Fershtater, G.B.; Montero, P.; Borodina, N.S.; Pushkarev, E.V.; Smirnov, V.N.; Bea, F. Uralian magmatism: An overview. Tectonophysics 1997, 276, 87–102.
2. Savelieva, G.N.; Sharaskin, A.Y.; Saveliev, A.A.; Spadea, P.; Pertsev, A.N.; Babarina, I.I. Ophiolites and zoned mafic-ultramafic massifs of the Urals: A comparative analysis and some tectonic implications. Mountain building in the Uralides: Pangea to the present. AGU Geophys. Monogr. 2002, 132, 135–153.
3. Savelieva, G.N. Gabbro-Ultrabasite Assemblage of Ural Ophiolites and their Analogues in Recent Oceanic Crust; Nauka: Moscow, Russia, 1987; pp. 1–45. (In Russian)
4. Savelieva, G.N.; Saveliev, A.A. Relationship between peridotites and gabbroic sequence in the ophiolites of the Urals and the Lesser Caucasus. Ofioliti 1992, 17, 117–137.
5. Garuti, G.; Fershtater, G.; Bea, F.; Montero, P.; Pushkarev, E.V.; Zaccarini, F. Platinum-group element distribution in mafic-ultramafic complexes of Central and Southern Urals: Preliminary results. Tectonophysics 1997, 276, 181–194.
6. Garuti, G.; Pushkarev, E.V.; Thalhammer, O.A.R.; Zaccarini, F. Chromitites of the Urals: Overview of chromite mineral chemistry and geo-tectonic setting (Part 1). Ofioliti 2012, 37, 27–53.
7. Koroteev, V.A.; de Boorder, H.; Necheukhin, V.M.; Sazonov, V.N. Geodynamic setting of the mineral deposits of the Urals. Tectonophysics 1997, 276, 291–300.
8. Tatarinov, P.M.; Krasnovsky, G.M. Alapaevsky Intrusion of Ultramafic Rocks in the Urals and Related Chromite Deposits; Gosgeoltehisdat: Moscow-Leningrad, Russia, 1940; pp. 1–43. (In Russian)
9. Zaccarini, F.; Pushkarev, E.V.; Garuti, G. Platinum-group element mineralogy and geochemistry of chromitite of the Kluchevskoy ophiolite complex, central Urals (Russia). Ore Geol. Rev. 2008, 33, 20–30.
10. The State Geological Map of Russian Federation. Scale 1:1000,000, 3rd ed.; Urals Series; O-41-Ekaterinburg; Explanatory Letter; VSEGEI; UGSE: St-Petersburg, Russia, 2011; p. 492.

11. Zoloev, K.K.; Chemyakin, V.I.; Shmaina, M.Y.; Medvedeva, T.N.; Mardirosyan, A.N.; Bashta, K.N. Bazhenovskoe Deposit of Chrysotile-Asbestos; Zoloev, K.K., Popov, B.A., Eds.; Nedra: Moscow, Russia, 1985; p. 271. (In Russian)
12. Petrov, G.A.; Ronkin, Y.L.; Maslov, A.V.; Lepikhina, O.P. Vendian and Silurian Ophiolite Formation Stages on the Eastern Slope of the Middle Urals. Dokl. Earth Sci. 2010, 432, 570–576.
13. Fershtater, G.B.; Borodina, N.S.; Rapoport, M.S.; Osipova, T.A. Orogenic Granitoid Magmatism of the Urals; IGG Ural Branch of RAS: Miass, Russia, 1994; p. 250. (In Russian)
14. Shilova, T.A. Chromspinel of the Alapaevsky Massif. In Mineralogical Digest No. 13: Mineralogy and Geochemistry of Ultramafites of the Urals; Uralian Scientific Center: Sverdlovsk, Russia, 1977; pp. 33–45. (In Russian)
15. Chashchukhin, I.S.; Votyakov, S.L.; Pushkarev, E.V.; Anikina, E.V.; Mironov, A.B.; Uimin, S.G. Oxythermobarometry of ultramafic rocks from the Ural Platiniferous Belt. Geochemistry 2002, 8, 846–863. (In Russian)
16. Perevozchikov, B.V.; Bulykin, L.D.; Popov, I.I.; Orfanitskiy, V.L.; Andreev, M.I.; Snachev, V.I.; Danilenko, S.A.; Cherkasov, V.I.; Chentsov, A.M.; Zharikova, L.N.; et al. The Registry of Chromite Deposits in Alpine-Type Ultramafites of the Urals; Perevozchikov, B.V., Ed.; KamaRICSDI: Perm, Russia, 2000; p. 474. (In Russian)
17. Chashchukhin, I.S.; Votyakov, S.L. Ultramafites of the Alapaevsky massif (Central Urals): Petrology, Geochemistry, Chromite resources. Litosphere 2012, 4, 140–157. (In Russian)
18. Thayer, T.P. Chromite segregations as petrogenetic indicators. Geol. Soc. S. Afr. Spec. Publ. 1970, 1, 380–390.
19. Arai, S. Chemistry of chromian spinel in volcanic rocks as a potential guide to magma chemistry. Miner. Mag. 1992, 56, 173–184.
20. Ferraio, A.; Garuti, G. Platinum-Group Minerals in Chromite-Rich Horizons in the Niquelandia Complex, (Central Goias, Brazil); Prichard, H.M., Potts, P.J., Bowles, J.F.W., Cribb, S.J., Eds.; Springer: Rotterdam, The Netherlands, 1988; pp. 261–272.
21. Chashchukhin, I.S.; Votyakov, S.L.; Scianova, Y.V. Crystal Chemistry of Chrome Spinel and Oxygen Thermobarometry in Ultramafic Rocks; Russian Academy of Sciences, Ural Division, Institute of Geology and Geochemistry: Ekaterinburg, Russia, 2007; p. 312. (In Russian)
22. Melcher, F.; Grum, W.; Simon, G.; Thalhammer, T.V.; Stumpfl, F.E. Petrogenesis of the ophiolitic giant chromite deposits of Kempirsai, Kazakhstan: A study of solid and fluid inclusions in chromite. J. Petrol. 1997, 38, 1419–1438.
23. Garuti, G.; Zaccarini, F.; Moloshag, V.; Alimov, V. Platinum-Group minerals as indicator of sulfur fugacity in ophiolitic upper mantle: An example from chromitites of the Ray-Iz ultramafic complex (Polar Urals, Russia). Can. Miner. 1999, 37, 1099–1115.
24. Anikina, Y.V.; Moloshag, V.P.; Alimov, V.Y.; Kononkova, N.N. Type chemistry of laurite from Polar Ural Alpine-type hyperbasites. Dokl. Russ. Acad. Sci. Earth Sci. Sect. 1996, 345, 410–412.
25. Britvin, S.N.; Kolomensky, V.D.; Boldyreva, M.M.; Bogdanova, A.N.; Krester, Y.L.; Boldyreva, O.N.; Rudashevsky, N.S. Nickelphosphide (Ni,Fe)3P—The nickel analog of schreibersite. Zap. Vserossi. Mineral. Obschch. 1999, 128, 64–72.
26. Skala, R.; Drabek, M. Nickelphosphide from the Vicenice octahedrite: Rietveld crystal structure refinement of a synthetic analogue. Miner. Mag. 2003, 67, 783–792.
27. Pratesi, G.; Bindi, L.; Moggi-Cecchi, V. Icosahedral coordination of phosphorus in the crystal structure of melliniite, a new phosphide mineral from the Northwest Africa 1054 acapulcoite. Am. Miner. 2006, 91, 451–454.
28. Britvin, S.N.; Murashko, M.N.; Vapnik, Y.; Polekhovsky, Y.S.; Krivovichev, S.V. CNMNC Newsletter No. 19, February 2014. Miner. Mag. 2014, 78, 165–170.
29. Uysal, I.; Zaccarini, F.; Garuti, G.; Meisel, T.; Tarkian, M.; Bernhardt, H.J.; Sadiklar, M.B. Ophiolitic chromitites from the Kahramanmaras area, southeastern Turkey:

Their platinum group elements (PGE) geochemistry, mineralogy and Os-isotope signature. Ofioliti 2007, 32, 151–161.
30. Irvine, T.N. Chromian spinel as a petrogenetic indicator. Part I. Theory. Can. J. Earth Sci. 1965, 2, 648–672.
31. Irvine, T.N. Chromian spinel as a petrogenetic indicator. Part II. Petrological applications. Can. J. Earth Sci. 1967, 4, 71–103.
32. Dick, H.J.B.; Bullen, T. Chromian spinel as a petrogenetic indicator in abyssal and alpine-type peridotites and spatially associated lavas. Contrib. Miner. Petrol. 1984, 85, 54–76.
33. Stowe, C.W. Compositions and tectonic settings of chromite deposits through time. Econ. Geol. 1994, 89, 528–546.
34. Barnes, S.J.; Roeder, P.L. The range of spinel composition in terrestrial mafic and ultramafic rocks. Contrib. Miner. Petrol. 2001, 42, 2279–2302.
35. Kamenetsky, V.S.; Crawford, A.J.; Meffre, S. Factors controlling chemistry of magmatic spinel: An empirical study of associated olivine, Cr-spinel, and melt inclusions from primitive rocks. J. Petrol. 2001, 42, 655–671.
36. Uysal, I.; Tarkian, M.; Sadiklar, M.B.; Zaccarini, F.; Meisel, T.; Garuti, G.; Heidrich, S. Petrology of Al- and Cr-rich ophiolitic chromitites from the Muğla, SW Turkey: Implications from composition of chromite, solid inclusions of platinum-group mineral, silicate, and base-metal mineral, and Os-isotope geochemistry. Contrib. Miner. Petrol. 2009, 158, 659–674.
37. Llanes Castro, A.I.; Proenza, J.A.; Zaccarini, F.; Garuti, G.; Pacheco Sarlabous, M.S.C. Al- and Cr-rich chromitites from the Eastern Havana-Matanzas ophiolites (Western Cuba). Episodes 2015, 38, 334–343.
38. Leblanc, M.; Nicolas, A. Ophiolitic chromitites. Int. Geol. Rev. 1992, 34, 653–686.
39. Zhou, M.-F.; Bai, W.-J. Chromite deposits in China and their origin. Miner. Depos. 1992, 27, 192–199.
40. Economou-Eliopoulos, M. Platinum-group element (PGE) distribution in chromite ores from ophiolite complexes of Greece: Implications for chromite exploration. Ofioliti 1993, 18, 83–97.
41. Leblanc, M. Chromitite and ultramafic rock compositional zoning through a Paleotransform fault, Poum, New Caledonia. Econ. Geol. 1995, 90, 2028–2039.
42. Proenza, J.A.; Gervilla, F.; Melgarejo, J.C.; Bodinier, J.L. Al and Cr rich chromitites from the mayari-Baracoa Ophiolitic Belt, (eastern Cuba): Consequence of interaction between volatile-rich melts and peridotite in suprasubduction mantle. Econ. Geol. 1999, 94, 547–566.
43. Kocks, H.; Melcher, F.; Meisel, T.; Burgath, K.P. Diverse contributing sources to chromite petrogenesis in the Shebenik Ophiolitic Complex, Albania: Evidence from new PGE- and Os-isotope data. Miner. Petrol. 2007, 91, 139–170.
44. Grammatikopoulos, T.; Kapsiotis, A.; Tsikouras, B.; Hatzipanagiotou, K.; Zaccarini, F.; Garuti, G. Spinel composition, PGE geochemistry and mineralogy of the chromitites from the Vourinos ophiolite complex, northwestern Greece. Can. Miner. 2011, 49, 1571–1598.
45. Zaccarini, F.; Garuti, G.; Proenza, J.A.; Campos, L.; Thalhammer, O.A.R.; Aiglsperger, T.; Lewis, J. Chromite and platinum-group-elements mineralization in the Santa Elena ophiolitic ultramafic nappe (Costa Rica): Geodynamic implications. Geol. Acta 2011, 9, 407–423.
46. Kapsiotis, A. Genesis of chromitites from Korydallos, Pindos Ophiolite Complex, Greece, based on spinel chemistry and PGE-mineralogy. J. Geosci. 2013, 58, 49–69.
47. Ismail, S.A.; Kettanah, Y.A.; Chalabi, S.N.; Ahmed, A.H.; Arai, S. Petrogenesis and PGE distribution in the Al- and Cr-rich chromitites of the Qalander ophiolite, northeastern Iraq: Implications for the tectonic environment of the Iraqi Zagros Suture Zone. Lithos 2014, 202, 21–36.
48. Miura, M.; Arai, S.; Ahmed, A.H.; Mizukami, T.; Okuno, M.; Yamamoto, S. Podiform chromitite classification revisited: A comparison of discordant and concordant

chromitite pods from Wadi Hilti, northern Oman ophiolite. J. Asian Earth Sci. 2012, 59, 52–61.
49. Miura, M.; Arai, S.; Ahmed, A.H.; Tamura, A. Formation of discordant chromitite as the initiation of sub-arc mantle process: Observations from northern Oman ophiolite. J. Miner. Petrol. Sci. 2014, 109, 38–43.
50. Zaccarini, F.; Idrus, A.; Garuti, G. Chromite Composition and Accessory Minerals in Chromitites from Sulawesi, Indonesia: Their Genetic Significance. Minerals 2016, 6, 46.
51. Li, C.; Thakurta, J.; Ripley, E.M. Low-Ca contents and kink-banded textures are not unique to mantle olivine: Evidence from the Duke Island Complex, Alaska. Miner. Petrol. 2012, 104, 147–153.
52. Arai, S. Dunite-harzburgite-chromitite complexes as refractory residue in the Sangun-Yamaguchi zone, Western Japan. J. Petrol. 1980, 21, 141–165.
53. Lago, B.; Rabinowicz, M.; Nicolas, A. Podiform chromite ore bodies: A genetic model. J. Petrol. 1982, 23, 103–125.
54. Arai, S.; Yurimoto, H. Podiform chromitites of the Tari-Misaka ultramafic complex, Southwest Japan, as mantle–melt interaction products. Econ. Geol. 1994, 89, 1279–1288.
55. Zhou, M.F.; Robinson, P.T. High-Cr and high-Al podiform chromitites, Western China: Relationships to partial melting and melt/rock reaction in the upper mantle. Int. Geol. Rev. 1994, 36, 678–686.
56. Zhou, M.F.; Robinson, P.T.; Bai, W.J. Formation of podiform chromites by melt–rock interaction in the upper mantle. Miner. Depos. 1994, 29, 98–101.
57. Zhou, M.F.; Robinson, P.T.; Malpas, J.; Zijin, L. Podiform chromites in the Luobusa Ophiolite (Southern Tibet): Implications for melt–rock interaction and chromite segregation in the upper mantle. J. Petrol. 1996, 37, 3–21.
58. Arai, S. Control of wall-rock composition on the formation of podiform chromitites as a result of magma/peridotite interaction. Res. Geol. 1997, 47, 177–187.
59. Johan, Z.; Dunlop, H.; Le Bel, L.; Robert, J.L.; Volfinger, M. Origin of chromite deposits in ophiolitic complexes: Evidence for a volatile and sodium-rich reducing fluid phase. Fortschr. Miner. 1983, 61, 105–107.
60. Ballhaus, C. Origin of podiform chromite deposits by magma mingling. Earth Planet. Sci. Lett. 1998, 156, 185–193.
61. Matveev, S.; Ballhaus, C. Role of water in the origin of podiform chromitite deposits. Earth Planet. Sci. Lett. 2002, 203, 235–243.
62. Baumgartner, R.J.; Zaccarini, F.; Garuti, G.; Thalhammer, O.A.R. Mineralogical and geochemical investigation of layered chromitites from the Bracco-Gabbro complex, Ligurian ophiolite, Italy. Contrib. Miner. Petrol. 2013, 165, 477–493.
63. Britvin, S.N.; Rudashevskii, N.S.; Krivovichev, S.V.; Burns, P.C.; Polekhovsky, Y.S. Allabogdanite, (Fe,Ni)2P, a new mineral from the Onello meteorite: The occurrence and crystal structure. Am. Miner. 2002, 87, 1245–1249.
64. Britvin, S.N.; Vapnik, Y.; Polekhovsky, Y.S.; Krivovichev, S.V. CNMNC Newsletter No. 15, February 2013. Miner. Mag.2013, 77, 1–12.
65. Buseck, P.R. Phosphide from meteorites: Barringerite, a new iron-nickel mineral. Science 1969, 165, 169–171.

CHAPTER 15

Structural Controls on the Geochemistry and Output of the Wells in the Olkaria Geothermal Field of the Kenyan Rift Valley

Ruth N. Wamalwa[1], Christopher M. Nyamai[2], Willis J. Ambusso[3], Josephat Mulwa[2], Aaron K. Waswa[2*]

[1]*Kenya Electricity Generating Company—KenGen, Naivasha, Kenya*

[2]*Department of Geology, University of Nairobi, Nairobi, Kenya*

[3]*School of Pure and Applied Sciences, Kenyatta University, Nairobi, Kenya*

ABSTRACT

The Olkaria geothermal field is located in the Kenya Rift valley, about 120 km from Nairobi. Geothermal activity is widespread in this rift with 14 major geothermal prospects being identified. Structures in the Greater Olkaria volcanic complex include: the ring structure, the Ol'Njorowa gorge, the ENE-WSW Olkaria fault and N-S, NNE-SSW, NW-SE and WNW-ESE trending faults. The faults are more prominent in the East, Northeast and West Olkaria fields but are scarce in the Olkaria Domes area, possibly due to the thick pyroclastics cover. The NW-SE and WNW- ESE faults are thought to be the oldest and are associated with the development of the rift. The most prominent of these faults is the Gorge Farm fault, which bounds the geothermal fields in the northeastern part and extends to the Olkaria Domes area. The most recent structures are the N-S and the NNE-SSW faults. The geochemistry and output of the wells cut by these faults have a distinct characteristic that is the N-S, NW-SE and WNW-ESE faults are characterized by wells that have high Cl contents, temperatures and are good producers whereas the NE-SW faults, the Ring Structure and the Ol'Njorowa gorge appear to carry cool dilute waters with less chloride concentration and thus low performing wells. Though the impacts of these faults are apparent, there exists a gap in knowledge on how wide is the impact of these faults on the chemistry and performance of the wells. This paper therefore seeks to bridge this gap by analysis of the chemical trends of both old wells and newly drilled ones to evaluate the impacts of individual faults and then using buffering technique of ArcGis estimate how far and wide the influence of the faults is. The data was obtained after the sampling and analysis of discharge fluids of wells located on six profiles along the structures cutting through the field. Steam samples were collected with a stainless steel Webre separator connected between the wellhead and an atmospheric silencer on the discharging wells whereas the analysis was done in house in the KenGen geochemistry laboratory. The results indicates that Olkaria field has three categories of faults that control fluid flow that is the NW-SE trending faults that bring in high temperature and Cl rich waters, and the NE-SW trending Olkaria fracture tend to carry cool temperature waters that have led to decline in enthalpies of the wells it cuts through. The faults within the Ol Njorowa gorge act to carry cool, less mineralized water. Though initially, these effects were thought to be in shallow depths, an indication in OW-901 which is a deeper at 2200 m compared to 1600 m of OW-23 well that proves otherwise. This is, however, to be proved later as much deeper wells have been sited.

Keywords: Structural Controls, Geochemistry, Productivity, Buffering, Olkaria Geothermal Field, Kenya

1. INTRODUCTION

Development of geothermal resources in the Olkaria area, a high temperature field, started in the early 1950s. Subsequent years saw numerous expansions with additional power plants being installed in Olkaria [1] . These included a binary plant at Olkaria South West (Olkaria III) in 2000, with a capacity of 110 Megawatts (MW), a condensing plant at Olkaria North East (Olkaria II) in 2003, with a capacity of 105 MWe and another binary plant at Olkaria North West (Oserian) in 2004, with a capacity of 2 MWe (Figure 1). Plans are underway to commission 280 Mw within Olkaria 1 and Olkaria Domes. Geochemistry in such a setting, given in terms of concentrations of various chemical constituents in both brine and steam discharged from geothermal wells is vital. This is through the understanding that various chemical constituents in both brine and steam discharged from geothermal wells, provide valuable quantitative information that can be used to note the reservoir condition and changes in response to steam exploitation. This is by grouping the elements into two i.e. tracer/conservative elements to trace the origin of the fluids and geo-indicators to reflect the specific environments of equilibrium. The reduction in the concentration of conservative elements will mean dilution occurring in the reservoir leading to cooling hence reduction in the steam production; steam meant to run the turbines for electricity generation. The geo-indicators which are chemically reactive reflect specific environments of equilibrium. For example Sodium, Silica and Potassium are to be used as solute Geothermometers. Such studies apart from being used directly in reservoir modeling and simulations also assist with calibration of the other reservoir models as trends in physical and chemical parameters discharged by the wells have a distant source and therefore have had lengthy interactions with the formations. The movement through the formation may have led to boiling thereby increasing the amount of CO_2 in both liquid and vapor phase. The deep casing therefore was to case off shallow aquifers and feeders of steam heated waters.

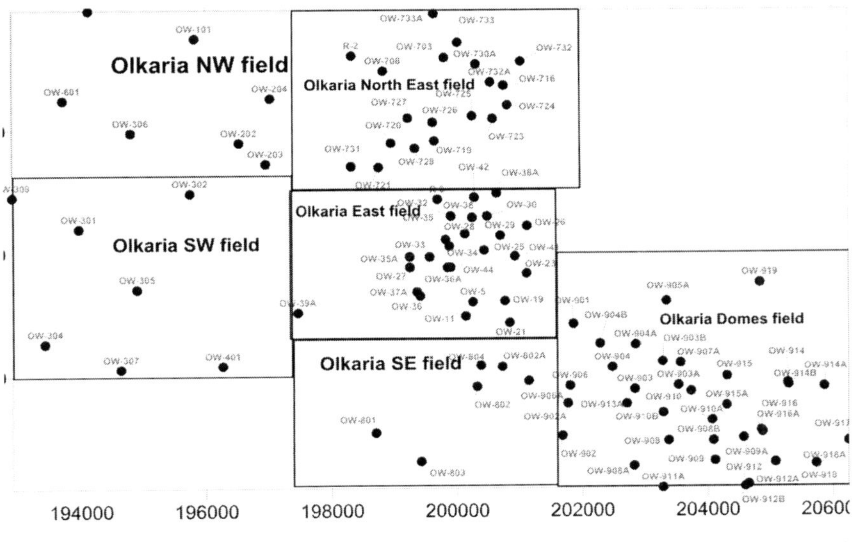

Figure 1. Geothermal fields within the Greater Olkaria geothermal area [2] .

2. STRUCTURAL GEOLOGY, CHEMICAL DATA AND PROFILES SELECTION

Geological structures within the Greater Olkaria volcanic complex include; the ring structure, rift fault systems, the Ol-Njorowa gorge, and dykes swarm (Figure 2). The faults are trending ENE-WSW, N-S, NNE-SSW, NW-SE and WNW-ESE. These faults are more prominent in the East, Northeast and Olkaria West fields but are elusive in the Olkaria Domes area, possibly due to the thick cover of the surface by pyroclasts [3] . The NW-SE and WNW-ESE faults are thought to be the oldest fault system and they link the parallel rift basins to the main extensional zone [4] . Gorge Farm fault, bounding the geothermal fields in the north eastern part and extending to Olkaria Domes area is the most prominent of these faults. The most recent structures are the N-S (Ololbutot eruptive fissure) and the NNE-SSW faults [5] . Dyke swarms exposed in the Ol-Njorowa gorge with strike direction in the NNE further attest to the recent reactivation of faults with this trend. Development of the Ol-Njorowa gorge was initiated by faulting along the trend of the gorge but the feature, as it is seen today, was mainly formed due to a catastrophic outflow of Lake Naivasha during its high stands [6] . Volcanic plugs (necks) and felsic dykes occurring along the gorge further attest to the fault control in the development of this feature. Hydroclastic craters located on the northern edge of the Olkaria Domes area mark magmatic explosions, which occurred in submerged country [5] . These craters form a row along, which the extrapolated caldera rim trace passes. Study considered 6 profiles along the structures cutting through the field as shown in Figure 4.

3. CHEMICAL DATA COLLECTION

The wells located in the production are usually monitored through geochemical sampling programs. The east production field, for example, is monitored in a biannual program in which gas samples and water samples are collected. The North east production field is monitored on a quarterly basis under the program of clean development mechanism (CDM) in which gas and water samples is collected for analysis. Newly drilled wells in the various sectors of development or further development have been discharge tested fluid chemistry of the areas made available. The sampling and analysis methods are based on those of [7] where steam samples were collected with a stainless steel webre separator connected between the wellhead and an atmospheric silencer on the discharging wells. For the earliest wells in the Domes, only the steam in the two phase flow was sampled. The water samples were taken from the weir box down the stream from an atmospheric silencer. However this changed later. The later sampling started collecting both the steam and water from the two phase flow from the line between the wellhead and the atmospheric silencer using a webre separator. Another water sample was also obtained from the weirbox. The analysis is done in house in the KenGen geochemistry laboratory according to standard methods for geothermal fluid analysis as described by various authors [7] and [8] . It is desirable that the charge balance error of geothermal waters be within ±5%. Therefore data that never met this criterion was left out.

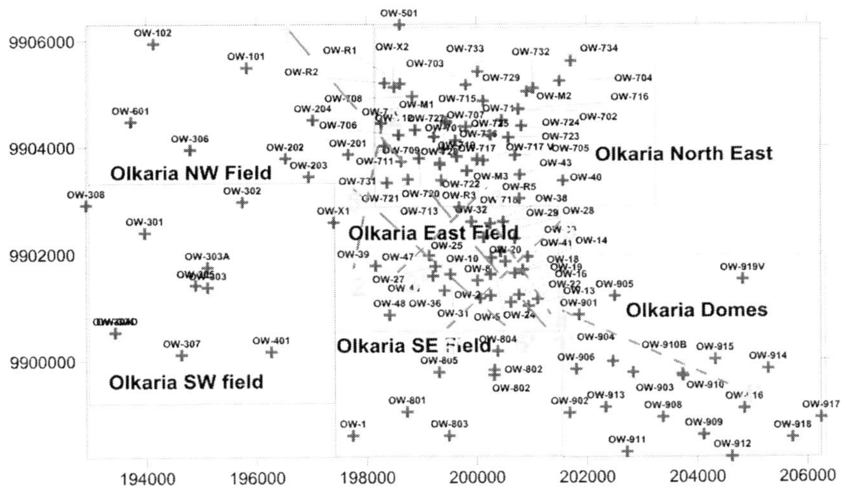

Figure 2. Olkaria geothermal sectors and selected profiles. The profiles allowed a distance of 1 km wide to cater for the buffering process.

4. GEOCHEMICAL RESULTS

The study considered one conservative element (Cl), one geoindicator (TNaK) and nitrogen gas to help in delineating the source of fluids, reservoir temperatures and signs of incursion of atmospherically modified fluids respectively. The results are in form of graphs of concentration variation with time as shown in Figures 3-5. Only profiles cutting at least 3 wells were considered.

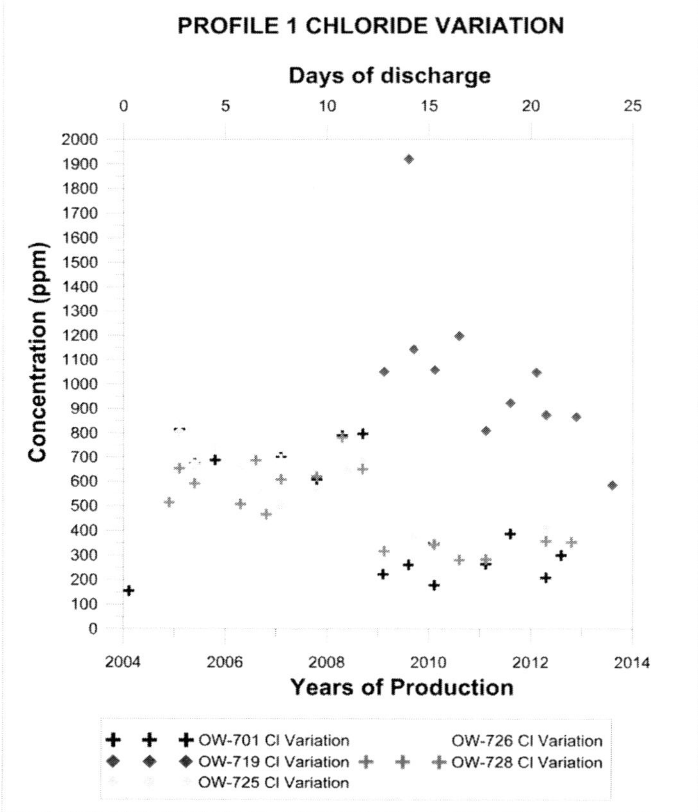

Figure 3. Chloride variation with time for profile 1.

5. DISCUSSIONS

This is in form of distribution maps plotted per profile (Figures 6-8). The Chloride and temperature distribution maps indicate the NW-SE trending faults bring in Cl rich and high temperature waters; these increase towards the central field a factor that has been attributed to boiling due pressure drop arising from production. Profile 5 and 6 give a clear distinguishing factor in that wells around the Ol Njorowa gorge, which has majorly NE-SW trending faults have low Cl content whereas those cut by NW-SE faults like OW-44 and OW-915 has high contents.

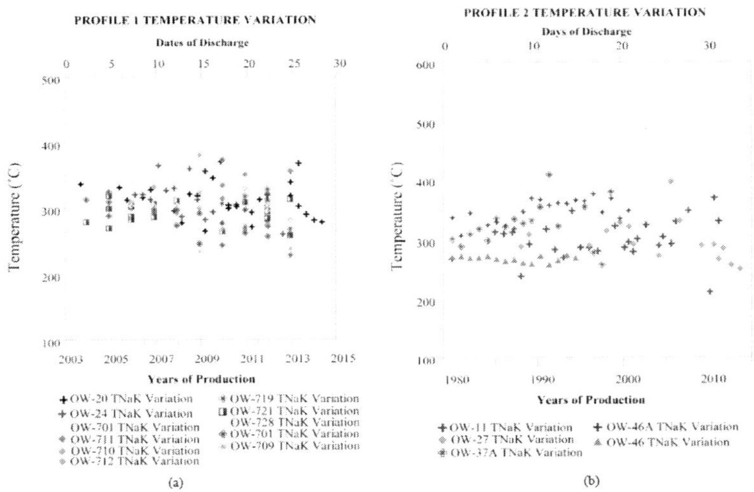

Figure 4. Temperature variation with time for profile 1 and 2.

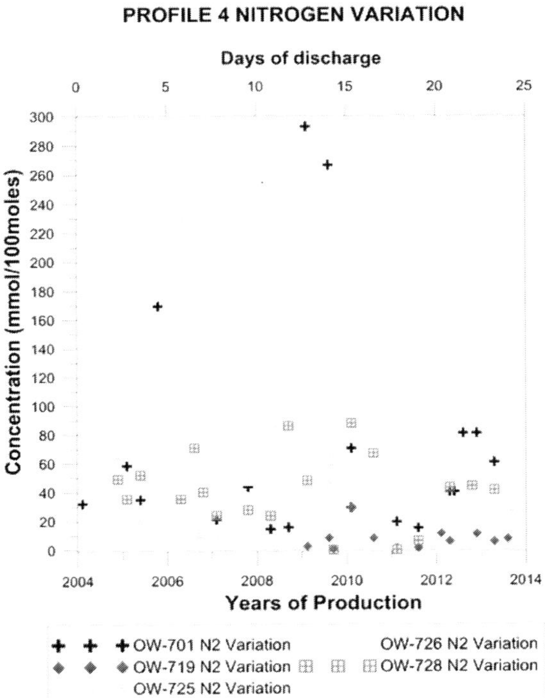

Figure 5. N_2 Variation with time for profile 4.

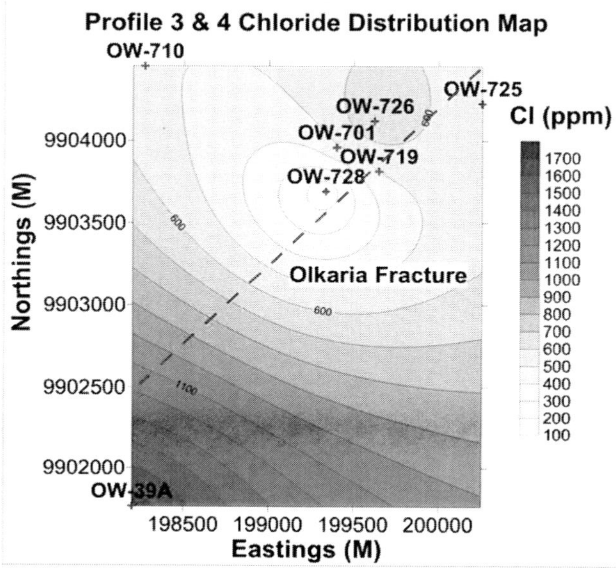

Figure 6. Cl distribution map for profiles 3 & 4.

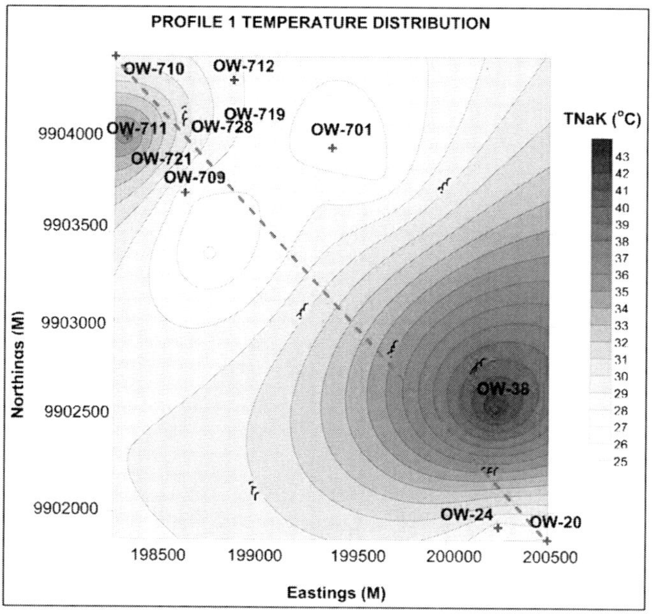

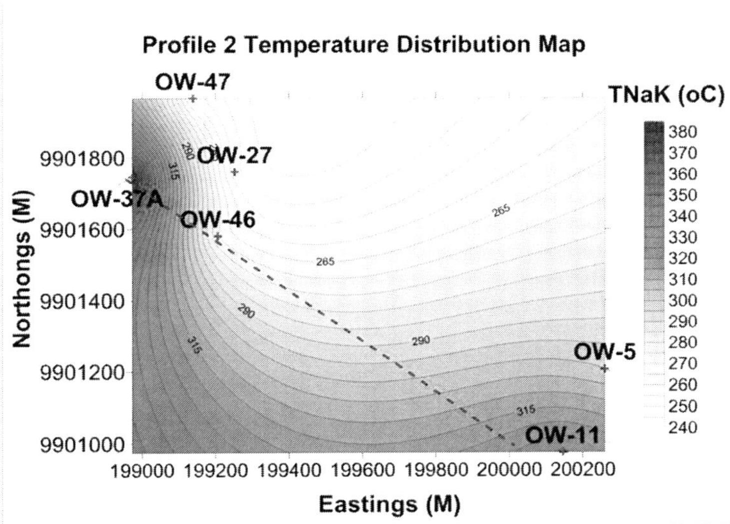

Figure 7. Temperature distribution per profile.

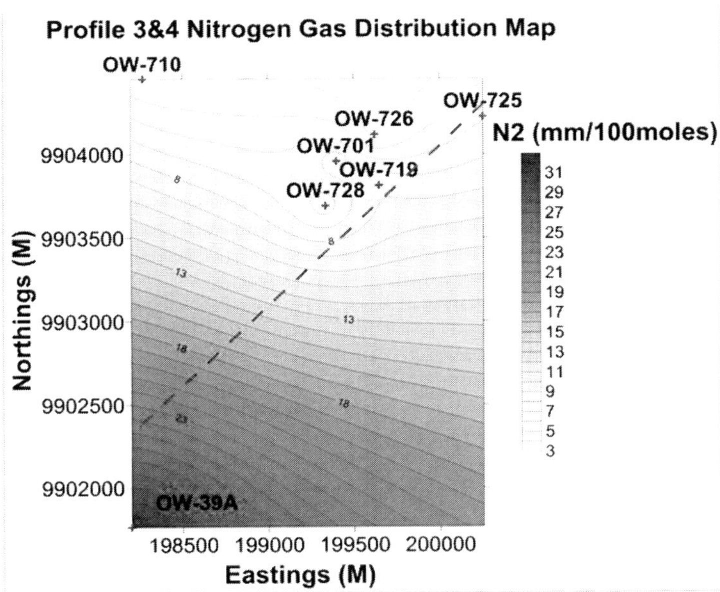

Figure 8. Nitrogen distribution per profile.

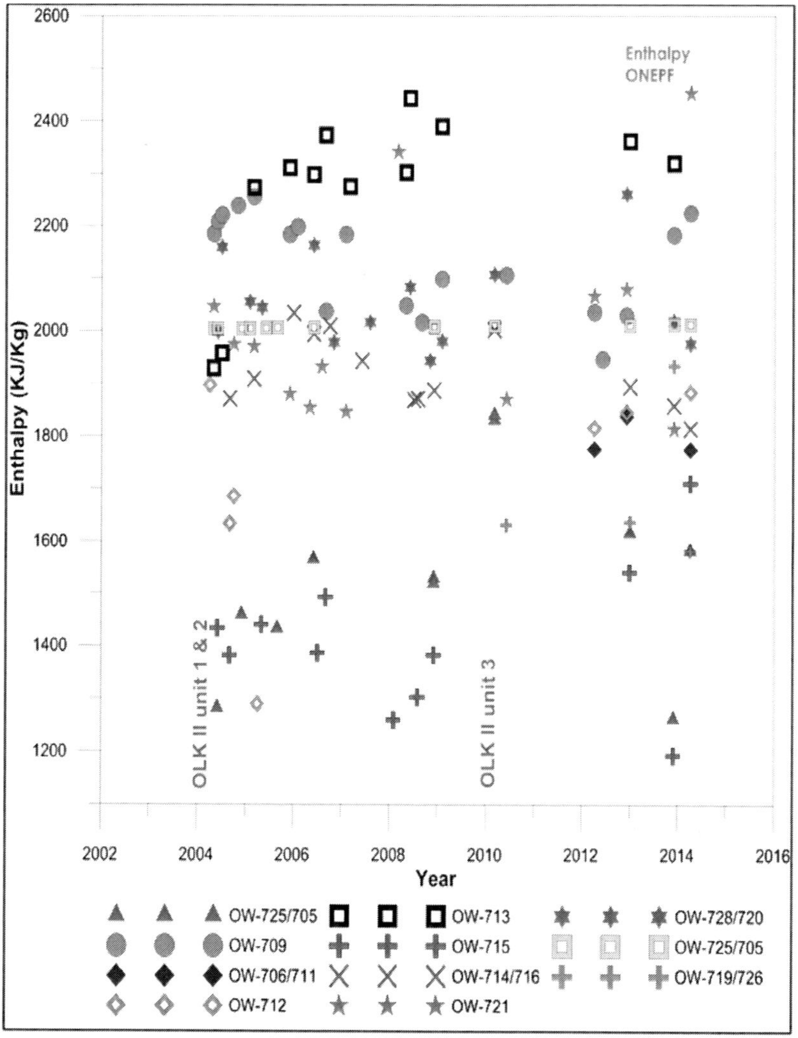
2600
2400
2200
2000
1800
1600
1400
1200
Enthalpy (KJ/Kg)
Enthalpy
ONEPF
OLK II unit 1 & 2
OLK II unit 3
2002
2004
2006
2008
2010
2012
2014
2016
Year
OW-725/705
OW-713
OW-728/720
OW-709
OW-715
OW-725/705
OW-706/711
OW-714/716
OW-719/726
OW-712
OW-721

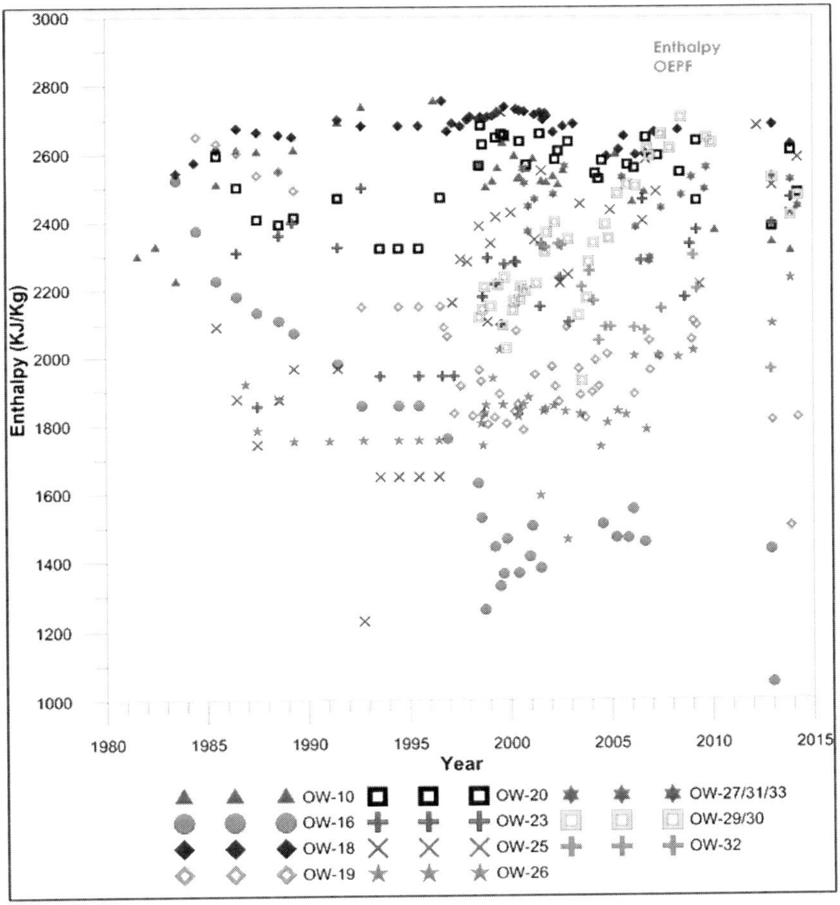

Figure 9. Enthalpy variation with time.

This study also considered N_2 as it is the best indicator of incursion of atmospherically modified reservoir fluid. The distribution map (Figure 9) shows low modifications associated with the NW-SE trending faults. As in case of profile 1, there is a sharp increase evident towards the SE corner; areas bordering the Ol Njorowa gorge. This therefore indicate that this feature act to recharge the reservoir.

The enthalpy variation has been used to summarize the wells output with time (Figure 9). The trending show that the NW-SE (Olkaria fracture), bringing cooler fluids has impacted on OW-725/OW-705 (see profile 3 & 4 Distribution map). These wells have the least combined enthalpies in the Olkaria North East field. The effect of the Ol Njorowa Gorge is also evident with the declining enthalpy and consequently output of OW-16, OW-19, OW-22 and OW-23 (see Profile 5 & 6 Distribution map).

6. CONCLUSIONS

After the evaluation of the chemical data, it was evident that Olkaria field has three categories of faults that control fluid flow:

1) The NW-SE trending faults bring in high temperature and Cl rich waters.

2) The NE-SW trending Olkaria fracture tends to carry cool temperature waters that have led to decline in enthalpies of the wells it cuts through.

3) The faults within the Ol Njorowa gorge act to carry cool, less mineralized water. Though initially, these effects were thought to be in shallow depths, an indication in OW-901 which is a deeper at 2200 m compared to 1600 m of OW-23 well that proves otherwise. This is, however, to be proved later as much deeper wells have been sited.

CITE THIS PAPER

Wamalwa, R.N., Nyamai, C.M., Ambusso, W.J., Mulwa, J. and Waswa, A.K. (2016) Structural Controls on the Geochemistry and Output of the Wells in the Olkaria Geothermal Field of the Kenyan Rift Valley. International Journal of Geosciences, 7, 1299-1309.

REFERENCES

1. Opondo, K.M. (2008) Fluid Characteristics of Three Exploration Wells Drilled at Olkaria Domes Field, Kenya. Proceedings of the 33rd Workshop on Geothermal Reservoir Engineering, Stanford University, Stanford, 6 p.
2. KenGen (1999) Conceptualized Model of the Olkaria Geothermal Field (Compiled by Muchemi, G.G.). Internal Report, The Kenya Electricity Generating Company, Ltd., Kenya, 46 p.
3. Lagat, J.L. (2004) Geology, Hydrothermal Alteration and Fluid Inclusion Studies of Olkaria Domes Geothermal Field, Kenya. MSc Thesis, University of Iceland, UNU-GTP, Iceland, Report 2, 71 p.
4. Wheeler, W.H. and Karson, J.A. (1994) Extension and Subsidence Adjacent to a "Weak" Continental Transform: An Example of the Rukwa Rift, East Africa. Geology, 22, 625-628.
5. Mungania, J. (1999) Geological Report of Well OW-714. Kenya Power Company Internal Report.
6. Clarke, M.C.G., Woodhall, D.G., Allen, D. and Darling, G. (1990) Geological, Volcanological and Hydrogeological Controls on the Occurrence of Geothermal

Activity in the Area Surrounding Lake Naivasha, Kenya, with Coloured 1:100,000 Geological Maps. Ministry of Energy, Nairobi, 138 p.

7. Arnósson, S., Stefánsson, A. and Bjanarson, J.?. (2007) Fluid-Fluid Interaction in Geothermal Systems. Reviews in Mineralogy & Geochemistry, 65, 258-312.

8. Arnósson, S., D'Amore, F. and Gerardo-Abaya, J. (2000) Isotopic and Chemical Techniques in Geothermal Exploration, Development and Use. International Atomic Energy Agency Publication, Vienna, 351 p.

CHAPTER 16

Geochemistry and Mineralogy of Tuff in Zhongliangshan Mine, Chongqing, Southwestern China

Jianhua Zou [1,2,*], Heming Tian [2] and Tian Li [2]

[1]*College of Geosciences and Survey Engineering, China University of Mining and Technology (Beijing), Beijing 100083, China*

[2]*Chongqing Key Laboratory of Exogenic Mineralization and Mine Environment, Chongqing Institute of Geology and Mineral Resources, Chongqing 400042, China*

ABSTRACT

Coal-bearing strata that host rare metal deposits are currently a hot issue in the field of coal geology. The purpose of this paper is to illustrate the mineralogy, geochemistry, and potential economic significance of rare metals in the late Permian tuff in Zhongliangshan mine, Chongqing, southwestern China. The methods applied in this study are X-ray fluorescence spectrometry (XRF), inductively coupled mass spectrometry (ICP-MS), X-ray diffraction analysis (XRD) plus Siroquant, and scanning electron microscopy in conjunction with an energy-dispersive X-ray spectrometry (SEM-EDX). The results indicate that some trace elements including Li, Be, Sc, V, Cr, Co, Ni, Cu, Zn, Ga, Zr, Nb, Cd, Sb, REE, Hf, Ta, Re, Th, and U are enriched in the tuff from Zhongliangshan mine. The minerals in the tuff mainly include kaolinite, illite, pyrite, anatase, calcite, gypsum, quartz, and traces of minerals such as zircon, florencite, jarosite, and barite. The tuff is of mafic volcanic origin with features of alkali basalt. Some minerals including florencite, gypsum, barite and a portion of anatase and zircon have been derived from hydrothermal solutions. It is suggested that Zhongliangshan tuff is a potential polymetallic ore and the recovery of these valuable elements needs to be further investigated.

Keywords: geochemistry; mineralogy; origin; tuff; Chongqing

1. INTRODUCTION

With the depletion of traditional rare metal deposits, coal deposits as promising alternative sources for rare metals have attracted much attention in recent years [1,2,3,4,5,6,7,8,9]. At present, germanium is the most successful rare metal element that has been extracted from coal ash [4,8,9]. The three well-known coal-bearing strata hosted Ge deposits include Lincang (Yunnan Province) and Wulantuga (Inner Mongolia) of China, and Spetzugli of Russia, are the main sources for the industrial Ge at present and for the foreseeable future [1,4,8,9]. The super-large coal-bearing strata hosted gallium deposit in the Jungar Coalfield (Inner Mongolia), China, is another typical example discovered in 2006 [10], which was considered as the third and the most outstanding discovery after the coal-bearing strata hosted uranium and germanium deposits [2,10]. Moreover, aluminum is also enriched in Jungar coalfield [1,10]. In 2010, another new type of coal-bearing strata hosted Nb (Ta)-Zr (Hf)-REE-Ga polymetallic deposit of volcanic origin was discovered in the late Permian coal-bearing strata of eastern Yunnan, southwestern China [11]. Similar polymetallic deposits have since been discovered in some coalfields from southern China [1,3]. Similar to most typical areas enriched in rare metals in coal-bearing strata, the tectonic controls on the localization of the metalliferous coal deposits and the mechanisms of rare-metal mineralization in south China and south Primorye of Russia have been studied comparatively in detail [3]. The possible recovery of rare earth elements from coal and its combustion products such as fly ash is an exciting new research area [2,12,13,14,15,16], because coal and its combustion derivation (fly ash) may have elevated concentrations of these rare metals.

The purpose of this paper is to discuss the mineralogical and geochemical compositions of tuff layer in late Permian coal-bearing strata of Zhongliangshan mine, Chongqing, southwestern China. It also contributes to the discussion on the origin and potential prospects of rare metals mineralization of the tuff.

2. GEOLOGICAL SETTING

The Zhongliangshan mine is located in the urban area of Chongqing, southwestern China (Figure 1). The coal-bearing sequence is the late Permian Longtan Formation (P_3l), which is composed of the light gray, gray, dark gray mudstone, sandy mudstone, siltstone, sandstone and coal seams. This formation is enriched in brachiopods, fern, cephalopods, bivalves, trilobite and other fossils. The Longtan Formation was deposited in a continental–marine transitional environment and has a thickness varying from 26.5 to 105.02 m, with an average of 71.08 m. It contains 10 coal seams, which are identified as K1 to K10 from top to bottom. The Changxin Formation conformably overlies the Longtan Formation and is mainly composed of thick layers of brown-gray, dark gray limestone intercalated with thin layers of mudstone and flint nodules. Some fossils including brachiopods, spindle dragonflies, sponges, corals, and trilobites are enriched in the Changxin Formation. The Maokou Formation disconformably underlies the Longtan Formation, which consists of thick layers of light gray to dark gray bioclastic limestone.

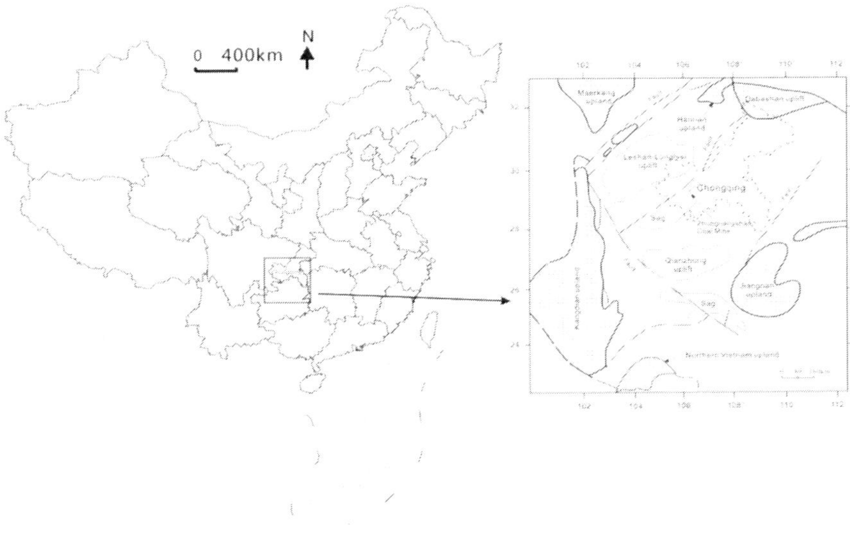

Figure 1. Location of the Zhongliangshan Mine, Chongqing, southwestern China.

The tuff layer, with a thickness mostly of 2–5 m, light-gray or light-gray–white in color, and a conchoidal fracture and a soapy feel, is located at the lowermost Longtan Formation. The K10 coal seam conformably overlies the tuff layer, which has a disconformable contact with the underlying Maokou Formation (middle Permian) (Figure 2). The tuff is enriched in pyrite and shows massive bedding structure. The tuff was derived from the basalt eruption and deposited directly on the weathered surface of the Maokou Formation limestone, and then was subjected to weathering, leaching, and eluviation [17,18]. It is usually described as bauxite or bauxitic mudstone during core sample identification or field lithological description [17,18].

3. SAMPLES AND ANALYTICAL PROCEDURES

A total of 21 bench samples were taken from the tuff layer in the Zhongliangshan mine, following the Chinese Standard GB/T 482-2008 [19]. Each tuff bench sample was cut over an area 10-cm wide, 10-cm deep and 10-cm thick. All collected samples were immediately stored in plastic bags to minimize contamination and oxidation. Large chips were selected at random from each sample for preparation of polished sections and also kept for later reference if required. The remainder of each sample was crushed and ground to pass through the 200-mesh sieve for analysis.

The loss of ignition (LOI) of each sample was determined according to ASTM standard D3174 [20]. All samples were analyzed by X-ray diffraction (XRD) using a D8 advance powder diffractmeter with Ni-filtered Cu-Kα radiation and a scintillation detector. The XRD pattern was recorded over a 2θ interval of 2.6°–70°, with a step size of 0.02°. X-ray diffractograms of the tuff samples were subjected to quantitative mineralogical analysis using Siroquant™ of China University of Mining and Technology (Beijing), a commercial interpretation software developed by Taylor [21] based on the principles for diffractogram profiling set out by Rietveld [22]. Further details indicating the use of this technique for coal-related materials are given by Ward *et al.* [23,24] and Ruan and Ward [25]. A Scanning Electron Microscope in conjunction with an energy-dispersive X-ray spectrometer (SEM-EDX, JEOL JSM-6610LV+OXFORD X-max, Tokyo, Japan), with an accelerating voltage of 20 kV, was used to study morphology and microstructure of minerals, and also to determine the distribution of some elements in tuff samples under a high vacuum mode in Chongqing Institute of Geology and Mineral Resources.

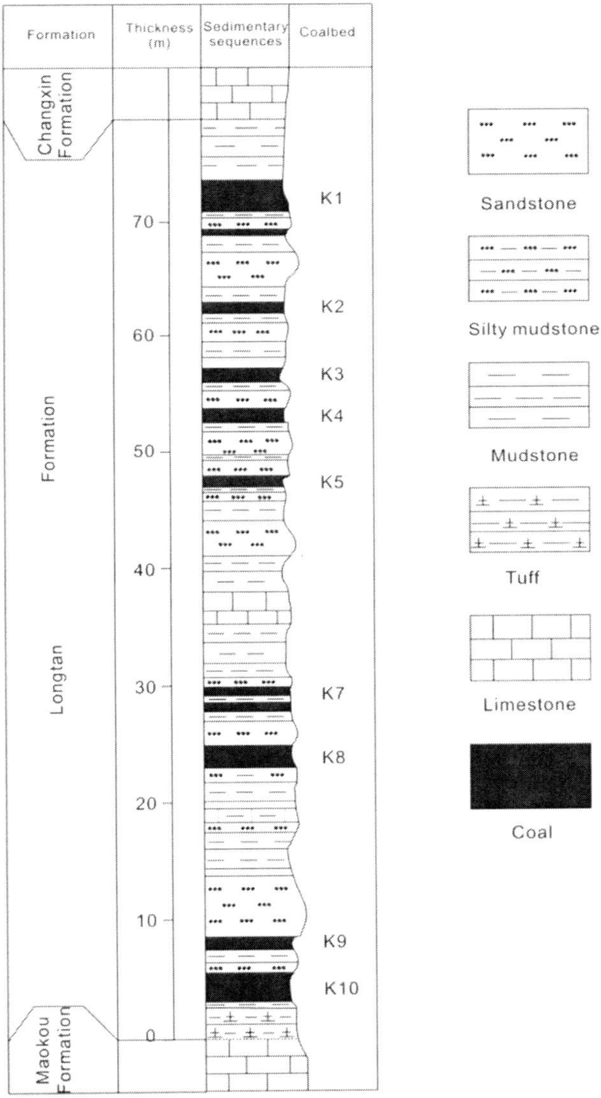

Figure 2. Generalized sedimentary sequence at the Zhongliangshan Mine, Chongqing.

Percentages of major element oxides including SiO_2, TiO_2, Al_2O_3, Fe_2O_3, MnO, MgO, CaO, Na_2O, K_2O, and P_2O_5 in the tuff samples were determined by X-ray fluorescence spectrometry (XRF) in Chongqing Institute of Geology and Mineral Resources. The contents of trace elements were determined by inductively coupled mass spectrometry (Thermo X series II ICP-MS, Thermo Fisher Scientific, Waltham, MA, USA) in Chongqing Institute of Geology and Mineral Resources. The procedures of ICP-MS were: weigh 0.25 g sample in a 50 mL Teflon beaker; add 20 mL HNO_3-$HClO_4$-HF (volume ratio of 4:1:5) and 2 mL H_2SO_4; place on a temperature controlled heating plate and heat to 230 °C until like wet salt; then heat to 280 °C and evaporate to dryness; turn off the heating plate to cool the sample for 3 min; add 8 mL concentrated aqua regia; incubate for 10 min; transfer the solution to a 25 mL plastic flask; mix and volume; take 5 mL of solution to a 25 mL volumetric flask and dilute to the mark; and study using high resolution inductively coupled plasma mass spectrometry.

4. RESULTS

4.1. Minerals

The proportion of each crystalline phase of the tuff identified by X-ray diffractometry plus Siroquant is given in Table 1. The minerals in Zhongliangshan tuff mainly include kaolinite, illite, pyrite, anatase, calcite, gypsum, and quartz. Some trace minerals such as zircon, florencite, jarosite, and barite, are observed under SEM-EDX.

Table 1. Mineral compositions of Tuff by XRD analysis and Siroquant (%).

Sample	Kaolinite	Illite	Pyrite	Anatase	Calcite	Quartz	Gypsum
S140SE7-1	83.2	-	13.6	2.7	-	0.5	-
S140SE7-2	85.1	-	9.3	3.5	-	2.1	-
S140SE7-3	65.4	-	11.6	8.2	-	14.8	-
S140SE7-4	69.9	-	20.2	6.7	-	3.3	-
S140SE7-5	53.7	-	38.2	2.8	-	-	5.3
S140SE7-6	61.4	-	30.8	3.2	-	-	4.6
S140SE7-7	61.9	-	19.8	4.1	-	-	14.2
S140SE7-8	78.7	-	14.8	3.3	-	0.1	3.1
S140SE7-9	80	-	10.6	7.5	-	0.3	1.7
S140SE7-10	65.8	-	26.5	4.7	-	0.3	2.7
S140SE7-11	72.4	-	16.1	8.8	-	0.3	2.5
S140SE7-12	73.4	6.7	2.8	14.3	-	0.4	2.4
S140SE7-13	49.9	18.6	10.6	9.7	6.1	0.3	4.8
S140SE7-14	11.4	3.8	28.8	0.8	45	-	10.3
S140SE7-15	52.9	21	3.1	8.4	10.7	0.5	3.5
S140SE7-16	81.5	4.4	1.4	6.9	3.8	0.1	1.8
S140SE7-17	86.4	4.4	1.1	6.9	-	-	1.2
S140SE7-18	80.7	6	2.4	7.2	0.4	0.4	2.8
S140SE7-19	34.5	31.3	1.9	3.6	24.7	0.3	3.7
S140SE7-20	67.2	18.6	1	10.5	0.1	0.5	2.2
S140SE7-21	56	35.3	0.4	7.2	0.3	-	0.7
Average	65.30	15.01	12.62	6.24	11.39	1.61	3.97

4.1.1. Kaolinite and Illite

Kaolinite is the dominant mineral of the tuff in Zhongliangshan (Figure 3). The average content of kaolinite is up to 65.3%, and all the samples are richer than 50% except for samples S140SE7-13, S140SE7-14, and S140SE7-19. Kaolinite occurs mainly as matrix material (Figure 4A), and to a lesser extent, as vermicular (Figure 4B) and individual massive (Figure 4C). Illite occurs at the lower part of the profile (Figure 3).

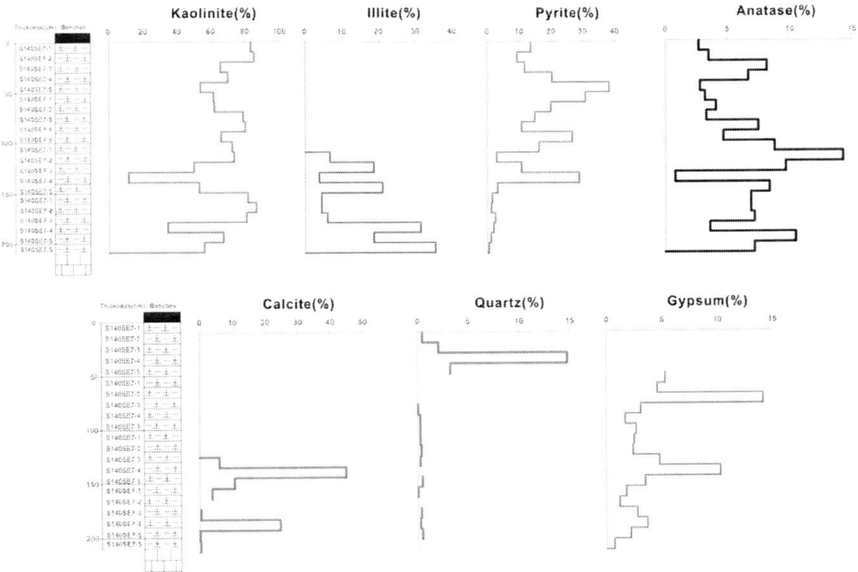

Figure 3. Vertical variations of minerals from the tuff in the Zhongliangshan mine.

4.1.2. Pyrite, Jarosite and Barite

Pyrite distributes widely in the tuff samples and its content varies from 0.4% to 38.2% (12.6% on average). Its content is gradually decreasing from top to bottom (Figure 3), suggesting that the upper portion has been more subjected to seawater. Pyrite mainly occurs as discrete particles (Figure 4A,D), lumps (Figure 4D), and in some cases, as cubic crystal and pentagonal dodecahedron (Figure 4E). Jarosite occurs as fracture-fillings (Figure 4F), indicating a weathering product of pyrite. Barite is located on the edge of jarosite (Figure 4F), which may be formed by the reaction of jarosite with the hydrothermal solution containing Ba.

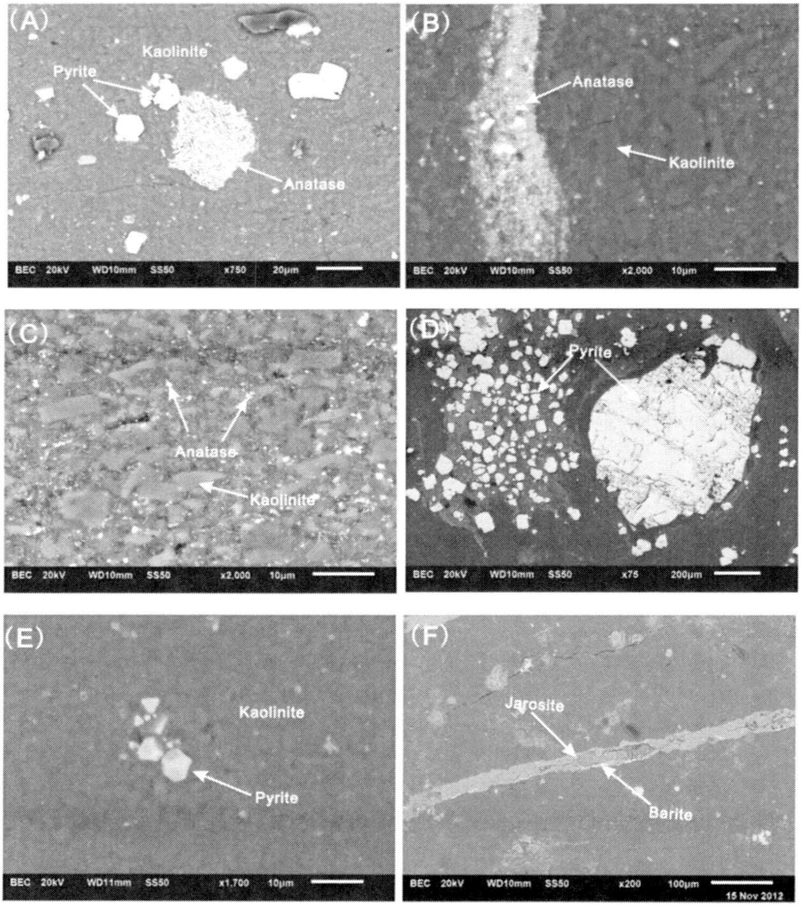

Figure 4. Back scattered images of minerals in the Zhongliangshan tuff: (**A**) kaolinite, pyrite and anatase in sample S140SE7-1; (**B**) kaolinite and anatase in sample S140SE7-8; (**C**) kaolinite and anatase in sample S140SE7-1; (**D**) pyrite in sample S140SE7-4; (**E**) pyrite and kaolinite in sample S140SE7-6; and (**F**) jarosite and barite in sample S140SE7-18.

4.1.3. Anatase

Anatase is present evenly in the tuff samples and varies from 0.8% to 14.3% with an average of 6.2%. The content of Nb in the anatase is up to 0.18% determined by SEM-EDX. Anatase occurs mainly as irregular fine particles (Figure 4B,C) or as colloidal (Figure 4A and Figure 5A) in the kaolinite matrix.

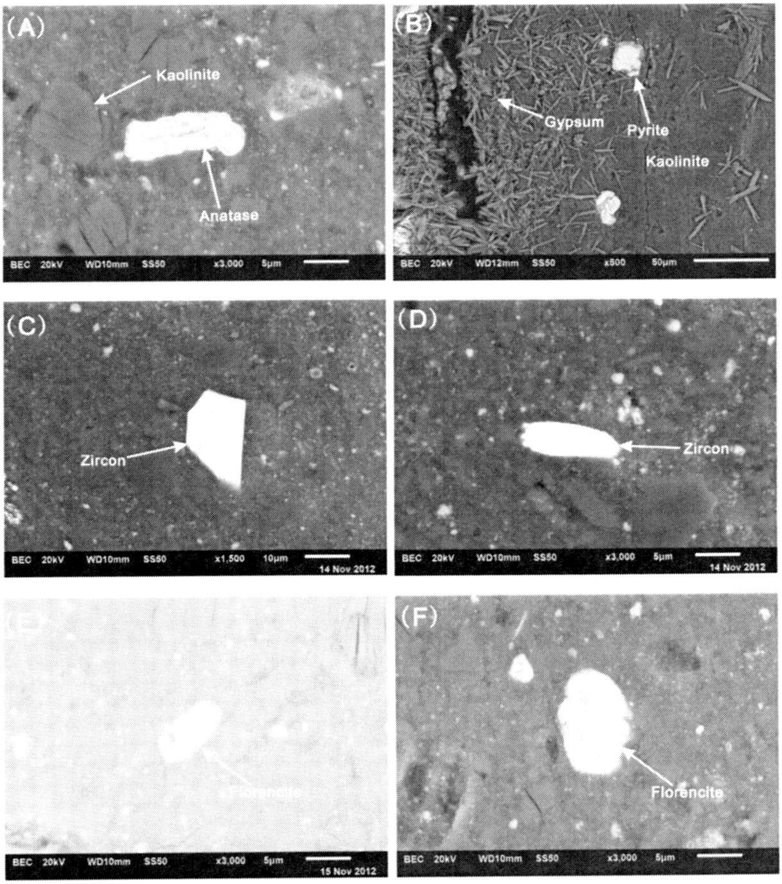

Figure 5. Back scattered images of minerals in the Zhongliangshan tuff: (**A**) kaolinite and anatase in sample S140SE7-4; (**B**) gypsum, pyrite, and kaolinite in sample S140SE7-10; (**C**) zircon in sample S140SE7-15; (**D**) zircon in the sample S140SE7-15; (**E**) florencite in sample S140SE7-21; and (**F**) florencite in sample S140SE7-4.

4.1.4. Calcite and Gypsum

Calcite distributes at the lower portion of the profile (Table 1, Figure 3), similar to that of illite. Gypsum occurs as radiating forms in the tuff and is present on the edge of fractures (Figure 5B), indicating an epigenetic origin.

4.1.5. Zircon and Florencite

Although zircon and florencite are at concentration below the detection limit of the XRD and Siroquant analysis, they have been observed under SEM-EDX in the tuff samples of the present study. Zircon occurs as subhedral (Figure 5C) and long axis (Figure 5D) in the kaolinite matrix. Florencite occurs as ellipsoidal form in kaolinite; however, minerals containing medium (M-REE) and heavy-rare earth elements (H-REE) have not been observed (M-REE include Eu, Gd, Tb, Dy, and Y; and H-REE include Ho, Er, Tm, Yb, and Lu [7]).

4.2. Major Elements

The loss of ignition of the tuff samples varies from 13.94% to 23.56%, with an average of 17.7%. The major element oxides are mainly represented by SiO_2 (35.3% on average) and Al_2O_3 (29.23%), followed by Fe_2O_3 (10.95%) and TiO_2 (3.82%) (Table 2). The ratio of SiO_2/Al_2O_3 is from 1.16 to 1.26 and averages 1.21, higher than the theoretical value of kaolinite (1.18). The ratio of TiO_2/Al_2O_3 is from 0.09 to 0.15, with an average of 0.13.

Table 2. Elemental concentrations in Tuff samples from the Zhongliangshan Mine (elements in μg/g, Oxides in %).

Sample	LOI	SiO_2	TiO_2	Al_2O_3	Fe_2O_3	MnO	MgO	CaO	Na_2O	K_2O	P_2O_5	FeO	SiO_2/Al_2O_3	TiO_2/Al_2O_3	Li	Be	Sc	V	Cr	Co	Ni	Cu
S140SE7-1	19.87	36.25	2.77	31.22	9.1	0.012	0.26	0.12	0.14	0.1	0.075	0.6	1.16	0.09	454	9.46	33.4	371	317	36.8	176	247
S140SE7-2	17.87	38.9	3.21	33.6	5.52	0.009	0.26	0.1	0.31	0.1	0.04	0.55	1.16	0.1	445	9.91	38	507	379	30.8	124	243
S140SE7-3	18.15	36.84	3.82	30.98	9.52	0.02	0.2	0.11	0.22	0.1	0.051	0.5	1.19	0.12	349	7.55	31.8	621	467	30.3	110	245
S140SE7-4	19.66	32.69	3.63	27.44	15.46	0.21	0.19	0.14	0.17	0.1	0.051	0.8	1.19	0.13	367	6.33	29.1	648	469	32.8	96.2	299
S140SE7-5	23.56	24.78	2.57	20.01	28.09	0.015	0.18	0.36	0.13	0.085	0.037	0.75	1.24	0.13	337	5.66	25.6	500	375	42.3	120	258
S140SE7-6	21.65	28.05	2.92	23.17	23	0.015	0.22	0.31	0.15	0.094	0.046	0.6	1.21	0.13	299	5.42	24	453	325	32.7	114	235
S140SE7-7	19.32	32.7	3.28	27.62	16.08	0.012	0.27	0.12	0.15	0.11	0.043	0.4	1.18	0.12	421	6.95	28.2	489	386	39.7	132	286
S140SE7-8	19.06	33.86	3.35	28.75	14.17	0.013	0.31	0.12	0.17	0.13	0.048	0.4	1.18	0.12	401	6.61	32.4	530	344	45	138	278
S140SE7-9	16.45	38.23	3.87	32.89	7.81	0.008	0.25	0.11	0.16	0.11	0.049	0.35	1.16	0.12	442	7.31	33.3	592	388	41.4	104	251
S140SE7-10	20.62	29.97	2.89	24.88	20.9	0.011	0.23	0.17	0.11	0.13	0.03	0.55	1.2	0.12	311	4.87	24.7	408	308	65.4	161	282
S140SE7-11	17.82	35.72	4.38	29.64	11.54	0.004	0.15	0.26	0.21	0.17	0.066	0.5	1.21	0.15	321	7.34	33.2	669	419	40.7	119	316
S140SE7-12	15.42	40.12	4.65	33.07	5.18	0.007	0.2	0.26	0.49	0.42	0.086	0.25	1.21	0.14	248	7.59	36.5	743	547	39.4	78.8	330
S140SE7-13	17.93	35.93	4.39	29.44	9.57	0.008	0.2	1.56	0.3	0.43	0.067	0.6	1.22	0.15	219	6.48	30.7	639	422	39.1	104	266
S140SE7-14	18.09	15.54	1.55	11.22	31.32	0.039	0.43	20.38	0.16	0.65	0.017	1.75	1.38	0.14	31.5	2.16	12.6	253	114	65.6	156	146
S140SE7-15	15.61	38.45	4.33	30.94	5.28	0.01	0.33	3.4	0.53	0.82	0.05	0.85	1.24	0.14	170	6.3	31.3	683	391	36.6	99.4	188
S140SE7-16	14.89	40.64	5.28	34.4	1.94	0.006	0.16	1.52	0.5	0.25	0.045	0.4	1.18	0.15	254	7.2	38.7	652	315	38.7	93.4	265
S140SE7-17	14.91	41.58	5.03	35.15	1.99	0.004	0.21	0.13	0.49	0.37	0.056	0.35	1.18	0.14	299	6.89	34.8	646	257	39.6	100	256
S140SE7-18	15.35	40.55	4.89	34.01	3.47	0.004	0.24	0.4	0.49	0.49	0.059	0.85	1.19	0.14	258	6.31	33.1	653	260	35.7	104	265
S140SE7-19	15.5	37.44	4.08	29.19	4.03	0.013	0.59	6.57	0.72	1.47	0.045	0.9	1.28	0.14	153	5.66	25	707	351	35.1	94.7	199
S140SE7-20	13.94	42.07	5.01	34.88	1.83	0.003	0.32	0.47	0.47	0.76	0.06	0.55	1.21	0.14	246	6.79	32.1	701	388	37.9	91.2	281
S140SE7-21	15.82	40.92	4.24	32.45	3.12	0.004	0.46	1.12	0.53	1.26	0.065	1.3	1.26	0.13	168	5.58	26.8	638	336	30.1	72	193
Average	17.7	35.3	3.82	29.28	10.9	0.02	0.27	1.8	0.31	0.39	0.05	0.66	1.21	0.13	295	6.58	30.1	576	360	39.8	114	252
UCC	nd	nd	nd	nd	nd	nd	nd	nd	nd	nd	nd	nd	nd	nd	20	3	11	60	35	10	20	25
CC	nd	nd	nd	nd	nd	nd	nd	nd	nd	nd	nd	nd	nd	nd	14.7	2.2	2.7	9.6	10.3	4	5.7	10.1

Sample	Zn	Ga	Rb	Sr	Zr	Nb	Mo	Cd	In	Sb	Cs	Ba	Hf	Ta	W	Re	Tl	Pb	Bi	Th	U
S140SE7-1	237	68.7	3.98	292	1898	213	3.4	3.13	0.615	4.22	1.17	146	57.6	15.9	4.02	0.042	0.11	58.1	1.41	50.3	25.7
S140SE7-2	190	64.5	2.86	190	1868	188	2.12	2.04	0.654	2.82	1.15	113	51.4	14.6	3.52	0.02	0.058	40.4	1.21	49.7	24.5
S140SE7-3	143	46.4	2.96	282	1487	134	1.73	0.952	0.587	3.51	1.48	141	37.7	8.88	4.44	0.018	0.085	39	1.62	33.4	9.83
S140SE7-4	123	34.9	3.52	256	1284	116	1.8	1.7	0.719	3.7	1.4	122	33.4	8.29	3.44	0.013	0.087	47.4	1.63	28.2	9.71
S140SE7-5	107	23.9	3.2	166	1046	91.6	1.56	1.46	0.457	4.13	1.07	92	26	6.91	3.16	0.014	0.207	61.3	1.22	23.8	8.11
S140SE7-6	100	30.7	2.03	191	1046	90.2	1.22	1.31	0.437	4.23	0.892	89.6	27.1	5.96	2.75	0.007	0.24	53.5	1.15	23.4	8.97
S140SE7-7	109	34.1	3.69	188	1206	109	0.988	1.22	0.494	3.41	1.1	101	30.7	7.6	2.83	0.006	0.103	49.3	1.35	27.9	11.4
S140SE7-8	124	35.9	3.1	232	1324	112	1.04	1.31	0.541	3.91	1.1	115	33.8	7.82	3.66	0.009	0.189	55.2	1.51	29	13.2
S140SE7-9	115	39.5	2.66	194	1392	114	1.03	1.5	0.519	2.23	1.05	99.3	36.9	8.57	3.67	0.006	0.058	34	1.48	30.4	16.6
S140SE7-10	103	35.4	2.46	154	978	77.7	1.45	1.83	0.516	6.79	0.927	79.1	25.4	5.68	2.8	0.01	0.09	75.2	1.14	21.7	12.6
S140SE7-11	192	38.5	5.34	287	1408	129	1.54	2.77	0.6	2.73	1.94	127	36.1	8.79	3.88	0.006	0.093	44	1.73	30	16
S140SE7-12	178	44.9	14.9	325	1467	132	1.68	4.45	0.572	1.71	5.03	141	37.9	8.98	4.05	0.013	0.118	23.6	2.06	31.8	17.7
S140SE7-13	161	37.6	12.5	341	1348	121	1.67	5.14	0.627	2.21	3.52	137	33.2	8.27	3.41	0.006	0.134	33.8	1.69	28.5	14
S140SE7-14	108	19	14.9	503	400	30.7	0.712	18.4	0.211	3.32	1.66	60.3	9.45	2.28	1.7	0.008	0.292	34.9	0.595	8.57	5.46
S140SE7-15	148	36.5	23.7	404	1342	113	1.37	6.96	0.501	2.08	5.27	143	34.9	7.88	3.69	0.006	0.156	26.9	1.32	28.7	14.2
S140SE7-16	170	32.6	6.83	279	1654	151	1.78	3.52	0.628	1.27	1.96	131	41.3	10.4	4.61	0.018	0.071	19.6	1.34	34	17.4
S140SE7-17	166	37	9.09	261	1631	143	0.996	4	0.567	1.68	2.37	131	41.4	10.1	3.76	0.025	0.08	23.8	1.36	32.2	18.5
S140SE7-18	147	35.1	12.1	286	1619	142	1.4	3.55	0.531	1.99	2.25	137	39.8	9.76	3.74	0.013	0.08	26.1	1.39	32	19.5
S140SE7-19	139	33.3	39.5	387	1153	106	1.49	6.92	0.505	2.42	5.33	160	28.8	7.07	3.48	0.036	0.287	21.8	1.39	27	20.1
S140SE7-20	174	35.6	19.8	322	1631	145	1.6	4.14	0.578	1.76	3.18	158	41.4	9.92	4.05	0.009	0.164	23	1.55	34	26.4
S140SE7-21	153	38.7	31.1	323	1405	120	1.48	3.47	0.526	1.5	4.24	165	35	8.39	3.43	0.015	0.236	25.2	1.27	29	23.8
Average	147	38.2	10.5	279	1361	123	1.53	3.8	0.54	2.93	2.29	123	35.2	8.67	3.53	0.01	0.14	38.9	1.41	30.2	15.9
UCC	71	17	112	350	190	25	1.5	0.1	nd	0.2	3.7	550	5.8	2.2	2	0.0004	0.8	20	nd	10.7	2
CC	2.1	2.2	0.1	0.8	7.2	4.9	1	38	nd	14.7	0.6	0.2	6.1	3.9	1.8	35.7	0.2	1.9	nd	2.8	7.9

UCC, the Upper Continental Crust; CC, concentration coefficient of trace elements in the tuff, normalized by average trace element concentrations in UCC [26]; nd, no data.

4.3. Trace Elements

Compared with the average concentration of the Upper Continental Crust (UCC) [26], some trace elements are enriched in the tuff samples from Zhongliangshan mine (Table 2). The concentration coefficients (CC, the ratio of the trace-element concentrations in investigated samples *vs.* UCC) of trace elements higher than 10 include Li, Cr, Cu, Cd, Sb and Re; whereas the elements with CC between 5 and 10 include V, Ni, Zr, Hf, and U. Elements Be, Sc, Co, Zn, Ga, Nb, REE, Ta, and Th, have a CC between 2 and 5. Elements Rb, Sr, Ba, and Tl are depleted, with a CC $<$ 0.5. Other trace elements have concentrations close to the UCC, with CC between 0.5 and 2.

4.3.1. Scandium

The average content of Sc in tuff samples is 30.1 μg/g, which is close to these of the tuffs from Songzao (29.8 μg/g), Nanchuan (26.3 μg/g) and the mafic rocks (29 μg/g, 1060 samples) [27]. Scandium is immobile during weathering and alteration and thus can be used as a reliable indicator for the source of tonsteins in coal-bearing strata system [28,29].

4.3.2. Vanadium, Cr, Co and Ni

The average contents of V, Cr, Co and Ni in the investigated samples are 576, 360, 39.8, and 114 μg/g, respectively, close to the tuff from Songzao (V, Cr, Co, and Ni being 576, 549, 37.9, and 164 μg/g, respectively) [1,3] and the normal detrital sediments (888 samples) in the south of Sichuan Province surrounding Chongqing (V, Cr, Co, and Ni being 442, 206, 31, and 61 μg/g, respectively) [30]. The contents of V and Cr have the same variations through the seam section, gradually increasing from top to bottom (Figure 6). However, the contents of Co and Ni are higher in the middle relative to the upper and lower portions (Figure 6). The terrigenous source of the inorganic matter in the late Permian coals and normal sediments in southwestern China is the Emeishan Basalt of the Kangdian Upland, which is enriched in V, Cr, Co, and Ni [31,32]. The values of tuff samples in the Zhongliangshan mine are close to those in normal sediments, indicating the normal sediments in southwestern China and tuff in Zhongliangshan have the same magmatic sources (the Emeishan basalt magma enriched in V, Cr, Co, and Ni). Dai *et al.* [1,3,18] suggested that some dark minerals such as basic plagioclase and pyroxene in the basalt rocks could be easily decomposed under weathering conditions and then transported into coal-bearing basin as complex anions.

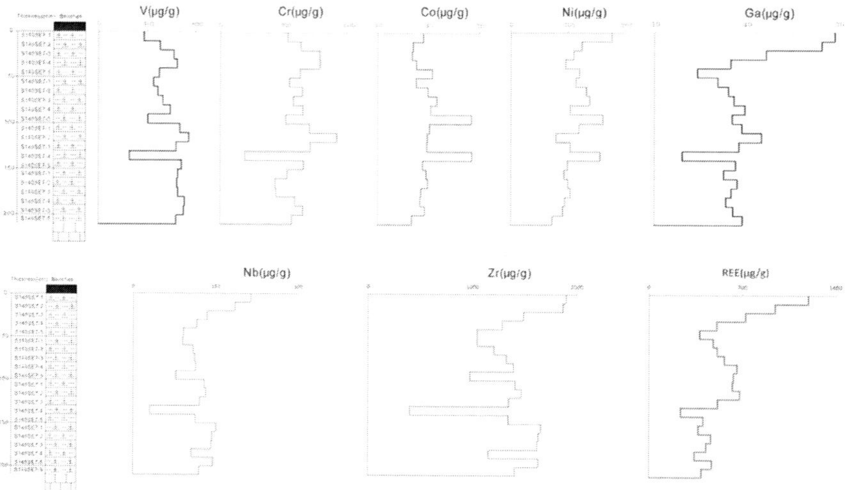

Figure 6. Vertical variations of selected trace elements of the tuff in Zhongliangshan mine, Chongqing.

4.3.3. Niobium, Ta, Zr and Hf

The average contents of Nb, Ta, Zr, and Hf of tuff in Zhongliangshan mine are 123, 8.67, 1361, and 35.2 μg/g, respectively, being close to those of the tuff from Songzao (Nb, Ta, Zr and Hf being 118, 9.46, 1377, and 41.5 μg/g, respectively). The Nb and Zr display a similar trend, both gradually decreasing from top to bottom (Figure 6).

The concentration of $(Nb, Ta)_2O_5$ of tuff in Zhongliangshan mine varies from 47 to 324 μg/g and averages 186 μg/g, lower than the concentration of the late Permian "Nb (Ta)-Zr (Hf)-Ga-REE" polymetallic deposit discovered in eastern Yunnan, southwestern China [11]. $(Zr, Hf)O_2$ varies from 551 to 2632 μg/g and averages 1880 μg/g, which does not meet the minimum industrial grade of the weathering crust type deposit (8000 μg/g) [33].

The common Nb-, Zr-, REE-, and Ga-bearing minerals have rarely been observed in the tuff, and thus it is suggested that these rare metals probably occur as absorbed ions [11,29]. However, Nb may occur as isomorph in the Ti-bearing minerals (Figure 4B,C) and Zr occurs as zircon (Figure 5C,D) in studied samples.

4.3.4. Gallium

The concentration of Ga in Zhongliangshan tuff varies from 19 to 68.7 μg/g and averages 38.2 μg/g, higher than the minimum industrial grade in bauxite (20 μg/g) and coal (30 μg/g) [34], but lower than the concentration of the late Permian "Nb(Ta)-Zr(Hf)-Ga-REE" polymetallic deposit in eastern Yunnan, southwestern China [11]. From top to bottom, the concentration of Ga gradually decreases, consistent with those of the Nb and Zr. Because the geochemical nature of Ga is similar to Al [1,8], it may occur as isomorph in Al-bearing minerals (e.g., kaolinite).

4.4. Rare Earth Elements (REE)

In this study, REE is used to specifically represent the elemental suite La, Ce, Pr, Nd, Sm, Eu, Gd, Tb, Dy, Y, Ho, Er, Tm, Yb, and Lu [35]. The abundances and geochemical parameters of REE in the tuff samples are listed in Table 3 and Table 4, respectively. The concentration of REE varies from 234 to 1189 μg/g, with an average of 548 μg/g. The concentration of REE gradually decreases from top to bottom, similar to that of Nb and Zr. Yttrium is closely associated with lanthanides in nature, because its ionic radius is very similar and its ionic charge is equal to that of Ho [7]. For this reason, yttrium is generally placed between Dy and Ho in normalized REE patterns [36]. Based on Seredin-Dai's classification [7], a three-fold geochemical classification of REE was used in the present study, including light (L-REE: La, Ce, Pr, Nd, and Sm), medium (M-REE: Eu, Gd, Tb, Dy, and Y), and heavy (H-REE: Ho, Er, Tm, Yb, and Lu) REE [7]. Accordingly, three enrichment types are identified, L-type (light REE; $La_N/Lu_N > 1$), M-type (medium REE; $La_N/Sm_N < 1$, $Gd_N/Lu_N > 1$), and H-type (heavy REE: $La_N/Lu_N < 1$), in comparison with the upper continental crust [7]. This classification has been widely adopted and used in recent years [2].

Table 3. Rare earth elements in the tuff samples collected from the Zhongliangshan Mine (μg/g).

Sample	La	Ce	Pr	Nd	Sm	Eu	Gd	Tb	Dy	Y	Ho	Er	Tm	Yb	Lu
S140SE7-1	266	372	55.7	212	34.3	5.38	30	4.92	28.3	142	5.19	15.7	2.38	13.2	1.88
S140SE7-2	180	314	41.8	174	32.2	4.79	25.9	4.22	22.9	111	4.79	12	2.08	10.8	1.67
S140SE7-3	134	313	26.8	93.9	16.9	2.99	15.4	2.94	16.2	76.8	2.99	8.08	1.23	8.02	1.25
S140SE7-4	103	224	17.8	61.1	8.7	1.83	10.4	1.87	10.9	52.7	2.11	5.66	0.929	5.45	0.897
S140SE7-5	74.6	152	12.8	48.1	8.55	1.79	8.32	1.76	9.73	46.7	1.81	5.28	0.843	5.4	0.837
S140SE7-6	91.5	201	18.2	66.4	13.7	2.58	11.2	1.97	9.95	49.5	1.73	5.21	0.796	5.23	0.763
S140SE7-7	95.7	237	16.5	52.9	9.71	2.09	9.78	1.93	9.99	56.6	2.02	5.97	1.03	6.63	0.996
S140SE7-8	108	265	17.6	59.6	10.7	2.05	10.2	2.03	11.1	59.1	2.15	6.5	1.21	7.37	0.986
S140SE7-9	111	338	19.8	67.4	11.9	2.5	11.7	2.24	11.8	62.2	2.27	7.23	1.22	7.69	1.19
S140SE7-10	77.7	293	22.9	107	29	5.19	18.7	2.42	10.9	50	1.87	5.5	0.895	5.65	0.866
S140SE7-11	103	297	22.3	79.7	14.7	2.47	11.4	2.23	12.4	58.3	2.42	7.58	1.28	8.05	1.23
S140SE7-12	106	326	24.4	87.8	15.7	2.91	12.5	2.33	13.6	62.5	2.65	8.38	1.42	9.21	1.43
S140SE7-13	76.5	251	17.7	60.4	10.4	2.46	9.64	2.04	12	52.7	2.26	6.79	1.18	7.91	1.12
S140SE7-14	18.9	101	9.12	45.4	12.7	2.56	7.86	1.23	5.44	22.9	0.907	2.45	0.367	2.45	0.326
S140SE7-15	57.3	184	13	47.7	9	2.22	8.96	1.98	10.7	53.5	2.16	6.45	1.13	7.54	1.04
S140SE7-16	46.2	163	10.7	40.3	9.95	2.44	9.47	2.13	12.2	54.7	2.17	6.58	1.12	7.07	1.06
S140SE7-17	67.2	222	15.4	55.7	9.3	2.59	9.83	2.01	11.9	50.8	2.2	6.38	1.13	6.99	1.08
S140SE7-18	61.8	201	13.7	50.2	9.63	2.64	9.67	1.96	10.5	48.9	2.09	6	1.01	6.68	1.04
S140SE7-19	50.7	158	11.1	37.9	8.35	2.43	7.86	1.45	8.4	41.1	1.56	4.65	0.799	5.21	0.714
S140SE7-20	67.3	241	14.3	49.6	10.3	3.03	10.7	1.73	10.1	46.8	1.86	5.59	0.939	6.37	0.913
S140SE7-21	56.1	197	12.2	43.3	8.68	2.79	8.27	1.46	8.61	40	1.67	4.98	0.817	5.35	0.827
Average	93.0	240	19.7	73.4	14.0	2.84	12.3	2.23	12.3	58.99	2.33	6.81	1.13	7.06	1.05

Table 4. Rare earth elements geochemical parameters of Zhongliangshan tuff.

Sample	REE (μg/g)	LREE (μg/g)	MREE (μg/g)	HREE (μg/g)	L/M	L/H	M/H	$(La/Lu)_N$	$(La/Sm)_N$	$(Gd/Lu)_N$	δCe	δEu
S140SE7-1	1189	940	211	38.4	4.46	24.51	5.49	1.41	1.16	1.26	0.70	0.77
S140SE7-2	942	742	169	31.3	4.40	23.68	5.39	1.08	0.84	1.22	0.83	0.76
S140SE7-3	721	585	114	21.6	5.11	27.10	5.30	1.07	1.19	0.97	1.19	0.85
S140SE7-4	507	415	77.7	15.0	5.34	27.56	5.16	1.15	1.78	0.92	1.19	0.88
S140SE7-5	379	296	68.3	14.2	4.33	20.89	4.82	0.89	1.31	0.78	1.12	0.98
S140SE7-6	480	391	75.2	13.7	5.20	28.47	5.48	1.20	1.00	1.16	1.12	0.96
S140SE7-7	509	412	80.4	16.6	5.12	24.74	4.83	0.96	1.48	0.78	1.36	0.99
S140SE7-8	564	461	84.5	18.2	5.46	25.30	4.64	1.10	1.51	0.82	1.39	0.90
S140SE7-9	658	548	90.4	19.6	6.06	27.96	4.61	0.93	1.40	0.78	1.64	0.97
S140SE7-10	632	530	87.2	14.8	6.07	35.83	5.90	0.90	0.40	1.70	1.58	1.02
S140SE7-11	624	517	86.8	20.6	5.95	25.13	4.22	0.84	1.05	0.73	1.41	0.88
S140SE7-12	677	560	93.8	23.1	5.97	24.25	4.06	0.74	1.01	0.69	1.46	0.95
S140SE7-13	514	416	78.8	19.3	5.28	21.60	4.09	0.68	1.10	0.68	1.56	1.13
S140SE7-14	234	187	40.0	6.5	4.68	28.79	6.15	0.58	0.22	1.90	1.75	1.18
S140SE7-15	407	311	77.4	18.3	4.02	16.98	4.22	0.55	0.96	0.68	1.54	1.14
S140SE7-16	369	270	80.9	18.0	3.34	15.01	4.50	0.44	0.70	0.71	1.67	1.15
S140SE7-17	465	370	77.1	17.8	4.79	20.79	4.34	0.62	1.08	0.72	1.57	1.24
S140SE7-18	427	336	73.7	16.8	4.57	20.00	4.38	0.59	0.96	0.73	1.58	1.26
S140SE7-19	340	266	61.2	12.9	4.34	20.57	4.74	0.71	0.91	0.87	1.52	1.38
S140SE7-20	471	383	72.4	15.7	5.29	24.41	4.62	0.74	0.98	0.93	1.77	1.33
S140SE7-21	392	317	61.1	13.6	5.19	23.25	4.48	0.68	0.97	0.79	1.72	1.51
Average	548	441	88.6	18.4	5.00	24.13	4.83	0.85	1.05	0.94	1.41	1.06

ΣREE, sum of La, Ce, Pr, Nd, Sm, Eu, Gd, Tb, Dy, Y, Ho, Er, Tm, Yb, and Lu; LREE, sum of La, Ce, Pr, Nd, and Sm; MREE, sum of Eu, Gd, Tb, Dy and Y; HREE, sum of Ho, Er, Tm, Yb, and Lu; L/M, ratio of LREE and MREE; L/H, ratio of LREE and HREE; M/H, ratio of MREE and HREE; $(La/Lu)_N$, ratio of $(La)_N$ and $(Lu)_N$; $(La/Sm)_N$, ratio of $(La)_N$ and $(Sm)_N$; $(Gd/Lu)_N$, ratio of $(Gd)_N$ and $(Lu)_N$; $\delta Ce = Ce_N/(La_N \times Pr_N)^{1/2}$; $\delta Eu = Eu_N/(Sm_N \times Gd_N)^{1/2}$; N, REE are normalized by Upper Continental Crust (UCC) [26].

Table 4 and Figure 7 illustrate that the tuff in the Zhongliangshan mine is mainly enriched in heavy REE. Only samples S140SE7-1, S140SE7-3, S140SE7-4, S140SE7-6, and S140SE7-8 are enriched in light REE; and samples S140SE7-2, S140SE7-10, and 140SE7-14 are enriched in medium REE. From top to bottom, the light REE enrichment only occurs in the upper portion of the profile, while the lower portion is enriched in heavy REE and the medium REE enrichment occasionally occurs in the middle portion.

The Ce-anomaly (expressed as δCe) values vary from 0.70 to 1.77, with an average of 1.41, indicating a well-pronounced Ce positive anomaly. The REE distribution patterns of the tuff display positive Ce anomalies, owing to the in-situ precipitation of Ce^{4+} in the process of weathering, leaching, and eluviation [35]. The Eu-anomaly (δEu) values varying from 0.76 to 1.51, with an average of 1.06, show a slight Eu positive anomaly, indicating the tuff and the Emeishan basalt have the same origin [35]. From top to bottom, δCe and δEu markedly increase. The distribution of REE of the tuff in the Zhongliangshan mine appears as a sawtooth shape, the portions of La-Sm and Gd-Lu occurring gentle and small slope, which indicates that the fractionation of REE is low.

Two reasons may be responsible for the H-REE enrichment of the tuff samples in the Zhongliangshan mine. First, L-REE can be easily leached by groundwater than H-REE; Second, L-REE can be easily adsorbed on the organic matter than the H-REE [37], which may be adsorbed by the coal seam overlying the tuff. The REE enrichment mode in the Zhongliangshan tuff is similar to that of Songzao Coalfield. Some studies have shown that L-REE are more easily to be leached by groundwater and are more apt to be adsorbed by organic matter [38,39,40,41,42].

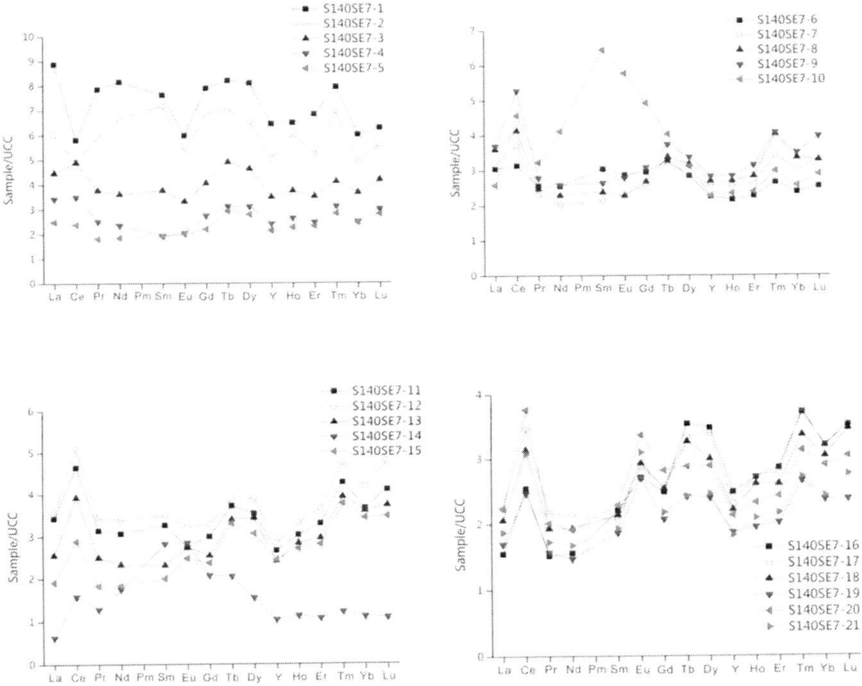

Figure 7. Distribution patterns of REE in the tuff samples from Zhongliangshan mine. REE are normalized by Upper Continental Crust [26].

5. DISCUSSION

5.1. Origin of Tuff

In the late Permian Age, the Dongwu movement, one of the most important tectonic events in southern China, caused the upper Yangtze basin uplifting and the subsequent sea regression, which led to an extensive erosion in the area. The upper part of Maokou limestone of Sichuan Basin had been subjected to a serious erosion, resulting in the formation of a vast weathering residual plain, where peat subsequently accumulated. Meanwhile, the Emeishan basalt volcano began erupting and reached a climax in the early stage of late Permian, leading to a tuff layer overlying the Maokou limestone [17].

Al_2O_3 and TiO_2 are both stable components in the rock and would be little altered during alteration, so the ratio of TiO_2/Al_2O_3 (KAT) would be kept constant and can frequently be used to study the origin of volcanic ash [29,42]. It is suggested that KAT values for silicic volcanic ash are <0.02, and those for mafic and alkali volcanic ashes are >0.08 and between 0.02 and 0.08, respectively [43,44]. The KAT ratios of the tuff in the Zhongliangshan mine are >0.08 (Figure 8), suggesting a mafic volcanic origin. In addition, the tuff samples fall in the area basalt to alkali basalt from the La/Yb-REE diagram (Figure 9), indicating a feature of alkali basalt.

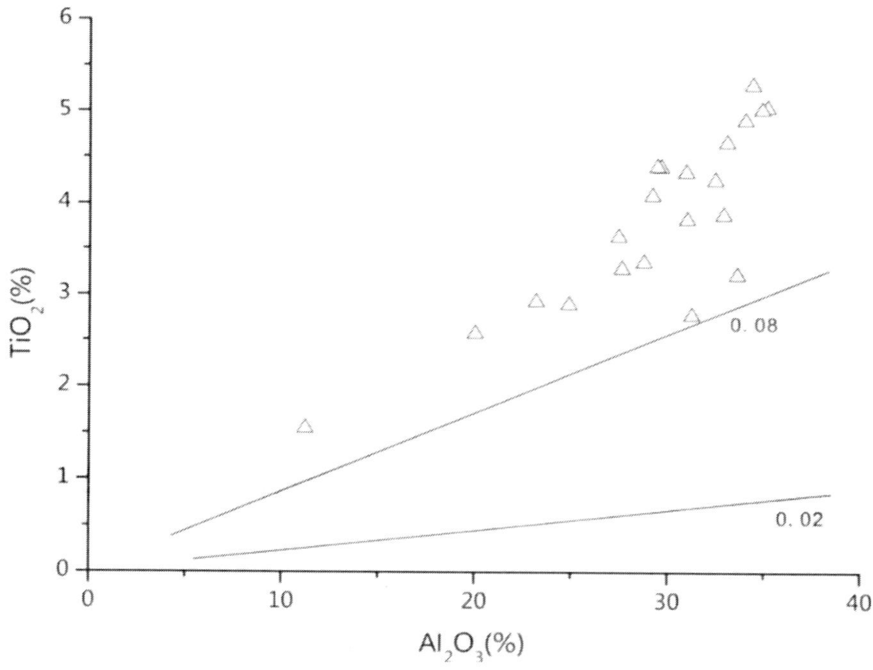

Figure 8. Plot for TiO_2 *vs.* Al_2O_3 of tuff samples in the Zhongliangshan mine, Chongqing.

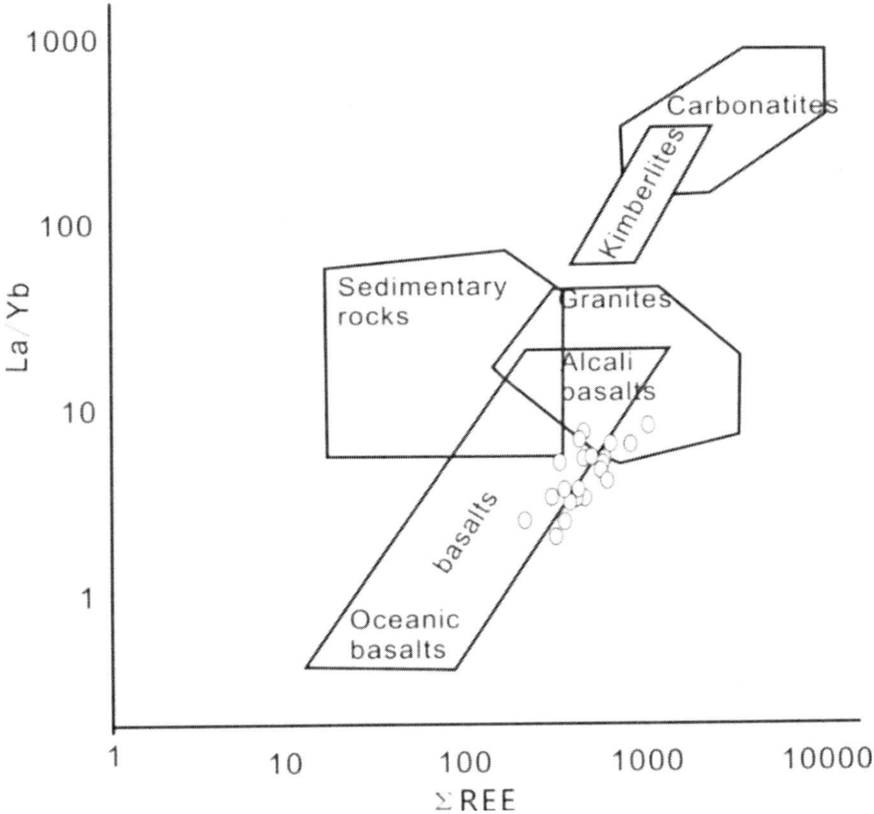

Figure 9. Relation between REE and La/Yb of tuff samples in the Zhongliangshan mine, Chongqing.

5.2. Hydrothermal Solution

Some researchers have shown that there have been activities of low-temperature hydrothermal solutions in the late Permian Age in southwestern China, which resulted in enrichment of trace elements and minerals in some coal [1,45,46,47,48,49,50,51]. Similarly, some minerals of tuff in Zhongliangshan are formed owing to the influence of hydrothermal solution.

In addition to the derivation from volcanic ash, anatase and zircon might have been derived from hydrothermal alteration in the Zhongliangshan tuffs. Anatase of various particle sizes is distributed in the kaolinite matrix (Figure 4B,C). Figure 4A and Figure 5A illustrate that part of anatase could be formed by hydrothermal alteration. Zircon from Figure 5D displays long axis and could be formed by the effect of hydrothermal alteration. Zircon in Figure 5D exclusively contains Zr, Si and O determined by the SEM-EDX. Finkelman [52] has demonstrated that Hf, Th, U, Y and HREE occur in the volcanogenic zircon, but were not identified in authigenic ziron, in accordance with the results of the Zhongliangshan tuff samples.

Florencite, the main carrier of REE in the Zhongliangshan tuff samples, occurs as ellipsoidal in the kaolinite matrix (Figure 5E,F), indicating a syngenetic or early diagenetic hydrothermal origin. Dai *et al.* [14] have also demonstrated that florencite is one of the important carriers of REE in the late Permian coals in southwestern China [3].

Gypsum (Figure 5B) and barite (Figure 4F) occur as crack-fillings, the former occurring as radiating and the latter on the edge of jarosite, indicating an epigenetic hydrothermal origin.

5.3. Preliminary Evaluation of Rare Metals

Coal and coal-bearing strata have recently become alternative sources for recovery of rare metals [2,3,7,8]. The U.S. Department of Energy's National Energy Technology Laboratory has selected 10 projects to receive funding for research in support of the lab's program on recovery of rare earth elements from coal and coal byproducts since 2015 [2,53].

Based on the Chinese industry standards [33], the required $(Nb,Ta)_2O_5$ concentrations for marginal and industrial grade Nb(Ta) ore deposits of weathered crust type are 80–100 and 160–200 μg/g, respectively; equivalent concentrations are 40–60 and 100–120 μg/g for Nb(Ta) ore deposits of river placer type. The concentration of $(Nb,Ta)_2O_5$ varies from 47 to 324 μg/g, with an average of 186 μg/g, higher than the marginal and industrial grade for weathered crust and placer deposit types. Concentration of TiO_2 varies from 1.55% to 5.28% and averages 3.82%, higher than the industrial grade of Chinese industry standard [54]. The average concentration of Ga (38.2 μg/g) is also up to the standards for industrial utilization in bauxite (20 μg/g) and coal mining (30 μg/g) [34]. In addition, the concentrations of REE vary from 234 to 1189 μg/g and averages 548 μg/g, higher than the cut-off grade of Chinese weathering crust ion adsorption type rare earth elements deposits (500 μg/g) [55].

The Nb, Ti, Ga, and REE all exceed their respective industrial grade of China in the tuff of the Zhongliangshan mine. It is considered that the Zhongliangshan tuff is a potential polymetallic ore worth in-depth study.

6. CONCLUSIONS

Compared with the Upper Continental Crust, some trace elements including Li, Be, Sc, V, Cr, Co, Ni, Cu, Zn, Ga, Zr, Nb, Cd, Sb, REE, Hf, Ta, Re, Th, and U are enriched in tuff from Zhongliangshan mine, Chongqing, southwestern China. The minerals mainly include kaolinite, illite, pyrite, anatase, calcite, gypsum, quartz, and traces of minerals such as zircon, florencite, jarosite, and barite. The tuff is of mafic volcanic origin with features of alkali basalt. The H-REE enriched in the tuff due to L-REE being leached easier by groundwater and adsorbed in the organic matter of the coal seam overlying the tuff. Some minerals including florencite, gypsum, barite, and a portion of anatase and zircon are precipitated from hydrothermal solution. It is suggested that Zhongliangshan tuff is a potential polymetallic ore and the opportunity for recovery of these valuable elements needs to be studied in depth.

ACKNOWLEDGMENTS

The authors wish to express their appreciation to Shifeng Dai for revision suggestions and English polishing. We thank Lei Zhao and Lixin Zhao, who helped to identify the minerals under SEM-EDX. We also thank Peipei Wang for the mineral quantitative analysis using Siroquant software. The authors are indebted to three anonymous reviewers for their careful reviews and constructive comments, which greatly improved the manuscript. This research was supported by the National Key Basic Research and Development Program (No.2014CB238902) and National Natural Science Foundation of China (No. 41502162).

AUTHOR CONTRIBUTIONS

Jianhua Zou and Heming Tian collected tuff samples in Zhongliangshan mine. Jianhua Zou conducted the determinations of major-element contents. Tian Li and Heming Tian were responsible for the analysis of trace-element concentrations. Jianhua Zou and Tian Li were responsible for the mineralogy investigation using XRD and SEM-EDX.

CONFLICTS OF INTEREST

The authors declare no conflict of interest.

REFERENCES

1. Dai, S.F.; Ren, D.Y.; Chou, C.-L.; Finkelman, R.B.; Seredin, V.V.; Zhou, Y.P. Geochemistry of trace elements in Chinese coals: A review of abundances, genetic types, impacts on human health, and industrial utilization. Int. J. Coal Geol.2012, 94, 3–21.
2. Hower, J.C.; Granite, E.J.; Mayfield, D.B.; Lewis, A.S.; Finkelman, R.B. Notes on Contributions to the Science of Rare Earth Element Enrichment in Coal and Coal Combustion Byproducts. Minerals 2016, 6, 32.
3. Dai, S.F.; Chekryzhov, I.Y.; Seredin, V.V.; Nechaev, V.P.; Graham, I.T.; Hower, J.C.; Ward, C.R.; Ren, D.Y.; Wang, X.B. Metalliferous coal deposits in East Asia (Primorye of Russia and South China): A review of geodynamic controls and styles of mineralization. Gondwana Res. 2016, 29, 60–82.
4. Dai, S.F.; Wang, P.P.; Ward, C.R.; Tang, Y.G.; Song, X.L.; Jiang, J.H.; Hower, J.C.; Li, T.; Seredin, V.V.; Wagner, N.J.; et al. Elemental and mineralogical anomalies in the coal-hosted Ge ore deposit of Lincang, Yunnan, southwestern China: Key role of N2-CO2-mixed hydrothermal solutions. Int. J. Coal Geol. 2015, 152, 19–46.
5. Johnston, M.N.; Hower, J.C.; Dai, S.; Wang, P.; Xie, P.; Liu, J. Petrology and Geochemistry of the Harlan, Kellioka, and Darby Coals from the Louellen 7.5-Minute Quadrangle, Harlan County, Kentucky. Minerals 2015, 5, 894–918.
6. Seredin, V.V.; Finkelman, R.B. Metalliferous coals: A review of the main genetic and geochemical types. Int. J. Coal Geol. 2008, 76, 253–289.
7. Seredin, V.V.; Dai, S.F. Coal deposits as potential alternative sources for lanthanides and yttrium. Int. J. Coal Geol.2012, 94, 67–93.
8. Seredin, V.V.; Dai, S.F.; Sun, Y.Z.; Chekryzhov, I.Y. Coal deposits as promising sources of rare metals for alternative power and energy-efficient technologies. Appl. Geochem. 2013, 31, 1–11.
9. Dai, S.F.; Seredin, V.V.; Ward, C.R.; Jiang, J.H.; Hower, J.C.; Song, X.L.; Jiang, Y.F.; Wang, X.B.; Gornostaeva, T.; Li, X.; et al. Composition and modes of occurrence of minerals and elements in coal combustion products derived from high-Ge coals. Int. J. Coal Geol. 2014, 121, 79–97.
10. Seredin, V.V. From coal science to metal production and environmental protection: A new story of success. Int. J. Coal Geol. 2012, 90–91, 1–3.
11. Dai, S.F.; Zhou, Y.P.; Zhang, M.Q.; Wang, X.B.; Wang, J.M.; Song, X.L.; Jiang, Y.F.; Luo, Y.B.; Song, Z.T.; Yang, Z.; et al. A new type of Nb (Ta)–Zr(Hf)–REE–Ga polymetallic deposit in the late Permian coal-bearing strata, eastern Yunnan, southwestern China: Possible economic significance and genetic implications. Int. J. Coal Geol. 2010, 83, 55–63.
12. Dai, S.F.; Seredin, V.V.; Ward, C.R.; Hower, J.C.; Xing, Y.W.; Zhang, W.G.; Song, W.J.; Wang, P.P. Enrichment of U-Se-Mo-Re-V in coals preserved within marine carbonate successions: Geochemical and mineralogical data from the Late Permian Guiding Coalfield, Guizhou, China. Miner. Deposita 2015, 50, 159–186.
13. Hower, J.C.; Eble, C.F.; O'Keefe, J.M.K.; Dai, S.F.; Wang, P.P.; Xie, P.P.; Liu, J.J.; Ward, C.R.; French, D. Petrology, Palynology, and Geochemistry of Gray Hawk Coal (Early Pennsylvanian, Langsettian) in Eastern Kentucky, USA. Minerals 2015, 5, 592–622.
14. Dai, S.F.; Zhao, L.; Hower, J.C.; Johnston, M.N.; Song, W.J.; Wang, P.P.; Zhang, S.F. Petrology, mineralogy, and chemistry of size-fractioned fly ash from the Jungar

power plant, Inner Mongolia, China, with emphasis on the distribution of rare earth elements. Energy Fuels 2014, 28, 1502–1514.
15. Zhuang, X.G.; Su, S.C.; Xiao, M.G.; Li, J.; Alastuey, A.; Querol, X. Mineralogy and geochemistry of the Late Permian coals in the Huayingshan coal-bearing area, Sichuan Province, China. Int. J. Coal Geol. 2012, 94, 271–282.
16. Dai, S.F.; Yang, J.Y.; Ward, C.R.; Hower, J.C.; Liu, H.D.; Garrison, T.M.; French, D.; O'Keefe, J.M.K. Geochemical and mineralogical evidence for a coal-hosted uranium deposit in the Yili Basin, Xinjiang, northwestern China. Ore Geol. Rev. 2015, 70, 1–30.
17. China Coal Geology Bureau. Sedimentary Environments and Coal Accumulation of Late Permian Coal Formation in Western Guizhou, Southern Sichuan and Eastern Yunnan, China; Chongqing University Press: Chongqing, China, 1996. (In Chinese).
18. Dai, S.F.; Liu, J.J.; Ward, C.R.; Hower, J.C.; French, D.; Jia, S.H.; Hood, M.M.; Garrison, T.M. Mineralogical and geochemical compositions of Late Permian coals and host rocks from the Guxu Coalfield, Sichuan Province, China, with emphasis on enrichment of rare metals. Int. J. Coal Geol. 2016.
19. Standardization Administration of the People's Republic of China. Sampling of Coal Seams; Chinese Standard GB/T 482–2008. Standardization Administration of the People's Republic of China: Beijing, China, 2008. (In Chinese).
20. ASTM International. Test Method for Ash in the Analysis Sample of Coal and Coke from Coal; ASTM D3174–11; ASTM International: West Conshohocken, PA, USA, 2011.
21. Taylor, J.C. Computer programs for standardless quantitative analysis of minerals using the full powder diffraction profile. Powder Diffr. 1991, 6, 2–9.
22. Rietveld, H.M. A profile refinement method for nuclear and magnetic structures. Appl. Crystallogr. 1969, 2, 65–71.
23. Ward, C.R.; Spears, D.A.; Booth, C.A.; Staton, I.; Gurba, L.W. Mineral matter and trace elements in coals of the Gunnedah Basin, New South Wales, Australia. Int. J. Coal Geol. 1999, 40, 281–308.
24. Ward, C.R.; Matulis, C.E.; Taylor, J.C.; Dale, L.S. Quantification of mineral matter in the Argonne Premium coals using interactive Rietveld-based X-ray diffraction. Int. J. Coal Geol. 2001, 46, 67–82.
25. Ruan, C.D.; Ward, C.R. Quantitative X-ray powder diffraction analysis of clay minerals in Australian coals using Rietveldmethods. Appl. Clay Sci. 2002, 21, 227–240.
26. Taylor, S.R.; McLennan, S.M. The Continental Crust: Its Composition and Evolution; Blackwell: Oxford, UK, 1985; p. 312.
27. Chi, Q.H.; Yan, M.C. Handbook of Elemental Abundance for Applied Geochemistry; Geological Publishing House: Beijing, China, 2007; pp. 1–148. (In Chinese)
28. Zhou, Y.P.; Ren, Y.L. Element gochemistry of volcanic ash derived tonsteins in late Permian coal-bearing formation of eastern Yunnan and western Guizhou, China. Acta Sedimentol Sin. 1994, 12, 123–132. (In Chinese)
29. Zhou, Y.P.; Bohor, B.F.; Ren, Y.L. Trace element geochemistry of altered volcanic ash layers (tonsteins) in late Permian coal-bearing formations of eastern Yunnan and western Guizhou provinces, China. Int. J. Coal Geol. 2000, 44, 305–324.
30. Sichuan Bureau of Coal Geology; Sichuan Institute of Coal Geology. Sedimentary Environment and Coal Accumulating Regulations of Late Permian Coal-Bearing Formation in Southern Sichuan; Guizhou Sceinece and Technology Press: Guiyang, China, 1994. (In Chinese)
31. Dai, S.F.; Chou, C.-L.; Yue, M.; Luo, K.L.; Ren, D.Y. Mineralogy and geochemistry of a Late Permian coal in the Dafang Coalfield, Guizhou, China: Influence from siliceous and iron-rich calcic hydrothermal fluids. Int. J. Coal Geol.2005, 61, 241–258.
32. Dai, S.F.; Ren, D.Y.; Zhou, Y.P.; Chou, C.-L.; Wang, X.B.; Zhao, L.; Zhu, X.W. Mineralogy and geochemistry of a superhigh-organic-sulfur coal, Yanshan Coalfield, Yunnan, China: Evidence for a volcanic ash component and influence by submarine exhalation. Chem. Geol. 2008, 255, 182–194.

33. DZ/T 0203–2002. Geology Mineral Industry Standard of P.R. China: Specifications for Rare Metal Mineral Exploration; Geological Press: Beijing, China, 2002. (In Chinese)
34. Mineral Resources Industry Requirements Manual Editorial Board. Mineral Resources Industry Requirements Manual; Geological Press: Beijing, China, 2010. (In Chinese)
35. Dai, S.F.; Graham, I.T.; Chou, C.-L.; Ward, C.R. A review of anomalous rare earth elements and yttrium in coal. Int. J. Coal Geol. 2016, 159, 82–95.
36. Bao, Z.W.; Zhao, Z.H. Geochemistry of mineralization with exchangeable REY in the weathering crusts of granitic rocks in South China. Ore Geol. Rev. 2008, 33, 519–535.
37. Dai, S.F.; Li, D.; Chou, C.-L.; Zhao, L.; Zhang, Y.; Ren, D.Y.; Ma, Y.W.; Sun, Y.Y. Mineralogy and geochemistry of boehmite-rich coals: New insights from the Haerwusu Surface Mine, Jungar Coalfield, Inner Mongolia, China. Int. J. Coal Geol. 2008, 74, 185–202.
38. Eskenazy, G.M. Rare earth elements in a sampled coal from the Pirin Deposit, Bulgaria. Int. J. Coal Geol. 1987, 7, 301–314.
39. Crowley, S.S.; Stanton, R.W.; Ryer, T.A. The effects of volcanic ash on the maceral and chemical composition of the C coal bed, Emery Coal Field, Utah. Org. Geochem. 1989, 14, 315–331.
40. Hower, J.C.; Ruppert, L.F.; Eble, C.F. Lanthanide, yttrium, and zirconium anomalies in the fire Clay coal bed, Eastern Kentucky. Int. J. Coal Geol. 1999, 39, 141–153.
41. Dai, S.F.; Ren, D.Y.; Chou, C.-L.; Li, S.S.; Jiang, Y.F. Mineralogy and geochemistry of the No. 6 coal (Pennsylvanian) in the Jungar Coalfield, Ordos Basin, China. Int. J. Coal Geol. 2006, 66, 253–270.
42. Dai, S.F.; Li, T.J.; Jiang, Y.F.; Ward, C.R.; Hower, J.C.; Sun, J.H.; Liu, J.J.; Song, H.J.; Wei, J.P.; Li, Q.Q.; et al. Mineralogical and geochemical compositions of the Pennsylvanian coal in the Hailiushu Mine, Daqingshan Coalfield, Inner Mongolia, China: Implications of sediment-source region and acid hydrothermal solutions. Int. J. Coal Geol. 2015, 137, 92–110.
43. Addison, R.; Harrison, R.K.; Land, D.H.; Young, B.R.; Davis, A.E.; Smith, T.K. Volcanogenic tonsteins from tertiary coal measures, East Kalimantan, Indonesia. Int. J. Coal Geol. 1983, 3, 1–30.
44. Burger, K.; Zhou, Y.P.; Ren, D.Y. Petrography and geochemistry of tonsteins from the 4th Member of the Upper Triassic Xujiahe Formation in southern Sichuan Province, China. Int. J. Coal Geol. 2002, 49, 1–17.
45. Dai, S.F.; Zhang, W.G.; Seredin, V.V.; Ward, C.R.; Hower, J.C.; Song, W.J.; Wang, X.B.; Li, X.; Zhao, L.; Kang, H.; et al. Factors controlling geochemical and mineralogical compositions of coals preserved within marine carbonate successions: A case study from the Heshan Coalfield, southern China. Int. J. Coal Geol. 2013, 109–110, 77–100.
46. Dai, S.F.; Zhang, W.G.; Ward, C.R.; Seredin, V.V.; Hower, J.C.; Li, X.; Song, W.J.; Wang, X.B.; Kang, H.; Zheng, L.C.; et al. Mineralogical and geochemical anomalies of Late Permian coals from the Fusui Coalfield, Guangxi Province, southern China: Influences of terrigenous materials and hydrothermal fluids. Int. J. Coal Geol. 2013, 105, 60–84.
47. Ren, D.Y.; Zhao, F.H.; Dai, S.F.; Zhang, J.Y.; Luo, K.L. Geochemistry of Trace Elements in Coal; Science Press: Beijing, China, 2006. (In Chinese)
48. Zhou, Y.P.; Ren, Y.L. Distribution of arsenic in coals of Yunnan Province, China, and its controlling factors. Int. J. Coal Geol. 1992, 20, 85–98.
49. Dai, S.F.; Chou, C.-L. Occurrence and origin of minerals in a chamosite-bearing coal of Late Permian age, Zhaotong, Yunnan, China. Am. Mineral. 2007, 92, 1253–1261.
50. Wang, X.B.; Dai, S.F.; Chou, C.-L.; Zhang, M.Q.; Wang, J.M.; Song, X.L.; Wang, W.; Jiang, Y.F.; Zhou, Y.P.; Ren, D.Y. Mineralogy and geochemistry of Late Permian coals from the Taoshuping Mine, Yunnan Province, China: Evidences for the sources of minerals. Int. J. Coal Geol. 2012, 96–97, 49–59.
51. Dai, S.F.; Tian, L.W.; Chou, C.-L.; Zhou, Y.P.; Zhang, M.Q.; Zhao, L.; Wang, J.M.; Yang, Z.; Cao, H.Z.; Ren, D.Y. Mineralogical and compositional characteristics of Late Permian

coals from an area of high lung cancer rate in Xuan Wei, Yunnan, China: Occurrence and origin of quartz and chamosite. Int. J. Coal Geol. 2008, 76, 318–327.
52. Finkelman, R.B. Modes of Occurrence of Trace Elements in Coal; US Geological Survey Open-File Report; United States Geological Survey: Reston, VA, USA, 1981; No. 81–99; p. 322.
53. The U.S. Department of Energy. Available online: http://www.energy.gov/fe/articles/doe-selects-projects-enhance-its's-research-recovery-rare-earth-elements-coal-and-coal.2015 (accessed on 2 December 2015).
54. DZ/T 0208–2002. Geology Mineral Industry Standard of P.R. China: Specifications for Placer (Metallic Mineral) Exploration; Geological Press: Beijing, China, 2002. (In Chinese)
55. DZ/T 0204–2002. Geology Mineral Industry Standard of P.R. China: Specifications for Rare Earth Mineral Exploration; Geological Press: Beijing, China, 2002. (In Chinese)

Index